A TEXTBOOK OF PLANT ECOLOGY

INCLUDING ETHNOBOTANY AND SOIL SCIENCE

(For B.Sc. & M.Sc Students and Aspirants of various Competitive Examinations)

Dr. R.S. SHUKLA
M.Sc., Ph.D.
Formerly from the Department of Botany
D.A.V. College, Kanpur

Dr. P.S. CHANDEL
M.Sc., Ph.D.
Formerly from the Department of Botany
Jai Narain P.G. College, Lucknow

S. CHAND
PUBLISHING

S Chand And Company Limited
(ISO 9001 Certified Company)

S Chand And Company Limited

(ISO 9001 Certified Company)

Head Office: D-92, Sector–2, Noida – 201301, U.P. (India), Ph. 91-120-4682700

Registered Office: A-27, 2nd Floor, Mohan Co-operative Industrial Estate, New Delhi – 110 044, Phone: 011-49731800

www.**schandpublishing.com;** e-mail: **info@schandpublishing.com**

Marketing Offices:

Chennai	:	Ph: 23632120; chennai@schandpublishing.com
Guwahati	:	Ph: 2738811, 2735640; guwahati@schandpublishing.com
Hyderabad	:	Ph: 40186018; hyderabad@schandpublishing.com
Jalandhar	:	Ph: 4645630; jalandhar@schandpublishing.com
Kolkata	:	Ph: 23357458, 23353914; kolkata@schandpublishing.com
Lucknow	:	Ph: 4003633; lucknow@schandpublishing.com
Mumbai	:	Ph: 25000297; mumbai@schandpublishing.com
Patna	:	Ph: 4011400; patna@schandpublishing.com

First Edition 1972
Subsequent Editions and Reprints 1974, 77, 80, 82, 85, 87, 89, 91, 93, 94, 96, 98, 2000, 2003, 2007, 2008, 2009, 2010, 2012
Twelfth Edition 2014
Reprints 2017, 2018, 2019, 2020, 2021, 2022, 2023

Reprint 2026

ISBN: 978-81-219-0548-0 **Product Code:** H6PLE68BOTN10ENAL14O

PRINTED IN INDIA

By Shri Shyamlal Printing Press Private Limited, Plot 20/4, Site-IV, Industrial Area Sahibabad, Ghaziabad – 201 010 and Published by S Chand And Company Limited, A-27, 2nd Floor, Mohan Co-operative Industrial Estate, New Delhi – 110 044.

PREFACE TO THE TWELFTH EDITION

In view of the changes introduced in the academic curricula of graduate and postgraduate programmes of Indian Universities, a compelling need was felt to revise the book to meet the need of students. Many additions and corrections have been made in the text of the previous edition. All the chapters have been supplemented with new facts, charts, tables, diagrams and photographs. The noteworthy additions in this edition are as follows:

- In the chapter on **Principles of Phytogeography, Climate, Vegetation and Botanical Zones of India,** the text of grassland vegetation has been supplemented with ecological status of grassland in India, grasslands of Madhya Pradesh, Uttar Pradesh and Uttarakhand, Assam and South India.
- In the chapter on **Environmental Pollution,** groundwater and marine pollution, effects of arsenic and a case study of Minamata disease caused by organic mercury have been incorporated. A table on diseases caused by polluted water, eutrophication, etc. has been added. In air pollution topic, the diseases caused by the various air pollutants and control measures of air pollution have been included. In agricultural pollution, a table of toxic pollutants and their effects has been included.

  Radiation pollution has been rewritten giving the details of radioactive isotopes of certain elements, types of nuclear radiation, natural sources, man-made sources of radiation pollution, consequences of radiation pollution, ionising and non-ionising radiations, diseases caused by radiation pollution and control of radiation pollution.

  In noise pollution, a table of intensities of noise, a brief summary of noise-induced diseases and control of noise pollution have been added.
- In the chapter on **Global Warming (Green House Effect),** a table giving details of green house gases from human activities, a chart on impact of global warming and action of green house gases on ozone layer have been added.
- In the chapter on **Conservation of Natural Resources,** discussions on wildlife resources, Red Data Book and bio-piracy have been given in conservation of wildlife resources.

  In the topic on water resources, the major problems of water management, rainwater harvesting, marine resources and coral reef conservation have been discussed at length. Under energy resources, wind energy and its harnessing, hydel energy, ocean energy, geothermal energy, nuclear energy, conservation of energy and future sources of energy have been discussed.
- The chapter on **Biodiversity** has been modified. To make the matter attractive and interesting, new photographs and tables providing latest data have been included.
- In this edition, two new chapters on current global ecological problems; **Waste and its Management, Disasters and their Impacts on Environment,** have been added.
- In the **Soil Science Section,** at the end of the chapter on **Soil Fertility,** an illustrated account of irrigation system has been added.

We hope that the present edition will meet the requirements of the students. Suggestions and letters of appreciation have been received in large numbers from the teachers and students and we are grateful to all of them. The valuable informations and photographs provided to us by Dr. Ratnesh Singh Chandel, Geologist, G.S.I., Lucknow are gratefully acknowledged.

We express our deep appreciation and sincere thanks to the management and the editorial team of S. Chand & Company Pvt. Ltd. for their continued interest and support.

Dr. R.S. SHUKLA
Dr. P.S. CHANDEL

CONTENTS

PLANT ECOLOGY & ETHNOBOTANY

SOIL SCIENCE

PLANT ECOLOGY & ETHNOBOTANY

1

Ecology

The things of the world are classified into two major groups namely the living or biotic component and the non-living or abiotic component. The biotic component includes all types of living organisms, both plants and animals and the abiotic component includes the non-living materials (soil, water, air, etc.) and the forces of nature (light, gravity and molecular energy). The living organisms exist in an environmental setting of which they are a part. Every aspect of life is influenced by the environment and the activities of organisms affect their environment. An aspect of biology which deals with the inter-relationship between biotic and abiotic components as well as the relationships among the individuals of the biotic component is called **ecology.** Organisms form interacting systems or communities, these communities are coupled to their environments by transfer of matter and energy and so the communities and environment are interrelated. A functional system formed by communities and their environment is called *ecosystem.* Thus ecology is a science of ecosystems or totality of reciprocal interactions between living organisms and their physical surroundings. (Clark, M.E. 1973).

Although the origin of the term Ecology is still uncertain, there is a general agreement that modern term Ecology is derived from "Oekologie". The word 'ecology' (oekologie), first proposed by Earnst Haeckel (1869), a German Biologist, is derived from Greek words, *oikos* meaning the dwelling place or home and *logos* meaning the discourse or study; thus, the word ecology literally means the study of living organisms, both plants and animals in their natural habitats or homes. It can also be defined as the study of life in relation to environment; the *environment* being the aggregate of all external conditions and influences which affect the life and development of organisms at a given spot. Recently, **Eudgene Odum (1963)** has defined ecology as *the study of the structure and the function of nature.* According to Charles J. Krebs (1972), Ecology is the scientific study of interactions that determine the distribution and abundance of organisms. Smith R.L. (1972) in his book Elements of Ecology and Field Biology has defined ecology as multidisciplinary science which deals with organisms and their environment, both biotic and abiotic.

ENVIRONMENTAL SCIENCE

The aggregate of all things; physical, chemical and biological and the influences which affect the life and development of organisms is called environment. This branch of science that seeks to determine the effects of environmental factors on the growth, distribution and migration of the organisms and also deals with some other aspects of relationship between organisms and those factors is called **Environmental Science.** The concept of environment is as old as human civilization. Early people considered science of nature as **Natural Philosophy.** Now environmental science has emerged as interdisciplinary subject comprising Biosciences, Physics, Chemistry, Geology, Geography and Social Sciences.

Ecology, like biology, has been subdivided into plant ecology and animal ecology. Plant ecology deals with the relationship between plants and their environments and animal ecology is concerned with the study of relationship between animals and their environments. Ethology is the term now generally used by biologists to denote the scientific study of animal behaviour with special reference to the behaviour of animal in its normal environment. In recent years there has grown an idea that in biological organisation the plants and animals are closely interdependent and they react with one another in many ways and at a particular place plants and animals share the same set of conditions and same environment. Therefore, these two subdivisions, plant ecology and animal ecology should be unified. In view of this reasonable fact, the authors of modern ecology have accounted for the study of both plants and animals in the environment.

ECOLOGY AND ITS DIVISIONS

Ecology may be divided into **autecology** and **synecology.**

Autecology. Autecology is concerned with the study of individual animal or plant species or its population throughout its life history in relation to the habitat in which it grows. In other words, it is a study of inter-relationship between individual species or its population and its environment.

Synecology. The other area of ecology which deals with systems of many species—whole communities or major fractions of communities and ecosystems is termed *Synecology* in English speaking countries, or *biocenology* or biosociology by many Europeans. (Whittaker, R.H. 1970). It is concerned with the structure, nature, development and causes of distribution of communities. To understand the ecology of plant communities, the ecological life cycles (autecology) of at least most important plant species of the communities must first be studied. Thus, autecology forms a basis for synecology.

The study of plant community structure is called phytosociology or plant sociology. The study of plant ecology merges with plant geography or phytogeography. This is a science which deals with the distribution of plants on or near the surface of earth and water and it also deals with the migration of species. Actually speaking there is no sharp line of distinction between plant ecology and plant geography. **W.B. Turrill** comments *that plant ecology is intensive while plant geography is extensive in outlook, but both are concerned with plants and in attempting to correlate observed structure and behaviour of plants with causes, both refer to the same sum total of environmental factors though the emphasis varies.*

DIFFERENT FIELDS OF ECOLOGY

Ecology involves study of organismal reactions *i.e.*, movement, morphological adaptations, behaviour patterns, physiological activities, growth and reproductive behaviour, social and economic structure, life expectancy of the organisms in their natural habitats or under domestication. The integrated approach of such studies provide proper insight for a better management of natural resources for the benefit of man. Ecological studies are either organism based or habitat based and are conducted at various levels. Different branches which are made to account for the various specific and detailed aspects of ecology are as follows :

I. Organism level:

1. **Autecology:** Ecology of species and individual
2. **Population ecology:** Study of population
3. **Community ecology:** synecology.

II. Habitat or Ecosm or Ecosystem level:

A. Terrestrial Ecosystems.

1. Forest ecology
2. Grassland ecology

3. Desert ecology
4. Wetland or Marsh ecology

B. Aquatic Ecosystems.

1. Marine ecology
2. Lagoon ecology
3. Estuarian ecology
4. Fresh water ecology or Limnology
   (*i*) Lotic waters (running water bodies e.g., rivers and streams)
   (*ii*) Lentic waters (standing water bodies like lakes and ponds)

III. Applied Ecology. It deals with the applied aspects of ecology, i.e., application of ecological concepts in human welfare including conservation of natural resources, forestry, wild life management etc. It includes the following:

1. Agricultural ecology or crop ecology.

2. Paleoecology. It is concerned with the organisms and geological environments of the past. Paleontology and radioactive dating have helped significantly in the study of Paleoecology.

3. Cytoecology. It deals with the cytological details of the species or populations in relation to different environmental conditions.

4. System ecology. It deals with the structure and working of ecological systems in relation to space and time and also with the analysis of components of ecosystem using applied mathematics, bioinformatics and statistics. In this, special emphasis is laid on the reciprocal relationship between living and non-living systems.

5. Conservation ecology or Resource ecology. It is concerned with the proper management of plant, animal, soil, water and mineral resources for human welfare.

6. Ecological energetics and Production ecology. These modern branches of ecology are still in developing stage. These deal with the mechanisms and quantity of energy conversion and flow of energy through organisms. Energy production processes, rate of increase in organic weights of organisms in relation to space and time are also discussed in this branch of ecology.

7. Landscape ecology.

8. Radiation ecology.

9. Ecophysiology.

10. Gene ecology.

PLANT ECOLOGY AND OTHER BRANCHES OF SCIENCE

Ecology is a synthetic branch of biological science which draws source materials from many other sciences. It is fundamentally related to morphology, taxonomy, physiology, biochemistry, cytology, genetics etc. Various other sciences, such as physics, chemistry, geology, geography, meteorology, climatology, hydrology, paleontology, Anthropology, sociology, mathematics, statistics are also being increasingly used in the study of ecological problems. Application of radioactive isotopes, use of many modern and advanced instruments like spectrometer, infrared gas analyser, flame photometer, computers in the analysis of data, calorimeters, phytotrons for culturing the plants in environment controlled chambers and many other equipments are common in ecological researches. Besides botany, zoology, chemistry and physics, the knowledge of climatology, geography, pedology and geology is also essential in the study of complicated problems of plant ecology.

APPLICATION OF PLANT ECOLOGY

The study of plants in their environment has yielded a large body of knowledge which provides aids to the science of conservation of natural resources. The knowledge of ecology is of great help in controlling soil erosion, reforestation, restoration of wild animals as well as grassland vegetation and flood control. Plant ecology is directly related to silvics and silviculture and other branches of forest biology. In *British Commonwealth Forestry Terminology* (1953) **silvics** has been defined as the study of general characteristics and life history of forest trees and crops with particular reference to environmental factors as the basis for practice of silviculture while the **silviculture** has been defined as the art and science of culturing forest trees and crops.

Every farmer or gardener is ecologist, since by such practices as cultivation, irrigation, artificial pollination and spraying, he affects the plant behaviour.

Knowledge of ecology is being applied in agriculture, food production and horticulture. The soil conservation practices are in use these days in agronomy. The modern ecology revolves round the biological production processes and ecological energetics. The International Biological Programme (IBP) was launched since July 1, 1967 to study the biological basis of organic productivity and conservation of natural resources in relation to human welfare. Launching of this programme has given impetus to the ecologists all over the world and over 70 nations including India have participated in the IBP at either national or international level. The future of ecology and indeed of biology is likely to be changed by some international programmes such as 'Man and Biosphere' (MAB).

The history of ecology in India is not very different from that of any other country in the world. Indeed, it has been much influenced by western school which provided the leadership. Publications of botanical explorations by Dudgeon (1920), Saxton (1922), Bor (1942), Osmaston (1926) and Champion (1936) provided enough opportunity for ecological investigation in India. Professor F.R. Bharucha, a student of Braun-Blanquet, established the first school of ecology at Bombay. This school contributed a great deal of informations on the biological spectra of different regions of India and on the phytosociology of grass and forest vegetation. The second school of ecology developed under the leadership of Professor R. Mishra first in Sagar and later at Varanasi. At present, many secondary schools of ecology have emerged at Ujjain, Ahmedabad, Pilani, Jodhpur, Pondicherry, etc. and ecologists in these centres are engaged in different fields of study.

QUESTIONS

1. What is ecology? How is it related with other branches of biology?
2. What are different branches of ecology? Why study of ecology is important for man?
3. Discuss the scope and objectives of ecology.
4. Write brief notes on:
   (*i*) Autecology
   (*ii*) Synecology
5. Why ecological study is importaht to man? (H.P. Univ. B.Sc. III, Bot. IX paper, 2010)
6. Explain:
   (*i*) Silviculture
   (*ii*) Applied ecology
7. Who coined the term ecology first of all ?

2

The Environment

The life containing and life supporting environment of the world is restricted to a very irregular layer (5 to 20 km thick) around the globe. This thin veil of living material on the earth is called *ecosphere or biosphere.* Thus, the **biosphere** is that part of earth in which life exists (Hutchinson, 1970). Biosphere is synonymous with biota and refers to the sum of all living organisms on the earth. Biosphere may be divided into *parabiosphere* and *eubiosphere.* The parabiosphere is that part of biosphere where environmental conditions are not entirely hospitable and this includes such broad areas as higher altitude, the polar regions, the deepest ocean troughs, the most extreme deserts and certain localised regions as volcanoes, geysers and heavily polluted areas of land and water. The remaining portion (the *eubiosphere*) is composed of three chief media—air, water and earth or land and accordingly it has been divided into three subdivisions :

(*i*) Atmosphere

(*ii*) Hydrosphere

(*iii*) Lithosphere or pedosphere

In ecology we study the reciprocal relationship between an organism or a group of organisms and its environment. The environment literally means the surrounding. The **environment** is the aggregate of all those things and set of conditions which directly or indirectly influence not only the development or growth and quality of life of individual organisms but also the communities at a particular place. It is comprised of a number of factors which interact with one another and also influence the responses of the organisms. Any external force, substance or condition affecting the organisms in any way is referred to as environmental factor. Soil, moisture, wind and temperature are, thus, factors and the environment thus is sum total of all such factors. The natural place where organisms or communities of organisms live is called *habitat.* The habitat implies a particular set of environmental factors in a given locality and is, therefore, generally used in a more concrete sense than the environment, as for example, water is the habitat of aquatic organisms and land is a habitat for numerous terrestrial organisms. Aquatic habitat comprises three major categories namely, *fresh water habitat, marine habitat and brackish habitat.* Biosphere can be divided into many major categories of land masses called **biomes**. *Biomes* are subdivisions of earth biota. They are distinct large areas of earth with relatively homogeneous climate and flora and fauna, as for example, deserts, forests, prairies etc. Biomes are subdivided into small units, each with its own particular set of physical conditions. These small units are called **Zones**. For example, a *forest biome* can be divided into *ground zone* and *canopy zone,* and likewise *desert biome* can be divided into *surface* and *subterranean zones.* The ocean can be regarded as single biome (the *marine biome*) which can be divided into *surface, abyssal* and *intertidal* or *littoral zones.*

Within each biome there may be numerous habitats, each characterised by a particular set of conditions and well adapted community of organisms. Within a particular habitat an individual is generally confined to restricted situation which is called *ecological niche.*

The environmental or the habitat factors influence the characters and composition of individual plants and plant communities. Any feature of an organism or its part which enables the organism to exist under conditions of its habitat is called an *adaptation.* An organism accumulates many adaptive features in it. Such features may ensure a degree of success either by allowing the organism to make full use of the amounts of nutrient, water, heat and light available to it or by providing a significant amount of protection against some unfavourable or adverse factors, such as very high or very low temperature, drought, parasitism and so on. The adaptive features of organism may be hereditary (*i.e.*, they may be genetically controlled) or they may be induced by the habitat factors.

In an ecosystem the species and its environment are taken as a complex *"whole"*. Environment, as pointed out earlier, in itself is a complex of factors acting, reacting and interacting with the organism complex. Thus, the organisms and their environment are wedded together and are in state of constant flux. Relationship between organisms and their environment are based on certain principles which are summarised as follows :

1. Everything influencing the life processes of an organism constitutes its environment.

2. Environment in a habitat may be considered into biotic and abiotic components and the activities of the organisms are influenced by the combined effects of various environmental factors.

3. An organism is a biotic component of the environment and the materials and energy required for the maintenance of the body and sustenance of life of organisms constitute the abiotic environment.

4. An organism cannot exist in vacuum.

5. Life is the energy exchange process between the organism and environment and death means cessation of the exchange process.

6. The environmental requirements of different organisms differ from individual to individual and also with age and need.

7. Life activities are influenced by that environmental component which occurs in minimum quantity. This is **Liebig's law of limiting factor**. Liebig stated that the growth of a plant is dependent on the amount of the food stuff which is presented to it in minimum quantity.

8. Life activities of an organism are influenced by the minimum or maximum quantity of the environmental components or factors, as for example, nutrients, light, temperature, moisture. Based on this principle Shelford founded the ***law of tolerance***.

9. Tolerance limits of an individual for different environmental factors may be different.

10. An organism may show different tolerance limit for a particular environmental factor in different habitats and at different age and stage of life history.

11. Organisms having wide tolerance limits for many environmental factors are widely distributed.

12. An organism is a product of nature (genetic set-up) and nurture (environmental upbringing). The inherited qualities are unfolded in proper environment.

13. Organisms react with the external stimuli caused by the environmental changes. The reactions may be exhibited by movements (migration) or adaptational changes in the body or physiological activities. All such adaptations have survival value.

14. Widely distributed species are adapted to various habitat conditions by evolving ecotypes.

15. Every habitat has potential to support a certain number of organisms. This is known as *carrying capacity* of the habitat. Knowledge of carrying capacity is essential for proper management of the habitat.

16. The biotic and abiotic components of the environment are in constant flux in any habitat which induce ecological succession with the passage of time and change in the environment.

17. Sun is the only source of energy in our earth's environment and the life depends on solar energy. Organisms live through exchange of energy.

18. Energy is neither created nor destroyed. It can be transformed from one form to the other.

19. During transformation of energy from one form to the other a large amount is dissipated into the environment mainly as heat energy.

20. Every living thing that we see has potential energy in huge quantity. One gm dry weight of body tissue contains 4—5 kcal of energy stored in it.

21. Solar energy reaching the earth's surface is not lost but only transformed from one form to the other. A very small amount (1 to 3 per cent) of solar energy is trapped by green plants and converted into chemical energy.

22. Energy flow from the sun to the plants, to the other organisms and then to the space is always unidirectional.

23. Energy and space relationships of the organisms cause niche differentiation within the habitat which brings about ecological stability in the community life. In any habitat a community is born, it grows with passage of time and through succession it is stabilized to form a climax community.

24. Life on the earth exists in a thin mantle or layer around the earth. This layer forms the biosphere.

25. The biosphere is not uniform structure and it consists of several life supporting habitats called *ecosms* or *ecosystems*.

26. An ecosystem has *producers* (plants), *consumers* (animals), and *decomposers* constituting the *biotic component,* and life supporting matter and energy which constitute the *abiotic component*.

27. In any ecosystem the stability is conditioned by :

(*i*) fixation and transfer of energy in the organisms at various levels.

(*ii*) conversion of abiotic components (nutrients and energy) into organic structures which adds to the biomass. The abiotic components are replenished through weathering of the lithosphere, atmospheric movements and biogeochemical cyclings of nutrients.

28. Energy flow, synthesis of matter and balanced relationship of biotic and abiotic components in an ecosystem are regulated through ecological processes.

ENVIRONMENTAL OR ECOLOGICAL FACTORS

Ecological factors are many and diverse and often intricately mixed and interdependent. These factors either singly or in combination influence and determine the presence or absence, vigour or weakness and relative success or failure of various plant communities in a particular habitat.

The environment of an organism may be subdivided into physical (abiotic) and biotic components. The physical or abiotic environment of all organisms on the earth includes three main media which also act like factors, viz., atmosphere (air), lithosphere (soil) and hydrosphere (water) and a variety of other physical factors like light, temperature, pH etc. The study of abiotic factors and substances has been described as **environmental ecology** by A.S. Boughey (1971). The environmental conditions which influence the life and development of organisms are grouped into four main classes which are as follows :

(1) Climatic factors (related to aerial environment) or atmosphere

(2) Edaphic factors (related to soil conditions);

(3) Physiographic (topographic) factors; and

(4) Biotic factors.

As already mentioned, these ecological factors are inter-related and intricately mixed. They work through one another acting and reacting together, as for example, the change in physiographic conditions at a place may bring about a change in local climate that, in turn, may affect the soil and competition impress.

CLIMATIC FACTORS

Climate is one of the important natural factors controlling the plant life. Its study is called climatology. The climate includes the following main factors :

(1) Light

(2) Temperature

(3) Precipitation and atmospheric humidity

(4) Air and atmosphere.

LIGHT

Light is the most essential abiotic factor without which no life can exist. It is non-lethal limiting factor both at the maximum and minimum levels. The chief natural sources of light are sunlight, moonlight, star light and the light producing or luminescent organisms. Of these the sunlight has the greatest ecological significance.

The radiant energy coming from the sun in the form of visible spectrum is called light or luminous energy. Radiation that penetrates the earth's atmosphere consists of electromagnetic waves of a wide range of wavelengths. A beam of light is pictured as a shower of particles called *photons*. Each photon carries a certain amount of energy called *quantum*. The energy varies inversely to the wavelength of the spectrum. The solar spectrum in the earth's atmosphere has been analysed on the basis of wavelengths into different radiations (Fig. 2.1). The solar radiations

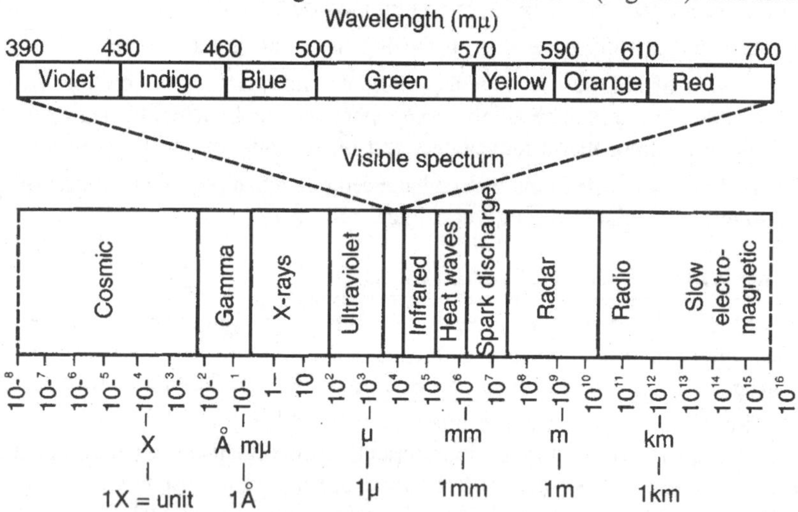

Fig. 2.1. Solar Spectrum

which penetrate earth's atmosphere consist of a band of visible (to man) light and a small proportion of ultraviolet and infrared radiations. It is not known whether the long radio waves have some ecological importance to plants. The visible light lies in the range of 400—750 mμ. When the visible sunlight is passed through a prism it is dispersed into a series of wavelengths exhibiting seven different colours—*violet, indigo, blue, green, yellow, orange,* and *red* (VIBGYOR). The wavelength of ultraviolet (2% of the radiation reaching the earth surface) is below 390 mμ and that of infrared above 750 mμ (1 millimicron or mμ = 1/1000 of a micron). Since a micron is 1/1000 of a millimetre, a milliμ (mμ) is equal to ·000001 mm or 10^{-7} cm). In working with ultraviolet and other rays of shorter wavelengths Angstrom unit is used (10 Å = 1 mμ). The photochemical activity is greatest at the violet end. Ultraviolet (UV) thus, has the greatest killing effect on the protoplasm. It helps in the synthesis of anthocyanin in leaves and inactivates the growth hormones and thus checks stem growth. Microbes injured by ultraviolet rays are often rejuvenated by exposure to visible light. X-rays and γ-rays show high degree of ionising effects. They cause mutation in the living systems and in high doses they are absolutely fatal. Infrared radiations are not so powerful as may stimulate biochemical reactions. These radiations have high heating effects. They exert influence on stem growth and germination. The radiations of wavelengths shorter than 290 mμ never reach to the earth surface.

Light is usually measured by an electric instrument called lightmeter or photometer which consists of a light sensitive photoelectric cell. The intensity of light is directly recorded on a dial of the photometer. It is measured in foot candle or lux. Photometers are never exposed to direct sunlight. Actually, light reflected by a thick white paper placed on the ground for surface of the plant is measured. For ecological purposes light intensity is measured in a number of habitats for different species. This will enable one to know whether the species are shade loving or light demanders.

Spatial Variations in Light Intensity

Light intensity differs from place to place. The spatial variation in light intensity may be caused by the following :

(*a*) **Effecting atmosphere.** When the light passes through the atmosphere of the earth, a small proportion of radiations of shorter wavelengths become absorbed by atmospheric gases, mainly nitrogen and oxygen. The places at high altitudes receive brighter light than those at lower altitudes (above the sea level) because at higher elevations the atmosphere is thinner than that at low elevation.

Atmospheric vapour exerts a powerful screening effect and for this reason the intensity of light is much greater in dry areas than in wet regions. In a cloudy day, the light may be reduced to 4% of the normal intensity. When the atmosphere is saturated with fogs and clouds, a relatively high proportion of light rays of longer wavelengths, such as infrared and other visible radiations, are absorbed by the atmospheric moisture and the light rays of shorter wavelengths and ultraviolet rays are absorbed by the gas molecules and vapour droplets and the light reaching the earth surface under such conditions is called diffused light or skylight. The diffused light on overcast day comprises upto cent per cent of the total light and on clear day it may comprise about 10-15% of the total light. Latitudinal variations in light intensities due to the height of the sun above the horizons are very important. At the equator, the light is most intense and contains highest proportion of direct sunlight. Towards the poles, the intensity of direct light decreases and the percentage of diffused light goes high.

(*b*) **Effect of water.** In water medium, the intensity of light is reduced and this decreases progressively with the increase of water depth. About 10% of the sunlight falling on the water surface is reflected and 90% of that penetrates water and is modified in respect of intensity, spectral composition, angular distribution (refraction) and time distribution. Phytoplankton, zooplankton and suspended particles either reflect or absorb the light rays. Submerged plants get weaker lights than the plants on the surface of water get. It is so because a proportion of light falling on the water surface is reflected and the major part of the penetrating light is absorbed by upper layers of water. When light reaches on the surface of water the major proportion of rays of shorter wavelengths. *i.e.,* violet, indigo, blue and green, are reflected and other lights are partly absorbed. This is why bodies of water appear bluish green in colour. Reflection of light is increased several times in the rough water surface. There is a selective absorption of light at various depths in water. The rays of longer wavelengths are absorbed near the surface and in general light rays of shorter wavelengths penetrate deepest layer. Thus, infrared rays are absorbed in upper layers of water (about 4 meters); red and orange rays are completely absorbed upto the depth of 20 metres; yellow rays may penetrate upto 50 m and green and blue rays penetrate upto 80 to 100 m depth Violet and ultraviolet rays penetrate upto 80 to 100 m depth Violet and ultraviolet rays penetrate beyond 100 m depth and no light penetrates beyond 200 m depth.

Depending upon the penetration of light, oceans are divided into the following three zones :

(*a*) Euphotic zone (upto 50 m depth)

(*b*) Disphotic zone (80 to 200 m depth)

(*c*) Aphotic zone (below 200 m of depth)

In the oceans, algae are distributed according to wavelengths of light rays to which their colours are best suited to absorb and utilize; green algae are found in intertidal zone, brown algae descend somewhat deeper and red algae are characteristic of deep oceanic water.

Light has got sufficient power of penetration. Photosynthesis may take place in some plants which are covered under as much as 40 cm thick snow.

(*c*) **Effect of suspended particles.** Dust, smoke and other solid particles dispersed in the air or water have great screening effect. In smoky industrial cities, the smoke may reduce 90% of light.

(*d*) **Effect of the layers of vegetation.** In complex plant community, for example forest, the tallest plants receive full sunlight, undershrubs receive subdued sunflecks or diffused light, herbs and epigeous cryptogams grow in still weaker light. In dense forest the leaves completely check the penetration of light and less than 1% of total sunlight reaches the surface.

Temporal Variations in Light

The light intensity during summer is much higher than that in winter. It is weak at sunset in winter and comparatively stronger during the mid-day hours. This periodical fluctuation in the intensity of light is due to the change in the angles of radiations reaching the earth. When the sun is at the horizon the solar rays travel through approximately 20 times the thickness of the air they have to penetrate when the sun is overhead in the noon. At the equator, the day light prevails 12 hours out of every 24 hours in both the seasons, summer and winter. In the polar regions, the day length (photoperiod) becomes longer than 12 hours in summer and shorter than 12 hours in winter. The skylight or diffused light available before sunrise and after sunset is of great ecological importance. Moonlight plays some important role in the plant life. It sometimes satisfies light requirements of certain seeds, promotes starch hydrolysis in the leaves, affects nocturnal leaf movement in legumes, and stimulates the sexuality in certain marine algae.

Importance of Light to Plants

Earth is a very small object in the solar system and it receives only 1—50 millionth of the total solar radiation reaching the universe. The journey of solar energy from sun to earth surface is very much moderated while it passes through stratosphere and the atmospheric envelope surrounding the earth. In the stratosphere lies the ozone regions which absorbs the harmful solar radiations of short wavelengths. There are various devices to measure the solar energy. The unit of measure is erg or calorie and energy falling at a place is expressed in terms of cal/cm^2/min. At any place on the earth surface this energy flux varies with season, time of day and atmospheric humidity and drought. In the northern hemisphere solar energy falling at the outer surface of the earth's atmosphere is 0.485 cal/cm^2/min. This is taken as hundred per cent solar radiation. Of this only 0.228 cal/cm^2/min reaches the earth surface, *i.e.,* only 47 per cent of the radiation reaches the biosphere and the rest 53 per cent is scattered and lost (Gate, 1965).

From the ecological point of view, the energy and its utilization is most important for the existence of life. Many of the environmental phenomena like movement of wind currents, rainfall, chemical action and organic behaviour are controlled by energy.

There are three possible pathways for light to take when it reaches the surface of a plant or animal; it may be reflected from the surface, it may be absorbed by the surface and it may be transmitted through the material (Moen, 1973), Fig. 2.2.

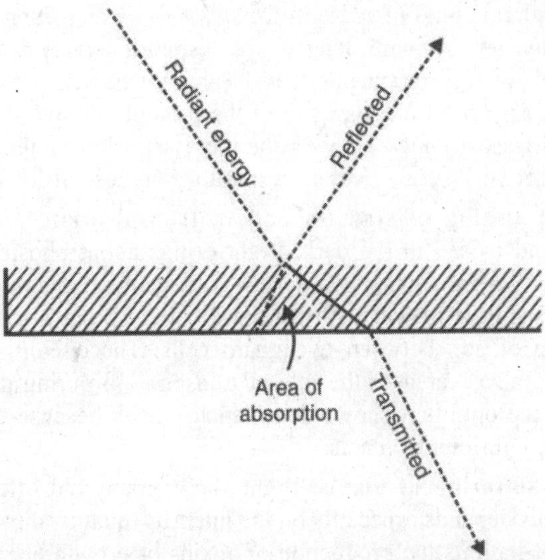

Fig. 2.2. Three pathways of solar radiation reaching the body surface.

Light affects many physiological activities of the plants. Light affects the following aspects of plant life :

1. **Photosynthesis.** Out of the total solar energy reaching the earth, only about 2% is used in photosynthesis and about 10% is used in other physiological activities. It has been estimated that a corn plant utilises only 0.13 per cent of light energy. All wavelengths of light are not utilised by a plant. Green light is completely reflected by green parts of the plants making them look green. Red and blue green are the two maxima of absorption of light. The green plants, the producers of ecosystem, synthesize their food (carbohydrates) from water and CO_2 in presence of sunlight. The solar radiations provide energy for this process. In this processes the radiant energy of sun available to the plants is converted into the chemical energy by chlorophylls. The chemical energy stored in food is utilised in various other biochemical activities of the plants. The rate of photosynthesis is greater in intermittent light than in the continuous light. The relationship of light intensity with photosynthesis in terrestrial as well as in aquatic plants follows the general pattern of a linear increase upto an optimum or saturation intensity followed by a decrease at high intensities (Rabinowitch, 1951; Thomas, 1955).

Light plays important role in the development of plastids and pigments. It has marked effect on the number and position of chloroplasts. The upper part of leaf which receives full sunshine has large number of chloroplasts which are arranged in line with the direction of light. At high intensity, the photo-oxidation of enzymes reduces not only the rate of carbohydrate synthesis but also that of protein synthesis. The protein synthesis is especially reduced by high intensity of light. High intensity of light, however, influences the formation of anthocyanin pigment. It is for this reason, alpine plants have beautifully coloured flowers.

2. **Respiration.** There is no direct effect of light on the respiratory activity in the plant body. Indirect effect is much important because in presence of light the respiratory substrates are synthesized. Under certain conditions, such as, in shade and under water, the light becomes a limiting factor and the photosynthesis is not sufficient for effective growth. Under such conditions, the rate of photosynthesis is just sufficient to meet the need of respiration. This is called *compensation point.* At this point, the dry weight of plant does not increase. The compensation point differs in different species and in different individuals of the same species at different ages. In many plants the respiratory rate increases with the increase in the light intensity. Ranjan and Saxena have studied the effect of light intensities on respiration rate in many plants and have

shown that respiratory rate could increase in *Canna, Nerium* and *Bougainvillea* with the increase in light intensity. However, in some other plants respiration rates decreased slightly in intense light. The rise and fall of respiration rate may be due to the effect of light on the permeability of plasma membrane, change in the viscosity of the protoplasm and photo-oxidation of enzymes. The permeability and viscosity increase with the increase in light intensity upto certain optimum. Light, however, has got very little effect on respiratory process of lower plants and thallophytes.

3. **Opening and closing of stomata and in transpiration.** Mostly the stomata remain opened in the light and closed in the dark. Light brings about phosphorylation and conversion of starch into soluble sugars in the guard cells and thereby increases their osmotic pressure which, in turn, causes inflow of water in the guard cells. The increase in the turgidity of guard cells causes widening of gap between two guard cells. The opening of stomata increases the gaseous exchange and also increases the rate of transpiration during day period. Increase in the light intensity above the optimum shows detrimental effects because the increased transpiration in the intense light is injurious to plants.

4. **Growth and flowering of plants.** Light shows manyfold effects on the growth of the plants. Growth of plants depends especially on the intensity, quality, duration and direction of light. High intensity of light inhibits the production of auxins or growth hormones and consequently it influences the shapes and sizes of plants. Plants growing in darkness or insufficient light produce maximum amount of growth hormones as a result of which they are elongated with slender pale yellow stem and small leaves. The plant growth is slow in the light of high intensity. Red light favours the growth. Lights of shorter wavelength, except violet, are detrimental to plant growth.

Duration of light is also very important. Actual duration or length of the day (photoperiod) has been shown to be important factor in the growth and flowering of wide varieties of plants. The controlling effect of the photoperiod, known as photoperiodicity, is currently an active field of physiological ecology. According to their response to photoperiods, the plants have been classified into three well defined groups :

(*i*) **Long day plants.** Plants which bloom when the light duration is more than 12 hours per day, as for example, radish, potato, spinach, etc.

(*ii*) **Short day plants.** Plants which bloom when the light duration is less than 12 hours per day, as for example, cereals, tobacco, cosmos, dahelia, etc.

(*iii*) **Day nuteral plants.** Plants which show little response to length of daylight, as for example, cotton, balsam, tomato plants.

Recently it has been shown that photoperiodic stimulus for flowering is also controlled by thermal points, Azzi (1957) has shown for the first time that initiation of flowering in a plant occurs at certain constant which is specific for a particular species. This constant is called Azzi's constant which is expressed as follows :

Azzi's constant = Total duration of light in hours + total mean temperature in °C.

Chinoy (1960) has confirmed this constant and called it **Photothermal quantum** requirement of a species. Thus, an increase in temperature will decrease the duration of light required for flowering. This information can be applied to advantage in autecological studies of certain crop plants.

Plants which receive direct sunlight are called *heliophytes* and those growing in the shades are called *sciophytes*.

Heliophytes exhibit the following features :

(*i*) Stem with short internodes and long lateral organs;

(*ii*) Roots numerous and profusely branched;

(*iii*) Thick cuticle;

(*iv*) Well developed palisade and weakly developed spongy tissue in the leaf;

(*v*) Well developed xylem with thick rays;

(*vi*) Small intercellular spaces in the tissues;

(*vii*) High respiration rate and much rapid transpiration;

(*viii*) Vigorous flowering and fruiting;

(*ix*) Early appearance of flowers;

(*x*) Low chlorophyll content; and

(*xi*) Proper development of mechanical tissues.

In the absence of light, the growth is very poor and plants show etiolation. The stem becomes tender, narrow, and long and the leaves become pale green, soft and small. Thus, light is essential for the normal and healthy growth of plants.

5. **Movement.** Light affects the movement of some plants. The stems, roots and leaves show different responses to light. The effect of sunlight on the plant movement is called heliotropic effect. The stems elongate towards the source of light (positively phototropic) and the roots are negatively phototropic. The leaves grow transversely to the path of light. In order to receive maximum sunlight, the leaves are oriented on the stem in such a way that they do not overlap one another.

6. **Germination of Seeds.** The seeds when moist are very sensitive to light. In some cases, the germination of seeds is retarded in light. The quantity of light needed for the stimulation of embryo varies in different seeds. In most cases, the red light promotes germination and far-red light inhibits germination. Investigations regarding the mechanism of seed germination in light have revealed that a pigment phytochrome is involved in this process. The pigment occurs in two reversible forms—Pr and Prf which develop under red and far-red light. The germination depends upon the balance between two forms :

$$\underset{\text{(Inactive)}}{\text{Pr}} \underset{730\text{ nm}}{\overset{660\text{ nm}}{\rightleftharpoons}} \underset{\text{(Active)}}{\text{Prf}}$$

Seeds of certain plants require blue light for germination. Germination of *Typha* seeds is promoted in yellow light. Yellow light counters the inhibitory effect of blue light.

Light is an important factor in the distribution of plants. Some plants grow in full sunlight, while others prefer to grow in shade. Bormann (1956) describes an interesting situation in certain species of pine in which young seedlings are shade adapted while older seedlings and young trees do not grow well in shades.

TEMPERATURE

Temperature is a variable factor which is influenced by time, season, latitude, altitude, slope, direction, soil texture, plant cover and human activities like urbanisation and industrialization. It penetrates every region of the biosphere and profoundly influences all forms of life by exerting its action through increasing and decreasing some of the vital activities, such as the metabolism, reproductive behaviour, embryonic development and growth. Temperature is a measure of intensity of heat. In terms of standard unit it is commonly expressed as degrees either in Fahrenheit or Celsius scale (Centigrade). Heat is a from of energy called thermal energy. Thermal energy is exchanged between organisms and environment by radiation, conduction, convection and evaporation. These four basic modes of transfer of heat energy occur within the organisms and in the interface between the organisms and their environment (Fig. 2.3).

The total amount of heat entering the biosphere from the sun is balanced by the amount lost per unit time. The flow of geothermal heat from the interior of the earth is small as compared to the amount of heat entering the biosphere from the sun. The estimate of thermal energy flow in the biosphere is referred to as *heat budget* (Vernberg and Vernberg, 1970).

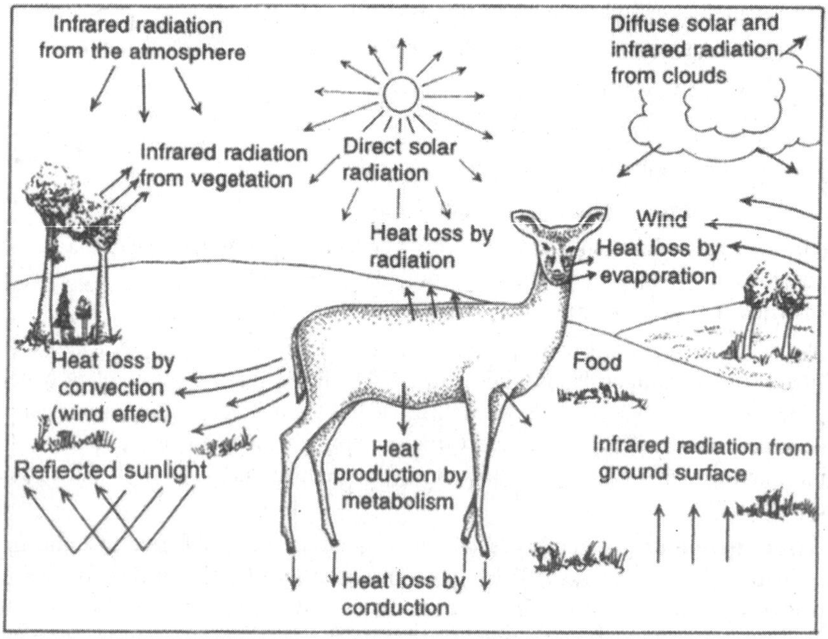

Fig. 2.3. Exchange of thermal energy between the organisms and their environment through radiation, conduction, convection and evaporation.

The most influential factors in the climate are temperature and moisture. The temperature affects the vegetation either directly or indirectly. Directly it affects in two ways :

(*i*) It affects the physiological processes of plants and consequently their growth and size; and

(*ii*) It determines which species can survive in a particular region. The different species of plants show a wide variation as regards their tolerance to temperature range and fluctuation.

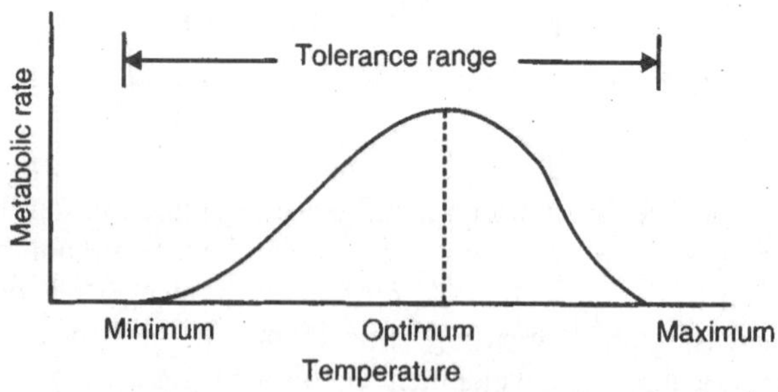

Fig. 2.4. Effects of temperature on the metabolic rate.

On our planet, organisms can carry on their life activities over a relatively narrow temperature range extending from 0°C to 50°C and every plant has a specific range of temperature requirement. This range differs from species to species. Plants do not thrive in places with higher or lower temperatures than their specific ranges.

Generally, at 40°C the protoplasm undergoes such changes as are minimal to plant life and it dies at temperatures above 60°C. At temperatures below freezing point plants generally die because of rapid crystallisation of protoplasmic water which results in mechanical injury. Air dried yeast can endure temperature as high as 114°C. Bacteria can endure temperature between 120°C and 130°C. A few fungi can withstand temperature upto 89°C. Air temperature above 32°C is most favourable for tropical plants. Existence of vegetation has been recorded between 26°C (some conifers) and 66°C (desert plants). Every organism has a definite temperature range. The temperature at which all metabolic processes necessary for life can only initiate and proceed with lowest pace is said to be **minimum temperature** and increase in the temperature above the *minimum level* increases the rate of metabolic activities until they reach the maximum level at a temperature which is called **optimum temperature**. Further rise in temperature beyond optimum level decreases the metabolic rate until it ceases at a temperature called **maximum temperature** (Fig. 2.4).

The general inability of protoplasm to endure high temperature can be ascribed in large part to the sensitivity of its enzymes to heat. Catalytic proteins in nearly all cases are irreversibly inactivated by exposure to high temperatures (usually greater than 50°C) for any length of time. They are, however, able to withstand lower temperatures even below the freezing point of water. The ability of cell to endure sub-freezing temperatures seems to depend principally upon the avoidance of ice formation. The appearance of ice crystals in cells is almost always associated with the death of cells, due in part to mechanical damage inflicted on the subcellular structures by the ice crystals themselves. Death may also be due to removal of water from the protoplasm during ice formation in the intercellular spaces, thus dehydrating the protoplasm.

According to the heat requirement of plants, Raunkiaer divided the gross vegetation into the following types :

(*a*) **Megatherms.** These are the plants of warm habitat which require high degree of heat throughout the year. They are found in areas with tropical climates *e.g.*, plants of deserts.

(*b*) **Mesotherms.** These are the plants of habitat which is neither very hot nor very cold. These plants cannot withstand extreme high or low temperatures and they are found in tropical and subtropical habitats.

(*c*) **Microtherms.** These are the plants of cold or temperate habitat and require low temperature for their growth. Such plants cannot tolerate high temperature. They may also be found in tropical and subtropical areas at high elevations where temperature conditions are less extreme.

(*d*) **Hekistotherms.** These are the plants of cold and alpine regions. They do not thrive well in heat and can withstand long and very severe winter.

Many plants are very sensitive to temperature. The sudden fall in temperature is injurious because plant tissues are badly affected by it. Forests suffer from night frost on the east side where the sun rays strike very early in the day. As an adaptation against frost, the starch of plants changes to fats or oils in the autumn. The fatty oils depress the freezing point and thus increase the power of resistance in plants against frost. The leaves of plants in the coldest lands store fats. Pentosans, mucilage and pectic substances which have high water retaining power are abundant in many plants. They decrease the danger of plants from desiccation and consequent death. Dried seeds and spores are not affected by freezing because there remains no liquid in them that can freeze. Due to removal of water from seeds, the cold resistance of seeds of certain plants increases upto the extent that their exposure for 3 weeks to —190°C does not diminish their germinability.

The temperature stimulates the growth of seedlings. The optimum temperature for seed germination ranges between 20°C and 27°C.

The absorption rate is retarded at low temperature. Photosynthesis operates over a wide range of temperature. Most of the algae require lower temperature range for photosynthesis than the higher plants. The photosynthesis continues even at 80°C in some desert plants.

The rate of respiration increases with the rise of temperature upto a certain level, but beyond the optimum limit the respiration rate shows marked decrease. The rate of respiration becomes doubled at the increase of 10°C above the optimum temperature provided other factors are favourable (*Vant Hoff's law*). High temperature generally favours the growth of plants, but for some crop plants low temperature is beneficial. If the temperature ranges of winter varieties are lowered upto 0°C to 5°C, the seeds sown in the spring season will grow luxuriantly and the plants will mature and flower at normal time. The process by which temperature range of plant is lowered in order to get early crop is called *vernalization*. This practice is very common in cold countries.

Temperature determines the growth of many plants. Cotton prefers high temperature. Potato gives highest yield in low summer temperature. Growth of plants is retarded at high temperatures.

Temperature in combination with humidity and other factors helps in the spread of diseases in plants. Low temperature and high humidity favour the rust attack. Low temperature, high humidity and cloudy weather favour the damping off, seedling blight, foot rot and root rot diseases of cucurbits, tobacco, papaya and ginger.

Temperature fluctuation in environment. The environmental temperature fluctuates both daily and seasonally. Temperature in any locality is governed by the brightness of the sun. It may vary from sunlight to shade and from daylight to dark. Surface temperature in soil may be 30°C higher in the sunlight than in the shade and upto 17°C higher during day than that in the night. In the desert the temperature may be 40°C higher during day than that during night. The Thar desert in south Rajasthan may show diurnal change of temperature to the extent of 20 to 30°C.

Latitudes also affect the temperature cycles. With the increase of every 150 metre altitude the temperature decreases by 1°C, Different habitats such as fresh water, marine and terrestrial environments show varying response to fluctuating temperature. Temperature fluctuations are less in the aquatic environment than in the terrestrial one. The increase in depth of aquatic medium increases the temperature fluctuation. Minimum temperature of the sea may be —3°C while temperature of fresh water pond never goes below 0°C. The maximum temperature of ocean generally reaches up to 36°C but it may go much higher in fresh water ponds and pools. In deep bodies of water heating or cooling is restricted to surface layer but deeper layers may also get heated or cooled as a result of vertical circulation wherein surface water is brought to the deeper regions and vice verse. Studies on the vertical fluctuations of temperature have led to the hypothetical classification of fresh water media into the following layers :

1. *Epilimnion*—This is the superficial layer of body of water which is constantly stirred by wind and is a layer of warmer water.

2. *Metalimnion* or *Thermocline*—This is the intermediate zone between the upper and bottom layers of body of water. This layer shows vertical temperature changes.

3. *Hypolimnion*—This is the bottom layer of stagnant water with little or no fluctuation in temperature. The process of differentiation of fresh water habitat on the basis of vertical changes of temperature into three strata is referred to as **thermal stratification**.

In terrestrial environment temperature fluctuations are varied and marked. Lowest temperature recorded for any land mass is —70°C (Siberia, 1957). Higher temperatures may often reach upto 85°C in deserts at noon. In Rajasthan the highest temperature exceeds 50°C. Water in hot springs and geysers may be as high as 100°C.

The temperature varies from place to place and likewise the vegetations of different areas also differ considerably. Desert plants grow in extreme heat, aquatic plants grow in low temperature range, and grasses prefer to grow in the area of moderate temperature.

Temperature in combination with moisture determines the general distribution of vegetation. Northern, southern, tropical and temperate vegetations depend solely upon temperature and moisture.

PRECIPITATION AND ATMOSPHERIC HUMIDITY

Water is one of the most important climatic factors. It affects the vital processes of all the living beings. It is the plenary agent that sets in motion the nutrients of the soil and makes them available to plants. It affects the morphology and physiology of the plants. It in combination with other factors regulates the structure and distribution of plant communities.

In nature, water may be found in vapour, liquid and snow or ice states. In the atmosphere, water is found in the form of vapour. The quantity of water retained in the atmosphere depends on temperature and wind. Vapour increases in the atmosphere if the temperature rises and pressure decreases. At certain temperature and pressure, the maximum water-laden air is called saturated atmosphere. At saturation point, if the temperature is lowered the water holding capacity of atmosphere is reduced which causes the condensation of water vapour in the form of rain drop, dew, frost, sleet, snow, etc. This is **precipitation**.

The water vapour present in unit volume of air is called **absolute humidity**. This is expressed in terms of percentage of water vapour present in unit volume of air at certain temperature. The amount of water vapour required to saturate the same unit volume of air under constant physical conditions is called **relative humidity**.

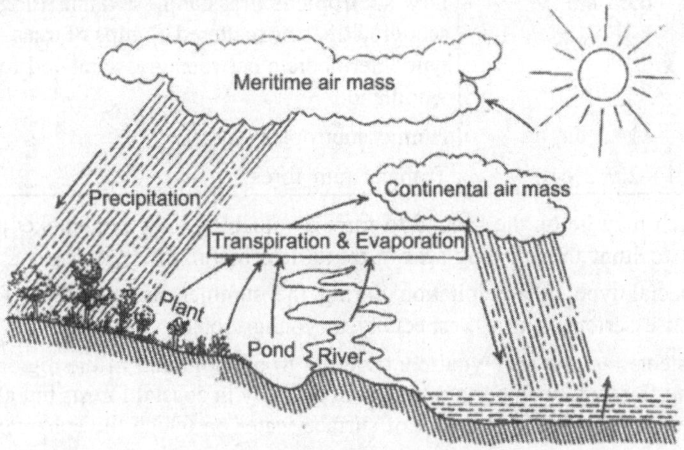

Fig. 2.5. Hydrologic cycle in nature

Water of atmosphere reaches to the earth's surface through precipitation and from earth's surface it reaches to the atmosphere through evaporation and transpiration (Fig. 2.5). Thus, a continuous circulation of water from earth to atmosphere and *vice versa* is maintained in nature. This is called **water cycle** or **hydrologic cycle**. It has been estimated that about 80,000 cubic miles of water from the oceans and 1,500 cubic miles of water from lakes and land surface evaporates annually. The total evaporation is equalled by total precipitation of which about 24,000 cubic miles of water falls on land surface. The main source of water on earth is rain water. Penck using precipitation-evaporation ratio, has classified the climate as follows :

(*i*) *Arid* — It is characterised by the condition in which evaporation is greater than precipitation.

(*ii*) *Arid-humid* — When evaporation is more or less equal to precipitation.

(*iii*) *Humid* — When evaporation is lesser than precipitation

The total rainfall, especially the distribution of rainfall throughout the year is one of the leading features of climate. Rainfall map of the world corresponds very closely with the distribution of great vegetational zones of the world.

Sudden and heavy rains are not so beneficial as are moderate and continuous rains because in the heavy rains a large amount of water is lost from the surface of soil as run off and the soil is eroded.

Rainfall is determined largely by geography and pattern of large air movements of weather systems. When the moisture laden winds blow from the oceans towards the high mountain they deposit most of the moisture on the ocean-facing mountain slopes with a resulting 'rain shadow', and produce desert on the other side. The higher the mountain the greater is the precipitation of moisture over it. This is the main reason why deserts are usually found behind high mountains. The deserts are also found along the sea coasts, where wind blows from large interior dryland areas rather than off the ocean.

The amounts of rainfall in different localities largely determine the nature of vegetation therein. The following tabulation gives a rough idea about the plant communities that may be expected in regions with different amounts of annual rainfall.

	Annual rainfall	Vegetation
1.	0 to 13.24 cm	Desert
2.	13.25—35.1 cm	Semi-arid grassland
3.	35.2—63.5 cm	Dry subtropical grassland, savanna (a grassland with scattered trees or scattered clumps of trees—a community type intermediate between grassland and forest) and open woodland
4.	63.6—114.3 cm	Humid subtropical forest
5.	114.4—203.2 cm	Tropical rain forest

Snow, which may lie on the ground to form a valuable protective blanket and also a reserve of water is apt to limit the growing season by its late melting.

Hail, a special type of precipitation during the summer season in the form of small ice pieces, may cause serious injuries, especially to young crops.

Dew and sleets make a very vital contribution to precipitation in the regions of low rainfall. Dew and ground fog may be important to plants not only in coastal forests but also in deserts near the sea coasts where they provide much of surface water on which the ephemeral plants depend.

The atmospheric humidity influences directly the form and structure of the plants. It directly affects the transpiration rate of the plants. In dry atmosphere transpiration rate increases and as a result of this the water content of leaf tissues decreases and the leaves wilt temporarily. Water requirements of different plant species differ considerably. Some species, on one extreme, thrive well in the region with an annual precipitation of 10 cm while some, on the other extreme, grow only when they are submerged in water. On the basis of their water requirements, the plants are grouped into three ecological groups :

(*i*) *Hydrophytes.* Plants adapted to aquatic environment.

(*ii*) *Xerophytes.* Plants adapted to grow in dry lands where water content is low.

(*iii*) *Mesophytes.* Plants living in the habitat that usually shows neither an excess nor a deficiency of water.

The actual effects of water on the plants may be complicated by other conditions, such as temperature and atmospheric humidity. The combination of temperature and precipitation plays vital role in determining the broad features of plant distribution on the earth surface.

The temperature exerts more limiting effect on the organisms when the moisture conditions are at extremes (*i.e.,* either very high or very low) than when such conditions are moderate. Likewise moisture also plays a more critical role at the extreme temperature. Some modern climatologists taking into consideration the quantitative measures, effectiveness and seasonal distribution of moisture and temperature have classified the climate into temperate, tropical, polar and high altitude climates. The characteristic features of these climates and peculiar type of vegetations restricted to them are given in the following chart :

	Climate	Characteristic features of climate	Vegetation
1	Temperate climate	Cold and moist with well marked seasonal and diurnal fluctuation, average annual rainfall more than 762 mm (30′′), warmest month temperature above 10°C.	Luxuriant vegetation in favourable situation. Trees and shrubs dominant, and herbs exceeding trees and shrubs in number.
2.	Tropical climate	Warm and widely humid with mean temperature of the coldest month usually 17.8°C and rainfall very heavy (200-400 cm per annum). Frost and snowfall are usually unknown. Little or no seasonal variation.	World's luxuriant rain forests in the regions of high rainfall. Scrubs, grassland, desert community in the regions of decreasing rainfall.
3.	Monsoon area	It is characterised by dry and wet seasons. Rainfall very heavy.	Deciduous trees and shrubs.
4.	Climates of polar regions and high altitudes.	Warmest month below 10°C. Precipitation mostly in the form of snow and widely distributed. Annual rainfall less than 254 mm (10′′). Owing to low temperature, the relative humidity is high and evaporation rate is low. In the regions of high altitudes there is continuous light in summer and darkness in winter.	Vegetation mostly low and scant. Dwarf shrubs and grasses dominant. Mosses and lichens are common.

ATMOSPHERE AND AIR

The earth is enveloped by a gaseous layer called atmosphere. Gaseous mantle forming atmosphere extends into outer space some 1000 km or so above the earth surface. It maintains contact with all the major types of environment of earth, interacting with them and greatly affecting their ability to support life. It is a reservoir of several elements essential to life. It serves many functions including the filtration of radiant energy coming from the sun, insulation from heat loss from the earth surface and stabilization of weather and climate owing to heating capacity of the air.

Structure of Atmosphere

There are five concentric layers within the atmosphere which can be distinguished on the basis of temperature (Fig. 2.6). These are as follows:

1. Troposphere
2. Stratosphere
3. Mesosphere
4. Thermosphere, and
5. Exosphere.

Troposphere. The lowest layer of atmosphere in which man and other living organisms live is called troposphere. Troposphere is about 20 km above the earth surface. It is thin in the polar regions (about 10 km thick). It is a mixture of several gases. The proportion of gases in the atmosphere is fairly constant. Troposphere is characterised by weather change and steady decrease in temperature with increasing amplitude and it may decrease upto —60°C in the upper layers. The composition of troposphere excepting water vapour and dust particles is presented in the table below :

Constituent	*Percentage by volume*
Nitrogen	78.0841
Oxygen	20.9476
Carbon dioxide	0.0318
Argon	0.02340
Neon	0.001818
Helium	0.000524
Methane	0.0002
Nitrous oxide	0.00005
Krypton	0.000114
Hydrogen	0.00005
Xenon	0.0000087
Ozone	0.000001

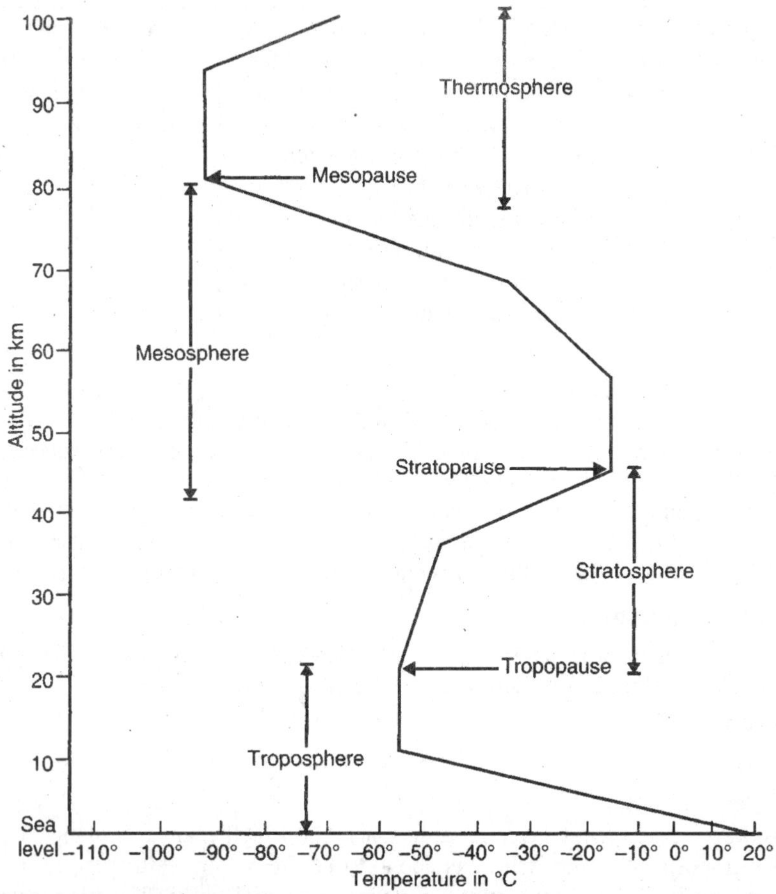

Fig. 2.6. Diagram showing the extent of different layers of atmosphere and temperature stratification of the atmosphere

The water vapour and dust particles occur in troposphere in variable concentrations. Concentration of water vapour may range from 0 to more than 4 per cent. The amount of water vapour in troposphere is maximum in the lowest level of atmosphere and it decreases gradually in the upper region and is entirely absent above 8 to 10 km. Dust is limited to lower levels and has no specific relationship with any other feature of atmosphere. Troposphere is the layer of sulphates and is the region of strong air movements, cloud formation, lightning, thundering etc. The upper layer of troposphere which gradually merges with the next zone or stratosphere is called is called *tropopause.*

Stratosphere. The second layer of air mass extending about 30 km above tropopause is called stratosphere. The uppermost layer of stratosphere is called *stratopause.* In this zone the temperature shows an increase from a minimum of about —60°C to a maximum of about 5°C. The increase in temperature is due to ozone formation under the influence of ultraviolet rays of solar radiation. Ozone is formed from oxygen by a photochemical reaction in which solar energy (symbolised as *hv*) splits the oxygen molecule to form atomic oxygen which then combines with oxygen molecule to form ozone

$$O_2 \xrightarrow{hv} 2O \quad \text{(atomic oxygen)}$$

$$O_2 + O \rightleftharpoons O_3 \quad \text{(ozone)}$$

The above reactions are reversible. Ozone content of stratosphere is constant which means that ozone is being produced from oxygen as fast as it is broken down to molecular oxygen.

The highest concentration of ozone (90%) in stratosphere approximately 20-25 km above, around the earth surface is known as ozone layer or ozonosphere. Ozonosphere is important because it absorbs ultraviolet radiations of the sun. The fact that upper region of stratosphere becomes warmer with increasing distance from the earth is due to transformation of absorbed ultraviolet rays into heat (Craig, 1968). The absorption of ultraviolet radiation by ozone umbrella is of paramount importance in the ecosystem because these radiations are prevented from reaching the earth surface where it would be lethal to living organisms. Ozonosphere also acts as a blanket that reduces the cooling rate of earth.

Mesosphere. It is the third layer of atmosphere next to stratopause. It is about 40 km in height. This region is characterised by low atmospheric pressure and low temperature. The temperature begins to drop from stratopause, goes on decreasing with the increase in the height and reaches a minimum of about —95°C at a level some 80 to 90 km above the earth surface The upper limit of the mesosphere is termed *mesopause.*

Thermosphere. Next to mesosphere is thermosphere which extends upto 500 km above the earth surface and is characterised by steady temperature increase with the height from mesopause. The thermosphere includes the regions in which ultraviolet radiations and cosmic rays cause ionization of molecules like oxygen and nitric oxide. This region is called *ionosphere.* In ionosphere, molecules of gases are so widely spaced that high frequency audible sound are not carried by the atmosphere.

Exosphere. The region of atmosphere above the thermosphere is called exosphere or outer space which lacks atoms except those of hydrogen and helium. This extends upto 32190 km from the earth. Exosphere has a very high temperature due to solar radiation. The earth's magnetic field becomes more important than gravity in distribution of atomic particles in the exosphere.

Air is an Ecological Factor

The gaseous mixture present in the troposphere is called air. Air moving from high pressure to low atmospheric pressure is called wind. It is an important ecological factor of the environment and it affects plant life mainly on flat plains, high altitudes in mountains and along the sea cost. Wind is directly involved in transpiration, causing various types of mechanical injuries and helps in dispersal of pollen, fruits and seeds. It also modifies the water relations and light conditions of a particular habitat through its effect on evaporation. It plays both positive and negative roles in atmosphere, for instance it has drying effect upon soil and may occasionally act in opposite direction bringing in moist air that reduces the transpiration and evaporation and may actually lead to the deposition and precipitation of moisture.

Effect of Air on Vegetation

The following are the important effects of wind on vegetation :

(*i*) Wind increases the water loss by constantly removing the air saturated with water vapour from the intercellular spaces of the leaves and bringing unsaturated air in contact with leaves and young shoots.

(*ii*) Mechanically, wind causes erosion of soil and abrasion of vegetation through removal of particles. Physiologically, it decreases the growth of plants by way of reducing the moisture content of air and reducing the turgidity of plant parts on which it impinges. Moist air promotes the growth of mesophytes.

(*iii*) In strong dry and hot winds, young parts of plants may become shrivelled and killed in a few hours and the surface of soil may become dry.

(*iv*) In open situations, *e.g.*, seashores and high mountain tops, where the strong winds blow all the year round in one direction, the trunks and branches of trees are twisted chiefly in the direction of prevailing wind. In such plants, generally, the growth of buds becomes checked on windward side (Fig. 2.7).

Fig. 2.7. Figure shows the effect of abrasive ice particles and strong winds on the branch growth. The buds on the windward side are either killed or they grow opposite the windward side.

(*v*) In strong winds, big trees are uprooted and small plants and grasses are affected but to a very little extent. By strong wind, weak plants like wheat, maize, sugarcane, jowar, etc., are bent against the ground. These lodged plants, if their stems are not too mature, may become partially erect. This is due to differential growth of the meristematic lower nodes of these plants.

(*iv*) Wind is an important agent for the dispersal of pollen grains, fruits, seeds and spores of the plants. In deserts, the strong wind carries the sand and seeds. Thus, it plays important role in local distribution of plant species or communities of plants. Some types are wind resistant but some may be totally dependent on wind for their dispersal. Many plants are unable to flourish or even exist in the exposed situations if they are brought to such places by wind. Strong wind causes injuries to plants growing at high altitudes. In desert, the storm results in big sand-dunes which cover the vegetation (Fig. 2.8).

Fig. 2.8. A view of sand dunes in the desert

(*vii*) In the areas subjected to strong winds, the leaves of plants become small and rolled. The transverse section of stem shows eccentrically developed secondary wood, *i.e.*, the diameter of the trunk in the direction of wind becomes greater than that at rightangle to it. The plants in such areas show extensive development of mechanical tissues which provide mechanical support and save the plants from wind injuries.

In India and many other countries of the world, unchecked winds have caused total disappearance of vegetation at certain places and rendered big areas deserted. Rajasthan desert in India is spreading eastward due to unchecked wind erosion.

MICROCLIMATE OR MICROENVIRONMENT

The important climatic factors, such as temperature, moisture, light and others show not only the regional variations but also the local, horizontal, and vertical differences. The organisms occupying the same general habitat may actually be living under varied conditions. The term 'microclimate' or 'microenvironment' refers strictly to combination of local atmospheric factors differing from the prevalent general climate (macroclimate) of the region. This is owing to the uneven topography, plant cover, etc. The microclimates are recognisable only in a very small area and they are sufficient to induce marked local variations in type and composition of flora. Thus, microclimate is of great ecological importance in the studies of plant and animal behaviour.

In one macroclimatic region there may exist a number of microclimates differing in several respects from the general climate, as for example, in a dense forest tall trees are exposed to high intensities of light and temperature, but shrubby plants growing in the shades of trees do not get the same climatic conditions or in other words those shrubby plants grow in the environment of dim light, high moisture and low temperature, and again at ground level the conditions are changed so much that only some peculiar shade loving plants can thrive in them. Thus, it becomes very clear that climatic conditions of a particular region vary at different levels, both vertically and horizontally.

Puri (1960) has summarised a good deal of informations on microclimates of plant communities. A number of physiological experiments can be conducted but nowadays emphasis is laid on work under controlled conditions. At some places phytotrons and climatotrons are used for such studies which constitute a series of chambers with devices to control different factors of climate. These experiments are useful in knowing the exact climatic requirements of a particular plant and defining the role of each factor. Although it is not easy to work out the effect of environment as a whole yet the results obtained by the study of individual factors if coordinated may provide important conclusions.

Importance of microclimate is realised in modern trends in agronomy and soil conservation. Interplay of variations in microclimate and soil conditions results in marked variations in agricultural potentiality within short distances.

SOIL

Earth is a spherical planet of the solar system. It is differentiated into three main layers—crust, mantle and core. The core is the central portion of the earth containing nickel and iron in molten or vapourised state. This is about 3470 km from the centre. The mantle is the middle layer of the earth which extends about 2900 km above the core. The mantle is in molten state. The crust is the outermost solid zone of the earth which is about 9 to 40 km thick. The solid component of earth is called **lithosphere** which is not a uniform layer. It is composed of different kinds of rocks, which as a whole has a composition and density of granite rock. The layer lining the floor of continents is granite layer and the layer of rock lining the floor of ocean is **basaltic.** The crust is very complex in composition and its surface is covered with the soil which supports rich and highly diversified biotic communities (Fig. 2.9). It contains the continents and the oceans, the soil and the sediments, minerals and water.

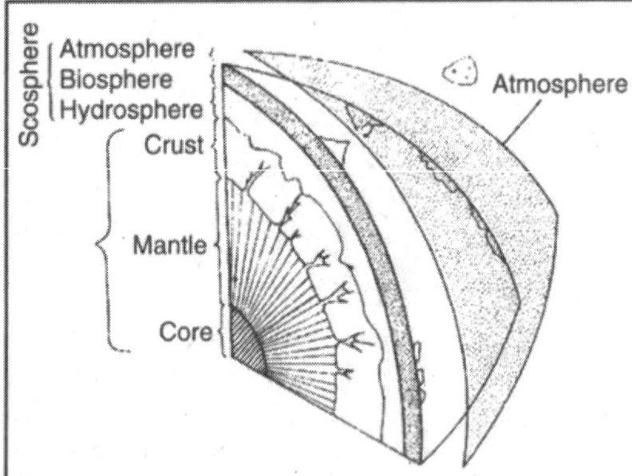

Fig. 2.9. Cross section of earth showing different regions.

The soil is one of the most important ecological factors called **edaphic factors**. It is the most characteristic feature of the terrestrial environment. It is the reservoir of biogenic salts and minerals which are essential for the living organisms. It is the outermost layer of earth's crust and is being constantly formed through weathering of rocks. It is not only a factor of the environment but also a product of the organic activities causing the biological weathering of rock.

Soil is defined as the unconsolidated top or superficial layer of earth's crust lying below any aerial vegetation and undecomposed dead organic remains and extending down to the limits to which it affects the plants growing about its surface. Beneath the soil, lie the subsoil and unweatherd rocks.

Edaphic factors are those which are dependent on the soil as such—on soil constitution, soil water, soil air, soil organisms and so forth. Soils at different places vary considerably in their structure, components and properties. These differences in the soils are often largely responsible for differences in vegetation within the same climatic region, and consequently they are of great significance in the distribution of plant communities.

Soil is a stratified mixture of inorganic and organic materials. The inorganic or mineral constituents of soil are derived from some parent materials, the soil forming rocks by fragmentation or break-down or weathering and the organic constituents of the soil are formed either by decomposition of dead remains of plants and animals or through metabolic activities of living organisms present in the soil.

Weathering of rocks is the initial steps of soil forming process which is brought about by (*i*) physical factors like temperature, water, ice, gravity and wind, (*ii*) chemical factors like solvent action, hydrolysis, oxidation, reduction, carbonation and hydration, and (*iii*) biological factors. Soil which is formed from the products of weathering processes is called embryonic or primary soil. It may mature into the following types of soils :

(*i*) **Residual or sedimentary soil.** This is the mature soil lying immediately over the parent rocks.

(*ii*) **Skeletal or immature soil.** This is only partly weathered material lacking maturation.

(*iii*) **Transported or secondary soil.** The transported soils are those in which the parent materials have been shifted to different places by the agency of glaciers, streams and rivers (**Alluvial soil**) the gravitational forces as landslides (**Colluvial soil**), wind (**Aeolian soil**), sand storms (**sand dunes**), standing water and wave action (**Lacustrine soil**) and oceanic waves (**marine soil**).

The second important process involved in the soil formation includes addition of organic matter, humification and mineralization. Organic matters are added to the embryonic soil by various living organisms. Plants and animals living in and on the soil after their death are decomposed and the decomposition product is mixed with the soil. All organic plant debris fallen recently to the ground is called ***litter***. The litter is composed of dead leaves, twigs, wood, dead roots and various plant products. Just below the fresh litter often occurs the material derived from preceding season's litter in which decay or microbial decomposition has set in. This is called ***duff***. The litter is decomposed by soil microbes such as bacteria, actinomycetes and other fungi. The products of decomposition include various types of inorganic and organic plant nutrients. They are all incorporated into mineral particles which then become dark in colour. The residual finally divided amorphous, incompletely decomposeds black coloured organic matter added to the mineral matter of the soil is called **humus** and the process of its formation is referred to as ***humification***. Humus includes two types of organic matter : the partially decomposed organic matter derived from litter and excreta of soil animals like centipedes, millipedes, earthworms, mites, grass-hoppers, etc. which feed on the litter of plant material. Gradually the humus is completely decomposed into simple compounds like carbon dioxide, water and minerals salts by a process called ***mineralisation***.

Depending upon the organic content, soils are generally classified as follows :

(*i*) *Mineral soils* which are rich in mineral particles.

(*ii*) *Peat and muck* which are rich in organic matter in wet areas.

(*iii*) *Mors* which are low in basic minerals.

(*iv*) *Mulls* which are rich in base contents.

Muller (1879, 1884) has recognised two kinds of humus : *mor* and *mull*. **Mor humus** is acidic, abounds in fungi and is low in bacterial content. Fungul mycelia may help to bind together particles of humus and decomposing litter into matted layers. A well developed mor can be differentiated into the following three layers :

1. **Surface litter or L-layer** composing undecomposed leaves and twigs.

2. **Fermentation or F-layer** in which decomposition has proceeded some way towards development of humus. This layer is just below the L-layer.

3. **Humus or H-layer** which is found just below the F-layer. In this layer degraded humus fraction is accumulated. ***Mor*** is deficient in calcium and it develops on sandy soil under conifers. ***Mull*** is, on the other hand, slightly alkaline or neutral and abounds in bacteria. It does not show layer differentiation because of abundance of earthworms which promote mixing of organic and mineral materials. Intermediate between two types of humus is ***moder*** which has a richer and varied fauna.

Soil Profile

Soil profile is the term used for the vertical section of mature soil upto the parental material to show different layers of soil.

Soils commonly become stratified into layers or horizons at different depths. The layers of soil at different depths show different compositions and natures. Normally three main horizons or groups of horizons can be recognised. These are the upper horizons or **'*A*' horizon**, the middle horizon or **'*B*' zone** and **'*C*' horizon** (Fig. 2.10). A and B horizons collectively constitute true soil. Some workers recognised an additional horizon called **organic** or **'*O*' horizon** above the *A* horizon, and **D horizon** below C horizon: These horizons are often divided by the use of appropriate subscripts.

'*O*' horizon. The upper most layer of soil profile is called *O* horizon. It is also called litter zone or Organic zone. *O* horizon is present in forest soil but absent in the soil of deserts, grasslands and cultivated fields *O* horizon can be divided into two sublayers normely *Aoo* or O_1 horizon and *Ao* or O_2 horizons.

***Aoo* or O_1 horizon.** It is the top layer formed by fallen leaves and things on the forest floor which are undecomposed.

***Ao* or O_2 Zone.** *Ao* or i_2 horizon lies below the *Aoo* or O_1 horizon. It is a layer of partly decomposed organic matter; the upper portion which is partially decomposed is called the **duff** and the lower part is completely decomposed and the organic structures have lost their identity to form black amorphous material called **humus.**

'*A*' horizon designates the top stratum which is subjected to marked leaching. It is a layer of greatest biological concern as the plant roots, small animals, and microflora and fauna are found here most densely. In this zone the concentration of the organic matter is highest, and hence it is the dominant reservoir of plant nutrients.

It is divided into three subzones which are denoted by A_1, A_2 and A_3. These can be distinguished by distinct colours, textures and organic contents.

A_1 horizon. In this rich humus is mixed with minerals. It contains fungi & bacteria in abundance. It is dark coloured.

A_2 Horizon. It is light coloured zone with less humus and maximum leaching of silicates, oxides of iron and aluminium.

A_3 Horizon. It is transition zone between A and B horizons, slightly darker than *O* horizon.

The ***B* horizon** or the **subsoil** lying under *A* horizon has little organic matter, very few plant roots and a sparse microflora and fauna. In it, iron and aluminium compounds are often accumulated. It is also divided into B_1, B_2 & B_3 zones. *A* and *B* horizons collectively represent the true soil.

***C* horizon**. At the bottom of *B* horizon is the *C* horizon which contains the parent materials of the soil. In this layer the organic matters are present in small amounts and little or no life is noted.

The essential components of most garden soils are as follows :

(*i*) Mineral particles of variable sizes obtained by weathering or breakdown of the rocks : 40—50% by volume and 95% by weight.

(*ii*) Soil water : 5.30%;

(*iii*) Soil air or soil atmosphere : 25–30%;

(*iv*) Organic matter or soil humus arising from the death and decay of the parts of plants and animals or added manures : 1 to 5% by weight.

(*v*) Soil organisms, including both soil flora and soil fauna, as for example, protozoa, nematodes, earthworms, bacteria, fungi, algae, etc. Fig. 2.11 gives a rough idea about the percentages of various components of garden soil.

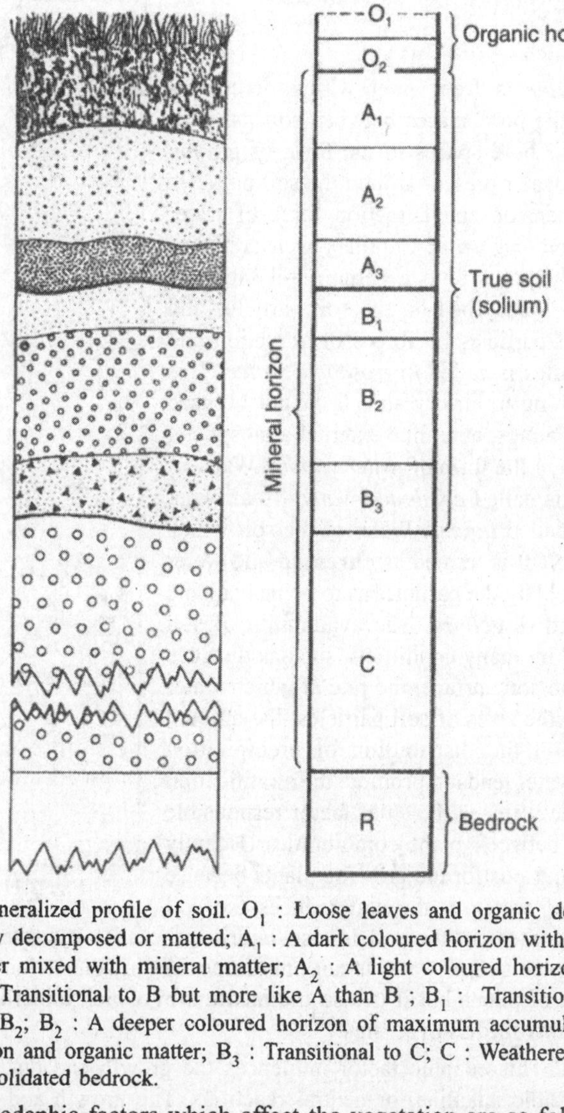

Fig. 2.10. A generalized profile of soil. O_1 : Loose leaves and organic debris; O_2 : organic debris partly decomposed or matted; A_1 : A dark coloured horizon with a high content of organic matter mixed with mineral matter; A_2 : A light coloured horizon of maximum leaching; A_3 : Transitional to B but more like A than B ; B_1 : Transitional layer but more like A than B_2; B_2 : A deeper coloured horizon of maximum accumulation of clay minerals or of iron and organic matter; B_3 : Transitional to C; C : Weathered material (regolith) : R : Consolidated bedrock.

The important edaphic factors which affect the vegetation are as follows:

(*i*) Soil moisture;

(*ii*) Soil reactions;

(*iii*) Soil nutrients;

(*iv*) Soil temperature;

(*v*) Soil aeration or soil atmosphere; and

(*vi*) Biotic components of the soil.

(*i*) **Soil moisture.** Plants absorb a small quantity of rain water and dew directly but they take a large quantity of water from the soil. Water held in the soil is found in the following forms (Fig. 2.12) :

(*i*) Gravitational water;

(*ii*) Capillary water;

(*iii*) Hygroscopic water;

(*iv*) Water vapour;

(*v*) Combined water.

Gravitational water is free water which percolates downwardly through the pore spaces between soil particles and accumulates in the pore spaces in the form of *ground water*. The amount of water present around the soil particles and held by surface tension and attraction force of water molecules is called *capillary water.* Capillary water remains readily available to the roots upto a certain soil moisture tension. Water which is adsorbed on the soil particles and held on the surface of particles by forces of attraction and cohesion of its molecules is called *hygroscopic water*. This is the moisture remaining in air-dry soil. It cannot be used by the plants. The soil atmosphere, like external atmosphere, also contains moisture in the form of *water vapour.* Water of chemical compounds is called *combined water.* Total water content of the soil is called ***holard***. Water of the soil which is easily available to plant is termed as **chresard** and water which is so strongly held by the particles as to be unavailable to the plants is termed as ***echard***. The availability of soil moisture is influenced by many conditions, such as distance of water table from the soil surface, the rate at which water percolates downward, the sizes of soil particles, the amount of annual rainfall, and the distribution of precipitation throughout the year. Water tends to promote the stratification of soil. Soil's available water is the chief factor responsible for local differences between plant communities. Heavily water logged soil is injurious for the growing plants because heavy accumulation of water in the soil reduces the soil aeration. Low water content in the soil is also injurious because it causes either temporary or permanent wilting of plants. Soil water is not only important in connection with direct fulfilment of water requirements of plants but it is also the medium by which mineral salts essential in the nutrition enter the plants in dissolved state.

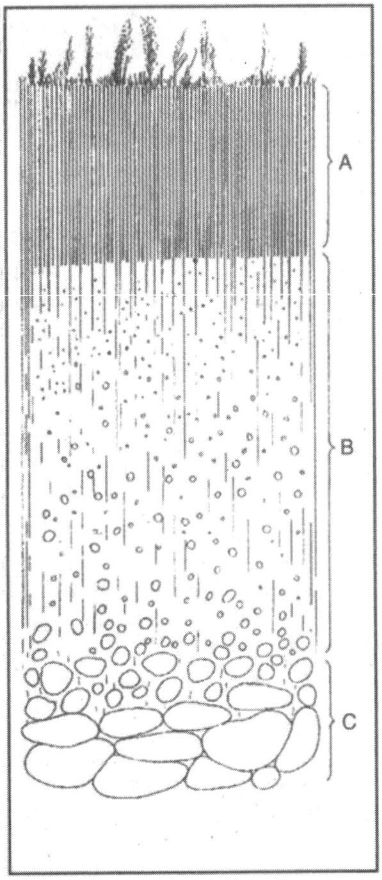

Fig. 2.11. Soil components

(*ii*) **Soil reaction.** This edaphic factor influences the growth and distribution of the plants. The soil may show acidic, alkaline, or neutral reactions. The growth and productivity of many species of plants are critically related to soil acidity. Species of *Rhododendron,* cranberris are acid loving. Most of the field crops, such as barley, maize, soybeans, tomato, rye, potato, flourish in slightly acidic soils. Many ferns and beech trees thrive best in slightly alkaline soils. Soil acidity is important in many ways; in the behaviour of soil solutes, and in the relation of roots to the soil. Soil acidity affects the availability of iron, manganese, phosphate and other ions. In the acid soils, iron and manganese are available in appreciable quantities, but in the neutral or alkaline soils they are available in meagre quantities to green plants.

The accumulation of calcium, sodium and magnesium salts in the soil results in alkalinity. The reaction of soil influences the absorption of water and soil solutes by roots through its direct effect on inhibition and permeability of root cell membranes.

(*iii*) **Soil nutrients.** The nature and availability of soil solutes are fundamentally important from the standpoint of plant nutrition. Normally, inorganic solutes are absorbed by the plants in the ionic forms. Different species of plants, no doubt they require the same ions for their normal development, require them in varying quantities. In saline soil where the percentage of salt is high, only halophytes (salt loving plants) grow. Some plants require lime and grow in calcium rich soils. Such plants are called *calcicoles* or *calciphytes*. Some plants do not thrive well when they grow in calcium rich soil. They are called *calcifuges* or *oxylophytes*.

Humus, a dark amorphous substance formed by partial degradation of dead organic remains is an important source of mineral and organic nutrients of plants. The fertility of soil is usually correlated with its humus content. Humus is the main source of nutrients for soil micro-organisms and green plants.

(*iv*) **Soil temperature.** Soil temperature in combination with other edaphic factors influences the properties of soil itself and plants as well. Low temperature reduces the rate of water and solute absorption by roots. Root injury due to low temperature in the winter is more common in sandy soil than in clay. Soil temperature is also important in the sense that it plays important role in determining the geographical distribution of plants on the earth. Soil temperature is affected by air temperature, the intensity of sunlight, the angles at which sun rays strike on the surface of soil, daily duration of sunlight, the amount of soil moisture and many other factors.

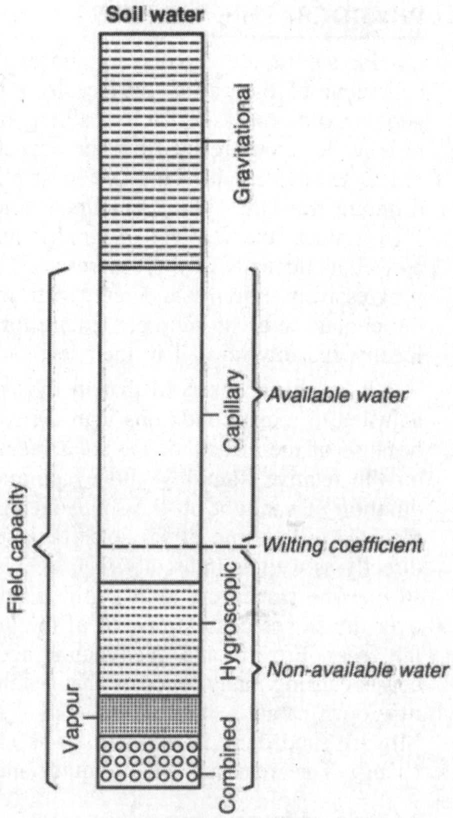

Fig. 2.12. Soil water.

(*v*) **Soil atmosphere.** In the soil, the spaces left between soil particles are called **pore spaces**. These spaces contain air. Soil air contains slightly lower proportion of oxygen and higher one of CO_2 than atmospheric air contains. Water logged soils are deficient in oxygen. Normally, plenty of oxygen for the soil is necessary for the life of micro-organisms and other soil inhabitants. It is also necessary for the respiration of underground parts of higher plants. Oxygen content of the soil is also an important factor in seed germination. The germinating seeds respire rapidly and usually they require large amount of oxygen.

(*vi*) **Soil organisms.** The plants, animals and microbes inhabiting the soils show marked effects on the soil fertility. Decomposing agents, such as bacteria, fungi, and many others convert dead organic matters into humus, free organic compounds and organic ions and thus make the nutrients available to plants. Some soil organisms secrete essential or beneficial substances including growth hormones and some of them secrete toxic substances in the soil which show marked effects on the growth and distribution of plants. Some of the bluegreen algae, like *Nostoc, Anabaena, Cylindrospermum* are beneficial to the higher plants because they fix atmospheric nitrogen into nitrogenous compounds that are utilised by the higher plants. Singh (1961) found that algal films that developed in paddy fields of Uttar Pradesh and Bihar were active nitrogen fixing bluegreen algae. The animals of burrowing habit also play important role in the soil by turning over the soil. Earthworms increase the fertility of soil by adding excretory matters to it and also by making it loose. Local distribution of plant communities in different regions also brings about changes in the composition and fertility of the soil in those regions.

PHYSIOGRAPHIC FACTORS

Physiographic factors are those which are introduced by the structure, conformity and behaviour of the earth's surface by topographic features such as elevation and slopes by the geodynamic processes, such as silting and erosion and consequently by local geology. Mountain, hills, valleys, etc. result from the irregularities in the earth's surface. These strong topographical reliefs tend to produce marked local climates or microclimates. The summits, for example, are different from the sides of mountains and narrow valleys are different from the open plains. At high altitude, the velocity of wind is high, temperature and air pressure decrease, and humidity as well as intensity of light increases. Thus the climatic factors show variations and they become progressively extreme and rapid with increasing altitudes. Major topographic features influence the climate to a considerable extent, for example, the mountain ranges may cause heavy rainfall locally and low rainfall in their lees.

It has been observed that in the northern hemisphere north facing slopes tend to be more adjusted to moist conditions than the south facing ones at similar altitudes. This is so, probably, because of the effects of the solar radiations on the air and soil temperatures and consequently, on the relative humidity and evaporation and through them on the local water situation. The duration of sunlight on the south facing slope is longest. The vegetation on north facing slopes receives little or no direct solar radiations. The slope of the soil surface affects the vegetation directly as well as indirectly. The steepness of slope accelerates the circulation of soil water. The steeper the slopes the more rapid is downward movement of surface water. The water moving over the slopes causes erosion of the top soil and as a result of this vegetations disappear from the areas. Erosion and geodynamic agencies are active in mountain regions and about the sea coasts causing many types of changes in the topography. Such activities often cause formation of new open habitats, both at the sites of erosion and at the places where eroded soil is deposited. Moving sand dune in desert may be cited as example of geodynamic source of physiographic change. The areas subjected to quick and continuous erosion are completely devoid of vegetation.

BIOTIC FACTORS

The biotic factors include the influence of living organisms, both plants and animals upon the vegetation. Any activity of the living organism which may cause marked effects upon vegetation in any way is referred to as **biotic effect.** The biotic effect may be both direct and indirect. It may be beneficial to the plants in some respects but detrimental in other respects.

The plants live together in a community and influence one another. In the forest there are many plant communities, such as trees, shrubs, herbs, mosses, lichens. These communities interact with one another and adjust according to environmental conditions. Trees cast their shadow on many shade-loving plants which grow around or beneath them. The micro-organisms, such as bacteria, algae, fungi, and viruses affect the life of plants of a given area in many ways. Besides these, the decomposition of dead parts of plant bodies causes significant addition of organic compounds and humus to soil. In this way, vegetation modifies the habitat to a considerable extent. Similarly, animals which are in close association with plants also affect the plant life in one or several ways. Many animals use plants as their food and for shelter as well. Besides animals, the man is most significant agent for modifying the vegetation.

The biotic effects modifying the vegetation can be discussed in the following heads :

(1) Interactions between the plants and local animals and man.

(2) Interaction among plants growing in a community.

(3) Interaction between plants and soil micro-organisms.

1. Interaction between Plants and Local Animals and Man

These can be described under the following heads :

(*i*) Effects of grazing and browsing by animals.

(*ii*) Role of animals in the pollination.

(*iii*) Role of animals in the dispersal of seeds and fruits.

(*iv*) Insects and carnivorous plants.

(*v*) Effects of human activities on vegetation.

(*vi*) Myremecophily.

(*vii*) Miscellaneous effects.

(*i*) **Effects of grazing and browsing.** *Grazing* means eating away of unharvested herbs as forage by animals (Fig. 2.13), as for example, eating away of grasses by goats whereas *browsing* refers to a similar use of shrubs or trees by animals (Fig. 2.14), as for example, eating away of leaves and small twigs of *Margosa* (Neem) by camels. The animals destroy a large part of vegetation by grazing and browsing. Some animals prefer to graze and browse on some particular plant species, *i.e.,* they show selective grazing and browsing, *e.g.,* sheep normally prefer forbs, horses and cattle prefer grasses and goats and deer prefer woody and leafy parts of plants. Small annual plants become uprooted and disappear after being grazed. In browsing, taller plants such as trees and shrubs are little affected.

Fig. 2.13 Grazing of grasses by buffalo.

Fig. 2.14 Grazing and Browsing by animals.

Various other effects of grazing and browsing are summarised briefly in the following points :

(*a*) The grazing and browsing adversely affect the aeration of soil and make it compact and hard and finally render the soil unfit for the growth of trees and shrubs. Forests open to cattle are changed first into shrubby vegetations and finally into grasslands. Excessive grazing and browsing may thus change the pattern of vegetation and finally lead the area to develop into desert. Murphy (1951) is of the opinion that Sahara desert developed as a result of unchecked and excessive grazing by goats, sheep and by introduction of camels in the area. Experimental works near the periphery of Sahara by Kassas (1956) have shown that many vegetational changes took place when grazing by animals was completely checked. Unrestricted grazing and browsing are the main causes for the eastward spread of desert in the part of Punjab, Delhi and Rajasthan.

(*b*) The grazing and browsing reduce greatly the frequency of photosynthetic organs (leaves and apical green parts of stem) and thus curtail the assimilation.

(*c*) The grazing and browsing reduce the vegetation from the surface of earth to a considerable extent and thus expose the soil for erosion.

(*d*) The most important effect of grazing and browsing is the trampling. In the trampling complete destruction of small and weak annual herbs is caused by the hoofs, paws and feet of animals, but the shrubs and trees are little affected. Usually trampled area becomes inhabited by special type of plants which can withstand the mechanical effect of trampling. These plants propagate vegetatively and are not dependent upon the seed for their propagation. Some grasses, *Heylendia latibrosa* for example, survive better in the grazed habitats than in non-grazed areas. Ecological differentiation due to grazing has been reported in certain other grasses and weeds. *Dichanthium annulatum* develops underground vegetative buds for reproduction in response to grazing of above ground parts. Plants show certain protective measures such as bitter taste of leaves, stems and fruits, prickles and spines on the surfaces of stems and leaves, which keep the herbivores away from them.

(*e*) In grazed pasture and meadows, dung avoiding (coprophobic) plants disappear giving place for the colonization of non-coprophilous vegetation.

(*ii*) Role of animals in pollination. A large number of plants depend on insects, birds and a number of animals for their pollination. These plants develop coloured flowers. The flowers possess scents, nectar, sap, edible pollens and many other characteristic structures for attracting insects towards them. Insects, birds and other pollinators visit the flowers in search of honey and edible pollens. Flowers in the families Rosaceae, Compositae, Leguminosae, Rutaceae, Umbelliferae, Euphorbiaceae, Cruciferae, Ranunculaceae are pollinated by insects. Some plants are specialized in their pollination by particular type of animals, for example, *Rafflesia* is pollinated by elephants and birds, bilipped flowers of *Salvia* are pollinated by bees, entomophilous flowers of orchids, *Ficus* and *Calotropis* are pollinated characteristically by insects. It is observed that different types of flowers and their pollinators generally live together in the same biotic communities and affect each other's life. R.F. Daubenmire says that "Extreme specialisation is as dangerous in connection with pollination as with other functions in the biological world, for not only must the ranges of two symbionts coincide but also the extinction of the one heralds the doom of the other."

Besides insects, birds, bats and some other animals, man too is taking active part in pollinating artificially one plant with the pollen of some other plant species. The artificial pollination is being used by man for the production of high yielding and disease resistant plant varieties.

(*iii*) Role of animals in the dispersal of fruits and seeds. Many animals, such as birds, bats, monkeys, act as important agents for disseminating the seeds, fruits and spores and thus they play important role in the migration of plants.

The seeds of many plants are very hard. Such seeds along with fleshy parts of fruits are swallowed by animals. While passing through the elementary canals of animals hard seeds are not affected by digestivejuices.

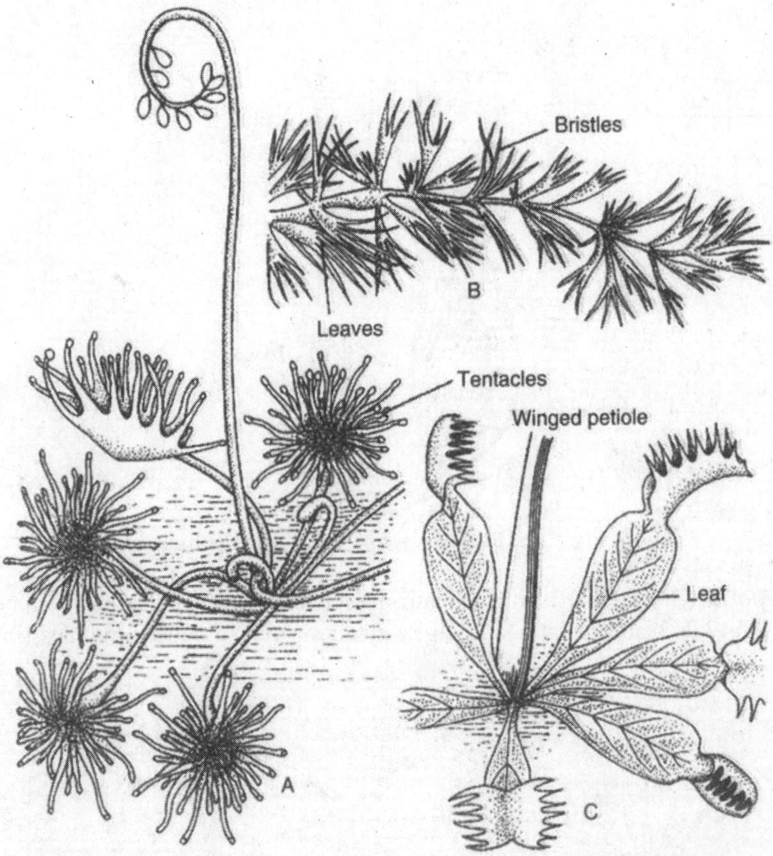

Fig. 2.15. Insectivorous plants A—*Drosera,* B—*Aldrovanda,* C—Venus fly trap (*Dionaea*)

When the animals leave faecal matter, the uninjured seeds present in it germinate. Passing of seeds through the digestive tracts sometimes facilitates their germination in certain cases. The seeds of tomato, tobacco, guava and many other plants are dispersed in this way.

The hairy, spiny, hooked and sticky fruits and seeds of some plants get entangled with the bodies of birds and other animals and with the clothes of man and are brought to distant places. When the animals clean their bodies at some places the seeds are dropped there. Seeds and fruits of *Xanthium, Andropogon, Plumbago, Aegle marmelos* are dispersed in this way.

Ants are good agents for transporting oily seeds and small grains of cereals.

(*iv*) Insects and Carnivorous plants. Semi-autotrophic insectivorous plants, as for example, pitcher plant, *Drosera* (Fig. 2.15A) *Aldrovanda* (Fig. 2.15B), *Dionaea* (Fig. 2.15C), bladderwort, etc., grow in the habitats which are deficient in nitrogenous compounds. These plants have some specialised organs and mechanisms for trapping and assimilating the preys.

Pitcher plants have leaf pitchers containing liquid and enzymes inside (Fig. 2.16). When the insects are trapped down in the pitcher they are digested and assimilated by it. In *Drosera* spathulate leaves are covered with sensitive glandular hairs which shine in the sunlight and attract insects and small flies. When the insects are entangled in the glandular hairs of leaves, digestive enzymes are secreted immediately which kill and digest the bodies of insects. The digested parts of insects are absorbed by the surface cells of the leaves.

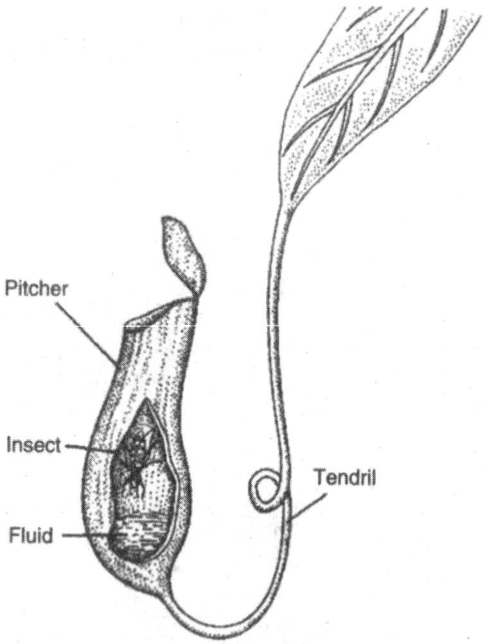

Fig. 2.16. Pitcher plant. A leaf pitcher with insect.

Utricularia (Bladderwort) is an aquatic insectivorous plant which is commonly found in ponds of India. Its bladders catch and digest small swimming aquatic animals (Fig. 2.17).

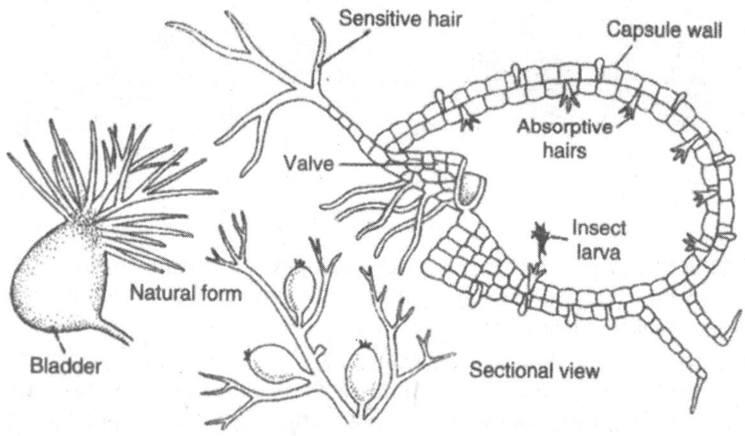

Fig. 2.17. *Utricularia.*

(*v*) **Effects of human activities on vegetation.** Man affects vegetation in the following ways :

(*a*) By cutting, felling and replanting the forest trees.

(*b*) **Cultivation.** Besides the old methods of cultivation, man has adopted a number of advanced methods for cultivation of plants. Cutting, budding, grafting and other methods used by man are proved beneficial for certain plants. Now at various research stations men are performing cross breeding experiments to evolve new varieties of plants that give high yields and are disease resistant. In cultivation, the destruction of weeds by man eliminates the competition among the plants. Proper spacing of plants during cultivation also checks the competition among them for food.

(*c*) **Fire.** Fire is a biological factor rather than a physical factor because it is mostly caused by man's activity. Lightning initiated fires have destroyed plants and animals since their early appearance on earth. In some countries, especially America and Africa, a lot of work has been done on the effects of fire on different ecosystems. A large body of information has developed on the effects of fire on glasslands and forests as well as on the use of fire in land management. The branch of ecology which deals with the effects of fire on ecosystem is called "**Fire ecology**" or **Ecopyrology**. Plants having ability to withstand fire with little or no damage are referred to as **Pyrophytes**. A number of pyrophytes are known to occur in Siwalik hills. Important examples of pyrophytes are *Cochlospermum religiosa, Combretum nanum, Grewia sapida* etc. These small plants are supposed to have become permanently dwarf by annual jungle fires. Pyrophytes are mostly woody plants with thick bark.

Fires caused by man's activity are responsible for complete destruction of vegetation at certain places resulting in temporary or permanent alterations in the characters of vegetations. In some parts of tropics and subtropics, especially in Africa, the burning of grassland has been a regular practice for the last many centuries. Generally in fires the aerial parts of plants are destroyed completely but their roots, rhizomes or other underground parts may sometimes remain unaffected which under favourable conditions may grow and produce new plants. Fire generally makes the area suitable for the growth of grasses and thus improves the quality of forage. Post-burn plants are preferred by herbivores. Animals grazing on burnt grasslands are found to gain weight more rapidly than those grazing on unburnt grasslands. Fire removes harmful plant and animal parasites and pests.

Litter accumulations physically prevent the healthy production and growth of some plant species in grassland. Fire not only removes the choking litter accumulation but also reduces the organic debris to ash. It affects the productivity by stimulating both the above and below ground growth, increases flowering in forbs and seed production in grasses, increases certain species like legumes and improves nutrient contents of the grassland species. Burning in normal course does not affect the grassland soils adversely and generally improves them. Mineral salts of calcium, magnesium, potassium and phosphorus increase with burning. Excessive burning may reduce the humus content and the fertility of the soil. Soil acidity increases and erosion is accelerated. Annual burning coupled with continuous heavy-grazing will have detrimental effects on the health of grassland. The grazing-burning rotation is very useful in the management of Pampas or Savannahs. In the ancient times man used fires to clean forest vegetations and also for lightening purposes. Vansteen (1936) is of the opinion that many deserts in the old world are man-made. Recent works in Australia by Spech *et al.* (1958), Coaldrabe (1951) and in Canada by Schmidt (1960) have thrown light on the significance of fire in determining the type and composition of plant communities.

(*d*) Man also clears the vegetation for making houses, roads, etc.

(*e*) In ancient times many human invasions took place in India which caused great destruction of vegetation. Alexander (330 B.C.), Muslim invasion after 850 A.D. Gori and Gaznavi, and Rajput invasions destroyed dense forests and converted them into deserts. Mohanjodaro and Harappa are examples which are supposed to have become deserted as a result of human invasions. The excavations of Mohanjodaro and Harappa indicate that shrubby plants were abundant in the desert areas of Punjab and Sindh.

(*vi*) **Myremecophily.** Sometimes ants take their abode or shelter on some trees such as Mango, Litchi, Jamun, South American *Acacia* (*Acacia sphaerocephala*) (Fig. 2.18) and so on. These ants act as body guards of the plants against any disturbing agent. In lieu of this defence, the plants provide food and shelter to these ants. This phenomenon is known as **myremecophily**.

(*vii*) **Miscellaneous effects.** The animals also affect the plant life in many other ways. Some animals, as for example, barkeater, rodents may kill a large number of trees. Juice suckin insects, woodpeckers, bud eating birds, sparrow, squirrel and other animals cause great harm to the vegetation. Elephants detach the branches of the trees and sometimes uproot the gigantic trees.

The insects, birds, squirrels, mice and rodents eat abundant seeds. Some animals eat and destroy seeds at the sowing time. Fishes, ducks and other aquatic animals depend upon aquatic plants for food and shelter.

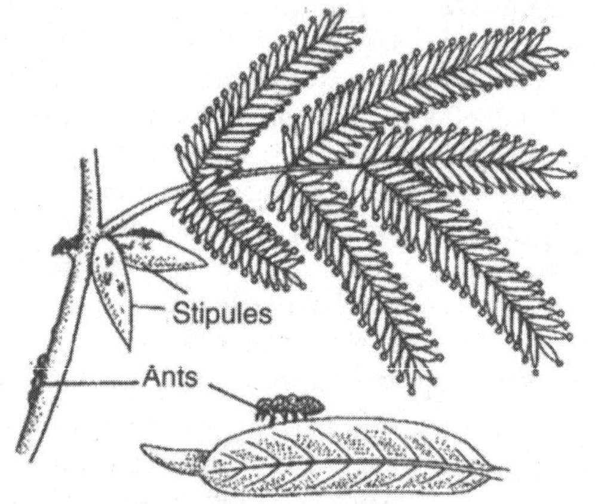

Fig. 2.18. Myremecophily

Termites are chief agents for destroying seeds, seedlings and standing crop plants in the fields. Termites and their termitaria serve important ecological function in African Tropical wet-dry soils which has been appreciated and utilized by indigenous African agriculturists in northern Malawi and neighbouring south-western Tanzania. Howard W. Mielke (Tropical Ecology. Vol. 19, No. 1, 1978) has highlighted the importance of termites and termitaria in Tropical wet dry Africa. Termites and termitaria contribute to the development of soils. The enhancement of the mineral content in termite mounds as compared to adjacent soils has been observed by many workers (Hesse, 1955; Stoops, 1964; Lee and Wood, 1971; Watson, 1975; Milne, 1974; Waston, 1969; Bouillon, 1970). This enhancement in mineral contents of termite mounds has been attributed to the interaction of several factors. Such as (1) Termite behaviour in aggregating finer soil particles within their mounds; (2) termite behaviour in collecting and storing significant amount of plant matter and accompanying base minerals in their mounds, and (3) relatively low rate of leaching occurring on the mounds due to their dome shape and concomitant propensity to shed water. Nitrogen fixing bacteria have been found associated with some species of termites and as a result of this association some species of termites facilitate the fixation of significant amounts of nitrogen. Termites are important food in the human diet of Africans of the Tropical dry-wet climatic regime. Termitaria soils are used as a raw material for the production of cement, mortar, plaster and bricks and are used in the construction of everything from buildings to roads (Lee and Wood, 1971). Termites and shifting cultivation also provide nutrients and habitats for migratory Palaearctic birds.

2. Interaction Among Plants Growing in a Community

Various plants in a community react with one another in several ways for

(*i*) Water,

(*ii*) Essential minerals and organic compounds, and

(*iii*) Light and air.

The taller plants modify the habitat for the plants growing around and underneath them by casting shadow, protecting them from injuries by strong wind, by increasing the atmospheric humidity, and by determining the humus content of the soil.

The most interesting instances of interactions among plants growing in a community are as follows :

(*i*) Action of lianas;

(*ii*) Effects of some epiphytes;

(*iii*) Effects of parasitic plants.

(*i*) **Action of lianas.** Lianas are woody vascular plants growing on the ground, maintaining, more of less, autotrophic mode of life and growing upward taking support of some trees and other objects. The woody stems of these plants have well developed alternating vertical columns of secondary xylem and parenchymatous tissues which enable them to twist around the supporting objects (Fig. 2.19). In tropical evergreen forests, lianas grow at the top of the trees and form the top layer of the forest canopy. This habit enables these lianas to get sufficient light. The lianas affect other plants also because they cast their shadow and check the light from reaching to the plants of lower storeys.

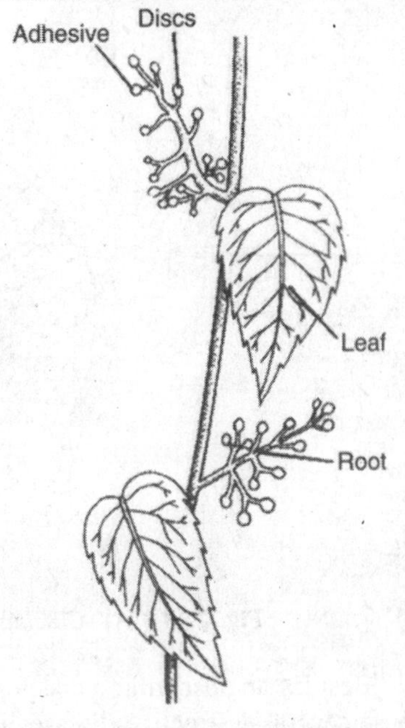

Fig. 2.19. A liana.

(*ii*) **Effects of some epiphytes.** The epiphytes grow on the leaves and stems of other plants. They are autotrophic and are dependent on other plants only for support. Epiphytes differ from parasites in not taking food from the hosts and also differ from lianas in not having any permanent connection with the soil. The examples of epiphytes may be found in the families Orchidaceae, Asclepiadaceae, Bromeliacease. Cactaceae, etc. *Dischidia, Tillandsia* are most common examples. Epiphytes are found in humid climates. The two main problems for these plants are (*a*) maximum absorption of water from the atmosphere and from the bark surface of the supporting plant and (*b*) maximum economy in the water consumption. These plants develop two types of roots, namely the aerial and clinging roots. The aerial roots are thick and have special thin walled porous absorptive tissue, the 'velamen' on their surface. These roots absorb rain water and moisture from the atmosphere. The clinging roots fix the epiphytes on the surface of supporting plants. Because the epiphytes are autotrophic, they do not affect the supporting plants to any considerable extent.

(*iii*) **Effects of parasitic plants.** Some plants are heterotrophic and are dependent on other plants for their food requirements. They are called parasites. These are of the following two types :

(*i*) Ectoparasites (external parasite); and

(*ii*) Endoparasites (internal parasites).

The endoparasites are more destructive than the ectoparasites. Because the parasites take their food from host plants, they check the growth and ultimately cause the death of their hosts. *Cuscuta, Loranthus, Orobanche, Rafflesia,* and sandal wood tree (*Santalum album*) are important parasitic angiosperms which may grow either on roots or on stems and some time even on the leaves of the higher plants. The parasites may be either obligate or facultative.

Cuscuta (Fig. 2.20 A & B) is an obligate stem parasite on *Acacia, Zizyphus* and a number of other angiospermic plants. *Loranthus* is a partial stem parasite on mango. *Orobanche* grows very commonly on the roots of crucifers and solanaceous plants as obligate roots parasite. Other important parasites are *Rafflesia* on the roots of *Vitis Viscum album* on coniferous trees. *Striga,* one of the smallest angiospermic parasites grows on the grasses. *Arceuthobium minutissimum,* an interesting smallest parasitic dicot, is an obligate stem parasite of *Pinus excelsa.*

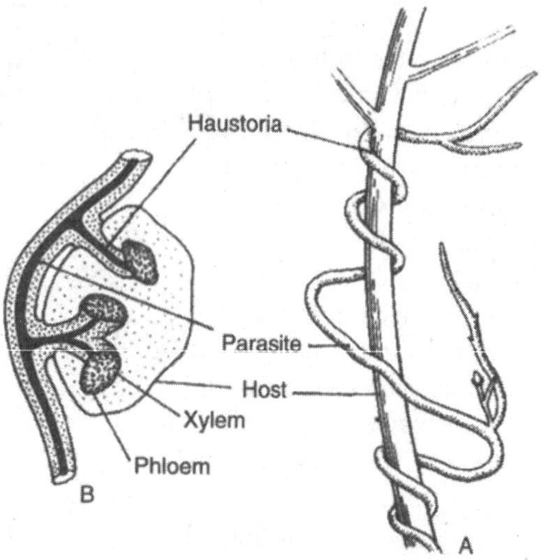

Fig. 2.20. A—*Cuscuta* on the host. B—T.S. of host along with *Cuscuta.*

Besides angiospermic parasites, fungi and bacteria are also known to parasitise plants and cause several destructive diseases in them.

Puccinia graminis is a parasitic fungus causing rust diseases in wheat. Mildews, smuts,, white rust, damping off, and blight diseases are generally caused by parasitic fungi.

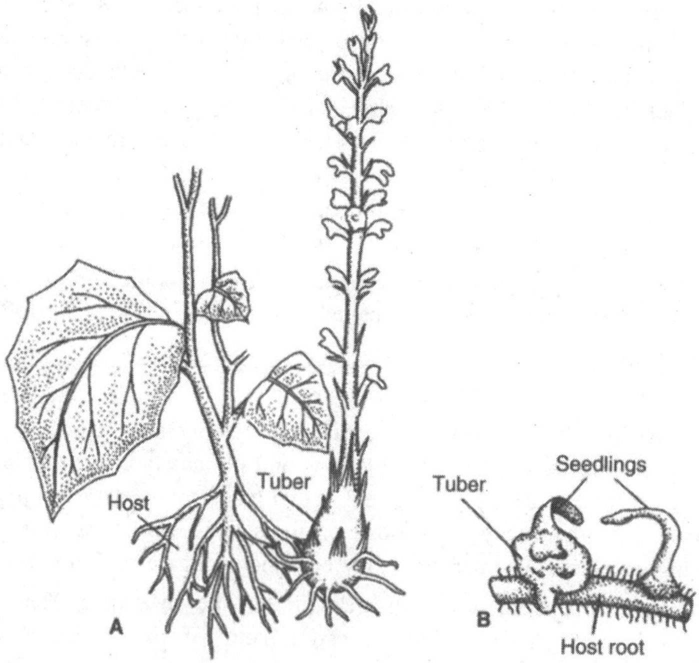

Fig. 2.21. A & B-*Orobanche*. A—Parasite growing on the root of brinjal.
B—A root of host with growing seedlings of parasite.

Some algae are also known to parasitise several plants and cause diseases in them. Most common example is *Cephaleuros* which is found on a number of angiosperms.

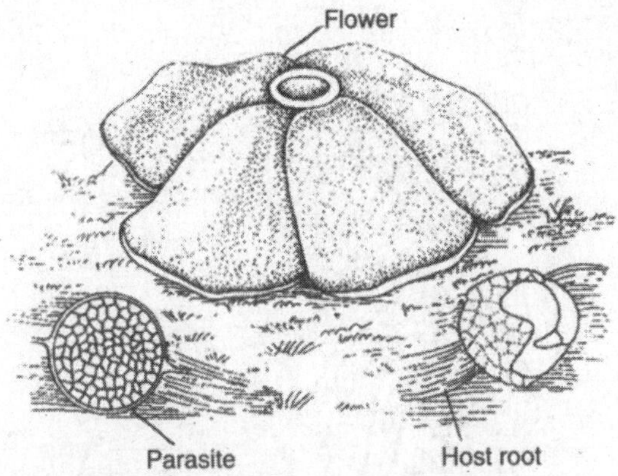

Fig. 2.22. *Rafflesia* on the root of *Vitis*.

3. Interaction between Plants and Microorganisms

Various kinds of bacteria, protozoa, algae, fungi, worms, nematodes and other soil microbes act as important agents which alter the physical and chemical properties of the soils and increase or decrease their fertility. These changes in the soil properties have great impact on the nature and growth of vegetation. Very often soil microbes, such as nematodes, bacteria and fungi cause many diseases in the underground parts of plants. Viruses too cause several mosaic and other diseases in many plants, as for example, the curling of tomato leaves, mosaic patterns in papaya and lady's finger (bhindi), bean mosaic, tobacco mosaic, etc.

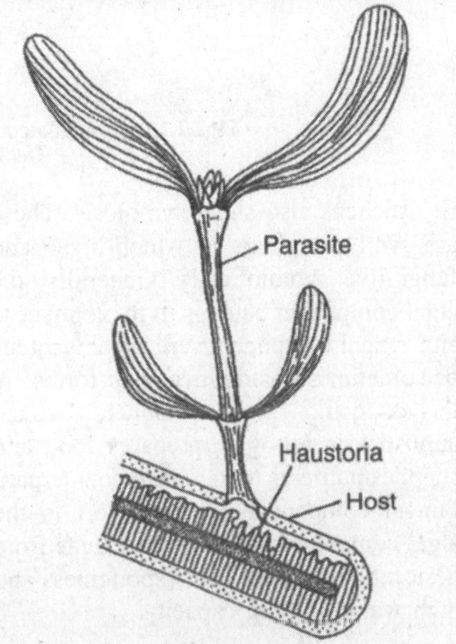

Fig. 2.23. Mistletoe (*Viscum*), a partial stem parasite.

Some microbes secrete growth stimulating substances in the soil which induce the growth of plants.

Besides above effects, the soil micro-organisms show symbiotic activities and many soil fungi form mycorrhizal association with the roots of higher plants. These two phenomena, *i.e.*, symbiosis and mycorrhizal associations are described below.

Symbiotic influence. Some soil microbes live in close association with plants, both benefiting each other. In this association both the organisms are interdependent and they do not harm each other. This mutual relationship between two organisms is known as symbiosis and the interdependent organisms are called symbionts.

Many cases of symbiosis in plants are known. The nodulated roots of legumes contain nitrifying bacteria (*Rhizobium*) (Fig. 2.24). These bacteria fix atmospheric nitrogen into nitrogenous compounds and benefit the legumes by supplying nitrogenous compounds in usable form. The leguminous plants, in return, provide nutrients, water and shelter to bacteria. When the roots of these plants die, decomposition of dead remains takes place in which the nitrogenous substances; nitrite, nitrate and ammonium ions are absorbed by the plants. *Nostoc* and *Anabaena* living symbiotically in the coralloid roots of *Cycas* also fix atmospheric nitrogen.

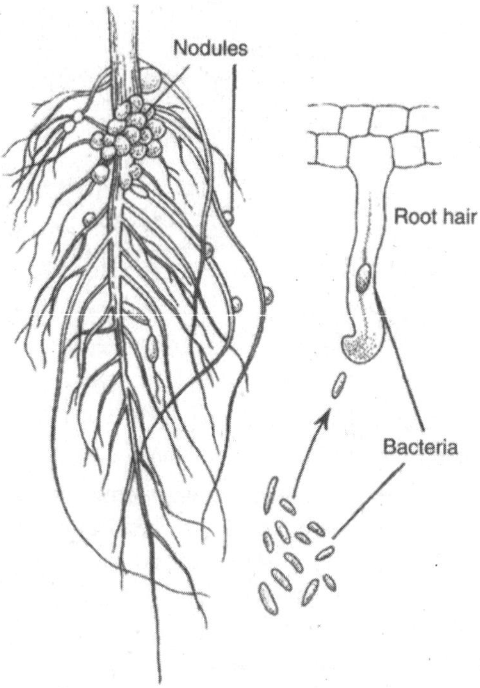

Fig. 2.24. A nodulated root, soil bacteria and infected root hair.

Lichens also show symbiosis. These are synthetic plants in which algae and fungi live symbiotically. Generally, the algal component belongs to myxophyceae and fungal components are ascomycetous or sometimes basidiomycetous forms. In this association, algal component fixes the atmospheric nitrogen, prepares food and supplies nutrients to its fungal counterpart. Fungal component gives support to the algal component. It also saves algae from desiccation because of its sponginess and high water holding capacity.

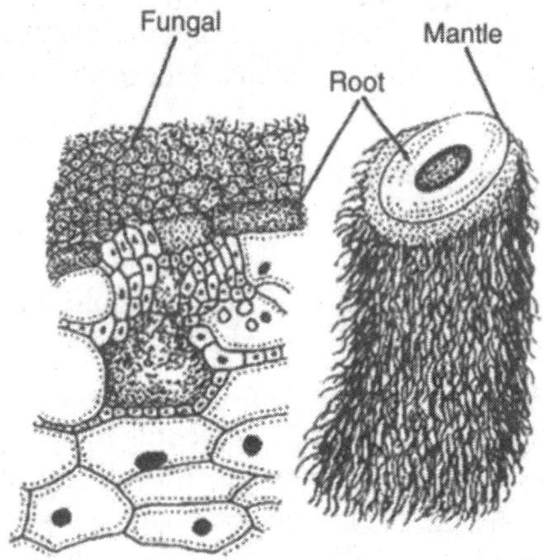

Fig. 2.25. Ectotrophic mycorrhiza or the root fungus.

Mycorrhizal association. Sometimes fungi grow on the surface or inside the roots of higher plants. They are called mycorrhizae. Micorrhizae are of two types.

(*i*) *Ectotrophic mycorrhiza.* Fungus lives on the surface of roots of higher plants (Fig. 2.25).

(*ii*) *Endotrophic mycorrhiza.* Fungus penetrates the deeper tissues of the roots and rhizomes of higher plants (Fig. 2.26).

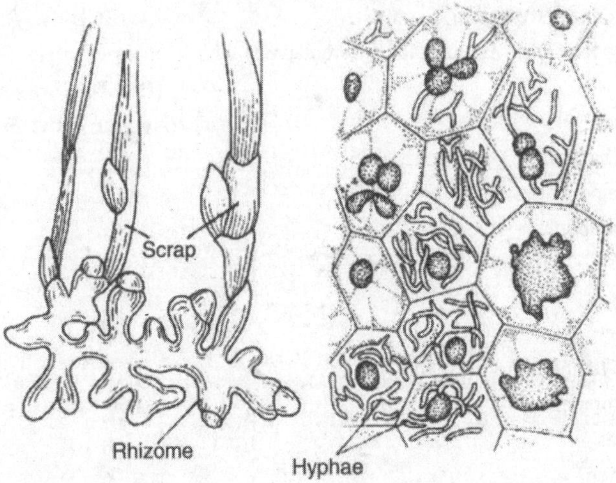

Fig. 2.26. A swollen rhizome with endophytic mycorrhiza.

The roots with mycorrhizae are unbranched and without root-caps and root hairs. Fungal hyphae in this association act like root hairs, absorb water and minerals from the soil and supply them to the roots. The roots in return, provide food and shelter to the mycorrhizae. It has been estimated that some species of plants in about 80% families of seed plants have mycorrhizal association. Blue berries cannot grow without mycorrhiza. Orchid seedlings do not develop properly unless they become infected with mycorrhizae.

Recent researches have shown that mycorrhizae play very important role in the root tissues. They regulate acidity and sugar content in the root tissues and enable the roots grow vigorously.

QUESTIONS

1. What is environment? What are various environmental factors which influence the plant life?
2. What are the ecological effects of light on vegetation? What is the significance of light to plants?
3. What are the effects of biological factors on plants? **(H.P. Univ. B.Sc. III, Bot. IX, 2010)**
4. Write short notes on :
   (*i*) Biotic factors
   (*ii*) Mycorrhiza
   (*iii*) Soil profile
   (*iv*) Soil water
   (*v*) Effect of wind on vegetation.
5. Write on:
   (*i*) Gaseous composition of atmosphere **(Nagpur Univ. B.Sc. III, Bot. II, 2011)**
   (*ii*) Physico-chemical properties of soil
6. Fill in the blanks
   Environmental factors which deal with the structure of soil are termed as ....
   (*i*) Topographic factors
   (*ii*) Climatic factors
   (*iii*) Biotic factors
   (*iv*) Edaphic factors **(H.N.B. Garhwal, Univ., B.Sc. II-2010)**

7. Explain humus and its types in detail. **(H.P. Univ., B.Sc. III, Bot. IX, 2010)**
8. What is ecological niche? Explain its different types with suitable examples. **(H.N.B. Garhwal Univ., B.Sc. II-2010)**
9. Write short notes on: **(Nagpur Univ., B.Sc., III. Bot. II. 2011)**
   (*a*) Global radiation
   (*b*) Biotic components
   (*c*) Abiotic components
   (*d*) Sciophytes
   (*e*) Grazing
   (*f*) Ozonosphere
   (*g*) Effects of temperature on plants
   (*h*) Soil development
   (*i*) Heliophytes
   (*j*) Vernalization
   (*k*) Soil profile

3

Ecosystem

An organism is always in the state of perfect balance with the environment. The environment literally means the surroundings. The environment refers to the things and conditions around the organisms which directly or indirectly influence the life and development of the organisms and their populations. Organisms and environment are two non-separable factors. Organisms interact with each other and also with the physical conditions that are present in their habitats. "The organisms and the physical features of the habitat form an ecological complex or more briefly an ecosystem." (Clarke, 1954).

The concept of ecosystem was first put forth by A.G. Tansley (1935). Ecosystem is the major ecological unit. It has both structure and functions. The structure is related to species diversity. The more complex is the structure the greater is the diversity of the species in the ecosystem. The functions of ecosystem are related to the flow of energy and cycling of materials through structural components of the ecosystem.

According to Woodbury (1954), ecosystem is a complex in which habitat, plants and animals are considered as one interesting unit, the materials and energy of one passing in and out of the others.

According to E.P. Odum, the ecosystem is the basic functional unit of organisms and their environment interacting with each other and with their own components. An ecosystem may be conceived and studied in the habitats of various sizes, *e.g.,* one square metre of grassland, a pool, a large lake, a large tract of forest, balanced aquarium, a certain area of river and ocean. All the ecosystems of the earth are connected to one another, *e.g.,* river ecosystem is connected with the ecosystem of ocean, and a small ecosystem of dead logs is a part of large ecosystem of a forest. A complete self-sufficient ecosystem is rarely found in nature but situations approaching self-sufficiency may occur.

STRUCTURE OF ECOSYSTEM

The structure of an ecosystem is basically a description of the organisms and physical features of environment including the amount and distribution of nutrients in a particular habitat. It also provides information regarding the range of climatic conditions prevailing in the area. From the structure point of view, all ecosystems consist of the following basic components :

1. Abiotic components
2. Biotic components

ABIOTIC COMPONENTS

Ecological relationships are manifested in physico-chemical environment. Abiotic component of ecosystem includes basic inorganic elements and compounds, such as soil, water, oxygen, calcium carbonates, phosphates and a variety of organic compounds (by-products of organic activities or death). It also includes such physical factors and ingredients as moisture, wind currents and solar radiation. Radiant energy of sun is the only significant energy source for any ecosystem. The amount of non-living components, such as carbon, phosphorus, nitrogen, etc. that are present at any given time is known as **standing state** or **standing quantity**.

Biotic Components

The biotic components include all living organisms present in the environmental system. From nutrition point of view, the biotic components can be grouped into two basic components :

(*i*) Autotrophic components, and

(*ii*) Heterotrophic components

The autotrophic components include all green plants which fix the radiant energy of sun and manufacture food from inorganic substances. The heterotrophic components include non-green plants and all animals which take food from autotrophs. So biotic components of an ecosystem can be described under the following three heads :

1. Producers (Autotrophic components),
2. Consumers, and
3. Decomposers or reducers and transformers

The amount of biomass at any time in an ecosystem is known as **standing crop** which is usually expressed as fresh weight, dry weight or as free energy in terms of calories/metre^2.

Producers (Autotrophic elements). The producers are the autotrophic elements—chiefly green plants. They use radiant energy of sun in photosynthetic process whereby carbon dioxide is assimilated and the light energy is converted into chemical energy. The chemical energy is actually locked up in the energy rich carbon compounds. Oxygen is evolved as by-product in the photosynthesis. This is used in respiration by all living things. Algae and other hydrophytes of a pond, grasses of the field, trees of the forests are examples of producers. Chemosynthetic bacteria and carotenoid bearing purple bacteria that also assimilate CO_2 with the energy of sunlight but only in the presence of organic compounds also belong to this category.

The term producer is misleading one because in an energy context, producers produce carbohydrate and not energy. Since they convert or transduce the radiant energy into chemical form, E.J. Kormondy suggests better alternative terms *'converters'* or *'transducers'*. Because of wide use the term producer is still retained.

Consumers. Those living members of ecosystem which consume the food synthesized by producers are called consumers. Under this category are included all kinds of animals that are found in an ecosystem. There are different classes or categories of consumers, such as,

(*a*) Consumers of the first order or primary consumers,

(*b*) Consumers of the second order or secondary consumers,

(*c*) Consumers of the third order or tertiary consumers, and

(*d*) Parasites, scavengers and saprobes.

(*a*) ***Primary consumers.*** These are purely herbivorous animals that are dependent for their food on producers or green plants. Insects, rodents, rabbit, deer, cow, buffalo, goat are some of the common herbivores in the terrestrial ecosystem, and small crustaceans, molluscs, etc. in the aquatic habitat. Elton (1939) named herbivores of ecosystem as "key industry animals". The herbivores serve as the chief food source for carnivores.

(*b*) ***Secondary consumers.*** These are carnivores and omnivores. Carnivores are flesh eating animals and the omnivores are the animals that are adapted to consume herbivores as well as plants as their food. Examples of secondary consumers are sparrow, crow, fox, wolves, dogs, cats, snakes, etc.

(*c*) ***Tertiary consumers.*** These are the top carnivores which prey upon other carnivores, omnivores and herbivores. Lions, tigers, hawk, vulture, etc. are considered as tertiary or top consumers.

(*d*) Besides different classes of consumers, the **parasites**, **scavengers** and **saprobes** are also included in the consumers. The parasitic plants and animals utilize the living tissues of different plants and animals. The scavengers and saprobes utilize dead remains of animals and plants as their food.

Decomposers and transformers. Decomposers and transformers are the living components of the ecosystem and they are fungi and bacteria. Decomposers attack the dead remains of producers and consumers and degrade the complex organic substances into simpler compounds. The simple organic matters are then attacked by another kind of bacteria, the *transformers* which change these organic compounds into the inorganic forms that are suitable for reuse by producers or green plants. The decomposers and transformers play very important role in maintaining the dynamic nature of ecosystems.

FUNCTION OF ECOSYSTEM

An ecosystem is a discrete structural, functional and life sustaining environmental system. The environmental system consists of biotic and abiotic components in a habitat. Biotic component of the ecosystem includes the living organisms; plants, animals and microbes whereas the abiotic component includes inorganic matter and energy. Abiotic components provide the matrix for the synthesis and perpetuation of organic components (protoplasm). The synthesis and perpetuation processes involve energy

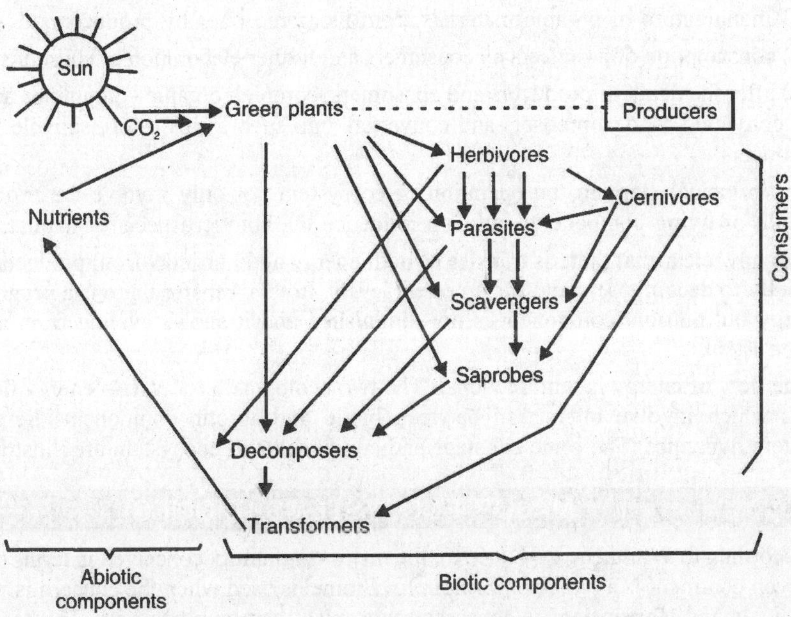

Fig. 3.1. Different components of ecosystem.

exchange and this energy comes from the sun in the form of light or solar energy. Thus, in any ecosystem we have the following functional components :

(*i*) inorganic constituents (air, water and mineral salts)

(*ii*) organisms (plants, animals and microbes), and

(*iii*) energy input which enters from outside (the sun)

These three interact and form an environmental system. Inorganic constituents are synthesized into organic structures by the green plants (**primary producers**) through photosynthesis and the solar energy is utilized in the process. Green plants become the source of energy for renewals (herbivores) which, in turn become source of energy for the flesh eating animals (carnivores). Animals of all types grow and add organic matter to their body weight and their source of energy is complex organic compound taken as food. They are known as **secondary producers**. All the living organisms whether plants or animals in an ecosystem have a definite life span after which they die. The dead organic remains of plants and animals provide food for saprophytic microbes, such as bacteria, fungi and many other animals. The saprobes ultimately decompose the organic structure and break the complex molecules and liberate the inorganic components into their environment. These organisms are known as **decomposers**. During the process of decomposition of organic molecules, the energy which kept the inorganic components bound together in the form of organic molecules gets liberated and dissipated into the environment as heat energy. Thus in an ecosystem energy from the sun, the input is fixed by plants and transferred to animal components. Nutrients are withdrawn from the substrate, deposited in the tissues of the plants and animals, cycled from one feeding group to another, released by decomposition to the soil, water and air and then recycled. The ecosystems operating in different habitats, such as deserts, forests, grasslands and seas are interdependent on one another. The energy and nutrients of one ecosystem may find their way into another so that ultimately all parts of the earth are interrelated, each comprising a part of the total system that keeps the biosphere functioning.

Thus the principal steps in the operation of ecosystem are as follows:

(1) reception of radiant energy of sun,

(2) manufacture of organic materials from inorganic ones by producers,

(3) consumption of producers by consumers and further elaboration of consumed materials; and.

(4) after the death of producers and consumers, complex organic compounds are degraded and finally converted by decomposers and converters into such forms as are suitable for reutilization by producers.

The principal steps in the operation of ecosystem not only involve the production, growth and death of living components but also influence the abiotic aspects of habitat.

It is now clear that there is transfer of both energy and nutrients from producers to consumers and finally to decomposers and transformers levels. In this transfer there is a progressive decrease of energy but nutrient component is not diminished and it shows cycling from abiotic to biotic and *vice versa.*

The flow of energy is unidirectional. The two ecological processes—energy flow and mineral cycling which involve interaction between biotic and abiotic components lie at the heart of ecosystem dynamics. The principal steps and components of ecosystem are illustrated in Fig. 3.1.

MAJOR ECOSYSTEMS (BIOMES)

According to Whittaker, R.H. (1970), the major community conceived in terms of physiognomy on a given continent is a *biome* or *formation.* 'Biome' is used when the concern is with both plants and animals and '**formation**' is concerned with plant communities only. These are the products of interaction of regional climate with biota and substrate. A biome is a grouping of terrestrial ecosystems on a given continent that are similar in vegetation structure (physiognomy), in the major features of environment to which this structure is a response and in some characteristics of their animal communities. The biome concept which is most widely applied to land ecosystems can also be applied in aquatic environment to such zones as, the worldwide structural types defined by major kinds of organisms.

In the broadest sense, there are two major types of ecosystems namely terrestrial and aquatic.

Terrestrial ecosystem operates on the land while the aquatic system operates in the aquatic habitat (Fig. 3.2).

Terrestrial ecosystem can further be divided into the following types:

(1) Forest ecosystem

(2) Grassland ecosystem

(3) Desert ecosystem

(4) Artificial ecosystems which are man-made, as for example, crop fields and gardens.

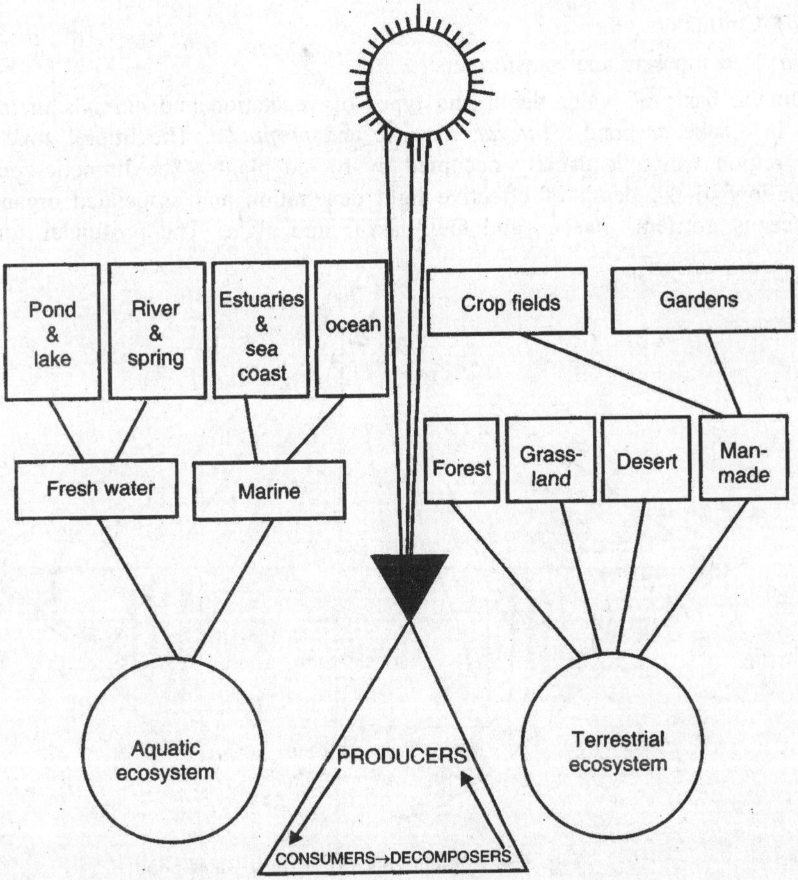

Fig. 3.2. Different types of ecosystems.

On the basis of salt contents in water, aquatic ecosystems can be divided into the following two types of minor ecosystems :

(1) Fresh water ecosystems

(2) Marine or oceanic ecosystems

The fresh water ecosystems include pond ecosystem, lake ecosystem, river ecosystem and spring ecosystem.

SOME EXAMPLES OF ECOSYSTEMS

POND AND LAKE AS ECOSYSTEMS

Pond and lake are fresh water ecosystems in which, like other ecosystems, there are two main components :

(*A*) Abiotic component

(*B*) Biotic component

(A) Abiotic component. Abiotic component of pond consists of water, dissolved minerals, oxygen and carbon dioxide. Solar radiations are the main source of energy.

(B) Biotic component. It includes the following :

(*i*) Producers

(*ii*) Consumers

(*iii*) Decomposers and transformers.

On the basis of water depth and types of vegetation and animals there may be three zones in a lake or pond : *littoral, limnetic* and *profundal.* The littoral zone is the shallow water region which is usually occupied by rooted plants. The limnetic-zone ranges from the shallow to the depth of effective light penetration and associated organisms are small crustaceans, rotifers, insects, and their larvae and algae. The profundal zone is the deep water parts where there is no effective light penetration. The associated organisms are snails, mussels, crabs and worms (Fig. 3.3).

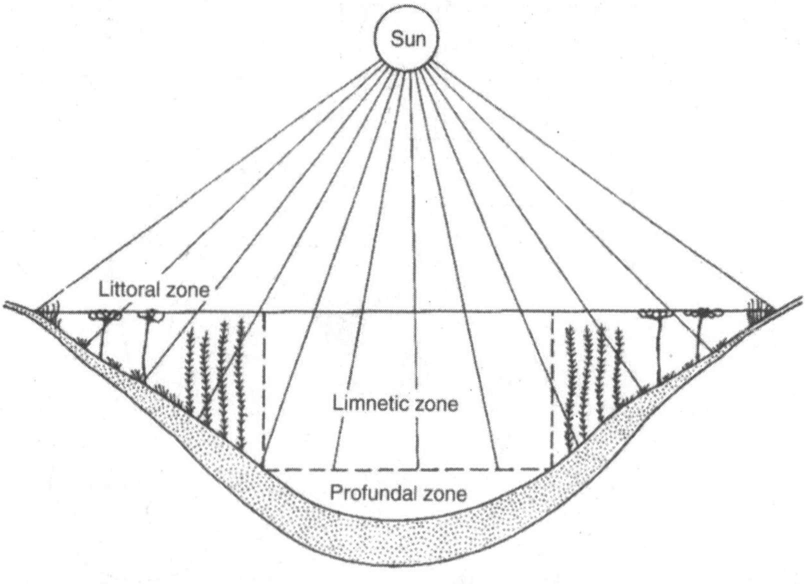

Fig. 3.3 Different zones of a fresh water pond.

(*i*) **Producers.** The main producers in pond or lake ecosystem are algae and other aquatic plants, such as *Azolla, Hydrilla, Potamogeton, Pistia, Wolffia, Lemna, Eichhornia, Nymphaea, Jussiaea,* etc. These are either floating or suspended or rooted at the bottom. The green plants convert the radiant energy into chemical energy through photosynthesis. The chemical energy stored in the form of food is utilised by all the organisms. Oxygen evolved by producers in photosynthesis is utilised by all the living organisms in respiration.

(*ii*) **Consumers.** In a pond ecosystem, the primary consumers are tadpole larvae of frogs, fishes and other aquatic animals which consume green plants and algae as their food. These herbivorous aquatic animals are the food of secondary consumers. Frogs, big fishes, water snakes, crabs are secondary consumers. In the pond, besides the secondary consumers, there are consumers of highest order, such as water-birds, turtils, etc.

(*iii*) **Decomposers and Transformers.** When aquatic plants and animals die, a large number of bacteria and fungi attack their dead bodies and convert the complex organic substances into simpler inorganic compounds and elements. These micro-organisms are called *decomposers*. The chemical elements liberated by decomposers are again utilized by green plants in their nutrition (Fig. 3.4).

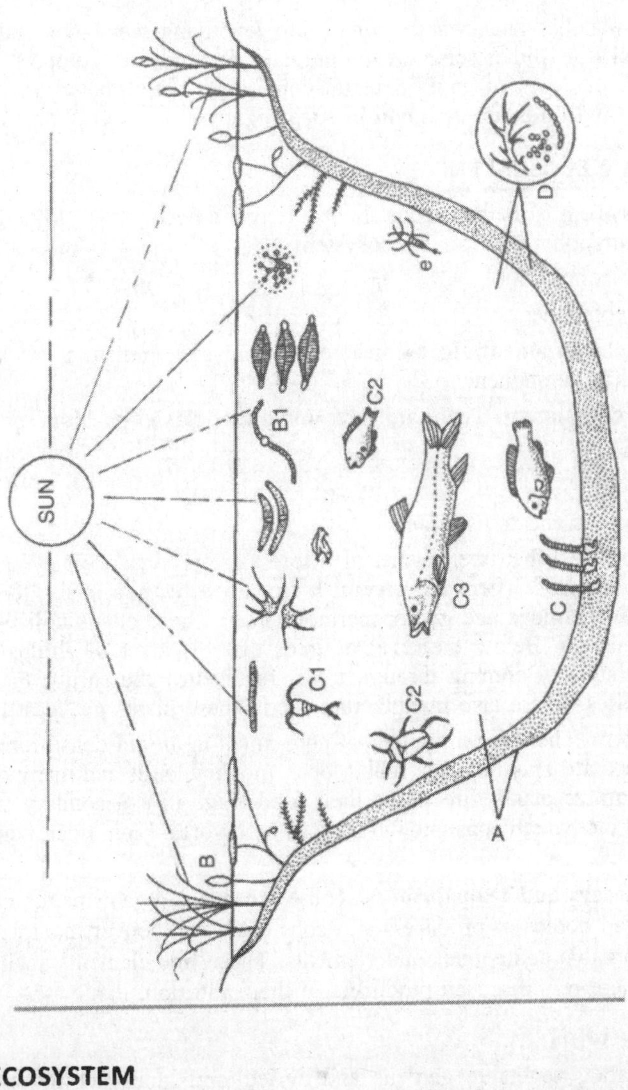

Fig. 3.4 A pond ecosystem.

A—Abiotic component; B and B1—Producers; C1—Primary consumers (herbivores); C2—Secondary consumers; C3—Tertiary consumers; D—Decomposers—saprophytic bacteria and fungi.

OCEAN AS AN ECOSYSTEM

The oceans constitute nearly 70 per cent of the earth's surface. The sea has unusual stability in temperature, salinity and gaseous content. The upper portion of the oceanic region through which light penetrates is called euphotic zone. The region of the ocean below the euphotic zone (below 200 metres) is called *benthal*. It is also known as aphotic zone. This zone is completely free of photosynthetic organisms. Thus, different sets of environmental conditions are present in different zones of the ocean and each zone has its own peculiar life form. Planktonic organisms are algae, diatoms, protozoa, small crustaceans and their eggs. The seashore organisms live under variable physical conditions and they are constantly under the impact of waves. Most of them are covered and exposed twice daily by the rise and fall of the tides.

Streams and Rivers

In the rivers and streams or in the shallow water, the velocity of current is fast to keep the bottom clear. Filamentous algae and fishes are common examples of producers and consumers respectively. In the zone of deep water the velocity of water current is reduced and silt and other loose materials tend to settle at the bottom.

Wet Lands

These are marshes and swamps which are low lying wet lands and their ecosystems are suitable for ducks and other semi-aquatic animals. Swamps also support large trees and shrubs. They have various types of environmental conditions, both aquatic and semi-aquatic. Aquatic insects, reptiles and birds are common in such regions.

FOREST AS AN ECOSYSTEM

Forest ecosystem is the best example of a terrestrial ecosystem. Like other ecosystems, there are two main components of forest ecosystem :

(*A*) Abiotic component

(*B*) Biotic component.

(A) **Abiotic component.** In a forest ecosystem soil, moisture, air and sunlight form the abiotic or physical component.

(B) **Biotic component.** There are three important classes of biotic components :

1. Producers
2. Consumers
3. Decomposers and transformers.

1. **Producers.** All the green plants of a forest are producers. They are the main sources of food for all the animals. There are several layers of vegetation in the forest. The plants of top stratum are angiospermous and gymnospermous trees. These plants utilize radiant energy of sun to the greatest extent. Below the level of trees there is layer of shrubs which consume light energy of low intensity coming through trees. Just below the shrubs there are grasses, herbs, lichens and mosses. These also manufacture food. These plants get least light.

2. **Consumers.** There are a number of consumers in an old dense forest. Consumers of first order in the forest are grasshoppers, rabbit, deer, monkey, birds and many other wild herbivorous animals which utilize plants directly as their food (Fig. 3.5). Secondary consumers are wolves, pythons, jackals etc. which consume the flesh of herbivores. Lion, tiger, hawks are the consumers of top level.

3. **Decomposers and transformers.** These are micro-organisms, chiefly bacteria and fungi which attack dead bodies of producers and consumers and convert complex organic compounds into simpler inorganic compounds and elements. These free elements again return to the abiotic component and are re-utilised by producers in their nutrition.

Artificial Ecosystems

Gardens, parks, vegetable gardens and agricultural lands are the artificial or man-made ecosystems. A balanced aquarium is also an artificial ecosystem. The space capsule which lasts for limited period is also an artificial ecosystem. For longer journey the spacecraft includes all the basic components of the ecosystem, namely, producers, consumers, decomposers and abiotic components.

BIOSPHERE – 2 (AN ARTIFICIAL ECOSYSTEM)

Earth is referred to as Biosphere – 1. Biosphere – 2 is a unique self-sustaining man-made artificial ecosystem, which has been constructed near Tucson in the **Arizona desert** (USA). The building is airtight and encloses 3.15 acres area. It is totally independent except for sunlight. It has a variety of natural living systems, which include a tropical rainforest, lakes, a desert, streams, freshwater and saltwater wetlands and even a small ocean with a coralreef. The structure has been designed to imitate the earth's natural chemical recycling systems. Water that evaporated from its aquatic systems condensed as rainfall over the tropical rainforest. This water was supposed to provide freshwater for the living organisms before evaporating again.

Fig. 3.5 Forest ecosystem.

Initially, in 1991, eight scientists (four women and four men) were completely sealed into Biosphere-2, for two years from the outside world. Besides, more than four thousand species of organisms were introduced. The scientists were supposed to grow their own food, breathe air generated by the plants and drink water produced by the natural recycling processes. They succeeded in preparing 85% of their food. However, many problems were encountered from the very beginning. Large amount of oxygen disappeared inexplicably and additional oxygen had to be pumped from the outside. Many plant and animals species could not sustain themselves and became extinct.

Fig. 3.6 A self sustaining man-made ecosystem in USA.

Thus, an expenditure of 200 million dollar failed to maintain a life support system for only eight people. The earth *i.e.* Biosphere-1, on the other hand, supports billions of people, flora and fauna at no cost. So one can well imagine the value of different natural processes and the environment.

Ecological Niches. Species of plants and animals present in an ecosystem perform different functions. Role of each species is spoken of as its niche. In other words, the total role of a species in a community is the ecological niche. The ecological niche includes the species of organisms, environmental factors, the area in which the species live and the specialization of species population within a community. Every population has an ecological niche which determines the structure and adaptation of that population. Ecological niche is not a simple concept but it relates the concepts of population and community. The niche is the property of the community which represents the place of the population in the community structure. Different communities in ecosystem characterised by similar environments are often similar in their structure and they may contain one or more niches that are essentially identical. The adaptations of population inhabiting these niches may also be similar, even though they are entirely unrelated.

Odum (1971) has classified the ecological niches into following types:

(*i*) **Habitat niches.** These are concerned with habitats of the organisms.

(*ii*) **Trophic niches.** These are related to physiological modifications.

(*iii*) **Geographical niches.** These are related to geographical modifications.

PROCESSES WITHIN THE ECOSYSTEM

In any ecosystem the biotic and abiotic components interact with each other. Processes are the events which either change the form of life or move on the energy and materials within the biotic and abiotic components of ecosystem. The processes operating within the ecosystems may be categorised as under :

1. Metabolic processes of the organisms. These are assimilation, respiration, growth, reproduction, uptake of nutrients and water, transpiration, fixation of nitrogen by bacteria, litter decomposition and mineralisation.

2. Energy flow within the ecosystem.

3, Biogeochemical cycling of minerals, water from abiotic to biotic to abiotic state.

The above processes maintain the dynamic equilibrium between the biotic and abiotic components within an ecosystem. Through metabolic processes the ecosystems become rich in biotic composition, add biomass and store energy in the form of food and maintain an order within the system by pumping out disorders. The second process of energy flow involves transfer of energy from autotrophs to various components of heterotrophs and helps in maintaining the diversity and order within the ecosystem. The third aspect pertaining to the cycling of minerals and water keeps them circulating within the biotic and abiotic components of ecosystem.

ENERGY AND ITS FLOW IN ECOSYSTEM

ENERGY

Energy has been defined as the capacity to do work. Energy exists in two forms : *potential* and *kinetic*. Potential energy is the energy at rest (*i.e.*, stored energy) capable of performing work. Kinetic energy is the energy of motion (free energy). It results in work performance at the expense of potential energy. Conversion of potential energy into kinetic energy involves the imparting of motion.

The source of energy required by all living organisms is the chemical energy of their food. The chemical energy is obtained by the conversion of the radiant energy of sun. The radiant energy is in the form of electromagnetic waves which are released from the sun during the transmutation of hydrogen to helium. The chemical energy stored in the food of living organisms is converted into potential energy by the arrangement of the constituent atoms of food in a particular manner.

In any ecosystem there should be unidirectional flow of energy. This energy flow is based on two important Laws of Thermodynamics which are as follows :

(1) **The first law of Thermodynamics.** It states that the amount of energy in the universe is constant. It may change from one form to another, but it can neither be created nor destroyed. Light energy can be neither created nor destroyed as it passes through the atmosphere. It may, however, be transformed into another type of energy, such as chemical energy or heat energy. These forms of energy can not be transformed into electromagnetic radiation.

(2) **The second law of Thermodynamics.** It states that non-random energy (mechanical, chemical, radiant energy) cannot be changed without some degradation into heat energy. The change of energy from one form to another takes place in such a way that a part of energy assumes waste form (heat energy). In this way, after transformation the capacity of energy to perform work is decreased. Thus, energy flows from higher to lower level.

Main source of energy is sun. Approximately 57% of sun energy is absorbed in the atmosphere and scattered in the space. Some 35% is spent to heat water and land areas and to evaporate water. Of the approximately 8% of light energy striking plant surface, 10% to 15% is reflected, 5% is transmitted and 80 to 85% is absorbed; and an average of only 2% (0.5 to 3.5%) of the total light energy striking on a leaf is used in photosynthesis and rest is transformed into heat energy.

Energy flow in Ecosystems. Living organisms can use energy in two forms : *radiant* and *fixed energy*. Radiant energy is in the form of electromagnetic waves, such as light. Fixed energy is potential chemical energy bound in various organic substances which can be broken down in order to release their energy content.

Organisms that can fix radiant energy utilizing inorganic substances to produce organic molecules are called *autotrophs*. Organisms that cannot obtain energy from abiotic source but depend on energy-rich organic molecules synthesised by autotrophs are called *heterotrophs*. Those which obtain energy from living organisms are called consumers and those which obtain energy from dead organisms are called *decomposers* (Fig. 3.7).

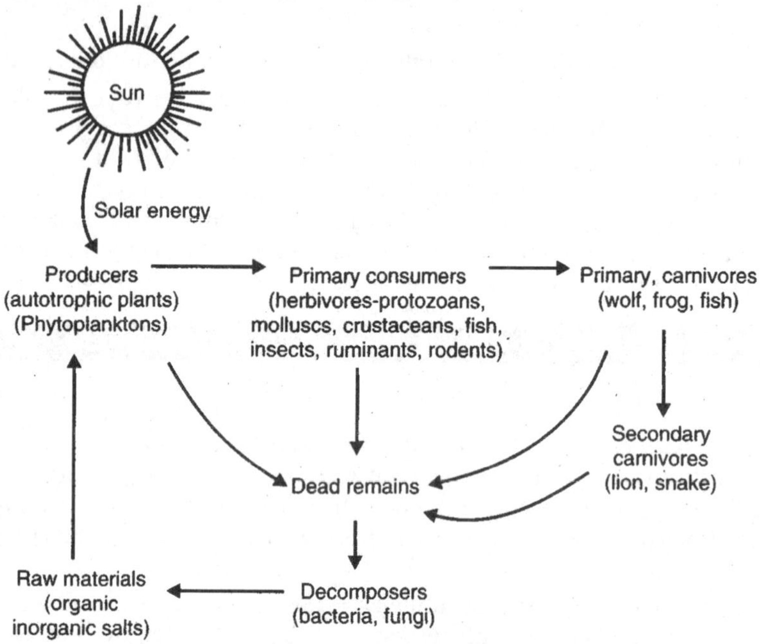

Fig. 3.7. Flow of energy at different levels of ecosystem.

When the light energy falls on the green surfaces of plants, a part of it is transformed into chemical energy which is stored in various organic products in the plants. When the herbivores consume plants as food and convert chemical energy accumulated in plant products into kinetic energy, degradation of energy will occur through its conversion into heat. When herbivores are consumed by carnivores of the first order (secondary consumers) further degradation will occur. Similarly, when primary carnivores are consumed by top carnivores, again energy will be degraded.

Trophic level. The producers and consumers in ecosystem can be arranged into several feeding groups, each known as *trophic level* (feeding level). In any ecosystem, producers represent the first trophic level, herbivores present the second trophic level, primary carnivores represent the third trophic level and top carnivores represent the last level.

Food Chain

In the ecosystem, green plants alone are able to trap in solar energy and convert it into chemical energy. The chemical energy is locked up in the various organic compounds, such as carbohydrates, fats and proteins, present in the green plants. Since virtually all other living organisms depend upon green plants for their energy, the efficiency of plants in any given area in capturing solar energy sets the upper limit to long-term energy flow and biological activity in the community. The food manufactured by the green plants is utilized by themselves and also by herbivores. Animals feed repeatedly. Herbivores fall prey to some carnivorous animals. In this way one form of life supports the other form. Thus, food from one trophic level reaches to the other trophic level and in this way a chain is established. This is known as *food chain.* A food chain may be defined as the transfer of energy and nutrients through a succession of organisms

During transformation of energy, only 10% of the total energy goes to the next trophic level and 90% is released into the environment in the form of heat. This rule of energy transformation known as "The Ten Percent Law" is proposed by Lindemann (1942).

through repeated process of eating and being eaten. In food chain initial link is a green plant or producer which produces chemical energy available to consumers. For example, marsh grass is consumed by grasshopper, the grasshopper is consumed by a bird and that bird is consumed by hawk. Thus, a food chain is formed which can be written as follows :

Marsh grass → grasshopper → bird → hawk

Food chain in any ecosystem runs directly in which green plants are eaten by herbivores, herbivores are eaten by carnivores and carnivores are eaten by top carnivores. Man forms the terrestrial links of many food chains.

Food chains are of three types :

1. Grazing food chain
2. Parasitic food chain
3. Saprophytic or detritus food chain

1. **Grazing food chain.** The grazing food chain starts from green plants and from autotrophs it goes to herbivores (primary consumers) to primary carnivores (secondary consumers) and then to secondary carnivores (tertiary consumers) and so on. The gross production of a green plant in an ecosystem may meet three fates—it may be oxidized in respiration, it may be eaten by herbivorous animals and after the death and decay of producers it may be utilized by decomposers and converters and finally released into the environment. In herbivores the assimilated food can be stored as carbohydrates, proteins and fats, and transformed into much more complex organic molecules. The energy for these transformations is supplied through respiration. As in autotrophs, the energy in herbivores also meets three routes : respiration, decay of organic matter by microbes and consumption by the carnivores. Likewise, when the secondary carnivores or tertiary consumers eat primary carnivores, the total energy assimilated by primary carnivores or gross tertiary production follows the same course and its disposition into respiration, decay and further consumption by other carnivores is entirely similar to that of herbivores. Thus, it is obvious that much of the energy flow in the grazing food chain can be described in terms of trophic levels as outlined below :

Autotroph → Herbivore → Primary carnivore → Secondary carnivore etc.

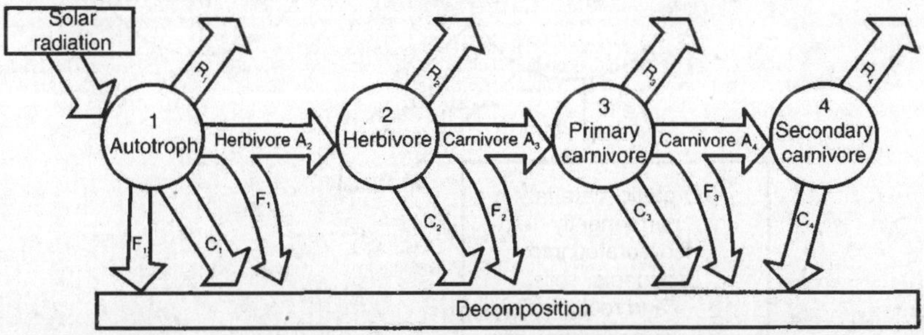

Fig. 3.8. Diagrammatic representation of a grazing food chain showing input and losses of energy at each trophic level. Trophic levels are numbered and used as subscripts to letters indicating energy transfer. A—assimilation of food by the organisms at the trophic level; F—energy lost in the form of faeces and other excretory products; C—energy lost through decay; and R—energy lost to respiration.

A schematic representation of grazing food chain showing input and losses of energy has been presented in Fig. 3.8.

2. **Parasitic food chain.** It goes from large organisms to smaller ones without outright killing as in the case of predator.

3. **Detritus food chain.** The dead organic remains including metabolic wastes and exudates derived from grazing food chain are generally termed **detritus**. The energy contained in detritus is not lost in ecosystem as a whole, rather it serves as a source of energy for a group of organisms

called detritivores that are separate from the grazing food chain. The food chain so formed is called **detritus food chain** (Fig. 3.9).

In some ecosystems more energy flows through the detritus food chain than through grazing food chain. In detritus food chain the energy flow remains as a continuous passage rather than as a stepwise flow between discrete entities. The organisms in the detritus food chain are many and include algae, fungi, bacteria, slime moulds, actinomycetes, protozoa, etc. Detritus organisms ingest pieces of partially decomposed organic matter, digest them partially and after extracting some of the chemical energy in the food to run their metabolism, excrete the remainder in the form of simpler organic molecules. The waste from one organism can be immediately utilized by a second one which repeats the process. Gradually, the complex organic molecules present in the organic wastes or dead tissues are broken down to much simpler compounds, sometimes to carbon dioxide and water and all that are left are humus. In a normal environment the humus is quite stable and forms an essential part of the soil. Schematic representation of detritus food chain is given in Fig. 3.9.

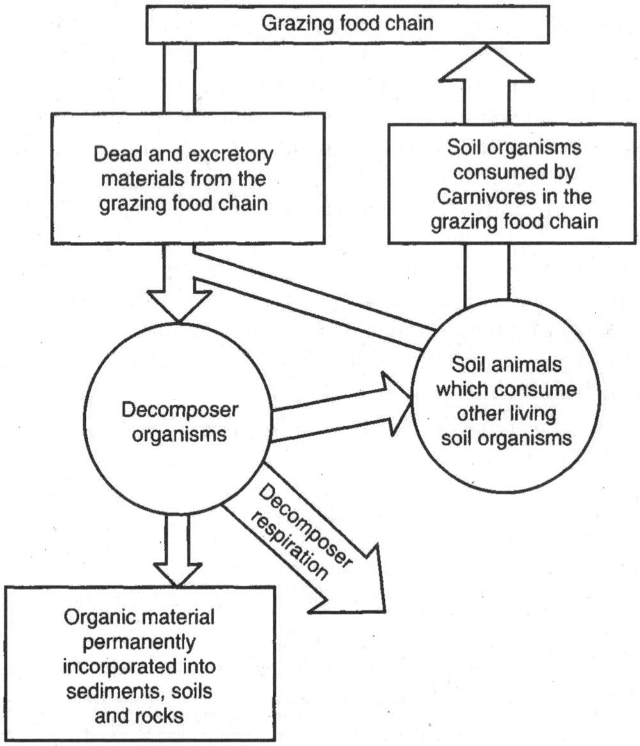

Fig. 3.9. Diagrammatic representation of the detritus food chain showing energy transfers between it and the grazing food chain, as well as energy losses to the detritus food chain.

Food web. Many food chains exist in an ecosystem, but as a matter of fact these food chains are not independent. In ecosystem, one organism does not depend wholly on another. The resources are shared specially at the beginning of the chain. The marsh plants are eaten by variety of insects, birds, mammals and fishes and some of the animals are eaten by several predators. Similarly, in the food chain grass→mouse→snakes→owls, sometimes mice are not eaten by snakes but directly by owls. This type of interrelationship interlinks the individuals of the whole community. In this way, food chains become interlinked. A complex of interrelated food chains makes up a **food web**. Food web maintains the stability of the ecosystem. The greater the number of alternative pathways the more stable is the community of living things. Fig. 3.10. illustrates a food web in ecosystem.

Fig. 3.10. Food web in an ecosystem.

Ecological pyramid. The trophic structure of an ecosystem can be indicated by means of ecological pyramid. At each step in the food chain a considerable fraction of the potential energy is lost as heat. As a result, organisms in each trophic level pass on lesser energy to the next trophic level than they actually receive. This limits the number of steps in any food chain to 4 or 5. Longer the food chain the lesser energy is available for final members. Because of this tapering off of available energy in the food chain a pyramid is formed that is known as **ecological pyramid**. The higher the steps in the ecological pyramid the lower will be the number of individuals and the larger their size.

The idea of ecological pyramids was advanced by C.E. Elton (1927). There are different types of ecological pyramids. In each ecological pyramid, producer level forms the base and successive levels make up the apex. Three types of pyramidal relations may be found among the organisms at different levels in the ecosystem. These are as follows :

1. Pyramid of numbers,
2. Pyramid of biomass (biomass is the weight of living organisms), and
3. Pyramid of energy.

1. **Pyramid of numbers.** It depicts the numbers of individuals in producers and in different orders of consumers in an ecosystem. The base of pyramid is represented by producers which are the most abundant. In the successive levels of consumers, the number of organisms goes on decreasing rapidly until there are a few carnivores. The pyramid of numbers of an ecosystem indicates that the producers are ingested in large numbers by smaller numbers of primary consumers. These primary consumers are eaten by relatively smaller number of secondary consumers and these secondary consumers, in turn, are consumed by only a few tertiary consumers (Fig. 3.11, 3.12a). This type of pyramid is

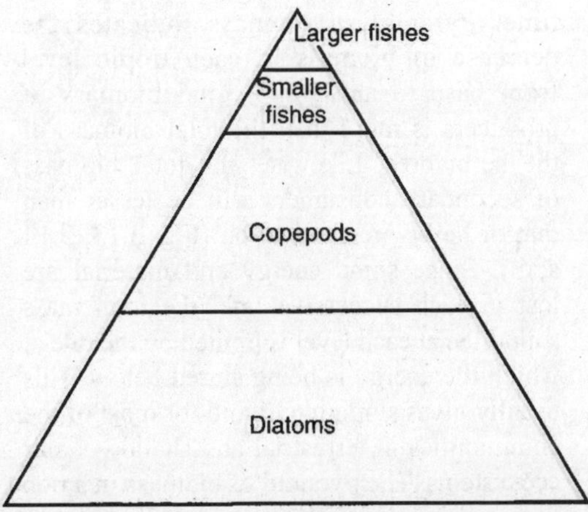

Fig. 3.11. A pyramid of numbers of a lake ecosystem.

best presented by taking an example of lake ecosystem. In this type of pyramid the base trophic level is occupied by producer elements—algae, diatoms and other hydrophytes which are most abundant. At the second trophic level come the herbivores or zooplanktons which are lesser in number than producers. The third trophic level is occupied by carnivores which are still smaller in number than the herbivores and the top is occupied by a few top carnivores. Thus, in the ecological pyramid of numbers there is a relative reduction in number of organisms and an increase in the size of body from base to apex of the pyramid. In parasitic food chain starting from tree, the pyramid of numbers will be inverted (Fig. 3.12)

Fig. 3.12 (*a* & *b*). Pyramids of number

(*a*) Fig. 3.12 a & b. Up-right Pyramids of numbers in a grassland and cultivated field

(*b*) Pyramid of numbers (inverted) of diseased tree (Parasitic ecosystem)

2. **Pyramid of biomass of organisms.** The living weights or *biomass* of the members of the food chain present at any one time form the pyramid of biomass of organisms. This indicates, by weight or other means of measuring materials, the total bulk of organisms or fixed energy present at one time. Pyramid of biomass indicates the decrease of biomass in each tropic level from base to apex, *e.g.,* total biomass of producers is more than the total biomass of the herbivores. Likewise, the total biomass of secondary consumers will be lesser than that of herbivores and so on (Fig. 3.13, 3.14 a, b). Since some energy and material are lost in each successive link, the total mass supported at each level is limited by the rate at which the energy is being stored below. This usually gives sloping pyramid for most of the communities in terrestrial and shallow water ecosystems. The pyramid of biomass in a pond ecosystem will be inverted as shown in Fig. 3.13 b.

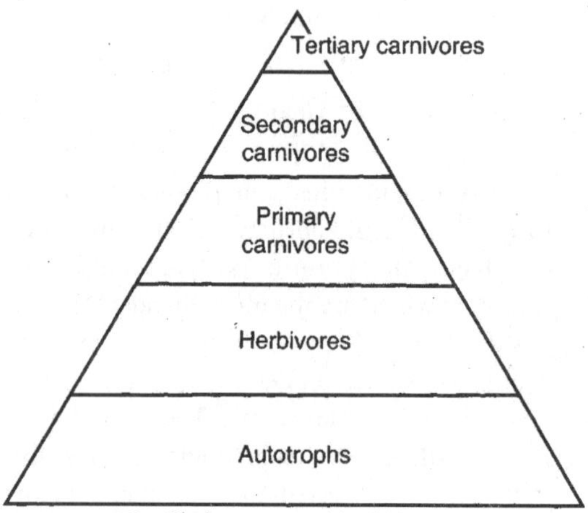

Fig. 3.13. A pyramid of biomass

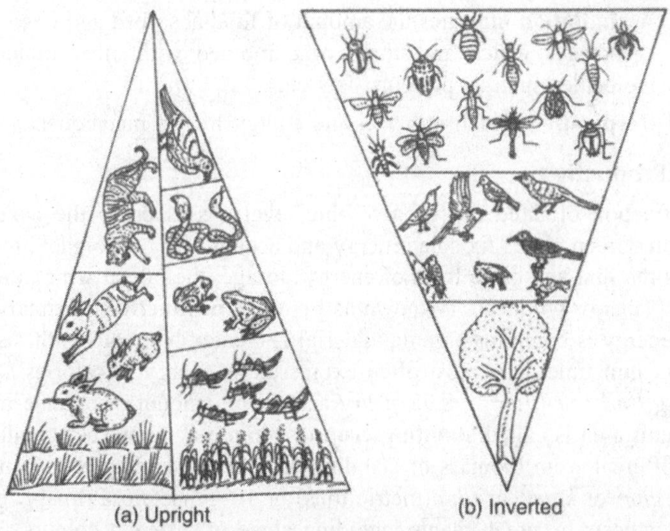

Fig. 3.14. (*a* & *b*). Pyramids of biomass
(*a*) A grassland ecosystems showing upright-triangular
(*b*) Inverted pyramid of biomass of an aquatic ecosystem.

3. **Pyramid of energy.** This depicts not only the amount of total energy utilised by the organisms at each trophic level of food chain but more important, the actual role of various organisms in transfer of energy. At the producer level the total energy will be much greater than the energy at the successive higher trophic level. Some producer organisms may have small biomass but the total energy they assimilate and pass on to consumers may be greater than that of organisms with much larger biomass. Higher trophic levels are more efficient in energy utilization but much heat is lost in energy transfer. Energy loss by respiration also progressively increases from lower to higher trophic states (Fig. 3.15).

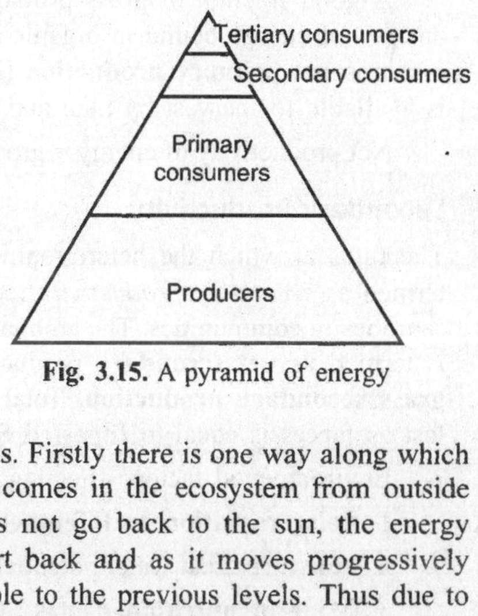

Fig. 3.15. A pyramid of energy

In the energy flow process, two things become obvious. Firstly there is one way along which energy moves *i.e. unidirectional flow of energy*. Energy comes in the ecosystem from outside source *i.e.* sun. The energy captured by autotrophs does not go back to the sun, the energy that passes from autotrophs to herbivores does not revert back and as it moves progressively through the various trophic levels, it is no longer available to the previous levels. Thus due to unidirectional flow of energy, the system would collapse if the supply from primary source, the sun is cut off. Secondly, there occurs a progressive decrease in energy level at each trophic level which is accounted largely by the energy dissipated as heat in metabolic activities.

PRODUCTIVITY

The relationship between the amount of energy accumulated and the amount of energy utilised within one trophic level of food chain has an important bearing on how much energy from one trophic level passes on to the next trophic level in the food chain. The ratio of output of energy to input of energy is referred to as *ecological efficiency*. Different kinds of efficiencies can be measured by the following parameters :

(*i*) **Ingestion** which indicates the quantity of food or energy taken by trophic level. This is also called **exploitation efficiency.**

(*ii*) **Assimilation** indicates the amount of food absorbed and fixed into energy rich organic substances which are stored or combined with other molecules to build complex molecules such as proteins, fats etc.

(*iii*) **Respiration** which indicates the energy lost in metabolism.

Primary Productivity

The fraction of fixed energy a trophic level passes on to the next trophic level is called **production**. Green plants fix solar energy and accumulate it in organic forms as chemical energy. Since it is the first and basic form of energy storage, the rate at which the energy accumulates in the green plants or producers is known as **primary productivity.** Primary productivity is the rate at which energy is bound or organic material is created by photosynthesis per unit area of earth's surface per unit time. It is most often expressed as energy in calories / cm^2 / *yr* or dry organic matter in $g / m^2 / yr$ ($g/m^2 \times 8.92 = lb$ / acre). The amount of organic matter present at a given time per unit area is called **standing crop** or **biomass** and as such productivity, which is a rate, is quite different from biomass or standing crop. The standing crop is usually expressed as dry weight in g/m^2 or kg/m^2 or *t/ha* (metric tons) or 10^6 *g*/hectare. Primary productivity is the result of photosynthesis by green plants including algae of different colours. Bacterial photosynthesis or chemosynthesis, although of small significance may also contribute to primary productivity. The total solar energy trapped in the food material by photosynthesis is referred to as **gross primary productivity (G.P.P.).**

A good fraction of gross primary production is utilized in respiration of green plants. The amount of energy bound in organic matter per unit area and time that is left after respiration in plants is **net primary production (N.P.P.)** or **plant growth.** Only the net primary productivity is available for harvest by man and other animals.

Net productivity of energy = gross productivity—energy lost in respiration.

Secondary Productivity

The rates at which the heterotrophic organisms resynthesize the energy-yielding substances is termed as *secondary productivity.*Secondary productivities are the productivities of animals and saprobes in communities. The amount of energy stored in the tissues of consumers or heterotrophs is termed as **net secondary** production and the total plant material ingested by herbivores is **grass secondary production.** Total plant material ingested by herbivores minus the materials lost as faeces is equal to **Ingested Secondary Production.**

Environmental factors affecting the production processes in an ecosystem are as follows :

1. **Solar radiation and Temperature**

2. **Moisture.** Leaf water potential, soil moisture and precipitation fluctuation and transpiration.

3. **Mineral nutrition.** Uptake of minerals from the soil, rhizosphere effects, fire effects, salinity, heavy metals, nitrogen metabolism.

4. **Biotic activities.** Grazing, above ground herbivores, below ground herbivores, predators and parasites, diseases of primary producers.

5. **Impact of human population.** Pollutions of different sorts, ionizing radiations like atomic explosions, etc.

There are three fundamental concepts of productivity :

1. Standing crop
2. Materials removed
3. Production rate.

1. **Standing crop.** It is the abundance of the organisms existing in the area at any one time. It may be expressed in terms of number of individuals, as biomass of organisms, as energy

content or in some other suitable terms. Measurement of standing crop reveals the concentration of individuals in the various populations of ecosystem.

2. **The materials removed.** The second concept of productivity is **the materials removed** from the area per unit time. It includes the yield to man, organisms removed from the ecosystem by migration, and the material withdrawn as organic deposit.

3. **The production rate.** The third concept of productivity is the **production rate.** It is the rate at which the growth processes are going forward within the area. The amount of material formed by each link in the food chain per unit of time per unit area or volume is the production rate.

All the three major groups of organisms—producers, consumers and reducers are the functional kingdoms of natural communities. The three represent major directions of evolution and are characterised by different modes of nutrition. Plants feed primarily by photosynthesis, animals feed primarily by ingesting food that is digested and absorbed in the alimentary canal and the saprobes feed by absorption and have need for an extensive surface of absorption. The principal kinds of organisms among saprobes are the unicellular bacteria, yeasts, chytrids or lower fungi and higher fungi with mycelial bodies.

In terrestrial communities as much as 90% of net primary production remains unharvested and are utilized as dead tissue by saprobes and soil animals. The saprobes have a larger and more essential role than animals in degrading dead organic matter to inorganic forms and in such ecosystems, secondary production by reducers (decomposers) should exceed that by consumers, though the former is even more difficult to measure than the latter. Biomass of decomposers with their microscopic cells and filaments embedded in food sources is also difficult to measure and that is small in relation to their productivity and significance for the ecosystem. Small masses of reducers degrade and transform larger masses of organic matter to inorganic remnants. In so doing decomposers disperse back to the environment the energy of photosynthesis accumulated in the organic compounds that are decomposed. Thus they have a major role in the energy flow of ecosystems. A community or ecosystem, like an organism, is an open energy system. The continuous intake of energy in photosynthesis replaces the energy dissipated to environment by respiration and biological activity and the system does not run-down through the loss of free energy to maximum entropy. If the amount of energy entrapped is greater than the energy dissipated, the pool of biologically useful energy of organic bonds increases. This results in increase of community biomass and consequently the community grows; such is the case in succession. If energy intake is lesser than energy dissipation, the community biomass will decrease and it must, in some sense, retrogress. If energy intake and loss are in balance, the pool of organic energy is in steady state; such is the case in climax communities. Three aspects of this steady state may be recognised :

(*i*) The steady state of population of climax communities in which equal birth and death rates in population keep the number of individuals relatively constant,

(*ii*) The steady state of energy flow,

(*iii*) The steady state of the matter of community, where addition of material by photosynthesis and organic synthesis is balanced by loss of material through respiration and decomposition.

Methods of Measuring Primary Production

There are several parameters for measuring primary production and the methods of measuring primary production are based on those parameters. The methods are discussed here as under :

1. **Harvest method.** It involves removal of vegetation periodically and weighing the material. For measuring above ground production, the above ground plant parts are clipped at ground level, dried to constant weight at 80°C and weighed. The dry weight in g/m^2 /year gives the ground production.

Below ground production is estimated by using frequent core sampling technique of Dahlman and Kucera (1965). It is expressed in terms of weight in gm per unit area per year. In terms of energy one gm dry weight of plant material contains 4 to 5 kcal. The limitations of harvest method are as follows :

(*i*) The amount of plant material consumed by herbivores and the food oxidised during respiration process of the plants is not accounted.

(*ii*) Root biomass is neglected.

(*iii*) Photosynthates translocated to underground parts of plants are not known.

In spite of these limitations the method is used allover for measuring net assimilation rate (NAR) and relative growth rate (RGR).

2. **Carbon dioxide assimilation method.** Utilization of CO_2 in photosynthesis or its liberation during respiration is measured by infrared gas analysis or by passing the gas through Baryta water $Ba(OH)_2$ and titrating the same. The CO_2 removed from incoming gas chamber is taken to be synthesized into organic matter by the green plants. Performing the experiment in light and dark chambers the net and gross production can be measured. In the lighted chamber photosynthesis and respiration take place simultaneously and the CO_2 coming out from the chamber is the unused gas of the atmosphere plus gas from the respiration of plant parts. In the dark chamber all CO_2 is due to respiration.

Net production = Gross production—Respiration

3. **Oxygen production method.** In the aquatic vegetation CO_2 gas analysis method is not used but oxygen evolution method is generally used. The light and dark bottle technique is employed for measuring primary production of aquatic plant. In this method two bottles, one transparent and the other opaque are filled with water at a given depth of lake, closed, maintained at that depth for sometime and then brought to laboratory for determination of oxygen content in the water. The decrease of oxygen in dark bottle is due to respiratory activity while increase of O_2 in light bottle is due to photosynthesis. The total increase of O_2 in light bottle plus the amount of O_2 decreased in dark bottle express gross productivity (O_2 value multiplied by 0.375 gives an equivalent of carbon assimilation). Recently, oxygen electrodes have been used for estimating oxygen content in water.

4. **Chlorophyll method.** Gesner (1949) pointed out that the amount of chlorophyll/m^2 is almost limited to a narrow range of 0.1 to 3.0 gm regardless of the age of individuals or the species present therein. There is direct correlation between the amount of chlorophyll and dry matter production in different types of communities with varying light conditions. The relation of total amount of chlorophyll to the photosynthetic rate is referred to as **assimilation ratio** or rate of production/gm chlorophyll. Total chlorophyll per unit area is greater in land plants as compared to that in aquatic plants. In marine ecosystem the rate of carbon assimilation is 3.7 g/hr/g of chlorophyll. The relationship between area based chlorophyll and dry matter production in terrestrial ecosystems has been worked out by Japanese ecologists Argua and Monsi (1963).

5. **Other methods.** Pandeya (1971), Sharma (1972) and several other ecologists have evolved correlation coefficients for evaluating biomass and productivity in forest trees by measuring their diameter at breast height (DBH), height, canopy cover, etc. Methods of establishing regression are as below :

(*i*) Diameter of trees in sample quadrats is measured at breast height and the height repeated is determined for each tree.

(*ii*) Different diameter and height classes are determined for each species.

(*iii*) A set of sample trees are cut and subjected to a detailed analysis for dry weight of stems, twigs, leaves and roots.

(*iv*) Regression values are computed for the sets of trees belonging to each girth class, relating the biomass of each fraction to the diameter at breast height.

(*v*) The regression values are used to compute the probable biomass and production of each tree in the sample area. These values for each species when pooled give biomass and production rate of trees per unit area in the forest.

Age of the trees markedly influences the annual net production.

BIOGEOCHEMICAL CYCLES

Nearly 30 to 40 elements are required for proper growth and development of living organisms. Most important of these are C, H, O, P, K, N, S, Ca, Fe, Mg, B, Zn, Cl, Mo, Co, I and F. These materials flow from abiotic to biotic components and back to the non-living component again in a more or less cyclic manner. This is known as the **biogeochemical cycle** or **inorganic-organic cycle**. The flow of these elements through the ecosystem must be cyclic, with matter being consistently reused. Because the flow involves not only the living organisms but also a series of chemical reactions in the abiotic environments, these cycles are called **biogeochemical cycles**.

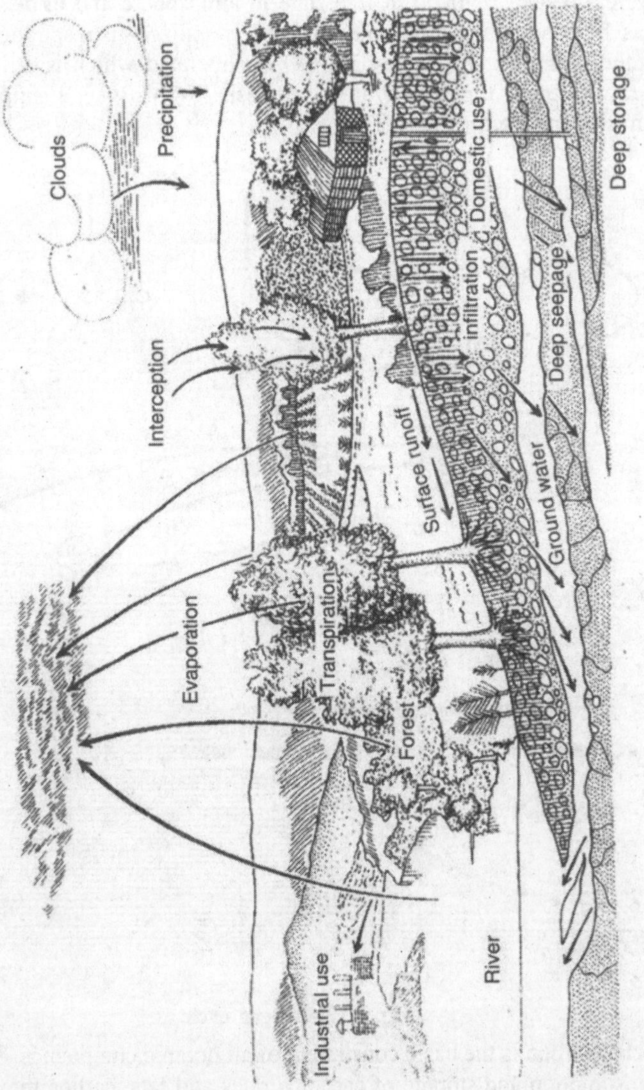

Fig. 3.16. The water cycle, showing the major pathways of water through the ecosystem.

There are three types of biogeochemical cycles :

(1) Hydrologic cycle or water cycle,

(2) Gaseous cycles, and

(3) Sedimentary cycles.

1. Hydrologic or Water Cycle

Interchange of water between atmosphere, land and sea and between living organisms and their environment is accomplished through water cycle. Water cycle or hydrologic cycle involves evaporation, transpiration, cloud formation and precipitation. Water of atmosphere reaches the earth surface through precipitation and from the earth surface it reaches the atmosphere through evaporation and transpiration. The amount of water available for evaporation is determined by the amount supplied by precipitation and condensation. Between rainfall input and evaporation output there lies a precarious water balance (Fig. 3.16).

2. Gaseous Cycles

Oxygen cycle. Oxygen is found in free state in atmosphere and in dissolved state in water. It is liberated as by-product of photosynthesis and is utilized in respiration by the plants and animals. When the living organisms respire, CO_2 is liberated which is utilized by green plants as an essential raw material for carbohydrate synthesis. In this way, a simple yet vital O_2 cycle is maintained in the ecosystem (Fig. 3.17).

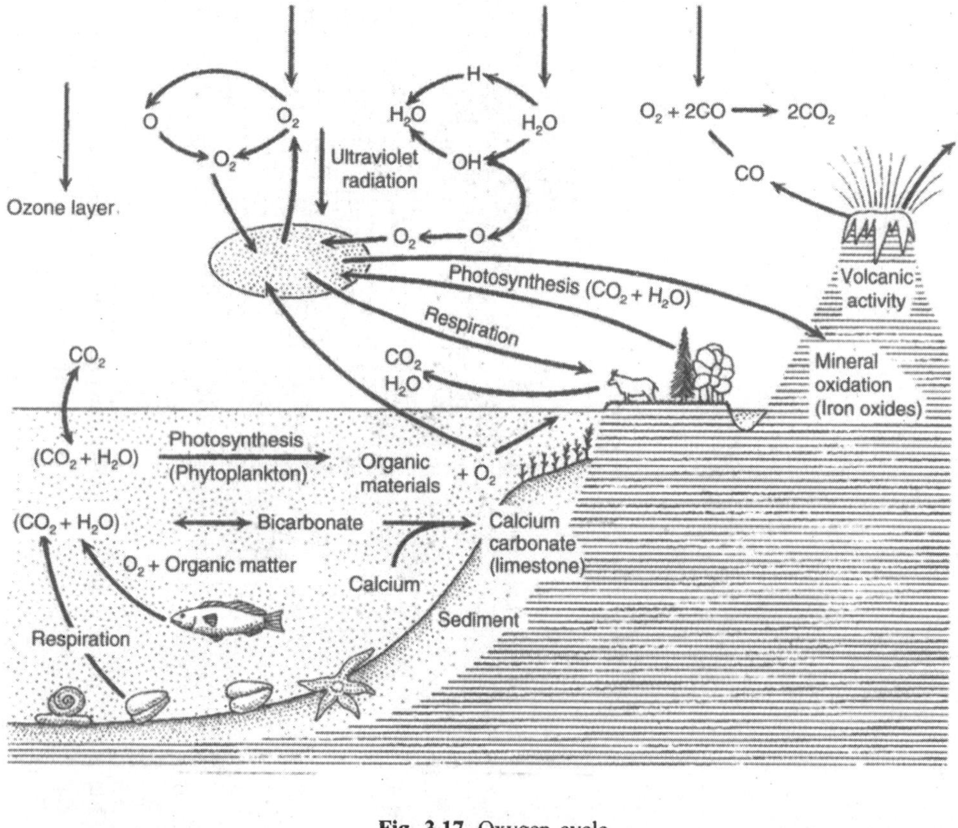

Fig. 3.17. Oxygen cycle

Carbon cycle. Carbon is the basic constituent of all organic compounds. Since energy transfer occurs in the consumption and storage of carbohydrates and fats, carbon moves to the ecosystem with flow of energy. The source of nearly all carbon found in the living organisms is CO_2 which is found in free state in atmosphere and in dissolved state in the water on the earth. Green plants (producers) use CO_2 through photosynthesis in the presence of sunlight and carbohydrate is formed. Later on, complex fats and polysaccharides are formed in plants which are utilised by animals. Flesh eating animals (carnivores) feed on herbivores and the carbon compounds are again digested and converted into the other forms. Carbon is released to the atmosphere directly as CO_2 in respiration of both plants and animals. Bacteria and fungi attack the dead remains of plants and animals. They degrade the complex organic compounds into simple substances which are then available for other cycles. Part of the organic carbon is incorporated into the earth's crust as coal, gas, petroleum, limestone and coral reef. Carbon from such deposits may be liberated after a long period of time (Fig. 3.18).

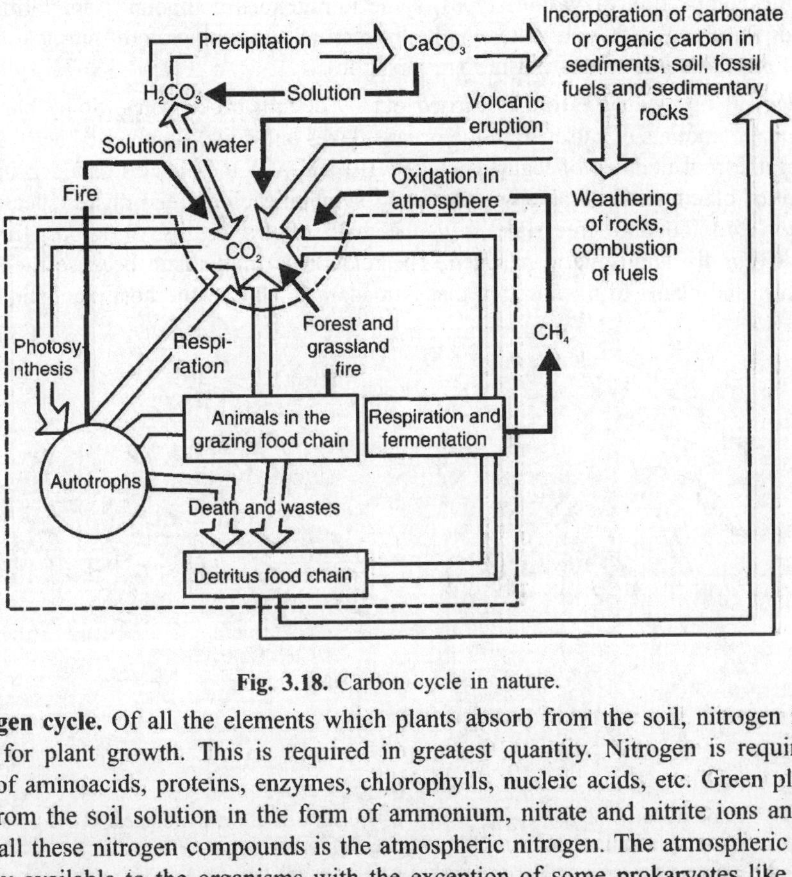

Fig. 3.18. Carbon cycle in nature.

Nitrogen cycle. Of all the elements which plants absorb from the soil, nitrogen is the most important for plant growth. This is required in greatest quantity. Nitrogen is required for the synthesis of aminoacids, proteins, enzymes, chlorophylls, nucleic acids, etc. Green plants obtain nitrogen from the soil solution in the form of ammonium, nitrate and nitrite ions and the main source of all these nitrogen compounds is the atmospheric nitrogen. The atmospheric nitrogen is not directly available to the organisms with the exception of some prokaryotes like blue green algae and nitrogen fixing bacteria. Nitrogen cycle consists of the following steps :

(1) Nitrogen fixation, (2) Nitrogen assimilation, (3) Ammonification, (4) Nitrification, (5) Denitrification, and (6) Sedimentation.

1. **Nitrogen fixation.** Conversion of free nitrogen of atmosphere into the biologically acceptable form or nitrogenous compounds is referred to as nitrogen fixation. This process is of two types :

(*a*) **Physicochemical** or non-biological nitrogen fixation

(*b*) **Biological nitrogen fixation.**

In physicochemical process of nitrogen fixation, atmospheric nitrogen combines with oxygen (as ozone) during lightning or electrical discharges in the clouds and produces different nitrogen oxides :

$$N_2 + 2\,(O) \xrightarrow[\text{Discharge}]{\text{Electric}} 2NO$$

$$2NO + 2\,(O) \longrightarrow 2NO_2$$

$$2NO_2 + (O) \longrightarrow N_2O_5$$

The nitrogen oxides get dissolved in rain water and on reaching earth surface they react with mineral compounds to form nitrates and other nitrogenous compounds :

$$N_2O_5 + H_2O \longrightarrow 2HNO_3$$

$$2HNO_3 + CaCO_2 \longrightarrow Ca(NO_3)_2 + CO_2 + H_2O$$

During combustion of various types, some nitrogenous compounds are formed which are washed down along with rain water. At high pressure and temperature, nitrogen and hydrogen react to form ammonia (industrial nitrogen fixation).

Biological nitrogen fixation is carried out by certain prokaryotes. Some blue-green algae fix significant amounts of nitrogen in the oceans, lakes and soils. Symbiotic bacteria (*Rhizobium*) inhabiting the root nodules of legumes (Fig. 3.19) and also the species of alder, buckbrush and a number of other non-leguminous genera and symbiotic blue-green algae (species of *Nostoc, Anabaena,* etc.) found in free state or in the thalli of *Anthoceros, Salvenia, Azolla,* coralloid roots of *Cycas* fix atmospheric nitrogen. The relation is mutualistic because the microbes use energy from the plants to fix nitrogen that is made available to the host plants and other plants

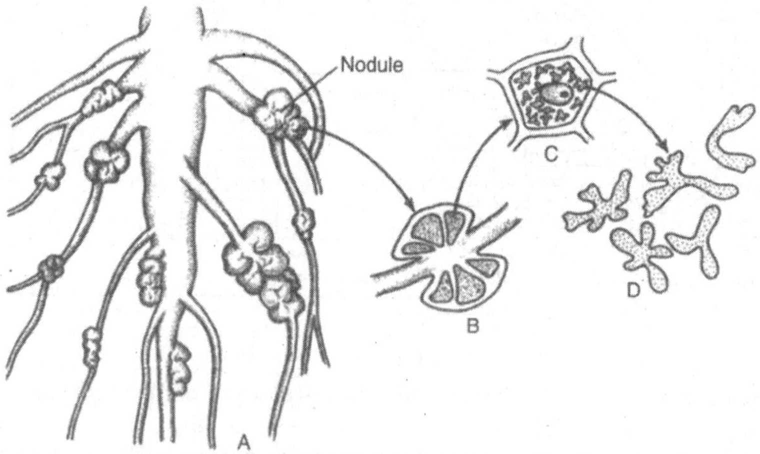

Fig. 3.19 Root nodules of a legume plant. A—Legume root with nodules. B—T.S. of root-nodule. C—Single cell of nodule. D—Nitrogen-fixing bacterium, *Rhizobium.*

of the community. Certain free living nitrogen fixing bacteria, such as *Azotobacter, Clostridium, Beijerinckia, Derxia, Rhodospirillium* also fix free nitrogen of atmosphere in the soil. *Frankia,* an actinomycetous fungus found in the roots of *Alnus, Percia, Casuarina,* etc. also fixes nitrogen. Nitrogen fixing organisms combine the gaseous nitrogen of atmosphere with hydrogen obtained from respiratory pathway to form ammonia which then reacts with organic acids to form aminoacids. Biological nitrogen fixation is the major source of fixed nitrogen upto 140—700 mg/m^2/year as against 35 mg/m^2/year by electrical discharge and photochemical fixation.

2. **Nitrogen assimilation.** Inorganic nitrogen in the form of nitrates, nitrites and ammonia is absorbed by the green plants and converted into nitrogenous organic compounds. Nitrates are first converted into ammonia which combines with organic acids to form aminoacids. Aminoacids are used in the synthesis of proteins, enzymes, chlorophylls, nucleic acids, etc. Animals derive

their nitrogen requirement from the plant proteins. Plant proteins are not directly utilized by the animals. They are first broken down into amino acids during digestion and then the amino acids are absorbed and manipulated into animal proteins, nucleic acids, etc.

3. **Ammonification.** The dead organic remains of plants and animals and excreta of animals are acted upon by a number of microorganisms especially actinomycetes and bacilli (*Bacillus ramosus, B. vulgaris, B. mesenterilus*). These organisms utilize organic compounds in their metabolism and release ammonia.

4. **Nitrification.** Certain bacteria, such as *Nitrosomonas, Nitrococcus, Nitrosogloea* and *Nitrospira* in oceans and soils convert ammonia into nitrites and then nitrites into nitrates. These bacteria primarily use the energy of dead organic matter in their metabolism.

$$2NH_4^+ + 2O_2 \longrightarrow NO_2^- + 2H_2O + \text{energy}$$

Conversion of nitrites to nitrates is brought about by several microbes like *Penicillium* species, Nitrobacter, *Nitrocystis* etc. *Nitrocystis oceanus* is the common marine autotroph which performs nitrification for obtaining energy.

$$2NO_2^- + O_2 \longrightarrow 2NO_3^- + \text{energy}$$

Some nitrates are also made available through weathering of nitrate containing rocks.

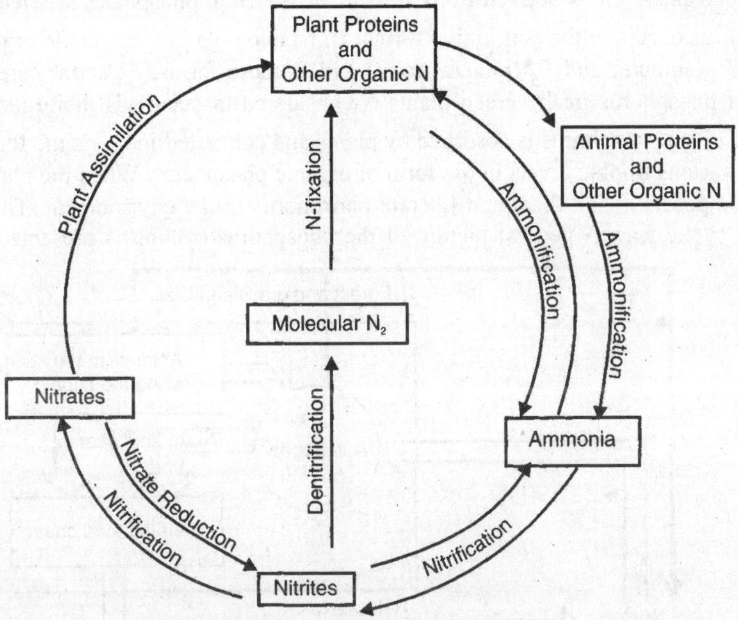

Fig. 3.20. The nitrogen cycle in ecosystem.

5. **Denitrification.** Ammonia and nitrates are converted into free nitrogen by certain microbes. This process is referred to as **denitrification.** *Thiobacillus denitrificans, Micrococcus denitrificans, Pseudomonas aeruginosa* are the common examples of denitrifying bacteria.

$$2NO_3^- \longrightarrow 2NO_2^- \longrightarrow 2NO \longrightarrow N_2O \longrightarrow N_2$$

6. **Sedimentation.** Nitrates of the soil are washed down to the sea or leached deep into the earth along with percolating water. Nitrates thus lost from the soil surface are locked up in the rocks. This is sedimentation of nitrogen. Nitrogen of rock is released only when the rocks are exposed and weathered.

Thus a large part of nitrogen is fixed up and stored in plants, animals, and microbes. Nitrogen leaves the living system in the same amount it is taken in from the atmosphere and the input and outflow of nitrogen are balanced in the ecosystem. The overall nitrogen cycle in nature is presented in Fig. 3.20.

3. Sedimentary Cycles.

Mineral elements required by living organisms are obtained initially from inorganic sources. Available forms occur as salts dissolved in soil water. Mineral cycles essentially consist of two phases : (*i*) the salt solution phase, and (*ii*) rock phase.

Mineral salts come directly from earth crust by weathering. Soluble salts then enter the water cycle. By movement of water minerals move from the soil to streams, lakes and ultimately to sea where they remain permanently. Other salts return to the earth's crust through sedimentation. They become incorporated into sediments or rock beds and after weathering of rocks they again enter the cycle. Plants and some animals take minerals in the form of mineral solution from their habitats. After the death of living organisms the nutrients return to the soil and water through the action of decomposers (bacteria and fungi) and transformers. Green plants at one end and decomposers at the other play very important role in circulation of nutrients.

Phosphorus cycle. Plants and animals obtain phosphorus from the environment. Phosphorus is a component of nucleic acids, ADP, ATP, NADP, phospholipids etc. It occurs in the soil as rock phosphate, apatite or calcium phosphate, fluorapatite [$Ca_{10}Fe_2$ $(PO_4)_6$], iron phosphate or aluminium phosphate. Soils derived from the rock beds rich in phosphates are rich in phosphorus.

Phosphorus occurs in the soil in five forms: P_1 (stable organic), P_2 (labile organic), P_3 (labile inorganic), P_4 (soluble) and P_5 (mineral form) and of these forms, P_3 and P_4 are in equilibrium and entry of phosphorus in the green plants is considered to occur via labile inorganic pool.

The dissolved phosphorus is absorbed by plants and converted into organic form. From plants it travels to various trophic levels in the form of organic phosphates. When the plants and animals die the decomposers attack them and liberate phosphorus to the environment. Thus, this process proceeds in cyclic way. A general picture of the phosphorus cycling is presented in Fig. 3.21.

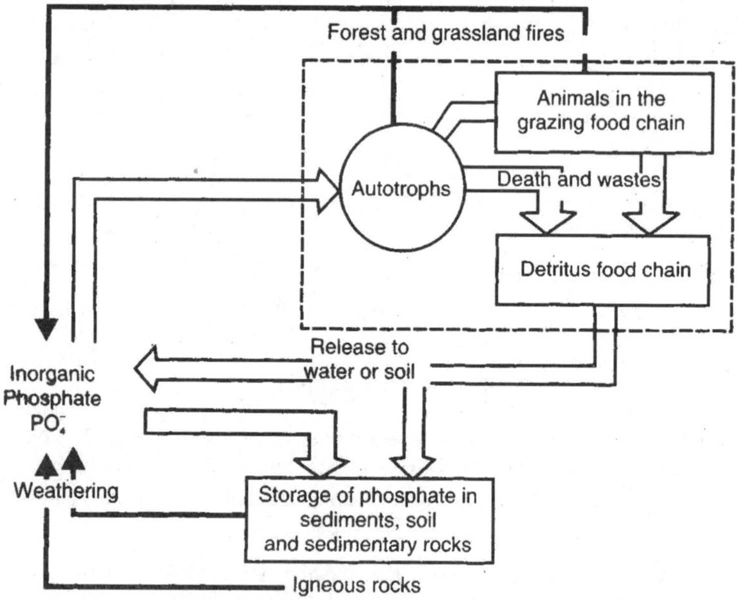

Fig. 3.21 Phosphorus cycle in ecosystem.

Phosphorus along with many other mineral elements reaches the oceans and settles down as sediment. A good proportion of phosphorus leaches down to deep layers of soil. In this way, major proportion of phosphate becomes lost to this cycle by physical processes, such as sedimentation and leaching. Biological processes such as formation of teeth and bones also keep phosphorus locked up for some time.

Sulphur Cycle. Sulphur cycle links soil, water and air. Sulphur occurs in the soil and rocks as sulphides (FeS, ZnS, etc.) and crystalline sulphates. In the atmosphere sulphur occurs in the form of SO_2 and H_2S. SO_2 gas is formed during combustion of fossil fuels or as a result of decomposition. H_2S or hydrogen sulphide gas is released to the atmosphere from water logged soils, continental shelf, lakes and springs. The organic and inorganic sulphur and SO_2 are formed through oxidation of H_2S in the atmosphere. A small amount of sulphur occurs in dissolved state in rain water and through rains it reaches earth surface. Except a few organisms which need organic form of sulphur as amino acids and cystein, most of the organisms take sulphur as inorganic sulphates. Most of the biologically incorporated sulphur is produced in the soil from aerobic breakdown of proteins by bacteria and fungi. Under an aerobic condition, however, sulphur may be reduced directly to sulphides, including H_2S.

$$2H_2S + O_2 \xrightarrow{\textit{Baggiatoa spp.}} 2S + 2H_2O$$

$$2S + 2H_2O + 3O_2 \xrightarrow[\text{(Thio-oxidation)}]{\textit{Thiobacillus}} 2H_2SO_4$$

Green and purple photosynthetic bacteria use hydrogen of H_2S as the oxygen acceptor in reducing carbon dioxide. Green bacteria are able to oxidise sulphide to elemental sulphur whereas the purple bacteria can carry oxidation to sulphate stage. In the ecosystems, sulphur is transferred from autotrophs to animals, then to decomposers and finally it returns to environment through the decay of dead organic remains (Fig. 3.22).

Sedimentary aspect of sulphur cycling involves precipitation of sulphur in presence of iron under anaerobic conditions. Sulphides of iron, copper, zinc, cadmium, cobalt are insoluble in neutral and alkaline water and consequently sulphur is bound to limit the amount of these elements. Thus, sulphur cycle affords an excellent example of interaction and complex biochemical regulation between the different mineral cycles.

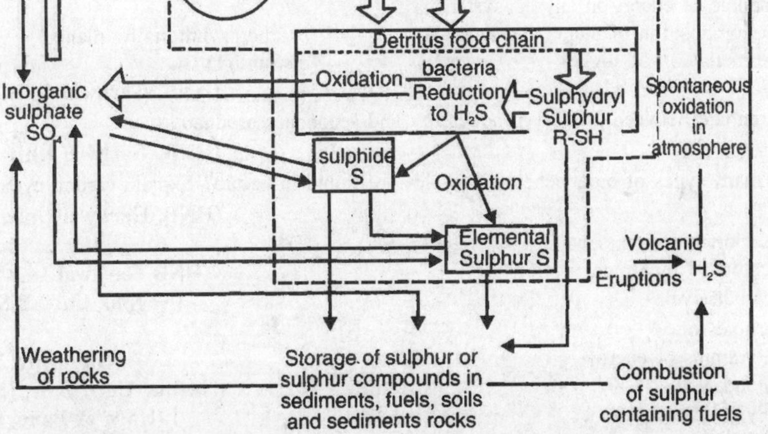

Fig. 3.22. Sulphur cycle. Organic phase is shown within broken lines.

The study of biogeochemical cycles in the ecosystem makes it clear that the abiotic components of ecosystem are transformed into biotic structures through metabolic processes and locked up in the biomass for some time depending upon the turnover rate. In lower plants with soft tissues the turnover rate is quicker than in higher plants and animals. The materials held up in the biomass are released to the environment by decomposing activities.

The nutrient cycle is not a close circuit within an ecosystem. The nutrients are continuously being imported as well as carried out of the ecosystem. Appreciable quantities of plant nutrients are brought to ecosystem by rain and snow. Small quantity of nutrients is carried to the forest by rains. The gain of nutrients to the ecosystem from precipitation, extraneous material and mineral weathering is offset by losses. Water draining away from forest carries with it more mineral matter than supplied through precipitation. Considerable quantities of nutrients in the forest are locked up in the trees and the humus layer. When trees and vegetation are removed, sufficient amounts of nutrient are removed. Intensive forestry and agriculture on some soils may reduce the nutrient reserves to such an extent that soils become unfertile. Ecosystem can remain productive only if the nutrients withdrawn are balanced by an inflow or replacement.

QUESTIONS

1. What is ecosystem? What are the various components of ecosystem? **(H.P. Univ., B.Sc. III, 2010)**
2. What are the major ecosystems of the world? Describe forest and pond ecosystems in detail.
3. What is meant by energy flow in an ecosystem? What are the laws of thermodynamics?
(Univ., of Pune, S.Y. B.Sc. Bot. I (New course 2010))
4. Write short notes on the following :
Producers, Food chain and Food web, **(Pune Univ., B.Sc. II, Bot. I-2010)**
Ecological pyramids, Hydrologic cycle, Nitrogen cycle, Productivity, Ecological niches.
(Univ. of Pune, B.Sc. II, Bot. I-2010)
5. Describe various Biogeochemical cycles occurring in nature. State Biotic components of ecosystem.
6. Which of the following are the main components of ecosystem?
(*a*) Biotic components (*b*) Climate
(*c*) Abiotic components (*d*) Abiotic & Biotic both
(HNB, Garhwal Univ. B.Sc. II, 2010)
7. Food chain is related to
(*a*) all the living organisms (*b*) only plants
(*c*) only animals (*d*) only man. **(HNB, Garhwal Univ. B.Sc. II, 2010)**
8. In a parasitic food chain starting from trees, the pyramid of number is
(*a*) upright (*b*) cylindrical
(*c*) Inverted (*d*) None of the above **(HNB, Garhwal Univ. B.Sc. II, 2010)**
9. The source of energy in any ecosystem is
(*a*) Decomposition of plants and animals (*b*) Photosynthesis by plants
(*c*) Fermentation of sugars (*d*) Sunlight
(HNB, Garhwal Univ. B.Sc. II, 2010)
10. Differentiate between primary productivity and secondary productivity.
(HNB, Garhwal Univ. B.Sc. II, 2010)
11. How many types of biogeochemical cycles are found in nature? Explain carbon cycle in detail.
(HNB, Garhwal Univ. B.Sc. II, 2010)
12. Write short notes on:
(*i*) Ecological pyramids, **(HNB Garhwal Univ. B.Sc. II 2010)**
(*ii*) Carbon cycle **(Nagpur Univ. B.Sc. Bot–I, 2010)**
13. Write notes on:
(*i*) Pyramids of energy **(H.P. Univ. B.Sc. III, 2010)**
(*ii*) Food web. **(Nagpur Univ. B.Sc. (F). Bot II-2011)**
(*iii*) Pyramids of Number **(Univ. of Pune, B.Sc. III. 2010)**

4

Autecology

An ecosystem usually contains numerous species of plants, animals and microbes. A **species** includes individuals which are usually alike and capable of interbreeding and producing fertile offsprings. A group of interacting and interbreeding individuals of a species found in a particular surrounding or locality is referred to as **population.** A branch of ecology which is concerned with the study of individual species at all stages of its life cycle in relation to its environment is called **autecology** or **species ecology.** This branch of ecology aims at the study of life cycle, distribution, differentiation, ecological relationships, adaptation and genetic variability of population of individual species in relation to its environment. Autecology provides a sound basis for synecology as it helps in explaining the structure and dynamics of communities by way of providing informations regarding the ecology of important species of communities. In the words of C.F. Harried II (1977), "the two types of study, autecology and synecology are inter-related, the synecologist painting with a broad brush the outline of the picture and autecologist stroking in the finer details." Much stimulus for autecological studies has come mainly from the economic importance of certain species and agronomy and silviculture are, indeed, applied aspects of autecology. Several plant species introduced from one country to the other countries where they never existed have become fully acclimatised and are creating serious problems in agriculture. The knowledge of autecology of those species may definitely be helpful in eradication or control of the species.

ECOLOGICAL LIFE HISTORY OF SPECIES

Several eminent workers have prepared detailed schemes for a systematic study of autecological life histories. Stevens and Rock (1952) have suggested scheme for autecological life histories of herbaceous plants, Pelton (1951) has given a scheme for the study of autecological life histories of trees, shrubs and stem succulents; Curtis (1952) for vascular epiphytes, and Cooke (1951) for fungi and they have emphasized the need for study of environment and performance of plants. The main purpose of such studies should be to uncover the response of various forms of life to stimuli and compulsions of environment, and also to throw light on the limits of its geographical distribution and causes related to that. The different aspects in the study of autecology of an individual species are outlined below :

(1) In any ecological or botanical study identification of plants is the first requisite. Taxonomy and nomenclature of the species under study are discussed. Under nomenclature scientific valid names, various synonyms and common names are discussed. The confirmation of identification is advisable in a standard regional or central herbarium. The botanical survey of India has developed a chain of herbaria; Central National Herbarium at Sibpur, Howrah, and Regional Herbaria at Poona, Coimbatore, Shillong, Dehradun and Allahabad. Besides, a small Economic plant herbarium exists at Central Botanical Laboratory, Calcutta. These herbaria maintain large collections of correctly identified and properly arranged plants specimens and run a plant identification service.

Associates. The other species associated with the plant species under study should be recorded in a number of habitat conditions in order to get a proper range of variations, Associate species indicate the sociability and adaptations of a plant.

(2) **Distribution and importance.** The range of distribution, altitudinal and latitudinal limits and its ecological importance in various regions and habitats are noted. The degree of dominance or subdominance of species in different types of vegetation; its place in succession, whether pioneer or seral member or a component of climax community; its economic importance for browse, timber, pulp, fruits, in erosion and medicine are emphasized.

(3) **Morphology of plants.** In this, the distinguishing features of various plant parts are described. Structural variations are also noted in the plants of same species growing on the different ecological habitats.

(4) **Cytogenetics of plant species.** Cytological features, such as the structure of cytoplasm, chromosome morphology and the number and behaviour of chromosomes during mitosis and meiosis are studied in detail. Sometimes plants of the same species growing geographically isolated become so much differentiated and morphologically changed that they are often supposed as separate species. In such cases, interbreeding experiments are made. If the individuals interbreed, it is supposed that they belong to the same species. Variations in nature and factors for segregation are also discussed.

(5) **Physiology of plant.** Various physiological processes of the plants of the particular species are studied in detail and the factors influencing the rates of those processes are also taken into consideration.

(6) **Environmental complex.** The life cycle of individual species is greatly influenced by a number of environmental factors operating in conjunction as environmental complex. The various stages of life cycle of plant species in nature remain completely embedded in the environmental complex. Different species differ in respect of their response to climatic factors at different stages of their life cycles. Each species has a definite time (month or season) in the year for seed germination, seedling growth, vegetative growth, flowering and fruiting etc. Study of all these processes of species in relation to different periods or seasons of the year is called its **phenology.** Since each species has a definite period for a particular stage of its life history, presence of species in that particular stage will indicate the time of the year. In other words, the phenological behaviour of the species and different environmental factors at different seasons of the year are so much interlinked that the species is said to be **biological** or **ecological clock**. In autecology of a species both biotic and abiotic aspects are measured quantitatively at different stages of plant growth at regular intervals in order to study the phenology (germination, leaf-fall, initiation of flower bud, etc.) in relation to different seasons of the year. The environmental complex which is made up of several factors in various combinations affects each stage of plant's life cycle. In the foregoing discussion correlations of phenology to the various environmental changes are discussed:

(1) **Flowering.** Various environmental factors, such as, light, temperature of the soil and atmosphere, percentage of nutrients in the soil, etc. affect the flowering process, of plants. In autecology of a species, the period of flowering as well as the specific roles of different environmental factors on flower initiation, is noted. The photoperiodic effects and temperature effects on the flowering are studied in detail.

(2) **Pollination.** During pollination, the environmental factors help the plant to a great extent. Some plants may be pollinated by wind, some by water and some others by biotic agencies, such as insects, birds and even man.

(3) **Fruiting.** The structure and number of fruits, number of seeds per fruit, season of fruiting, etc. are some of the important aspects in study of ecological life cycle of a species. The environmental factors by way of their influence on the fruiting of the plants determine the

success or failure of particular species of the community in regeneration and establishment. Biotic factors, especially certain disease causing fungi and insects, damage fruits and thus affect fruit formation to a great deal.

(4) **Seed output.** The number of seeds produced by a plant in one reproductive flux is known as *seed output*. The seed output is measured by the following methods:

Average number of seeds per fruit $= a$

Average number of fruits per plant $= b$

Average seed output $= (a \times b)$

Average number of fruits per branch (for trees and shrubs) $= b_1$

Average number of branches per plant $= b_2$

Average number of fruits per plant $= (b_1 \times b_2)$

$\therefore$ Average seed output for trees $= a\,(b_1 \times b_2)$

Counts for seed output are made from the plants growing in a number of diverse localities in order to get a correct assessment. In autecology the knowledge of seed output is of great importance. It is commonly observed that annuals reproduce only once in their life, whereas the perennial shrubs and trees produce seeds usually once in a year but many times in their life time. Every species has its own range of seed output. The range of seed output may be affected by many habitat factors, both biotic and abiotic. So while taking into consideration the seed output of a particular species, various factors which affect it are also considered.

(5) **Dispersal of seeds.** The shape, size, weight and volume of seeds are important characters which govern the mode and extent of dispersal. Distribution and success in the establishment of plant species on the habitats are governed directly by the distribution of their seeds. If all the seeds scattered around the mother plants are allowed to grow, the new plants developing from them will have little chance for survival and regeneration because of over-crowding. Movement of plant populations to wider areas ensures success in the establishment of species. Thus, the study of dispersal of seeds of plants is one of the important aspects of autecology. The dispersal mechanisms and the agencies causing the dispersal of seeds and other reproductive bodies are also studied in autecology.

(6) **Viability of seeds.** Seeds have life spans of their own. They lose their power of germination after some time if stored under favourable conditions for long period. The period right from their formation up to the time when they begin to lose their germination capacity is called '**viability period**'. This period varies from species to species. In some cases, the viability period is zero and the seeds are totally incapable of germination.

To test the viability, seeds are placed in petridishes on sawdust every week from the day of maturation and seedfall. The experiment is continued till the seeds do not germinate at all.

For calculating the percentage of viable seeds in a given seed sample, seeds are cut and dipped in dilute solution of TTC (tetra zolium trichloride) with a little sucrose. After 24 hours, if the embryos of dipped seeds become pink, such seeds are treated as viable.

The seeds of many crop plants remain viable for 5 to 10 years. In *Mimosa glomerata, Cassia bicapsularis* and *Astragalus massibiensis* the seeds are reported to be viable even after over one hundred years of storage. It is usually seen among in the seeds that cannot germinate immediately after formation. Viability is affected by the conditions prevailing in the particular place where the seed is stored. It has been observed that the conditions which reduce the metabolic activities are usually responsible for increasing the seeds longevity. Low temperature, low oxygen and high CO_2 contents in the atmosphere have marked effects in increasing the viability period of seeds.

(7) **Dormancy.** In some ca es the viable seeds do not germinate up to a certain length of time even if the conditions for germination are favourable. This period is called dormancy period and such seeds are said to be in dormant stage and this phenomenon is known as dormancy. Dormancy may be on account of physiological immunity for germination or due to impermeability of seed coat to water and gases or due to specific light requirements or due to the presence of some germination inhibiting substances. Recently caumarin and its derivatives have been found to be present in the dormant seeds of some species. The dormancy can be broken by the following methods :

(*a*) **Mechanical methods.** When the dormancy is caused by hard seed coats that are impermeable either to water or to oxygen or both, it can be broken by abrasion, removal or puncture of seed coats. *Coriandrum* seeds are pounded mechanically before sowing which gives high percentage of germination.

(*b*) **Temperature.** The following alternating temperatures are usually given to seeds for breaking dormancy : (*i*) 0°C and 10°C, (*ii*) 30°C and 45°C and (*iii*) 0°C and 45°C. High temperature may also break dormancy, *e.g.*, in *Cassia occidentalis*.

(*c*) **Irradiation of seeds.** Irradiation of seeds to red light breaks the dormancy and increases the germination percentage in some plants. X-rays and Gamma-rays have also been found to break dormancy.

(*d*) **Chemical treatments.** Many chemicals are known to affect germination of seeds. Dormancy of seeds can be broken by dipping them in conc. H_2SO_4 for a few minutes (say 2 to 5 minutes) followed by thorough washing with water. Sometimes, it can be broken by putting soaked seeds in solution of copper sulphate or potassium permanganate or hormones.

(*e*) **Removal of Inhibitors of germination.** When the dormancy is caused by the presence of some germination inhibiting substances, it can be removed by washing the seeds in running water for varying periods of time before placing them for germination.

(8) **Seed germination and Reproductive capacity.** All the seeds may or may not germinate; some of them are inviable, some do not get proper conditions for germination and some may be destroyed by living organisms and by excess water. According to Salisbury (1942), the percentage of germinated seeds in the average seed output is called 'reproductive capacity'.

$$\text{Reproductive capacity} = \frac{\text{Average seed output of a plant} \times \text{Percentage of germinated seeds}}{100}$$

The reproductive capacity and the seedling establishment are significant for physiognomy (outward appearance) and sociological structures of species population. The process of seed germination is controlled by several factors, such as, temperature, dormancy of seeds, availability of water, concentrations of oxygen and carbon dioxide, pH value of medium and so on. In autecology of a plant species the different factors which influence the seed germination are also taken into account.

(9) **Seedling and vegetative growth.** When mature and viable seeds get favourable conditions for germination, they start germinating and after some time they establish their seedlings. Proper development and growth of seedlings are important factors for the survival of plant. Growth performance of shoot and root in relation to climate, physiography, soil character (physical and chemical both), fire, root/shoot ratio, stomatal counts, their physiology of movements are considered. Besides, the aerial and underground productivity in various habitats and communities, water requirement (the amount of water required per gm of dry weight) are also studied. Several factors influence the growth and establishment of seedlings. Many seedlings of sal develop on moist leaf litter during rainy season but the onset of dry season even for a short period due to interruption of rains may kill many of them and unless the roots of seedlings have penetrated deep enough into the soil establishment of seedlings cannot be achieved. The aerial shoots of seedling of sal *die back* year after year but the root growth continues. Shoots regenerate every

year during or before the rains. This phenomenon of "die back" continues for 20 to 25 years and only after the roots have become established in deep strata of soil with good supply of water and nutrients the plant shoots up and establishes itself. In autecological studies, thus, the knowledge about seedling growth, establishment of young seedling in a particular area and climate is essential. The intensity of solar radiations, durations and quality of light, temperature, soil conditions, water, etc. affect vegetative growth of any species. Every species in a population or community has its own requirements for the environmental conditions and has its own ecological amplitude of tolerance to an extent of fluctuation towards higher and lower sides from the optimum. Interspecific and intraspecific competitions for space, nutrients, light, water, etc. may also wipe out or delay the establishment of seedlings. Dense shade of trees and absence of sufficient light may have lethal effects on seedlings of some species. Interspecific and intraspecific competitions may be studied by growing varying number of seedlings of the same species (intraspecific) and those of associated species (interspecific) in mixed condition in a unit area and recording growth performance (mainly dry weight) of individual plants under all conditions.

POPULATION ECOLOGY

A group of individuals of a particular species occupying a definite space in which the individuals interact, interbreed and exchange genetic information is referred to as *population.* Thus one can speak of population of a bird species, population of locust species in a crop field, and so on. Population is a unit of community through which energy flows and nutrients are cycled. It is a self-regulatory system that helps in maintaining the stability of ecosystem. A population can be divided into small subgroups called *demes* or local populations. The branch of ecology which deals with the characteristics, structure and regulation of population is called ***population ecology***.

Ecological Amplitude and Law of Tolerance. A range of environmental conditions within which a species shows its characteristic growth potentiality is called *ecological amplitude* or tolerance. The geographical distribution of a species is determined by the environmental factors limiting the ecological amplitude and different ecological forms or *ecotypes* of the species. According to McMillan (1959), the species with one or few ecotypes may have narrow ecological amplitude, as for example, *Stipa spartea,* but the species represented by several ecotypes, each having its own phenotypic expression, have wide ecological amplitude, *e.g., Andropogon.*

According to Shelford's (1913) *Law of Tolerance,* each environmental factor has two limits—*the maximum* and *the minimum* limits within which a species survives. They are called limits of tolerance. Maximum growth and vigour of a species is exhibited at the *optimum.* Since the environment is dynamic and it keeps on changing from time to time, it acts as a natural check on population. Physiological stress and loss of vigour begin to appear towards the limits of tolerance and the individuals of one species are unable to compete with the better adapted individuals of other species and consequently they become infrequent.

The subject of population ecology is very wide but only the following aspects of population study will be discussed here :

(*i*) Characteristics of population

(*ii*) Population structure including (*a*) analysis of population dispersal and (*b*) various types of interactions.

CHARACTERISTICS OF POPULATION

The population has the following characters :

1. Population density
2. Natality
3. Mortality
4. Population growth
5. Age distribution of population
6. Population fluctuations.

1. Population Density

Population density refers to the size of any population in relation to some unit of space. It is expressed in terms of the number of individuals or biomass per unit area or volume, as for example, 500 teak trees per hectare; 40 lions per 100 km^2, 5 million diatoms per cubic metre of water. Population density is seldom static and it changes with time and space. Population size can be measured by several methods :

(*i*) *Abundance*—Absolute number of individuals in population.

(*ii*) *Numerical Density*—Number of individuals per unit area or volume. It is expressed when the size of individuals in the population is relatively uniform, as in mammals, insects and birds.

(*iii*) *Biomass Density*—Biomass density is expressed in terms of wet weight, dry weight, volume, and carbon and nitrogen weight per unit area or volume.

Population density can be expressed in two ways :

(*i*) **Crude Population Density.** When the density is expressed with reference to total area at a particular time.

(*ii*) **Ecological Density.** When the density is expressed with reference to total area of habitat available to the species.

The distribution between crude density and ecological density becomes important because the patterns of distribution of individuals in nature are different and individuals of some species like *Cassia tora, Oplismenus burmanni* are found more crowded in shady places than in other parts of the same area. Thus population density calculated in total area would be crude density and the densities for the shade areas and open areas separately would be ecological densities.

Population density can be calculated by the following equation :

$$D = \frac{n/a}{t}$$

where D is population density; n is the number of individuals; a is area and t is unit time. Density of human population can be obtained by dividing the total number of persons in the area by the total land area of the region. Density of population of a country can be obtained by dividing the total number of persons living in the given region by total land area of that region. Average population density in developing countries is more as compared to those in developed countries. Netherlands is smaller than India but its population density is greater (319/km^2 in Netherlands and 168/km^2 in India). Area of India is 2.5% of the world but 15% population of the world lives in India alone. The population density of India is 4% higher than that of Europe and more than 7 times that of U.S.A.

Population density is affected by a number of environmental factors, such as geographical factors, mortality, natality, emigration and immigration and socio-economic factors.

2. Natality

Natality refers to the rate of reproduction or birth per unit time. It is an expression of the production of new individuals in the population by birth, hatching, germination or fission.

Natality is calculated by the following formula :

$$\text{Birth rate or Natality } (B) = \frac{\text{Number of births per unit time}}{\text{Average population}}$$

The maximum number of births produced per individual under ideal conditions of environment is called *potential natality*. It is also called *reproductive* or *biotic potential, absolute natality* or *maximum natality*.

Natality varies from organism to organism. It depends upon the population density and environmental factors. It is a general rule that if the population density is usually low, the birth rate is also low. This is so because the chances of mating between males and females are low. If population density is unusually high, the birth rate may also be low due to poor nutrition or physiological or psychological problems related to crowding. The maximum or absolute natality is observed when the species exists under ideal ecological and genetic conditions. The actual number of births occurring under the existing environmental conditions is much less as compared to absolute natality. It is referred to as *ecological natality* or *realised natality.* It is not constant for population and may vary with the size of population as well as with the time.

3. Mortality

Mortality refers to the number of deaths in population per unit time.

Mortality rate $= \frac{D}{t}$ where D is the number of deaths in the time t.

Mortality can be expressed in the following two ways:

(*i*) **Minimum or Specific or Potential Mortality.** It represents the minimum of theoretical loss of individuals under ideal or non-limiting condition. Thus, even under the best conditions individuals of a population would die of old age determined by their physiological longevity. So it is constant for a population.

(*ii*) **Ecological or Realised Mortality.** It refers to the death of individuals of a population under existing environmental conditions. Since it varies with environmental conditions, it is never constant. The maximum mortality occurs at the egg, larval, seedling and old age.

Mortality is affected by a number of factors, such as, density, competition, disease, predation and environment. Death rates vary among the species and are correlated with birth rates. When the rate of natality is equal to the rate of mortality the population is stationary.

A birth-death ratio $\left(\frac{\text{Births}}{\text{deaths}} \times 100\right)$ is called *Vital index*. For a population, the survival of individuals is more important than the death. The number of births in relation to the carrying capacity of the habitat is a fundamental factor influencing the mortality rate. When more youngs are born than the habitat can support, the surplus must either die or leave the area. Because the number of survivors is more important than the number of dying individuals, mortality is better expressed as survival or as *life expectancy.* The life expectancy refers to the average number of years the members of a population have left to live.

Life Tables and Survivorship Curve. The species differ in respect of birth rates, average life span and mortality rate. When sufficient informations about a species are available, *life-table* can be formulated which provides vital statistics of mortality and life expectancy for the individuals of different age-groups in the population. In such tables age is usually represented by the subscript index x which is some convenient fraction of species life span, such as, years or stage of development. The life table is set up on the basis of an initial cohort or group of 100, 1000, 10,000 10,00,00 individuals and the number of living in the beginning of each successive age interval is symbolised as lx. Plotting these data gives a survivorship curve for a species. The number of dying individuals within each age group is denoted as dx. The rate of mortality during each age interval (qx) is commonly expressed as the percentage of the number at the beginning of the interval.

$$qx = \frac{dx}{lx} \times 100$$

Survival rate is the difference between the mortality rate and 100 per cent (*i.e.*, 100—qx) and is denoted by δx. Life expectancy (ex), thus, is the mean time between any specified age and the time of death of all individuals in the age group.

Types of Survivorship Curve. If it could be assumed that all members of an original population have the same capacity for survival (environmental effects for the moment are ignored), plotting the number of surviving individuals against time would produce a survivorship curve in the form of a right angle. There are three general types of survivorship curves which represent different natures of survivors in different types of population (Fig. 4.1).

(*i*) **First Type or Highly Convex Curve.** Curve *A* in the Fig. 4.1 is the characteristic of the species in which the population mortality rate is low until near the end of life span under ideal environmental conditions. Thus, all the members born at the same time live out the full physiological life span characteristic of the species and all die at about the same time. Many species of animals as deer, mountain sheep, and modern man show such curves.

(*ii*) **Second Type or Diagonal Curve.** Survivorship curve *B* in the figure 4.1 is characteristic of organisms in which rate of mortality is fairly constant at all age levels, a more or less uniform percentage decrease in the number that survives.

(*iii*) **Third Type or Highly Concave Curve.** Survivorship Curve *C* in the Fig. 4.1 is characteristic of such species in which mortality rate is high during the early stage and constant in all other age-groups. Oyster, some birds, oak trees, etc. show this type of curves.

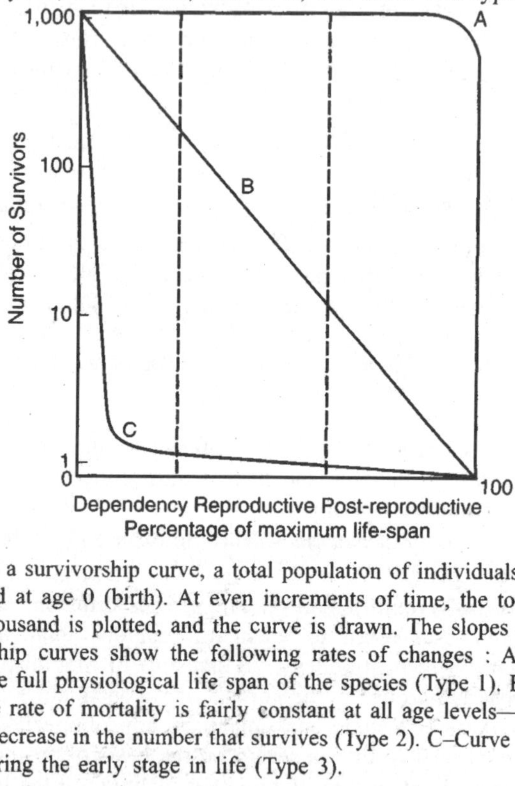

Fig. 4.1. To construct a survivorship curve, a total population of individuals, such as one thousand, is considered at age 0 (birth). At even increments of time, the total number of survivors from this thousand is plotted, and the curve is drawn. The slopes of the three basic types of survivorship curves show the following rates of changes : A—Curve for organisms living out the full physiological life span of the species (Type 1). B—Curve for organisms in which the rate of mortality is fairly constant at all age levels—a more or less uniform percentage decrease in the number that survives (Type 2). C–Curve for organisms with high mortality during the early stage in life (Type 3).

4. Population Growth

The growth is one of the dynamic features of species population. Population size increases in a characteristic way. When the number of individuals of population is plotted on the *y*-axis and the times on the *x*-axis, a curve is obtained that indicates the trend in the growth of population size in a given time. This curve is called *population growth curve.* There are two types of growth curves :

(*i*) **Sigmoid Curve.** When a few organisms are introduced in an area, the population increase is very slow in the beginning (*positive acceleration phase* or *lag phase*), in the middle phase, the population increase becomes very rapid (logarithmic phase) and finally in the last phase the population increase is slowed down (*negative acceleration phase*) until an equilibrium is attained around which the population size fluctuates according to variability of environment. The level beyond which no major increase can occur is referred to as *saturation level* or *carrying capacity.* In the last phase the new organisms are almost equal to the number of dying individuals and thus there is no increase in population size. In this way, one gets sigmoid or *S*-shaped growth curve (Fig. 4.2).

(*ii*) **J-Shaped Curve.** The second type of growth curve is J-shaped. Here in the first phase there is no increase in population size because it needs some time for adjustment in the new environment. Soon after the population is established in the new environment, it starts multiplying rapidly. This increase in population is continued till large amount of food materials exist in the habitat. After some time, due to increase in population size, food supply in the habitat becomes limited which ultimately results in decrease in population size. This will result in J-shaped growth curve rather than *S*-shaped (Fig. 4.2).

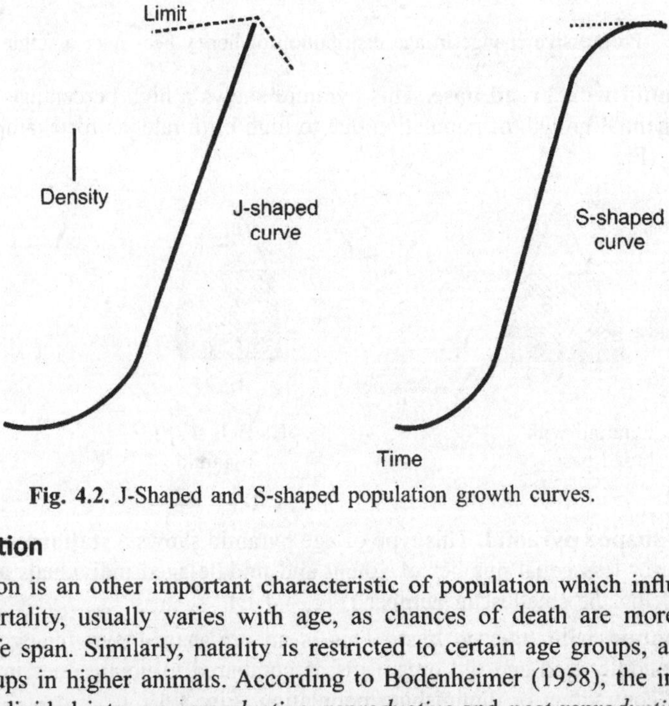

Fig. 4.2. J-Shaped and S-shaped population growth curves.

5. Age Distribution

Age distribution is an other important characteristic of population which influences natality and mortality. Mortality, usually varies with age, as chances of death are more in early and later periods of life span. Similarly, natality is restricted to certain age groups, as for example, in middle age-groups in higher animals. According to Bodenheimer (1958), the individuals of a population can be divided into *pre-reproductive, reproductive* and *post-reproductive* groups. The individuals of pre-reproductive group are young, those of reproductive group are mature and those in post-reproductive group are old.

The distribution of ages may be constant or variable. It is directly related to the growth rate of the population. Depending upon the proportion of the three age-groups, populations can be said to be *growing, mature* or *stable,* and *diminishing.* In other words, the ratio of various age groups in a population determines the reproductive status of the population. Rapidly increasing population contains a large proportion of young individuals, a *stable population* shows even distribution of individuals in reproductive age-group and a *declining population* contains a large proportion of old individuals.

Age Pyramids. Age pyramid is a model in which the numbers or proportions of individuals in various age groups at any given time are geometrically presented. In an age pyramid, the number of pre-reproductive individuals is shown at the base, that of reproductive age group in the middle and the number of post-reproductive individuals at the top. The shape of age-pyramid changes with the change in the population age distribution over a period of time (Fig. 4.3). The age pyramid indicates whether a population is expanding or stable or diminishing and accordingly three hypothetical age pyramids have been suggested. These are as follows:

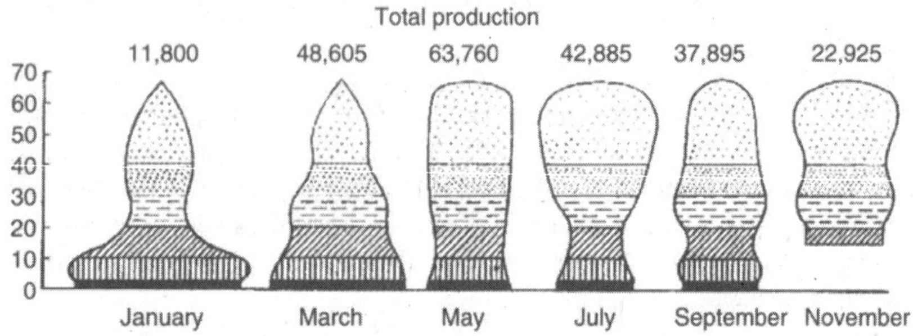

Fig. 4.3. Progressive change in age distribution of honey bees over a single season.

(*i*) **Pyramid with broad base.** This pyramid shows a high percentage of young individuals and an exponential growth of population due to high birth rate, as for example in yeast, housefly, *Paramecium* (Fig. 4.4A).

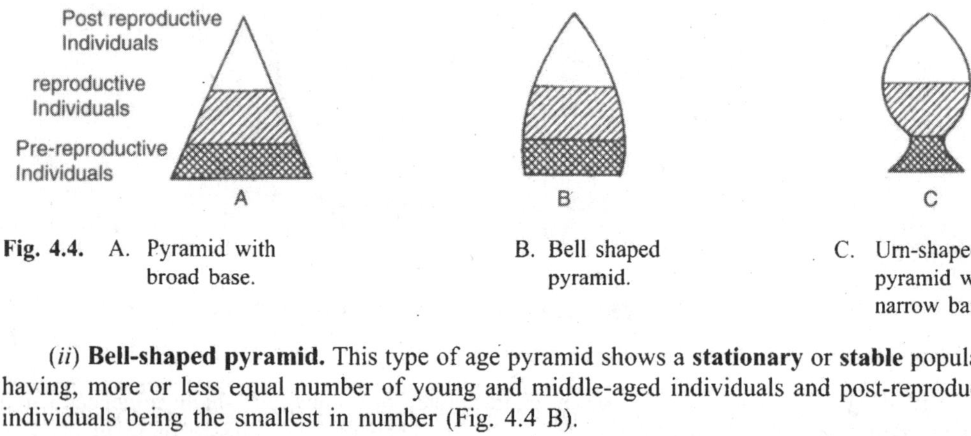

Fig. 4.4. A. Pyramid with broad base. B. Bell shaped pyramid. C. Urn-shaped or pyramid with narrow base.

(*ii*) **Bell-shaped pyramid.** This type of age pyramid shows a **stationary** or **stable** population having, more or less equal number of young and middle-aged individuals and post-reproductive individuals being the smallest in number (Fig. 4.4 B).

(*iii*) **Pyramid with narrow base.** This is an urn-shaped pyramid which shows increased numbers of middle aged and old organisms as compared to young ones in the population. It is indicative of contracting or diminishing population (Fig. 4.4 C).

6. Population Fluctuations

The size and density of natural population show a changing pattern over a period of time. This is called population fluctuation. There are three types of variations in the pattern of population change :

(*i*) **Non-fluctuating.** When the population remains static over the years, it is said to be non-fluctuating.

(*ii*) **Cyclic.** The cyclic variations may be (*i*) *seasonal*, and (*ii*) *annual*. Sometimes *seasonal changes* occur in the population and there are additions to the population at the time of maximum reproduction and losses under adverse climatic conditions. Common examples of seasonal variations are met in mosquitoes and houseflies which are abundant in particular season and so also the weeds in the field during the rainy season.

When the population of a species shows regular ups and downs over the years, it is called annual cyclic variation. It appears in the form of a sigmoid curve with regular drops in population after peaks.

(*iii*) **Irruptive.** When the change in population density does not occur at regular intervals or in response to any obvious environmental factor, it is said to be **irruptive fluctuation.** In this, there is a sudden exponential or logarithmic increase in population density in short time, followed by equally quick drop in population density due to deaths, and final return to normal level or even below that level.

BIOTIC POTENTIAL OF POPULATIONS

It is also known as maximum reproductive power. The constant percent growth rate of population under optimum environmental condition represents the biotic potential or reproductive power.

$$\text{Biotic potential } (r) = \frac{\Delta AN / \Delta t}{N}$$

Where, N = number, t = time, A = Area and Δ is a constant.

A biometrician A.J. Lotka (1925) has advanced this equation. According to him, the conditions altering biotic potential will in turn change the age structure of populations and when the conditions return to original state, the population returns to normal reproductive power and so does the age structure.

POPULATION STRUCTURE

It includes an analysis of population dispersion and various types of interactions.

1. Population Dispersal

Movement of individuals into and out of the population is called population dispersal. It plays important role in the geographical distribution of organisms even to the areas previously unoccupied by the members of population. Dispersal of organisms occurs for various reasons such as food, protection, prevention from overcrowding, action of wind and water, environmental factors, such as light, temperature, breeding behaviour, physiological reasons or for interchange of genetic materials between the populations.

Population dispersal occurs in nature in the following three ways :

(*i*) Emigration

(*ii*) Immigration

(*iii*) Migration.

(*i*) **Emigration.** It is one way movement of individuals out of a population. This movement is permanent and causes spread of a species to new areas. Emigration under natural condition occurs when there is over crowding in the population and is generally regarded as an adaptive behaviour that regulates the population on a particular site and prevents over-exploitation of the habitat. This type of dispersal offers new opportunity to the individuals of a population to interbreed with those of the other populations leading to more genetic heterozygosity and adaptability.

(*ii*) **Immigration.** Immigration is one way movement of individuals into a population. It leads to rise in density of population. It may result in decreased mortality among the immigrants or decreased reproductive capacity of the individuals.

(*iii*) **Migration.** Migration is two way mass movement of the entire population. It involves a periodic departure and return of the individuals of a population and occurs only in mobile organisms during unfavourable periods. It is shown by many birds, fishes and certain mammals.

In most cases, migration of population occurs for food, shelter or reproduction. Through this type of movement the chances of utilization of resources in the habitats not previously occupied by any organism are great. However, during migration of population, mortality of numerous individuals may occur due to various ecological hazards, such as temperature fluctuation, scarcity of food, predation etc. Migration has certain benefits for populations as it enables wider dispersion of populations. It also avoids intraspecific competition for food, shelter, etc.

2. Interactions Among Populations

Plants and animals exhibit a wide range of relationships. Individuals of one species interact with the other individuals of the same species or with those of other species. Various types of interactions are as follows:

1. **Neutralism.** When the presence of one species appears to have no effect on the second species (*i.e.*, no interaction), it is a state of neutralism.

2. **Commensalism.** It is one-sided relationship between two species in which one species is benefited but the other is neither benefited nor harmed. Some epiphytes, as for example orchids, are the best examples. Epiphytes depend upon the other trees for support and nutrients. They manufacture their own food but do not help supporting plant in any way.

3. **Protocooperation.** It is less extreme type of interaction in which two species interact favourably with each other, though both of them are able to survive separately.

4. **Mutualism.** It is an obligatory interaction that is beneficial to both species. The term **symbiosism** has also been used for this relationship. Mutualism is best demonstrated in lichens. The lichen is composed of two components—alga and a fungus. The fungus supports the alga while the alga supplies food to fungus. Green *Hydra* presents another example of mutualism. This animal has green photosynthetic alga in the protective ectoderm. The alga gives off oxygen benefitting the animal which, in turn, supplies CO_2 and N_2 to the plant. Root nodules of legumes containing *Rhizobium leguminasarum* bacterium provide another example of mutualism in which there is reciprocal beneficial relationship between the root and bacteria.

5. **Amensalism.** In this type of interaction between the two species, one species is harmed or inhibited and the other is neither benefitted nor harmed by the association.

Many algae produce extracellular toxic metabolites which inhibit the growth of other algae species. *Chlorella vulgaris* (green alga) produces chlorellin which is toxic to other species of algae.

6. **Parasitism.** When two organisms live together in which one derives nourishment at the expense of the other, the condition is called parasitism. In the parasitic association, the species which provides nourishment and support is called the **host** and the one which gets support and nourishment is called the **parasite.** Several species of plants and animals form parasitic associations with other organisms. A parasite usually parasitizes a host which is larger in body size than it and ordinarily it does not kill the host, at least until it has completed its reproductive cycle.

7. **Cannibalism.** This type of interaction is limited within a species in which the bigger individuals kill and feed on the smaller ones. It is a natural method of population control.

8. **Predation.** In this type of association and interaction one species (predator) kills and feeds on second species (prey). Predation is important process in the community dynamics. Predator is always larger than the prey. From population ecology point of view predation is the action and reaction in the transfer of energy from one trophic level to the other. It represents a direct and complex interaction between two or more species of eaters and eaten.

9. **Competition.** When in the association of two or more species each species is adversely affected by the presence of the other species in respect of food, shelter, space, light, etc., this phenomenon is termed competition. It is of two types :

(*i*) **Intraspecific competition.** When competition occurs between the individuals of the same species and their requirements are common, the process is called intraspecific competition.

(*ii*) **Interspecific competition.** In this type of competition, the individuals of different species compete for common materials and conditions.

QUESTIONS

1. What is autecology? Discuss the various aspects that are important in the study of autecological life history of a tree.
2. What do you mean by phenology? Discuss briefly the correlations between the phenology and the various environmental changes.
3. Write brief notes on the following :

   (*i*) Seed Dormancy (*ii*) Reproductive capacity of a species (*iii*) seed output (*iv*) Biological clock.
4. Define population. Discuss briefly the various characteristics of population.
5. Write short notes on the following:

   (*i*) Population growth (*ii*) Population dispersion and various types of interactions among population (*iii*) Population density (*iv*) Ecological amplitude.
6. Fill in the blanks:

   (*i*) The study of individual plant species is called ....

   (*ii*) Periodic departure and return of a population is called ....

   (*iii*) One way movements of individuals out of populations is ....

5

Ecological Genetics of Population or Gene Ecology

Species differ in their environmental requirements. They also differ in respect of their tolerance to environmental fluctuations. The species occur in several morphological forms in different habitat conditions. Gote Turesson (1922), a Swedish worker conducted a series of experiments on variations within a Swedish plant species *Plantago maritina.* He collected a group of 20 or more individuals of the same species from different areas and planted them under identical environmental conditions in his experimental garden at Akark. He noted that (*i*) several intergrading forms existed within the species which differed from one another in morphological or physiological features and (*ii*) some of differences might be unstable and induced only by the environmental factors while some others might be permanent (*i.e.,* genetically fixed). These observations led him to formulate the concept of gene ecology which brought about many revolutionary changes in plant taxonomy. The population sampling and subsequent procedures that Turesson followed led him to conclude that species differ from one another morphologically, physiologically and in habitat requirements and on the basis of such studies he proposed the following classes of morphological forms within a species :

1. Ecads or Ecophenes or Habitat forms,
2. Ecotypes,
3. Ecoclines
4. Ecospecies, and
5. Coenospecies.

ECADS OR ECOPHENES

These are also called epharmones or habitat forms which are environmentally induced variations. They belong to the same genetic stock or species and the variations in their morphology (in shape, size, number and reproductive capacity) are induced by the environmental influences. The variations are not fixed but are temporary, somatic and reversible. If one type of ecad is transplanted into environment of another type of ecad, the differences would disappear. These morphological variations are not permanent because the genetic composition of ecads is not affected and so they are identical in their genetic behaviour and interfertile. The variants or ecophenes may differ to such an extent that they can be treated as separate species. *Euphorbia hirta* plants growing in the grassland are prostrate and profusely branched while the plants of the same species growing on the footpaths are compact, small leaved and cushioned. When these forms are grown under identical habitat conditions, their differences disappear. In *Bothriochloa pertusa* and *Dichanthium caricosum* several distinct morphlogical forms of ecads have been noted under different habitats by Pandeya (1962). In both the species there are two types of ecads : (*i*) basket form habit shown by the plants growing in protected areas, and (*ii*) saucer-shaped habit of the plants growing in over-grazed areas. Grazing causes reduction in the size of erect stem, number of spikes per raceme, number of spikelets per spike and in length and breadth of lower glumes of spikelets. In addition to these morphological variations, some physiological variations are also noticed, as for example development of anthocyanine in the leaves and early initiation of flowers.

ECOTYPES

These are also called ecological or physiological races. The term ecotype was proposed by Turesson (1922) to the groupings of populations or ecological races or sub-species of a species in relation to different environmental or habitat conditions. These are interfertile forms or biotypes of a species which possess different genetic compositions or genotypes and arise due to mutations, hybridization and isolation. Though the different ecotypes of a species are morphologically and genetically distinct, yet because of their interfertility, they are put into one taxonomic species. Ecotypes are morphologically, physiologically and developmentally adapted to live under varying environmental conditions. Ecotype is the product of genetic response of a population to a habitat. In ecotypes adaptations are irreversible, *i.e.,* they retain their features even when planted in a neutral habitat. The differences in the ecotypes are so marked that some of them are treated as separate species by old taxonomists. Mishra and Shiva Rao (1948) found *Lindenbergia polyantha* and *L. urticaefolia* to be the ecotypes of the same species. The two ecotypes differ from each other in respect of their tolerance to high doses of lime. *L. polyantha* can tolerate high doses of lime. In another study. Ramakrishnan (1959, 60) has shown that *red* and *green populations* of *Euphorbia thymifolia* are two ecotypes. The red ecotype grows in calcium rich (**calcicol**) as well as calcium deficient soils and the green ecotype is a **calcifuge** *i.e.,* it cannot grow in calcium rich soil. The red ecotype possesses a pair of dominant alleles whereas the green type possesses a pair of recessive alleles. Ecotypes have been reported in several species of plants, such as *Euphorbia hirta, Cassia tora, Ageratum conyzoides, Cenchrus ciliaris.*

During last two decades, many investigators have studied either only a few populations or several widely separated populations and have shown that each population was an ecotype and the term is now being widely used below the level of species. Our increasing knowledge of genecology and population differentiation has shown that an ecotype may be the ecological unit but not an evolutionary unit. According to McMillan (1969), the concept of genetically based variation that is habitat correlated or population differentiation or micro evolution is valid, but ecotype as a word has limited value in its application.

Consideration of the Turesson's ecotype as a reliable ecological unit requires an assumption of environmental uniformity within climatic or altitudinal regions or habitat types. There is strong trend among evolutionary biologists to emphasize the importance of the local environment in shaping the population. Since the unit of study is the local population, it seems most appropriate if we present our information in terms of local population, rather than to imply that local population is an ecotype. If the term "ecotype" is used as equivalent to ecological race or a taxonomic sub-species, we cannot consider the ecotype to be a basic evolutionary unit. Since the evolutionary unit is the population, the use of ecotype or genetically based variation that is habitat correlated cannot contribute additional information but will create confusion. Ecotype is now used by botanists and ecologists to indicate almost any degree of genetic difference below the level of species. According to Philips (1971), the ecotype has been used as the word for an abstract concept of an ecosystem type.

Ecotype is still considered on the basis of morphological, physiological or ecological evidences. According to Davis and Heywood (1963), ecotype is based on the reaction between genotype and prevailing environment. They defined 'ecotype' as a population distinguished by morphological and physiological characteristics most frequently of a quantitative and physiological characteristics nature; inter fertile with other ecotypes or the ecospecies, but prevented from freely exchanging genes by ecological biomass.

Characteristics of Ecotypes

The characteristic features of ecotypes are mentioned below :

1. Ecotypes of a species, though genotypically distinct, are always interfertile.
2. They retain their original features when cultivated in a natural habitat.
3. Ecotypes are genetically fixed.
4. A species with wide ecological amplitude can be distinguished on the basis of morphological and physiological characters into different habitat forms or ecotypes.
5. They occur in distinct habitats.
6. Ecotypes are discrete entities with clear differences which separate one ecotype from another.
7. The differences are not due to plastic response to change in environment but are actually due to natural selection of locally adapted populations.

Formation or Origin of New Ecotypes

Now ecotypes can be produced by the following methods :

1. **Hybridization.** It is produced by the natural cross between two species. For example, when *Spartia stricta* is naturally crossed with *S. alterniflora,* the new hybrid *S. townsendii* results which eliminates both the parents from their natural habitats owing to its greater adaptability.

2. **Mutation.** Due to natural mutation and recombination small gene pools accumulate in a segregating population which make it better adapted to the particular habitat or environment. Some new ecotypes also arise by cultivation or protected growth as it eliminates competitive selection.

3. **Chromosomal changes.** Structural changes in the chromosomes such as translocation, inversions, and loss or addition of chromosome segments produce changes in genotypes and phenotypes resulting in the formation of new ecotypes. Polyploidy also leads to the formation of new ecotypes because polyploids hardly exhibit ecological tolerance as their parents do.

Kinds of Ecotypes

Ecotypes have been observed in a large number of species and the cause of ecotype differention may be the latitude, altitude, light, soil, biotic interference, physiological changes, etc. According to varying environmental conditions, ecotypes may be of following types :

1. **Climatic ecotypes.** Ecotypes which are produced due to varying climatic factors as light, temperature, water and wind are called climatic ecotypes. Turesson (1930) has recorded the climatic ecotypes in *Leontodon autumnalis.*

2. **Edaphic ecotypes.** Ecotypes which are produced due to differences in edaphic or soil factors, such as soil moisture, excess or deficiency of nutrients, change of soil pH, etc. are said to be edaphic ecotypes. Misra and Rao (1948) have studied *Lindenbergia polyantha* and Ramkrishnan (1961) has studied *Euphorbia thymifolia* and recorded several edaphic ecotypes in them.

3. **Climatic-Edaphic ecotypes.** The ecotypes produced due to the influences of both climatic and edaphic factors are called climatic-edaphic ecotypes. Panday and Jayant (1970) have reported climatic-edaphic ecotypes in *Cenchrus ciliaris.*

4. **Altitudinal and latitudinal ecotypes.** These ecotypes are produced due to change in altitude and latitude. Such ecotypes are found in *Cassia tora, Anagallis arvensis, Pinus* and many other gymnosperms.

5. **Physiological ecotypes.** These ecotypes are produced due to physiological changes as in photoperiod, water absorption, nutrient uptake, etc., for example in *Boutelona curtipendula,* there are two photoperiodic ecotypes—short day and long day which are morphologically alike.

Delimitation of Ecotypes

Ecotypes are not always based on morphological characters. Sometimes, single ecotype has several ecophenes which depend upon the habitats. Now a days following techniques have been applied to differentiate ecotypes:

(*i*) **Morphological features.** In this case morphological or physiological features of several individuals are studied at random in different populations of a species and one or few characters are considered and the results obtained are shown in the graph. Unimodal curve shows homogenous population, bimodal, trimodal and multimodal curves show two, three and many ecotypes in the population respectively.

(*ii*) **Anderson's Scatter Diagrams.** This was devised by Anderson (1940) to delimit ecotypes. For this some measurable characters are taken into consideration. One character (for example, leaf length) is plotted on one axis of the graph and other character (for example, leaf breadth) is plotted on another axis. In this way large number of characters are taken from different localities.

(*iii*) **Cytological behaviour.** In this case karyotypes and their behaviour are observed in different forms. The differences in cytological behaviour show the existence of distinct ecotypes.

(*iv*) **Transplantation experiment.** In this experiment plants from all the different localities are grown under uniform environmental conditions and their morpho-physiological characters are compared with plants growing in natural habitat. If the characteristic features are not changed in the neutral area, the existence of particular ecotypes is confirmed.

(*v*) **Breeding experiments.** In this case, crossing is done between different forms of variable nature followed by self-fertilization to determine the characteristic features of variable forms. If the characteristic features show segregation in the offsprings, the presence of distinct ecotypes can be proved.

Significance of Ecotypes

1. Cultivation of economically important plants has been made possible in different habitats.

2. New ecotypes of a species enable it to be adapted to climatically and edaphically different places.

3. Ecotypes help the species to extend its ecological range and spread to new areas.

4. Morphological variations can be marked in the species growing on varying habitats which lead to evolution.

ECOCLINES

Ecoclines are the intermediate forms which are found in the transitional regions between two adjacent regions occupied by separate ecotypes of a species. They are produced by hybridization between two related ecotypes and they survive under intermediate environmental conditions. Thus a continuous change in the characteristics of plants from one region to another is apparent.

ECOSPECIES

Ecospecies and coenospecies are two parallel terms. Gregor (1939) defined ecospecies and coenospecies purely on the basis of fertility-sterility criteria, eliminating any specific demand for morphological, physiological or cytological differences. According to him, ecospecies is a unit of gene ecological classification. An ecospecies has one or more ecotypes but ecotypes of one ecospecies do not produce viable offspring when crossed with the ecotypes of other ecospecies.

COENOSPECIES

Different ecospecies which can undergo occasional crossing but do not produce viable offspring are collectively brought under coenospecies. In other words, populations incapable of exchanging genes with other populations of the same species are referred to as coenospecies.

QUESTIONS

1. What is ecotype? Describe the origin and development of ecotypes.
2. What are characteristic features of ecotype? How is new ecotype evolved?
3. What are the different types of ecotype? Describe with suitable examples. Gives some Indian examples of ecotypes.
4. What is the difference between ecotypes and ecophenes? What are the significances of ecotypes?

6

Synecology or Study of Communities

No plant or animal lives as isolated individual. Plants and animals generally prefer to live in groups or colonies. Different plants and animals living in a habitat constitute a biotic community. When only assemblage of plants in a habitat is considered, it is called *plant community.* Similarly, assemblage of animals in a habitat is called *animal community.* In any biological organization plants and animals are very closely related and interdependent and at a particular place plants and animals share the same set of conditions and same environment. In view of these facts, modern biologists prefer use of biotic community to plant community or animal community. The study of the relationships of plants and animals making up a natural community is termed as ***community ecology*** or ***synecology***.

The basic unit of vegetation is called plant community or a plant association. The communities are not the random mixtures of species. The species living together in groups exhibit various degrees of adjustment among themselves and with their physical habitats as well. Each community consists of a set of many different species which persist year after year. In a community, each plant species is represented by innumerable individuals. A group of individuals of the same species is commonly known as *population.* Thus a population is a part of community and populations of different species may be intermingled in a community. Oosting defines community as "an aggregation of living organisms having mutual relationship among themselves and to the environment." In modern books of ecology, plant community is defined as *'uniform floristic composition.'* Earlier Gleason (1939) in his individualistic concept of plant association suggested that an association is a complex of slight irregularities, all of which blend into an entirety of apparent homogeneity. It is uniform either in space or in time and possesses definite limits of area only to a reasonable extent.

Pondeyar (1960, 1961) has discussed this subject and suggested that an association is a temporary phenomenon which is continuous in space and with slight irregularities, all of which blend into an entirety of apparent abstract homogeneity. According to him, the association is not a final vegetation but an assemblage of plants of any status. Misra and Puri (1954) are of the opinion that Clements' original concept which defined "community as any unit of vegetation, whether developmental or climax in status", should be accepted as such and they used the term *association* for *climatic climax.* According to them, the climatic climax community is in, complete equilibrium not only with climate but also with the whole complex of environment.

COMMUNITY COMPOSITION

The following points characterise the community :

(1) **Species diversity.** The biotic community is a natural assemblage of a large number

of plant and animal species in an area. Actually it is a part of a larger whole ecosystem in which living and non-living components interact and bring about circulation, transformation and accumulation of energy and matter. Various species of plants and animals living in a community exhibit species diversity. Of the total number of species in a biotic community a few species are abundant while a large percentage is rare. The common abundant species are called **dominants** whereas rare species determine species diversity. The ratios between the number of species, their biomass and productivity provide valuable information about community and are called **species diversity indices.** Species diversity is high in natural community and low in managed ecosystem such as croplands.

Because in any particular habitat there is no considerable variation in environmental conditions, the plants growing together in a community show unique uniformity in their behaviour. Vegetation, therefore, is reflection of a climate and, in general, widely separated areas having similar climate have similar aspects of landscape.

Some community areas have limits but more often the community boundaries are hard to define. A clearly distinguished area or a type of area with uniform habitat conditions and supporting characteristic type of vegetation is termed "biotope".

Each species of community has got definite range of tolerance towards the physical and biological environmental conditions of the habitat. The range of environment a species can tolerate is called its **ecological amplitude.** The nature of community of a particular habitat is determined by the species contents, ecological amplitudes of the species and physical and biotic influences prevailing in the locale of community.

(2) **Coexistence.** Species occurring in the particular habitat do not live in complete isolation as pure cultures, but they coexist in mutual adjustment. The coexisting populations are interrelated and they show some sorts of interaction. The relationships, between coexisting species may be obligatory in one direction or in both. The trees in a forest community can live just as good as without shrubs and herbs which grow under them. This relationship is obligatory in one direction only. The relationship between plants and animals which pollinate them is obligatory in both directions.

The nature of interaction between two coexisting species may fall into one of the following types :

(*i*) **Exploitation.** In this, one species lives at the expense of another.

(*ii*) **Mutualism.** In this, two or more coexisting populations benefit from the relationship but none suffers.

(*iii*) **Competition.** In this, two populations may compete for same resources of the habitat.

(*iv*) **Neutralism.** In this, two populations may be quite independent and neither population affects the other.

(3) **Interdependency.** All the members of a community have ability to live under the conditions of the habitat and they are interdependent upon one another to some extent. It is called *dependency.* Thallophytes, mosses, ferns and many shade loving herbs that are found on the forest floor are dependent on the forest trees because trees provide shadow and moist conditions. If the trees of forest are removed, the ground vegetation may disappear. Similarly, the fungi and saprophytes found in the forest depend upon the roots of plants and on the rich humus and some fungi form mycorrhizal associations with the plant roots. Some sort of relationship also exists between plants and the insects and other animals which pollinate them.

(4) **Species dominance.** Not all the species of a community are found in adundance. Only a few species are found in abundance, either in number or in biomass (living weight) while the majority are rare. The common species which are abundant and contain maximum biomass are considered to be **dominants**. Dominant individuals influence the associated individuals. In the

forest, tallest trees, for example, influence the understorey plants and ground vegetation not only by decreasing the intensity of light reaching the forest floor and increasing the moisture content of air but also by changing the soil structure and its chemical composition. The dominance in the community may be the result of co-action between two or more species. Different communities are generally recognised and named on the basis of dominant species occurring in them.

(5) **Stratification.** In a plant community, the plants, which have some sort of relationship among themselves, may be trees, shrubs, herbs, mosses, lichens and thallophytes. These plants form, more or less, distinct strata or layers or storeys on vertical as well as in horizontal planes. This is characteristically known as *stratification.* The individuals of different layers represent different *"life forms"*. Each layer of community may sometimes include individuals of different morphological classes, as for example, the top layer or canopy of forest may be formed by tallest trees and lianes (woody climbers). In order to overcome this objection, plants belonging to different morphological classes are put in 'sinusiae' (singular—sinusia), as for example, trees are put in sinusia of trees, epiphytes are put in sinusia of epiphytes, and so on.

In grassland, there are essentially three strata : (1) the root and rhizome layer, (2) the ground layer, and (3) herbaceous layer.

Fig. 6.1. A forest showing stratification of the plants.

In forest vegetation, the stratification reaches its greatest complexity. As shown in Fig. 6.1, five vertical subdivisions (units of homogeneous life forms and ecological relations) may be present—(1) *subterranean zone,* (2) *forest floor,* (3) *ground vegetation* nearly extending upto a metre or so, (4) *understorey tree* and *shrub layer* extending to the height of one to five metres and (5) *tree layer* or *top storey* extending to the extent of 5 to 15 metres in most of the forests, but sometimes it may extend upto 25-30 m in the coniferous forests and to about 40 or 50 m in rain forests. In *Sequoia* forest, the upper canopy may surpass 100 m. The different strata of plant community are determined by light penetration; the maximum light is available to plants of top storey and the minimum light reaches the ground vegetation. It is evident from Fig. 6.2.

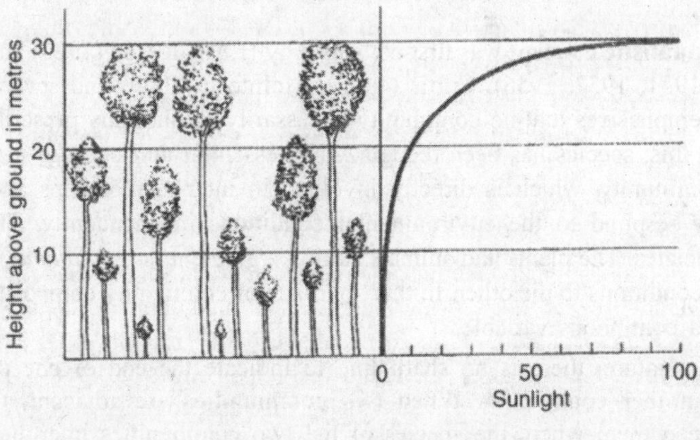

Fig. 6.2. Stratification of a forest and height levels.

The factors causing stratification of biotic communities are as follows:

1. Specific tolerance and adaptations.
2. Chemical reactions between the lay-products and physical stratification.
3. Changes in species structure with time. New organisms take up temporary or permanent position as a direct response to the presence of initial or preexisting population in the area.

(6) **Succession.** Interacting populations of community are characterized by continuous death and replacement and usually by immigration and emigation (one way movement from home range to other habitat) of their individuals. In this way, composition and shape of community remains in changing state. The changes in the community go on taking place until a complete balance is established between community and environment. This is called **succession**. At complete equilibrium state a stable community is established which is called ***climax community***. In the climax community very little or no change in the shape can be anticipated over a long period of time.

A mixture of species which live in a habitat and are held together by common ecological tolerances form a community. As indicated earlier, some communities have sharply defined limits and they are characterized by a, more or less, definite species contents, as for example, community of a pond or a small island or old field. Thus, they show *discontinuous* vegetational patterns.

There are, however, several instances of community types which do not have well defined limits and one community generally merges with another community. The species occurring in a community are not bound together into group of associates; and they respond independently to the physical and biotic factors of the environment. Thus with the change in the environment, the composition of community also tends to change. The sequence of communities showing a gradual change in composition is called ***continuum*** (Curtis 1959). Community in such a gradient can be recognized as a discrete point with uniform floristic composition in the *continuum.*

As regards the composition of community, there are two opposing philosophies :

(*i*) Organismic view, (*ii*) Individualistic view.

The **organismic view** was advanced by Clements (1916) and is still supported by his school of ecology. According to this view, the community is a sort of "*super organism*", the highest stage in the organisation of living world, rising from cell to tissue, organs, organ-systems, species, population and to community. The community is regarded as super organism because it grows, adjusts under some circumstances, reproduces itself, and functionally represents higher level of integration than individual plants and animals that make it up. The community acts as unit in the succession.

The **individualistic concept** was first advanced by H.A. Gleason (1926) and further supported by Whittaker (1951, 1952, 1956), Curtis (1958), McIntosh (1959) and several other ecologists. This approach emphasizes that no community necessarily reaches any prescribed composition or steady state. In this, species has been recognized as essential unit because it is only the species, and not the community, which is directly involved in inter-relationships and distribution. The species directly respond to the environmental conditions independently. They are not bound together in associates. The plants and animals exhibit a gradient or *continuum* from one extreme of environmental conditions to the other. In this approach of community composition the community is regarded as a continuous variable.

Ecotone. In nature, there is no sharp line to indicate the end of one community and the beginning of another community. When two communities are adjacent, there is a zone of transition between them where the species of the two communities intermingle. The transition or marginal zone between two major communities presenting a situation of special ecological interest is called an *ecotone* or *tension zone.* The transition zone between grassland and forest or the intermediate zone between any two major land or aquatic communities are the examples of ecotone. Strictly speaking, a transition zone is an ecotone only if tension exists between the bordering communities. The ecotone is colonized by the species that are commonly found in the communities on both the sides, as well as by some versatile species of plants and animals. As a rule, the ecotone contains more species and often denser populations than the bordering communities. This is known as *'principle of edges'* (Fig. 6.3). This edge effect is chiefly due to wider range of suitable environmental conditions.

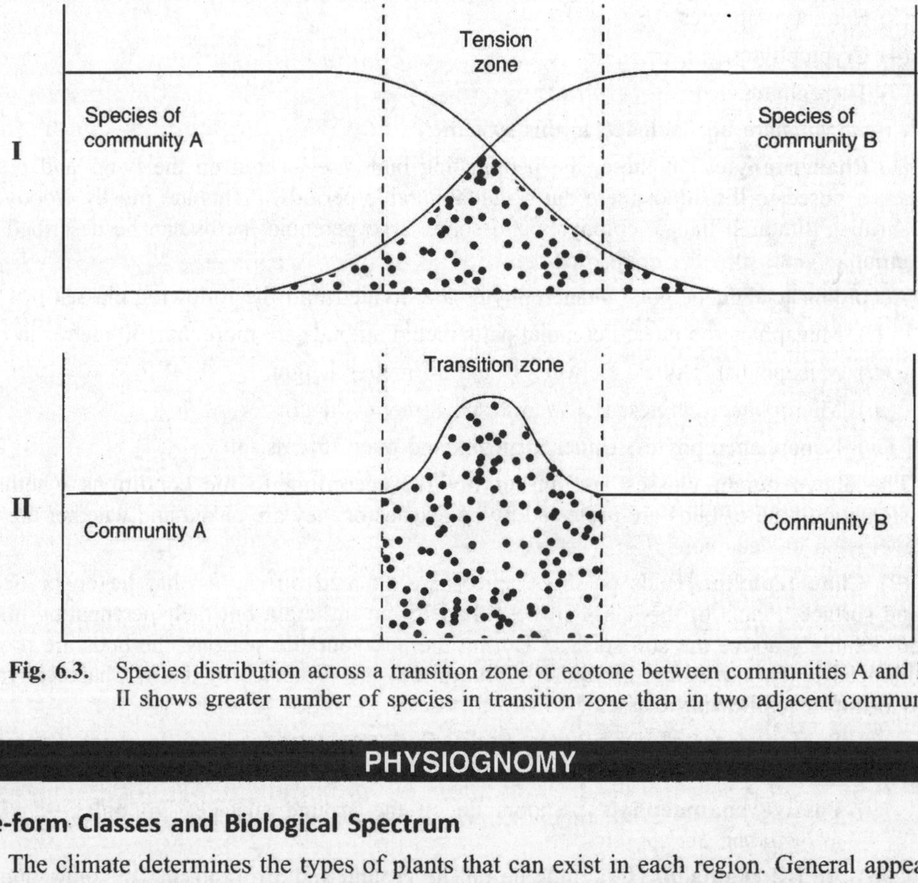

Fig. 6.3. Species distribution across a transition zone or ecotone between communities A and B. Fig. II shows greater number of species in transition zone than in two adjacent communities.

PHYSIOGNOMY

Life-form Classes and Biological Spectrum

The climate determines the types of plants that can exist in each region. General appearance of vegetation is referred to as **physiognomy.** It constitutes general stature, shape and life-forms of the species comprising the vegetation and actually the classification of vegetation types has been done on the basis of physiognomy. The species of a community can be grouped into several life-forms on the basis of general appearance and growth. The life-forms of the vegetation are to a certain extent indicators of the climate. This idea has been extensively developed by a Danish botanist Raunkiaer (1934). He considered that it is the unfavourable environmental condition which exerts a major control over the growth form and he stressed the significance of the adaptations of buds and shoot tips for overcoming adverse temperatues and survival in drought. A classification of plants on the basis of the types and kinds of perennating organs seems to be natural. Raunkiaer called the kinds and types of these organs as **life-forms.** Raunkiaer employed three guiding rules in the selection of life-form characteristics :

(1) The character must be structural and essential and must represent important morphological adaptation.

(2) The character must be sufficiently clear so that one can easily see in nature.

(3) All the life-forms employed must be of such nature as could constitute a homologous system and represent a single point of view or aspect of plants and thus enable a comparative statistical treatment of flora of different communities. On the basis of these principles, he recognised five life-form groups. These groups are as follows :

(1) Phanerophytes

(2) Chamaephytes

(3) Hemicryptophytes

(4) Cryptophytes

(5) Therophytes

Cryptogams are not included in this system :

(1) **Phanerophytes.** In these, the perennating buds are located on the twigs and branches and are exposed to the atmosphere during unfavourable periods. Plants are mostly woody (trees and shrubs), although lianas, epiphytes and some large perennial herbs can be described under this group.

According to their heights, phanerophytes are divided into the following classes :

(*i*) Megaphanerophytes; Perennial parts including buds are more than 30 metres in height.

(*ii*) Mesophanerophytes; Between 8 and 30 metres height.

(*iii*) Microphanerophytes; Between 2 and 8 metres height.

(*iv*) Nanophanerophytes; Under 2 metres and over 25 cms tall.

The above height classes are further divided according to the conditions whether the meristematic tissue of buds are protected by bud scales or they are naked and whether the plants are evergreen or deciduous.

(2) **Chamaephytes.** Buds or shoot apices are situated on shoots that lie on or near the ground surface. Plants of this class are less than 25 cm in height but their perennating buds are found definitely above the soil surface. During the unfavourable seasons, the buds are protected by the dead fallen leaves and on higher altitudes they are protected by snow. Chamaephytes are divided into the following classes :

(*i*) **Subfructicose chamaephytes.** Buds are protected by dead plant remains. Examples—Caryophyllaceae, Labiatae, Leguminosae, etc.

(*ii*) **Passive chamaephytes.** Shoots lie on the ground surface. Examples—*Saxifraga, Polygonum, Sedum,* etc.

(*iii*) **Active chamaephytes.** Buds lie on the ground and are protected by snow under the unfavourable conditions. Example—*Vinca.*

(*iv*) **Cushion chamaephytes.** They represent transition stage from chamaephytes to hemicryptophytes. In this group the shoots are closely packed, for example, *Azorella.*

(3) **Hemicryptophytes.** They are found on the soil surface and buds and shoots are protected by soil and dead leaves. They are herbaceous and spread on the surface horizontally as runners. The group is divided into protohemicryptophytes, partial rosette plants, and rosette plants.

(4) **Cryptophytes.** In this group, the buds and shoots of plants are underground. Reserved food material is stored in the subterranean perennating organs. The class is divided into the following categories :

(*i*) Geophytes

(*ii*) Helophytes

(*iii*) Hydrophytes

(*i*) **Geophytes.** They are land plants with subterranean perennating buds. They are of following types :

(I) Rhizome geophytes.
(II) Stem tuber geophytes,
(III) Bulb geophytes,
(IV) Root tuber geophytes,
(V) Root geophytes.

(*ii*) **Helophytes.** These are the plants of water saturated soil. The leaves and flowers emerge above the ground but the rhizomes are subterranean.

(*iii*) **Hydrophytes.** They are water plants, for example—*Eichhornia, Nelumbium,* etc.

(5) **Therophytes.** This group includes those annual summer plants which survive under unfavourable seasons in the form of seeds. In summer, all the pats of plants die except seeds and fruits. Thus, the perennating bud is actually the embryo of the seed.

Braun Blanquet's Classification

Braun Blanquet (1951) modified classification of Raunkiaer and proposed a new scheme of classification of vegetation. The scheme is as follows :

1. **Phytoplankton** (Microscopic floating plants)
   (*a*) Aeroplankton (float in air)
   (*b*) Hydroplankton (float in water)
   (*c*) Cryoplankton (float in ice and snow)
2. **Phytoedaphon** (Microscopic soil flora)
   (*a*) Aerophytobionts (aerobic)] Bacteria
   (*b*) Anaerophytobionts (anaerobic)
3. **Endophytes**
   (*a*) Endoxylophytes (Parasites of plants)
   (*b*) Endolithophytes (Algae, Fungi and Lichens)
   (*c*) Endozoophytes (Pathogens on animals)
4. **Therophytes**
   (*a*) Thallotherophytes (Myxomycetes and moulds)
   (*b*) Bryotherophytes (Liverworts and mosses)
   (c) Pteridotherophytes (Vascular cryptogams)
   (*d*) Entherophytes (annual seedplants) which may be creepers (climbers or erect).
5. **Hydrophytes** (All water plants except planktonic forms)
6. **Geophytes** (Earth plants)
   (*a*) Geophyta mycetosa (Root and tuber inhabiting fungi)
   (*b*) Geophyta parasitica (*e.g., Rafflesia,* a root parasite)
   (*c*) Eugeophytes with bulb, rhizome and root tubers.
7. **Hemicryptophytes** (Some thalloid algae, fungi, lichens and rooted creepers, climbers and Rosette plants).
8. **Chamaephytes** (Mosses, lichens, leaf succulents, cushion plants, hard grasses, semishrubs and trailing shrubs).
9. **Phanerophytes** (Succulent cacti, herbs, liana plants).
10. **Epiphyta arboriela** (Tree epiphyte).

BIOLOGICAL SPECTRUM OR PHYTOCLIMATIC SPECTRUM

Raunkiaer's system is unitary being based completely on the position and protection of perennating buds during unfavourable seasons. The characters are structural, essential and adaptive.

Further more, they provide simple basis for statistical treatment. The array of percentages of life-form classes of a flora of any area composing any floristic community is called the *biological spectrum* or *phytoclimatic spectrum.* Since each life-form class is related to the environment, the biological spectrum is direct indicator of its environment. In biological spectrum, the life-form classes are represented by their percentages. Raunkiaer prepared a normal spectrum based on sampling of world flora using one thousand entities. The normal spectrum provides a base line from which the departure of percentages of any class in any given flora can be ascertained. The normal spectrum has:

Phanerophytes 46%

Chamaephytes 9%

Hemicryptophytes 26%

Cryptophytes 6%

Therophytes 13%

Generally, the biological spectra are worked out and compared with Raunkiaer's normal spectrum.

The percentage of phanerophytes in different floras ranges from zero to over 74% in the tropical rain forest. The higher percentage of phanerophytes is indicative of phanerophytic climate. The higher percentage of therophytes (above 40%) indicates therophytic climate (desert). High percentage of chamaephytes (above 50%) indicates an extremely cold climate. High percentage of hemicryptophytes in temperate forest vegetation indicates the conditions favourable for the development of extensive grassland. Geophytes (cryptophytes) are abundant in the regions of mediterranean climate and in the broad leaf deciduous forest. The biological spectra of different areas differ from one another and there is no base line for comparison.

There are certain limitations to usefulness of biological spectrum as are indicative of climatic condition, because at certain places it does not indicate the environmental conditions, for example, therophytes are abundant in Indogangetic planes where phanerophytic vegetation should actually be dominant. Biotic agencies are the chief causes for changing the biological spectrum in a given floristic zone. Thus comparison of biological spectrum with normal spectrum may sometimes create confusions and may also lead to wrong conclusions.

As the leaves are essential part and are very much affected by climatic conditions, their shapes and sizes have been taken as important criteria in further classification of different life-forms. Raunkiaer used leaf size as important character in classification, and classified plants into the following leaf sizes classes :

Leaf size classes	Leaf size
1. Leptophyll	Upto 25 sq mm in area.
2. Nanophyll	From 25 to 225 sq mm in area, the upper limit being 9 × 25 sq mm.
3. Microphyll	Area from 225 to 2025 sq mm (20.25 sq cm or 9^2 × 25 sq mm).
4. Mesophyll	Area from 2025×18,225 sq mm (182.25 sq cm or 9^3 × 25 sq mm).
5. Macrophyll	Area from 18,225 to 164.025 sq mm (1,640.25 sq cm, or 9^4 × 25sq mm).
6. Megaphyll	Larger than macrophylls, area over 1,640.25 sq cm.

EVOLUTION OF PLANT COMMUNITY OR COMMUNITY DYNAMICS

Evolution of plant community on a bare area is quite a prolonged process. This involves a number of stages. Each stage is characterised by particular assemblage of plant populations and dominants. It is difficult to recognise these stages because of the fact that the process of community evolution is continuous. However, these stages can be defined on the basis of their characteristic vegetation. Evolution of plant community involves the following important steps and processes :

(1) **Nudation.** The development of bare areas is initial prerequisite. The naked areas develop either by emersion, submergence, glacial recession, erosion, deposits and climatic change or by biotic agencies.

(2) **Migration including initial colonisation.** When the area becomes bare some plants from the nearby localities move into it in the form of germules, propagules or migrules (structures or off-springs reaching from different places). This process is known as migration. Migration starts when germules leave their parent areas and terminates when they reach the final resting place. The movement between these two places may complete in one or two steps. It is only by migration, plants from an area are brought into new areas. Several agencies help in the migration of plants to new areas. They are wind, water, animals, man, glacier, etc. Migrules may jump into the new area from all the surrounding localities or from one side only.

(3) **Ecesis.** It is a process of establishment of immigrants. When the migrants enter a new area, they germinate, grow and reproduce there. It is not necessary that all the migrules reaching the new area must stabilize. The stabilization depends greatly on the conditions prevailing in that area. The first plants growing in the new area are known as *pioneer colonisers.* The germination may be affected by a number of external and internal factors. Dormancy may be a barrier in the germination. Viviparous germination is helpful in the establishment of halophytes in the saline marshy places as the saline habitat has marked inhibiting effect on germination.

(4) **Aggregation of germules.** In the beginning, pioneer plants may be present in very small number and they grow far from one another. These plants produce reproductive structures which will be dispersed in the open areas around them and after germination they form their family groups.

In the course of evolution, more new migrules reach the open areas and become stabilized there. This grouping together of colonising individuals in bare area after migration is termed as *aggregation.* Aggregation may be of two types :

(*i*) **Simple aggregation.** In this, the germules are aggregated in a group around the parent. This is independent of immigration. It increases the number of individuals of only one species, *e.g., Gloeocapsa, Tetraspora.* Falling down of fruits and seeds of the plant just below it is also aggregation.

(*ii*) **Mixed aggregation.** When the individuals from the family groups migrate away and some more new immigrants are brought by some means into the area under colonisation, it is called mixed aggregation.

(5) **Evolution of community relationships.** When the bare areas become occupied by the individuals of colonizing species, they become related with one another. The relationship may be of the following three types :

(*i*) **Exploitation.** In this, one species lives at the expense of another.

(*ii*) **Mutualism.** In this, one or both species benefit from the relationship but none suffers.

(*iii*) **Co-existence.** In this, the species live together in some measures of actual or potential competition for same necessities, such as light, moisture, space and nutrients.

Competition starts among the constituents of vegetation when the supply is in adequate to meet full requirements of all. Competition may be interspecific or intergeneric. It increases with the increase in the number of individuals in the populations. If the populations over the entire community range are habitually at levels producing interspecific competition, coexistence will be possible only for species best adapted to some recurring variants of the ecological pattern of the community. Competition is, perhaps, most acute in the early stages of growth when mortality is highest. Between individuals of the same species an equilibrium may be established so that over a wide range of population density the total productivity is very similar. The reactions vary greatly between the individuals of different species. In perennials, the mode of growth and rate of potential spread play important role in competition. Competition continues until the vegetation is fully established.

(6) **Invasion.** In the process of colonization, germules of aggressive and more adapted plants reach the adjacent area from time to time. There they grow and become established. This process is termed as *invasion* and the new aggressive and more adapted organisms are called *invaders.* Invasion may be intermittent (periodic) or continuous. The invaders establish themselves in the new area either temporarily (partial invasion) or permanently (permanent invasion). New invaders may come either from the areas adjacent to the locality under colonisation or they come from other places in the same locality. There are several barriers to check the invasions. Some of them are as follows :

(*i*) Topographic barriers : mountains, valleys, slopes, etc.

(*ii*) Physical barriers : oceans, lakes, rivers, deserts, etc.

(*iii*) Biotic barriers : man, animals, insects, etc.

(7) **Reaction.** This essentially involves the changes that are brought in the habitat conditions by the plants themselves. This is the effect of interactions between vegetation and habitat. Plants modify the environment particularly in two ways :

(*i*) by changing the nature and reaction of soil, and

(*ii*) by modifying the climate.

Acute competition among developing plant communities causes the disappearance of many individuals. The dead remains of colonisers are added to the soil in the form of humus. Humus changes the soil structure to a considerable extent. It also increases the water holding capacity, aeration and mineral contents of the soil. Shadow of the plants checks the rise of temperature and increases the humidity of the air.

Reaction is continuous process. This leads to the development of such conditions as are less favourable for the growth of previous colonising individuals and more favourable to the new invaders. In this way reaction plays important role in the replacement of pre-existing plants by the new invaders.

(8) **Stabilization.** Continuous competition and reaction bring about several marked changes in the environment and consequently introduce gradual change in the structure of vegetation. After a long time, some such individuals come and dominate in the area as are least affected with the new changes in the habitat. Climate at this stage plays principal role in determining the nature of community. This process is called stabilization.

(9) **Climax.** The final stage of vegetation development after the stabilization is called climax stage. The dominant species of climax community are nearly in complete harmony with its habitat and environment. Climax community is nearly stable and will not change so long as the climate and physiography remain the same. However, complete climax is impossible because the community and environment both are changing, *i.e.,* they are in dynamic state.

CLASSIFICATION OF COMMUNITY

Plant communities have been classified in several ways according to particular need or viewpoints. The following criteria have been widely used in the classification of communities :

(1) Physiognomy, (2) Habitat, (3) Species composition and dominance.

Physiognomy refers to the general appearance of plant community. Major plant communities of large area are classified into component communities on the basis of physiognomy. Component communities recognized on the basis of physiognomy are named after the dominant life forms as for example, forest, grass land, desert community, etc.

Major communities are sometimes divided into smaller segments on the basis of habitat, mainly on the basis of water contents in the habitats. Sometime, five different types of component communities can be delineated in a major community on the habitat basis :

(1) Wet land community, (2) Wet-mesic community, (3) Mesophytic community, (4) Dry-mesic community, and (5) Dry land community.

Major plant communities are often divided into smaller divisions on the basis of species composition and dominance. Such a classification requires the knowledge of species content of the community, frequency (the regularity with which a species is distributed throughout community), dominance, and fidelity (faithfulness of species to their community). If two areas have similar communities, the communities are named after dominant organisms or those that show high frequency, *e.g.*, *Betula-Rhododendron-Magnolia* association, Oak-hickory forest.

Clements recognized dynamic nature of community and he developed floristic classification with emphasis on succession, dominance, constancy, and diagnostic species. According to Clements, vegetation can be analysed into the following classificatory units in descending sequence:

1. Formation. According to Clements, the major unit of vegetation is plant formation. Plant formation is a great vegetational unit in a region determined by several dominant growth forms, as for example, the forests which are characterised by trees. Plant formation is a product of macroclimate and is controlled and delimited by climate alone. In other words, 'formation occurs in a natural area of essential climatic unity' (Weaver and Clements, 1938). Whittaker is of the opinion that plant formations are not distinct or concrete vegetational units determined solely by climate, but they are abstract groupings of communities of similar physiognomy (outward appearance of community) and environmental relations. The formation, according to Clements (1938), is a complex organism and as such it arises, grows, matures and reproduces. Mature vegetational unit which shows no sign of change in its climatic region is termed as *climax community* and changing vegetational units are termed as *seral communities.*

Dansereau (1958) considers the following basic formation types :

(*i*) **Forests.** These are characterised by trees. The forests may be evergreen and deciduous, high and low (mostly above 8 metres) tangled with lianas, heavy with epiphytes or sparse, long lived and some rather short lived. They may harbour little undergrowth or a dense carpet of mosses. Forests of 6 types have been recognized by Dansereau (1958). These are (*a*) tropical rain forest, (*b*) temperate rain forest, (*c*) tropical deciduous forest, (*d*) evergreen deciduous forest, (*e*) needle leaf evergreen forest, (*f*) evergeen hard-woods.

(*ii*) **Woodland.** It is open stand of trees which are as tall as trees growing in forest but scattered rather than clumped.

(*iii*) **Savanna.** It contains low, branched, often flat topped small trees (less than 10 m in height) and many other woody plants scattered regularly or assembled in small groves. The intervening spaces are often occupied by seasonal grasses (in tropics).

(*iv*) **Scrub.** It is an essentially continuous stand of medium sized, compact growing, busy plants separated by bare ground or herbaceous patches.

(*v*) **Prairie.** It is continuous stand of tall dense grasses which are more than 50 cm high. They are usually seasonal in distribution. Besides grasses, very few shrubs may be present there.

(*vi*) **Meadow.** It is continuous stand of herbaceous plants, the majority of them being graminoids. Meadows are devoid of woody plants.

(*vii*) **Steppe.** It is an open stand containing bunch of grasses interspersed with low shrubs. Steppe and scrubs or steppe and meadow or even prairie differ each other in respect to the plant coverage.

(*viii*) **Desert.** It is characterised by extremely low permanent coverage. The perennial vegetation (mostly woody or succulent and evergreen) is very sparse and during rainy season the ephemerals may cover the area.

(*ix*) **Tundra.** It consists of a very low woody vegetation, trailing shrubs and cushion plants intermingled with mosses.

(*x*) **Crusts.** These are formed by algae, fungi, lichens on the rocky surface or soils.

2. **Association.** Every climax formation consists of two or more subdivisions which are known as associations, each being marked by more than one dominant species that are peculiar to it. Association is regional vegetation in the formation. It is climax of subclimate within the general climate of the formation. Each association is similar throughout its extent in physiognomy, in its ecological structure and in general floristic composition (Weaver and Clements, 1938). Example of association is Oak-Beech association. Developing counterpart of association is called ***associes***. Now the association concept has fallen in disuse and in its place community continuum is gaining popularity. The vegetation is continuous, though differing from place to place. It cannot be categorised into well defined units because it is in changing state. Whittaker (1951, 56) asserts that association is not a concrete natural community. L.E. Braun too criticised association concept in a symposium organised by Ecological Society of America in August, 1956. The main points of objections are :

(*i*) That the communities do not have definite limits but intergrade with one another.

(*ii*) That the species which seem to characterise them may extend to other communities, although, probably in different proportions.

(*iii*) That two communities are not exactly alike.

(*iv*) That the vegetation (leaving aside only those places where habitats change abruptly) is continuous, though differing from place to place.

3. **Faciations.** Each association which fundamentally includes many dominant species consists of two or more subunits. These are called ***faciations***. Each faciation may be dominated by two or more dominants, but the total number of dominants in faciation will be lesser than that of *association*. Seral or developing counterpart of faciation is termed as ***facies***. Another local variation of the association is *location* which varies particularly in the composition of subdominants and influence.

4. **Consociation.** When there is only one dominant in climax community that is known as consociation. Consociations are smaller unit communities whose single dominants still have the life-forms characterising the formation. Such **eca** (vegetational units) are modified greatly by edaphic conditions, example—Oak or Beech consociation in Oak—Beech association. The developing counterpart of consociation dominated by single species is termed as ***consocies***.

5. **Societies.** Association and consociations can further be analysed into several minor communities which are under the direct influence of local variations of habitat. These minor communities are dominated by one or more species other than the dominants of associations and consociations. These smaller units are called **societies**. The dominants of societies are sub-dominants of higher **econ**. Society thus shows dominance within dominance whose dominant

species is (or are) subordinate when we consider consociation as a whole. Developing societies are called *socies*. If socies have got two or more invading species without evident association, they may be called colonies.

6. **Clans.** In each society there may be found two or more smallest units. These are called **clans**. Each clan is a small aggregation of single but very locally and overwhelmingly dominant species. The seral equivalent is the family derived from multiplication and gregarious growth of single immigrant.

STUDY OF PLANT COMMUNITY STRUCTURE

The special field of synecology which is concerned with the structure and classification of plant community is known as ***phytosociology***. The study of structure and composition of plant communities has been developed largely in Europe and Zurich-Montpether school of vegetational analysis led by J. Braun-Blanquet who has outlined several methods grouping them into phytosociology. Two sets of characters, *viz., analytical* and *synthetic* are studied in a community at the same time.

ANALYTICAL CHARACTERS

According to Hanson (1950) and Braun Blanquet (1932), analytical characteristics are those features of community which can be observed or measured directly in each stand. They include kinds and number of species, distribution of individuals, species vigour, form, number of individuals, height of plants, area volume, growth rate and periodicity, etc. There are two different aspects of vegetational analysis—namely ***quantitative*** characters which can be measured more readily than the others, and ***qualitative*** characters (which are described and not measured).

SYNTHETIC CHARACTERS

Those aspects of community which are based on analytical characteristics and utilise data obtained in the analysis of a number of stands.

A. Qualitative Structures of Plant Community

The qualitative structure and composition of plant community can be described on the basis of visual observations without any special sampling and measurement. In the qualitative characteristics floristic enumeration (species content), stratification, aspection, sociability, interspecific associations, life-forms and biological spectrum, etc. are studied in the field.

1. **Floristic composition or species content of community.** The study of species content in a community is of paramount importance. The species content of a community can be studied by periodic collection and identification of plant species for the whole year. This will show the tolerance of each species for different environmental conditions (Hanson, 1950).

2. **Stratification and aspection.** The number of strata or layers in a community can be determined by general observation of the vegetation. If one periodically observes the flora for the whole year, changes in the appearance of vegetation may be apparent with the change in the season. This is known as ***aspection***. For this phenology of species in relation to different seasons of the year is recorded. The term phenology was first used by shelford (1929) to mark the events associated with seasonal, succession in natural communities. Later ecological society of America defined phenology as the science dealing with the appearance of certain events during the life cycle of an organism in different seasons of the year. Lieth (1970) defined phenology as the art of observing the life cycle or periodicity of plant and animals. Phenology, thus forms a part of community periodism. It is applied to those ecosystems where seasonal changes are not well marked such as desert cosystem. On the basis of general observations of the vegetation a number of layers have been distinguished :

L_1—ground stratum like mosses, thallophytes, lichens, etc.,

L_2—herbaceous or ground flora,

L_3—middle layer or shrubby layer, and

L_4—top layer or canopy layer of trees.

In grasslands even two to three strata may be distinguished.

3. **Life-forms.** On the basis of general appearance and growth, the species of community are grouped into different life-form classes. The chief criteria for recognising life-form classes have already been given earlier in this chapter. On the basis of percentage values of different life-form classes, real nature of habitat and community can be understood.

4. **Sociability.** In a plant community, the individuals of species are not evenly distributed. Individuals of some species grow widely spaced while those of some other species are found in clumps or mats. The space relationship of plants is referred to as **sociability**. Individuals of some species when growing in clumps are either very weak or they tend to disappear due to hard competition and as such they cannot form big populations. Braun-Blanquet (1951) has recognised five sociability classes which accommodate different types of species.

Class 1. Shoots grow singly,

Class 2. Scattered groups or tufts of plants,

Class 3. Small scattered patches or cushions,

Class 4. Large patches or broken mats, and

Class 5. Very large mats of nearly pure population covering the entire area.

High degree of sociability is seen in those plants which produce large number of seeds with high germination percentage, show good survival of seedlings and mature plants and have many adaptive features. From gregariousness of species conclusions may often be drawn as to the nearness of approach to optimum conditions (Braun-Blanquet, 1932). This may explain the importance of the rough determinations of sociability values.

5. **Interspecific associations.** When the plants belonging to two or more different species grow near one another they form a community. This type of association is known as interspecific association. Interspecific association is possible if—

(*i*) the species can live in similar environment,

(*ii*) the species in question have similar geographical distribution,

(*iii*) the species belong to different life-forms (this reduces the competition), and

(*iv*) the plants of one species are related to the plants of other species. The relationship may be obligatory in one or both the directions.

Interspecific association can easily be observed in the field.

B. Quantitative analysis of Plant Community

The structure of sociological order in any plant community cannot be studied by observing each and every individual of plant species growing in a habitat. It is rather impossible. Therefore, rough estimate of species content of a habitat is made by observing the plant species at different places or sample areas, in the habitat. Several methods have been used by ecologists for this purpose which are as follows :

(1) Quadrat method,

(2) Transect method,

(3) Loop method, and

(4) Pointless or point method.

1. Quadrat Method of Sampling the Vegetation

The quadrat is a square sample plot or unit for a detailed analysis of vegetation. It is actually the sample-plot method of Clements (1898). It may be a single sample plot or may be divided into several subplots. In vegetational analysis, quadrats of any size, shape, number and arrangement may be used. In the study of a forest community quadrats of one-fifth acre are established to include maximum number of trees, while for studying shrubs and grass covers usually the quadrats of smaller sizes are used (Fig. 6.4). For grassland and low herbaceous community, the quadrats of one square metre size or 50 cm × 50 cm size or even 20 cm × 20 cm size may serve the purpose. The shape of quadrat is usually a square (Fig. 6.5 A and B) but rectangular, or even circular ones are also used. In some cases rectangular sample plots often give the best results. The ratio of breadth and length in rectangular plots is generally 1 : 2 or 1 : 4 or 1 : 8.

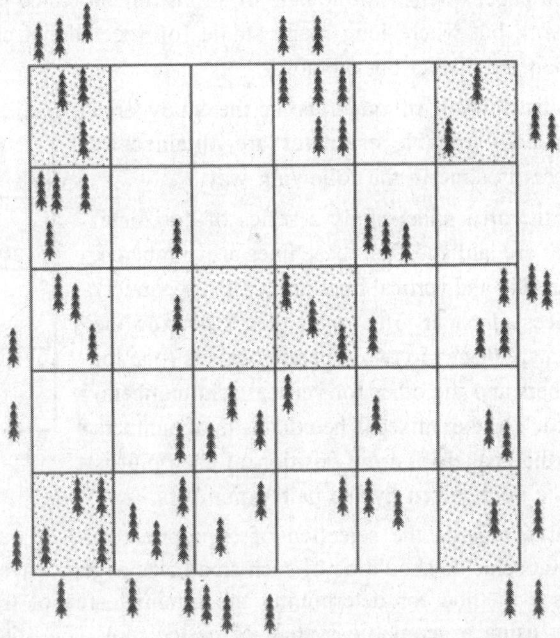

Fig. 6.4. Quadrat sampling method for population estimation.

Kinds of Quadrats. Quadrats are named according to the use. These are of the following types :

(*i*) **List quadrats.** When the organisms encountered in the sample plot are listed by their names, the quadrat is called *list quadrat.* It includes all the species botanically identified or otherwise. A series of list quadrats gives floristic analysis of the community. This is used for studying the frequency of different species.

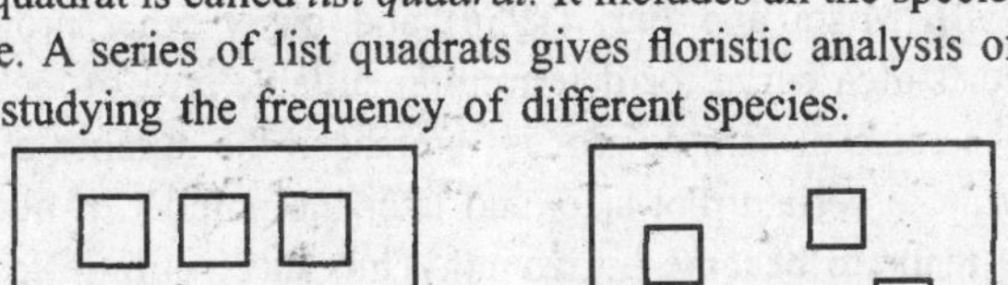

Fig. 6.5. A and B—Arrangement of quadrats in the study area

A—Systematically distributed quadrats. B—Randomized quadrats.

(*ii*) **Count quadrat or list-count quadrat.** When the species name and the number of individuals of each species found in the sample area are recorded, the sample plot is called *count* or *list-count* quadrat. This type of plot is usually used in forest survey work.

(*iii*) **Cover quadrat.** When the actual or relative coverage is recorded usually as percentage of ground area covered or shaded by vegetation, the sample area is known as *cover quadrat.*

(*iv*) **Chart quadrat.** Quadrats that are mapped to scales to show the location of individuals of species are called chart quadrats. Individuals plants are recorded on miniature quadrat on a graph paper often with the help of an instrument called plantograph. This is very tedious work but where long range studies of vegetational changes are being made, this method provides a big picture.

So far as the distribution of quadrats in the study area is concerned, statistically reliable estimates are obtained by randomized plots. This is done in the following way :

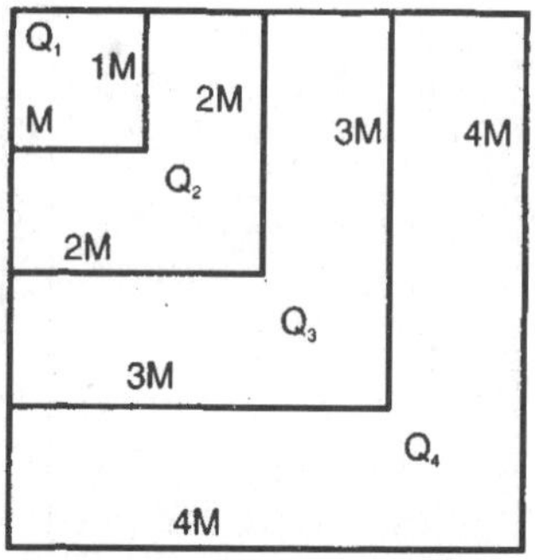

Fig. 6.6. Nested quadrats.

On the map of the area under study a series of horizontal and vertical grid lines are laid and then these lines are numbered. The numbers of horizontal and vertical lines are written separately on small square pieces of paper. The paper pieces bearing the numbers of grid lines are placed in two separate beakers (one for horizontal grid numbers and the other for vertical grid numbers) and paper-pieces in each beaker mixed. Then the pairs of numbers are drawn out from the beakers and the position of each plot is located by putting the point given by the paired numbers.

Species-area curve regards the selection of appropriate size of quadrat for sampling the vegetation, the quadrat selected should be of such small size as may cover the maximum number of species. There is a method for determining the minimal area of the sample plot. In this, sampling is done by using a geometric system of nested plots (Fig. 6.6). In plots of different sizes, number of species found in them are counted and recorded separately.

Then the numbers of species found in the plots of different sizes are plotted on vertical axis (O-Y axis) against sample plot sizes plotted on the horizontal axis (O—X axis). The resulting sigmoid curve will be obtained. This is called *species-area curve* (Fig. 6.7). Braun-Blanquet (1932), Oosting (1958), Misra and Puri (1954) and many other prominent ecologists of the world have used species-area curve for determining suitable area of the quadrat. Such a curve is obtained because as the size of sample plot is increased the number of species increases in the initial stage but only up to certain plot size, and later the number of new species added declines and finally the curve tends to become horizontal. Thus there is little to be gained by increasing the plot size. The desirable minimal quadrat size is determined by locating the point on curve where line takes horizontal course and joining it to the horizontal axis will indicate the minimal size of small plot.

In a well stratified community, the study of different strata is done with the help of different sized quadrats. For the tree and shrub strata large sized quadrats are taken but for ground vegetation small sized quadrats are used. The minimal sizes of quadrat are determined by "species-area curve" method.

In the same way, the minimum number of quadrats to be taken is also determined.

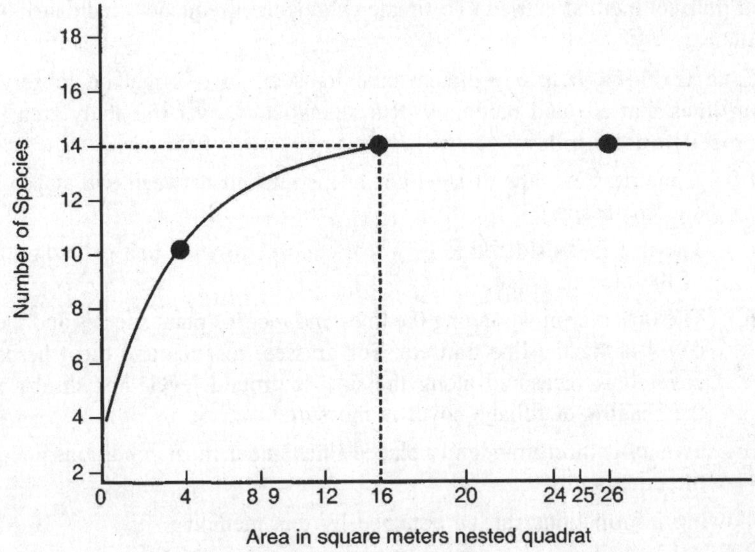

Fig. 6.7. Species-area curve.

2. Transect Method

A cross-section of an area used as a sample for recording, mapping or studying vegetation is called ***transect.*** It may be a strip, belt or a line across the area of study. The species occurring along these strips or lines are recorded. Because transect is continuous through the study area, it can be applied in studying the gradual and continuous changes in the vegetation along the line or strip with the change in environment. On the slopy area, the transect is laid between two points at different altitudes.

The transects are of two types :

(*i*) Belt transect, (*ii*) Line transect.

(*i*) **Belt transect.** It may be established as follows :

(*a*) The total area of the site to be studied is divided by 5 or 10 to obtain the total number of sample areas.

(*b*) A series of belt transects of predetermined width and length are laid and the belts are divided into equal sized segments (Fig. 6.8). These segments are sometimes called quadrats, but they differ from true quadrats in that each of them represents one observation point. Each segment within a belt is a part and the belt as a whole is one sampling unit.

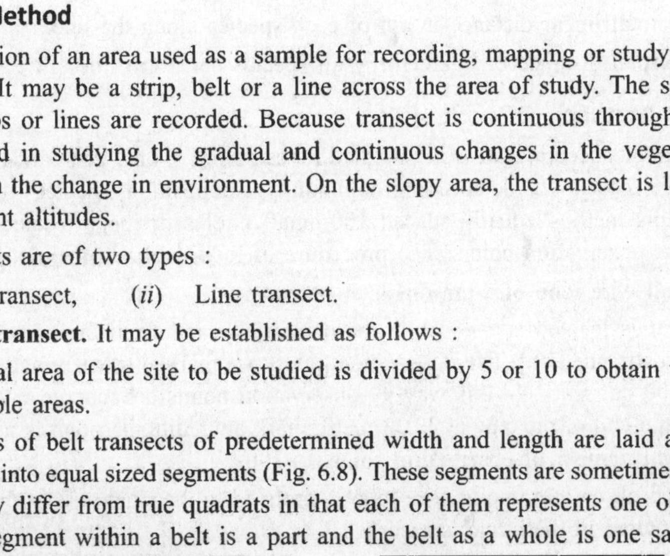

Fig. 6.8. Transects

(*c*) Names of species and number of individuals of each species in each unit are recorded.

The belt transect method is used to estimate abundance, frequency, and distribution of species in the community.

(*ii*) **Line transect.** It is one dimensional transect. In this method, observation is taken on lines that are laid randomly or systematically over the study area (Fig. 6.8). The procedure is as follows :

(*a*) A metric steel tape or steel chain is stretched between two stalks 33.5 metres or one chain apart.

(*b*) The line is considered to be a one centimetre wide belt extending along one side of the tape or chain.

(*c*) The observer moves along the lines and records plant species and the distance they cover along the line transect. For grasses, rosettes and dicot herbs, the distance covered is measured along the line at ground level. For shrubs and tall herbs, the shadow or foliage cover is measured.

(*d*) Twenty or thirty randomly placed lines under most conditions adequately sample the community.

The following informations can be obtained by this method :

(*i*) The number of times each individual species appears along the transect.

(*ii*) The occurrence percentage for each species in relation to the total species.

(*iii*) The total linear distance in cm of each species along the line.

(*iv*) The total distance of intercept by all species per 30 m line.

3. The Loop Method

This is a simple, accurate and quick method for sampling of only grassland and low herbaceous communities. It is used for determining community composition, species frequency and range condition. In this method equally spaced 100 small circles or loops located along a stretched line are used as observation points. The procedure of loop method of sampling is as follows:

(*a*) A small wire loop of 2 cm diameter is made.

(*b*) A point is located at random in the community and from the point 33.5 m long steel tape is stretched out. The observation points are marked in each metre at 33 cm, 66 cm, and 100 cm mark. In this way 99 observation points are marked in 33 metre distance. Near the end of the tape at 33.33 metre mark an additional point is marked. This brings the total number of observation points to 100.

(*c*) At each observation point wire loop is dropped to the right side of the tape and species encountered in the circle is recorded.

In this method 20 to 30 transects under most conditions adequately sample the community. Since 100 observation points are used in each line, the sum-totals are read as percentage. By this method species contents and cover are easily computed.

4. Pointless or Point Method

In this method of sampling observations are taken on the point in the study area where a nail or set of nails touch the ground on grid lines or at random places. There are several point methods of sampling but here only the following two methods will be discussed :

(*i*) Point frame method. (*ii*) Point centre method.

(*i*) **Point frame method.** This method was introduced by Levy and Maiden (1933). It is done with the help of point frame. This consists of a scale like frame, supported by a pair of legs. The frame bears 10 equidistant holes having 60 cm long pointers or pins (Fig 6.9). It is placed one after another at several observation points in the study area and the plant species that are hit by the pointed end of the pointers or nails are recorded. Besides this, the number of times the species are hit, and the total number of points taken are noted. From these values the quantitative structure of community is explained. This method is employed in the study of grassland and low herbaceous communities even on uneven ground.

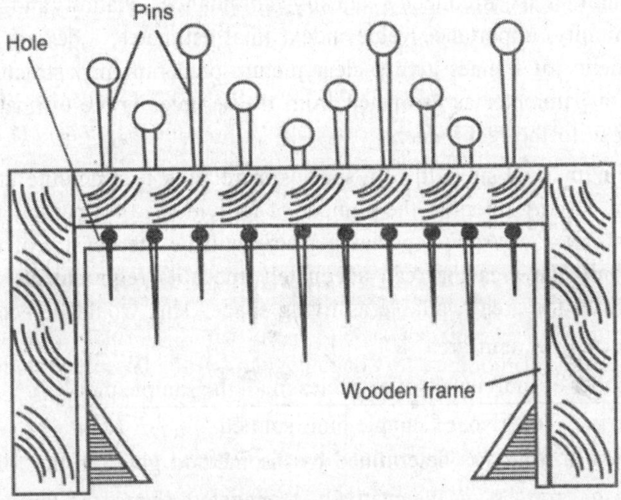

Fig. 6.9. Point frame apparatus.

(*ii*) **Point Centre or Quarter method.** In this method of sampling four measurements are taken at each observation point. The observation points can be mechanically placed along a straight line or they can be located at random. Quarter method was first described by Cotton and Curtis (1956). In this method an easy instrument is used which consists of a brass needle or a nail fitted with rubber cork and compass on the top (Fig. 6.10). The procedure of sampling is as follows :

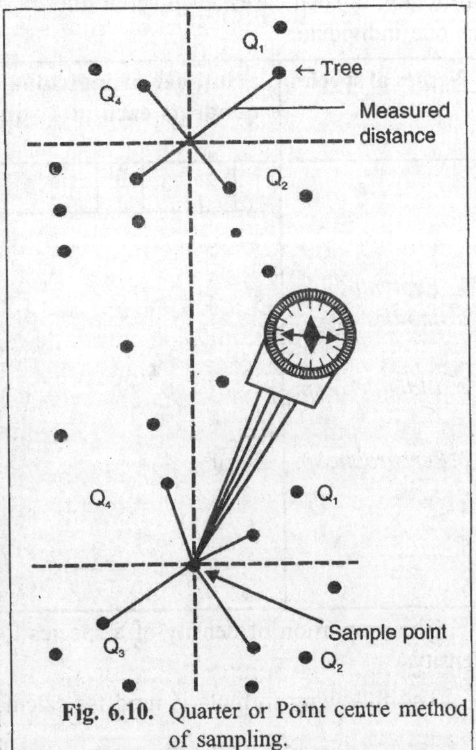

Fig. 6.10. Quarter or Point centre method of sampling.

(*a*) At each observation point the needle is fixed in the ground. This is central point.

(*b*) The working area is divided into four quarters or quadrats by visualizing two grid lines predetermined by the compass at right angles. Both the lines should cross each other at the central point.

(*c*) Now in each of the four quarters plant nearest to the central point is spotted and species recorded. The distance of each plant from the central point and also the basal diameter of the plant are measured.

(*d*) Tally at least 50 such points.

Quantitative Structure of Plant Community

Coexistence and competition both are affected directly by the number of individuals in the community. Therefore, it is essential to know the quantitative structure of community. To characterise the community as a whole certain numerical constants called parameters are used. The total count of individuals of each species, mean value of individuals of a species per plot, for example, are parameters. Frequency, density, abundance, shadow and area coverage of species in the community, importance value index, total estimates, index of association, index of similarity, and fidelity of species give a clear picture of community structure in quantitative terms. The value of a parameter as estimated from the samples is the estimate which is hoped to be accurate or close to the real value.

1. **Density.** The numerical strength of a species in relation to a definite unit space is called its density. The *crude density* refers to the number of individuals of a particular species per unit area, *e.g.*, 2000 plants of *Peristrophe* species per acre will be the density of this species. Each organism occupies only the area that can adequately meet its requirements. Thus the density of an organism refers to the area available as living space. This would be **ecological density.**

Density of a species per unit area

$$= \frac{\text{Total number of individuals of a species in all the sample plots}}{\text{Total No. of sample plots studied}}$$

Density of species in a field is determined by the method given in the following table :

In some cases, *e.g.*, grasses and vegetatively propagated plants, the term individual creates difficulty. In such cases, each aerial tiller or shoot arising out from the soil is generally regarded as one individual.

Names of species	**Number of individuals in different quadrats each of 1 square metre size**										**Total No. of Individuals**	**Density**
	1	2	3	4	5	6	7	8	9	10		
1. *Evolvulus alsinoides*	5	4	7	×	1	3	9	2	8	5	44	$\frac{44}{10}$ = 4.4 plants/m^2
2. *Indigofera sp.*	×	7	6	9	2	4	×	1	5	×	34	$\frac{34}{10}$ = 3.4 plants/m^2
3. *Peristrophe sp.*	3	1	4	×	×	×	1	2	×	7	18	$\frac{18}{10}$ = 1.8 plants/m^2
4.												
5.												
6.												

The proportion of density of a species to that of stand as a whole is referred to as relative density.

The following formula is used for calculating relative density of a species:

$$\text{Relative density of a species} = \frac{\text{Total no. of individuals of a species}}{\text{Total no. of individuals of all species}} \times 100$$

2. **Frequency.** In the community, the individuals of all the species are not evenly distributed. Individuals of some species are widely spaced while those of some other species are found

in clumps or mats. The distribution patterns of individuals of different species indicate their reproductive capacity as well as their adaptability to the environment. Frequency refers to the degree of dispersion in terms of percentage occurrence. In order to study the frequency of species in an area, the study area is sampled by any sampling method at several places in desired pattern or at random and only the names, not the numbers, of individual species encountered in each sample are listed. The frequency of a species is determined with the help of the following formula:

$$\text{Frequency} = \frac{\text{Total no. of quadrats in which the species occur}}{\text{Total no. of quadrats studied}} \times 100$$

Suppose, species 'A' occurred in 4 quadrats out of total ten quadrats studied, the frequency of species A will be

$$\frac{4}{10} \times 100 = 40\%$$

If the line or belt transect method is used for sampling then each line or belt is recognised as one quadrat for the purpose of frequency calculation.

If point frame method of sampling is used, the frequency is calculated by the following formula :

$$\text{Frequency of a species} = \frac{\text{Total no. of hits the species secured} \times 100}{\text{Total hits made}}$$

For calculating the frequency, the following pattern is adopted :

species	Number of sample plots										Frequency Percentage
	1	2	3	4	5	6	7	8	9	10	
1. *Evolvulus alsinoides*	×	×	–	–	×	×	–	–	×	×	$\frac{6}{10} \times 100 = 60\%$
2. *Peristsophe bicalyculata*	–	–	×	×	×	×	–	×	–	–	$\frac{5}{10} \times 100 = 50\%$
3. *Cynodon sp.*	×	×	×	×	×	×	×	×	×	–	$\frac{9}{10} \times 100 = 90\%$
4. *Euphorbia sp.*	×	–	–	–	–	×	×	×	–	×	$\frac{5}{10} \times 100 = 50\%$
5. *Boerhaavia sp.*	–	–	–	–	–	–	–	×	–	×	$\frac{2}{10} \times 100 = 20\%$
6.											

(In the above table, the occurrence of species is marked by ×)

If the individuals of a species are evenly distributed over the area they may occur in all the sample plots and thus the frequency of species will be 100%. Poorly dispersed species will occur only in a few quadrats and their frequency will be low. This indicates that higher the frequency value of a species in the area the greater will be the uniformity in the spread.

Raunkiaer recognized five frequency classes of plant species in the community on the basis of their frequency percentages. These are as follows:

Class A—1 to 20% frequency

Class B—21 to 40% frequency

Class C—41 to 60 % frequency

Class D—61 to 80% frequency

Class E—81 to 100% frequency

Raunkiaer (1934) suggested that the number of species in frequency class A is greater than that of class B; B is greater than in class C, class C is greater, or equal or lesser than class D; and D is lesser than class E.

$A > B > C \overset{>}{\underset{<}{=}} D < E$. This is also read as Raunkiaer's *'frequency law'*. From the above frequency law it is apparent that the species with low frequency value are higher in number than the species with higher frequency value in most natural communities.

The dispersion of species in relation to that of all the species is termed as relative frequency of a species. Relative frequency is determined by the following formula :

$$\text{Relative frequency of a species} = \frac{\text{Frequency of the species in stand } x}{\text{Sum of the frequencies for all species in stand } x} \times 100$$

3. **Abundance.** The estimated number of individuals of a species per unit area is referred to as abundance. To determine abundance, the sampling is done by quadrat or other methods at random at many places and the number of individuals of a species is added for all the quadrats studied. The abundance is determined by the following formula :

$$\text{Abundance of of a species} = \frac{\text{Total number of individuals of the species in all quadrats}}{\text{Total number of quadrats in which the species occurred}}$$

The abundance is usually expressed by assigning the species to one of the following abundance classes:

Abundance Classes

Classes	Stalks per square metre quadrat
Rare	1 to 4
Occasional	5 to 14
Frequent	15 to 29
Abundant	30 to 90
Very abundant	100 +

Abundance refers actually to *density* of population in those quadrats in which a given species occurs. In low form of vegetation like grasslands abundance can be recorded by uprooting the plants.

4. **Cover.** The cover implies the area covered or occupied by the leaves, stems and flowers, as viewed from the top. The coverage is studied at the *canopy level* and the *basal region.* In forest, where several strata are well marked, each layer of vegetation is considered separately for measuring the coverage. Basal cover is best expressed as the basal area, the ground actually covered by the crowns or by stems penetrating the soil. In forest the basal area is the cross

section area of a tree measured at 4.5 feet above the ground (cross section area of a tree at breast height). It is estimated by point method of sampling (quarter method). In grasslands, estimate of total spread of foliage has little meaning. Basal area in such cases refers to the coverage of ground one inch above the ground surface by stems and leaves. It is also called *herbage cover* (Fig. 6.11). The coverage can be measured by quadrat method, transect method and point method of sampling. Basal coverage of tree is measured at breast height. Basal area of tree is calculated by the following formula :

$$\text{Basal area per tree} = \frac{\text{Total basal area}}{\text{Number of trees}}$$

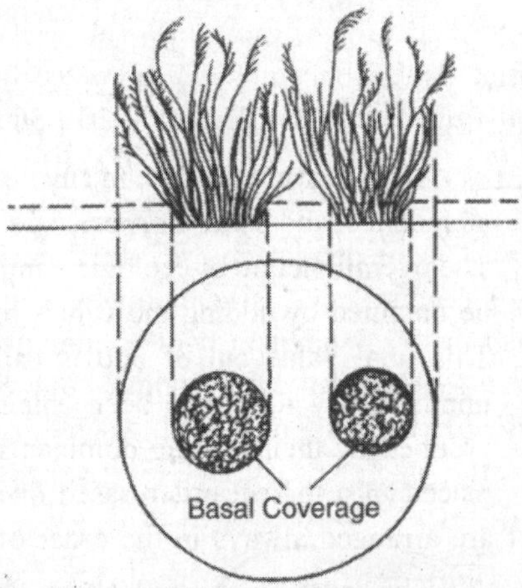

Fig. 6.11. Herbage cover of two clumps of grasses. Basal coverage of each clump one inch above the ground level is shown by shaded smaller circles and total coverage by large circle.

The area of coverage is used to express the dominance. The higher the coverage area the greater is the dominance. The average basal area is calculated out of average cross section areas of stems penetrating the soil. The average area of one stem multiplied by the density (no. of individuals per unit area) gives the basal cover per unit area.

Relative dominance (R.D.) is the proportion of the basal area of a species to the sum of the basal coverage of all the species in the area.

$$\text{Relative dominance of a species} = \frac{\text{Total basal area of the species in all the quadrats}}{\text{Total basal area of all the species in all the quadrats}} \times 100$$

5. Total estimate. Although abundance and coverage have their own importance in the community structure, yet they can be combined in a community description as total estimate. It is probably the best method for obtaining a complete general picture of a plant community. Total estimate (abundance plus coverage) scale as suggested by Braun-Blanquet is as follows:

+ Individuals of a species very few; coverage very poor.

1. Individuals of a species in plenty; but coverage small.
2. Individuals numerous if small and a few if large; coverage 5% of the total area.
3. Individuals few or many; coverage 25 to 50% of the total area.
4. Individuals few or many; coverage 50 to 75% of the total area.
5. Plant species over 75-100% of the total area.

6. **Association Index and Index of similarity.** The inter-specific association can be evaluated by association index and also by calculating the index of similarity. The index of similarity is utilized to compare two coexisting groups.

Suppose, out of 100 quadrats studies species A is encountered in 90 quadrats and species A in association with another species B is found only in 40 quadrats, the association index of species A is calculated by dividing the total number of quadrats in which A occurred in association with B by the total number of quadrats in which species A is found $\left(\frac{40}{90} = 0.44\right)$.

Index of similarity is calculated as follows :

Suppose that in one group of coexisting species the number of plant species is 30 and in the other group the number of plant species is 20 and in first and second groups 15 species are common:

$$\text{Index of similarity} = \frac{2 \times \text{no. of common species}}{\text{Total number of species in both associations}} \times 100$$

$$= \frac{2 \times 15}{20 + 30} \times 100 = \frac{30 \times 100}{50} = 60$$

7. **Importance value.** In any highly heterogeneous plant community, data of frequency, density, abundance, and cover of species do not yield total picture of ecological importance independently. The overall picture of ecological importance of a species in relation to the community structure can be obtained by adding the values of relative density, relative dominance, and relative frequency. This total value out of 300 is called *Importance Value Index* (IVI) of the species. Once the importance values have been obtained for the species within the stands, the stands can then be grouped by their leading dominants according to the importance values and the groups are then placed in a logical order based on relationships of several predominant species. The dominants are arranged always in the order of decreasing importance values.

The importance of IVI was first pointed out by Curtis and McIntosh (1951). The IVI, as pointed out earlier, gives complete picture of sociological character of a species in the community but it does not give the dimension of relative density, relative dominance and relative frequency. The individuals as well as combined aspects of the position of a species in the community structure can be shown with the help of phytographs. In one type of phytograph a circle is made and then the circle is divided into four equal quarters by two diagonal lines lying at right angles to each other. Three radii from centre to circumference are divided into 100 segments and the fourth radius is divided into 300 parts (Fig. 6.12). On radius A is marked the value of relative frequency, on radius B is marked the value of relative density, on radius C is marked the value of relative dominance, and on D IVI on 0—300 scale. All these points on different radii are joined by lines. Thus, a phytograph illustrating the sociological characters and IVI of individual species is obtained.

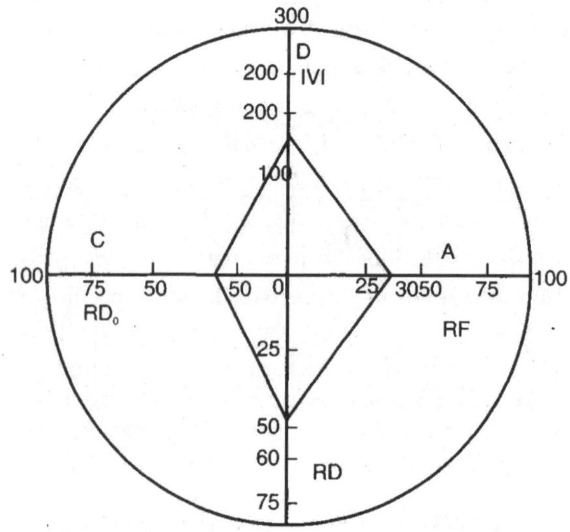

Fig. 6.12. Phytograph of species with Relative frequency (RF) 30%, Relative density (RD) value 60%, Relative dominance (RD) value 40% and Importance value index 130.

Synthetic Characters

Synthetic characters describe the make-up of a community. The chief synthetic characters used and advocated by Braun-Blanquet (1932), Cain (1932) and Nichols (1930) are presence, constance and fidelity.

Fidelity. The term fidelity refers to the faithfulness of a species to its community. In the community, there are different types of species. Some plant species are confined to one particular community and they are called indicator species. Some can flourish in several communities. According to Hanson (1950), it is a measure of ecological amplitude. Pandeya (1960) observes that characteristic species with high fidelity value has *low ecological* amplitude. **Ecological amplitude of a species** or its **tolerance** (Good, 1947) is the capacity of growing and reproducing within a certain range of environmental conditions. Fidelity of a species is expressed in relation to a particular community. A species may have high fidelity for one community and a low fidelity for another. Braun-Blanquet and Pavillard (1925), and Oosting (1956) recognised the following classes of species on the basis of their fidelity.

Fidelity Classes	Characters
Class—1	**Strange species** or **accidentals** which are either intruders from other community or relics from other successional stages.
Class—2	**Indifferent species** or companions which have no preference or affinities for any community.
Class—3	**Preferential species** which may be found in several communities but have affinity for only one community.
Class—4	**Selective species** which may occur rarely in other communities but have strong affinity to one particular community.
Class—5	**Exclusive species** which occur exclusively or almost exclusively in one community.

Species of 3rd, 4th and 5th fidelity classes are called **characteristic or key species of the community**.

Presence. It indicates the presence of a species in a stand. It is generally expressed in a scale of 1 to 5.

(*i*) *Rare,* which occur in 1% to 20% of stands examined

(*ii*) *Seldom present,* which occur in 21-40% of stands examined

(*iii*) *Often present,* which occur in 41-60% of the stands examined

(*iv*) *Mostly present,* which occur in 60 to 80% of stands examined

(*v*) *Constantly present,* which are present in 81 to 100% stands

Constance. It is the degree of "presence" in a unit area (sample area) instead of the whole stand. It actually improves the method of *presence* study, otherwise there is no fundamental difference between *presence* and *constance.* It is generally determined from frequency and the following 5 constance classes have been recognised.

Con. 1—1 to 20% frequency

Con. 2—21 to 40% frequency

Con. 3—41 to 60% frequency

Con. 4—61 to 80% frequency

Con. 5—81 to 100% frequency

QUESTIONS

1. What is plant community? Write an essay on the composition of plant community.
2. Discuss the structure of plant community.
3. Describe Raunkiaer's life-form classes of plant community.
4. Explain how a plant community has evolved.
5. Describe the various classificatory units of plant community.
6. Describe the various methods of studying plant community structure.
7. Write short notes on the following :

   Synecology, Ecotone, Biological spectrum, Ecesis, Consociation, Evolution of plant community, Density and frequency.
8. Write breif notes on the following:

   (*i*) Species diversity

   (*ii*) Stratification of plant community

   (*iii*) Phenology

   (*iv*) Ecological amplitude of a species
9. What do you mean by quadrat method of sampling of vegetation? Discuss the various types of quadrats used for the study of vegetation. How do you determine the minimal area of the sample plot?

7

Plant Succession

"Vegetation is dynamic, an ever-changing complex, now appearing quiescent and in complete equilibrium with the habitat, now displaying an obvious evidence of change" says E. Lucy Braun (1956). Observation of the natural changes in vegetation long ago resulted in the concept of succession. Under natural conditions the vegetation occupying a given habitat is called plant community. Since the community is not stable, it passes through many developmental stages in definite sequence and in definite direction generally from simple to complex and rarely from complex to simple. The gradual replacement of one type of plant community by the other is referred to as *plant succession.* According to E.P. Odum, "plant succession is an orderly process of community change in a unit area." Salisbury defines plant succession as follows :

"Plant succession is a competitive drift in which at each phase, until the climax, the constituent species render the habitat more favourable to their successors than to themselves."

Clements gave a very simple definition of plant succession. According to him, succession is a natural process by which the same locality becomes successively colonised by different groups or communities."

Succession is a complex universal process which begins, develops, and finally stabilizes at the climax stage. The climax is the final mature, stable, self-maintaining and self-reproducing stage of vegetational development in a climatic unit. Succession is generally progressive and thus it brings about—

(1) progressive changes in the soil conditions or habitats. These changes bring the habitat from extreme to optimum conditions for plant growth, and

(2) progressive changes in the life forms or phiads.

CAUSES OF SUCCESSION

The main causes of succession are as follows :

(1) Climatic causes,

(2) Topographic causes, and

(3) Biotic causes.

1. **Climatic causes.** Plants cannot adjust with the long range variations in the climate. The fluctuating climate sometimes leads the vegetation towards total or partial destruction and, as a result, the bare area develops which becomes occupied by such plants as are better adapted for changed climatic conditions. Drought, heavy snowfall, hails and lightnings are some of the important factors for the destruction of vegetation.

Sometimes new bare ground is formed by emersion of land from the bodies of water (ponds, rivers etc.).

2. **Topographic causes.** These are concerned with the changes in the soil. The following are two important soil factors which bring about changes in the habitat.

(*i*) **Erosion of the soil.** Sometimes surface soil is removed by a number of agents, such as wind, water currents, and rainfall. This process is known as soil erosion. In the process of erosion new and bare area is exposed in which new plant communities begin to appear one after another.

(*ii*) **Soil deposition.** It is one of the important causes that initiates succession. Soil deposition results owing to heavy storms, glaciers, snowfalls and landslides. If the deposition of soil takes place over an area already covered with vegetation, the plants occurring over there may be suppressed and destroyed. Deposition results in a new bare area on which succession of vegetation starts.

3. **Biotic causes.** Many biological or living agencies also affect the vegetation in many respects. Grazing, cutting, clearing, cultivation, harvesting, and deforestation, all caused by living agencies, are directly responsible for vegetational change. The parasitic plants and animals also affect the vegetation and destroy it.

SUCCESSION AND CLIMAX CONCEPT

Plant succession is an orderly change of vegetation. It involves gradual and successive replacement of one plant population by the other. Any concrete example of plant succession taking place on a particular habitat is termed as *sere,* its various intermediate stages are called the *seral stages* and communities representing these stages are called the *seral communities.* Though the seral communities are not clearly distinct, yet they are recognised only because of some dominant plant species growing in them. The first plants which appear on the bare habitat are called pioneer plants. Actually speaking, plant succession is not a series of steps or stages but is continuous, and very slowly changing complex. It is dynamic process. The replacement of vegetation takes place individual by individual. There is no jump from one dominant community to an other. Dominant species of one community may persist along with some new migrants for several generations in a given area and bring about several changes in the habitat by their dense shades and leaf litter. When the habitat becomes extremely non-tolerable for the existing plants then the plants that are well suited to that habitat will come and become dominant. After several such changes, a stage may come when the habitat becomes occupied by most tolerant species that can reproduce and perpetuate well. Thus, the process leads to establishment of **climax community;** a mature, dominant, self-maintaining and slow changing plant community. Climax dominants are the species best adjusted to habitat and are able to take possession of the habitat and hold it against the new invading species. The treatments of climax and succession in Clementsian ecology imply the following assumptions about the orderliness of vegetation :

(1) The succession is an orderly growth process. Succession beginning in different environmental conditions finally reaches similar climaxes.

(2) The climax is determined only by climate, consequently climax and climatic region must correspond.

(3) The vegetations consist of climaxes and their successional conditions are clearly distinguishable.

Whittaker (1953) says that "successional processes give an impression of relative irregularities and disorderliness in detail together with a degree of orderliness in general pattern and trend."

Clements (1928) considered the climax formation as adult organisms, the fully developed community of a region, of which all other communities are the stages of development. Since the climate alone determines the climax formation, there is one true or climatic climax in a climatic region. The climatic climax is achieved where physical conditions of the substratum are not so extreme as to modify the effects of the prevailing regional climates. Sometime the climax is greatly modified by the physical conditions of soil, such as its topography and water content. Such a climax is known as *edaphic climax.* Relatively stabilized vegetation other than climatic or true climax may be produced in a given region because of distinctive soils or other habitat characteristics. These are referred to as developmental communities.

Sometimes the vegetation is prevented from reaching to the actual climax stage by the factors other than climate, as for example, fire, cutting, grazing, flooding, etc. Thus, the vegetation which is in the imperfect stage of development, is held indefinitely in stages preceding to real climax either by natural or by artificial factors. This type of imaginary climax is termed as *sub-climax.* Such plant communities would tend to reach real climax state if the causative factors are removed. Sometimes many disturbances cause modification or replacement of the true climax and consequently a modified sub-climax is formed which is termed as *disclimax.* If temporary change of the climate stops the development of vegetation before it has reached the expected climax, it forms *preclimax.* Sometimes the environment changes in such a way that the vegetation progresses beyond the expected climax stage. The new climax, thus formed, is termed *post-climax.*

MONOCLIMAX AND POLYCLIMAX THEORIES

As regards the number of climaxes in a given habitat or climatic region, there are two different schools of thought. These are as follows :

(*i*) **Monoclimax theory.** According to Clementsian school, *there develops only one true climatic climax in a particular climatic region.* This concept is generalised as monoclimax theory.

(*ii*) **Polyclimax theory.** Tansley (1935) proposed this theory. The second school of thought holds the view which is opposite to monoclimax concept. *It defines climaxes as the stabilized and self-maintaining plant communities and considers that a number of climaxes may exist in a given area and further classfied them as edaphic climax, topographic climax," biotic climax, etc.*

"The essential difference between monoclimax and polyclimax approaches seems to be the relative emphasis," says Whittaker (1963). According to him, those who support the monoclimax theory emphasize the essential unity of climax vegetations in a given area with allowance also for stabilized plant communities other than climax, but the exponents and supporters of polyclimax concept emphasize the inherent complexity of climatic climax with allowance also for prevailing climax community which characterizes the vegetation of given area and expresses its relations to climate."

Clements' monoclimax theory has been severely criticized in recent years on the ground that in the theory regionally prevailing undisturbed vegetation occupying the large part of the land surface was regarded as real climax and other stabilized plant communities in the same area were recognised as subclimaxes which only theoretically could be replaced by the climax.

Braun, L.E. does not believe in monoclimax concept and comments *"to me, monoclimax seems impossible."* He supports polyclimax idea but that too seems questionable to him. While working on the 'Deciduous forests of Eastern-north America' in 1950, Braun found many intergrading communities among which some were in the developing state and the others had reached the climax stage, some were very local in occurrence and extent (regional climaxes) and others recurred frequently over a big geographic extent (edaphic or topographic climaxes). Thus, his observations fully supported the polyclimax idea. Again in the year 1956, in the paper "Development of Association and Climax Concept", Braun supported the polyclimax idea and said, "the prevalence of any particular climax type may be due to control of climate; it may be related to the history of erosion cycles and past climate; it may be due to state of development of climax communities or more likely due to a combination of all.

The monoclimax theory, of course in modified form, has also been expressed by Dansereau (1954) and Walter (1954). Henry J. Oosting, of Duke University in his book, "Study of plant Communities", mentions that polyclimax theory is more practical. Nichols (1923) also expressed the same view. British ecologists under the leadership of Tansley maintained a general skepticism of monoclimax concept. Russian ecologists also have similar belief.

Smithusen (1950), Whittaker (1951; 53) and many other American ecologists support **climax pattern** or polyclimax concept and all believe in the fact that the climax state is determined by the environments of individual plant communities and not by regional climate, the latter identifies the climax as a community steady state and substitutes '**prevailing climax**' for climatic climax.

KINDS OF SUCCESSION

Depending upon the nature of bare area on which it develops, the succession may be of two kinds : *primary* and *secondary*.

1. **Primary succession.** When the succession starts on the extreme bare area on which there was no previous existence of vegetation it is called *primary succession* or *presere.* This succession starts in the areas of extreme conditions and the process terminates after a long series of intermediate stages.

2. **Secondary succession.** This type of succession starts on the secondary bare area which was once occupied by original vegetation but later became completely cleared of vegetation by the process called denudation. This denudation process is brought about by the destructive agencies, such as fire, cultivation, strong winds, and rains. When the existing vegetation of an area becomes completely destroyed by destructive agents, it becomes naked or bare. Such a bare area is termed denuded area or secondary bare area. The succession progressing on such an area is also termed ***subsere***. Secondary succession has fewer stages than the primary succession and the climax is reached very quickly in the secondary succession. The primary and secondary successions may be of the following types :

(*a*) **Hydrosere.** The plant succession which starts in the aquatic environment is called "hydrarch." A series of changes taking place in the vegetation of hydrarch is called hydrosere.

(*b*) **Halosere.** It is special type of sere which begins on a salty soil or in saline water.

(*c*) **Xerosere.** When the vegetational succession develops in xeric or dry habitat, it is called xerarch or xerosere. Xerosere may be of two types :

(*i*) **Psammosere.** It refers to the vegetational succession that begins on the sandy habitat.

(*ii*) **Lithosere.** It refers to the succession that occurs on the rock surface.

(*d*) **Serule.** It refers to the miniature succession of micro-organisms, such as bacteria and different types of fungi on the fallen logs of decaying wood, tree bark, etc.

Irrespective of the types of succession, all the seres generally lead the plant communities towards the climax stage of mesic community (mesophytic or mixed mesophytic forests).

HYDROSERE

It is succession occurring in the aquatic environment. Such a type of succession does not necessarily lead the aquatic communities toward the development of land communities. If the body of water is large and very deep or very strong wave action and other powerful physical forces are at work, the succession results in a stable aquatic community in which any considerable further change is hardly recognisable. Succession is recognisable only if the colonisation of plant communities takes place in artificial small and shallow ponds, lakes, etc. where wave action speeds up the process by allowing the erosion of soil towards edge regions. In this way, the filling process also speeds up quickly and consequently the body of water disappears within few years time. Fig. 7.1 illustrates the different types of vegetation at different depths in a pond; floating plants in the central region; rooted hydrophytes in shallow region, amphibious plants in the marginal mud and trees developing in dry haibtat.

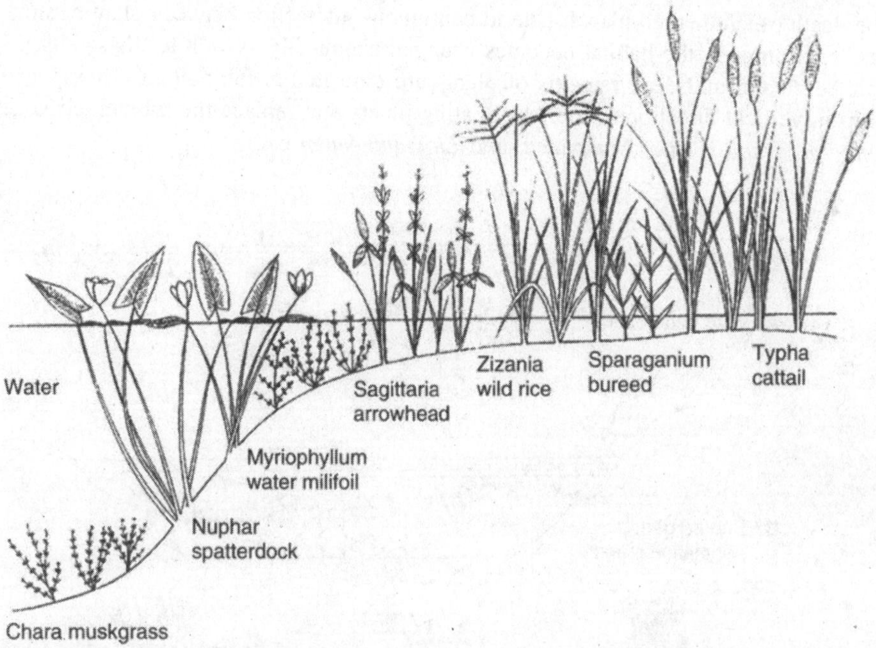

Fig. 7.1. Zonation of aquatic vegetation (hydrophytes) along a pond and along river banks. Note the changes in vegetation with water depth.

In a new and virgin pond hydrosere starts with the colonisation of phytoplankton and finally terminates into a forest (the climax community). The process of aquatic succession completes in the following stages (Fig. 7.2) :

1. **Phytoplankton stage.** In the initial stage of succession algal spores are brought in the body of water. The simple forms of life like bacteria, algae and many other aquatic plants (phytoplankton) and animals (zooplankton) floating in water are the pioneer colonizers. All these organisms add large amount of organic matter and nutrients due to their various life activities and after their death, they settle at the bottom of pond to form a layer of muck.

2. **Submerged stage.** The phytoplankton stage is followed by submerged plant stage. When a loose layer of mud is formed on the bottom of the pond, some rooted submerged hydrophytes begin to appear on the new substratum. The submerged aquatic vegetation develops in the regions of ponds or lakes where water depth is about 10 feet or more. The pioneers are *Elodia, Potamogeton, Myriophyllum, Ranunculus, Utricularia, Ceratophyllum, Vallisneria, Chara,* etc. These plants form tangled mass and have marked effects upon the habitat. When these plants die their remains are deposited at the bottom of the ponds or lakes. The eroded soil particles and other transported materials are also deposited at the bottom. This gradually raises the bottom of the ponds and lakes up. As this process of stratification progresses, the body of water becomes more and more shallow, consequently the habitat becomes less suited for the submerged vegetation but more favourable for other plants.

3. **Floating stage.** When the depth of water reaches about 4 to 8 feet, the submerged vegetation starts disappearing from its original place and then the floating plants make their appearance gradually in that area. In the beginning the submerged and floating plants grow intermingled but in the course of time the submerged plants are replaced completely. The most tolerant species in the area are able to reproduce and perpetuate. Their broad leaves floating on the water surface check the penetration of light to deeper layer of water. This may be one of the main causes responsible

for the death of submerged plants. Due to continuous interaction between plant communities and aquatic environment, the habitat becomes changed chemically as well as physically. More water and air borne soil and dead remains of plants are diposited at the bottom. Thus, the substratum rises up in vertical direction. Important floating plants that replace the submerged vegetation are *Nelumbium, Trapa, Pistia, Nymphaea,* and *Limnanthemum* etc.

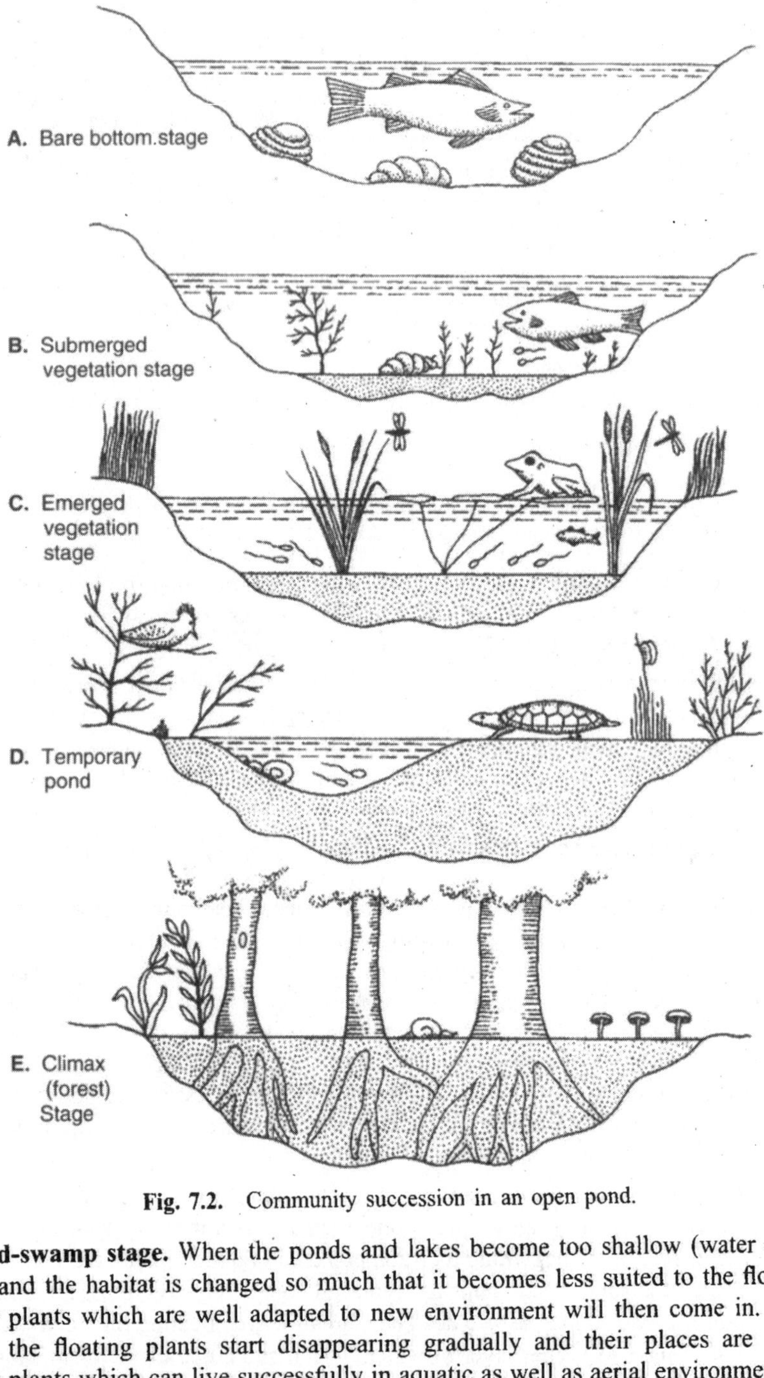

Fig. 7.2. Community succession in an open pond.

4. **Reed-swamp stage.** When the ponds and lakes become too shallow (water depth one to three feet) and the habitat is changed so much that it becomes less suited to the floating plants, some other plants which are well adapted to new environment will then come in. Under these conditions, the floating plants start disappearing gradually and their places are occupied by amphibious plants which can live successfully in aquatic as well as aerial environment. Important

examples are *Bothrioclova, Typha, Phragmites* (Reed), etc. The foliage leaves of such plants are exposed much above the surface of water and roots are generally found either in mud or submerged in water. The foliage leaves form a cover over submerged and floating plants and thus they cut off light from the plants underneath them. Under such conditions neither submerged nor floating plants can survive. Further deposition of soil and plant debris at the bottom reduces the depth of water and makes the habitat less suitable for the pre-existing plants. When the bottom reaches very close to the water surface many secondary species, such as those of *Polygonum, Sagittaria,* etc. make their appearance. Later, they also bring about such reactions by which the habitat becomes less suitable for most of the existing species, and consequently new successional step follows.

5. **Sedge Marsh or Meadow stage.** The filling process finally results in a marshy soil which may be too dry for the plants of pre-existing community. Now the plants well adapted to new habitat begin to appear in the pre-existing community in mixed state. Important plants that are well suited to marshy habitat are the members of cyperaceae and gramineae. The species of sedge (*Carex*) and rushes (*Juncus*), species of *Themeda, Iris, Dichanthium, Eriophorum, Cymbopogon, Campanula, Mentha, Caltha, Gallium, Teucrium, Cicuta,* etc. are the first invaders of marshy area. As these plants grow most luxuriantly in the marshes, they modify the habitats in several ways. They absorb and transpire a large quantity of water and also catch and accumulate plant debris and wind and water borne soil particles. Consequently a dry habitat results which may be totally unfit for the growth of normal hydrophytes. Gradually the mesophytes start appearing and after some time the sedge vegetation is totally replaced by them.

6. **Woodland stage.** In the beginning some shrubs and later medium sized trees form open vegetation or woodland. These plants produce more shade and absorb and transpire large quantity of water. Thus, they render the habitat more dry. Shade loving herbs may also grow under the trees and shrubs. The prominent plants of woodland community are species of *Buteazon, Acacia, Cassia, Terminalia, Salix, Cephalanthus,* etc.

7. **Climax forest.** After a very long time the hydrosere may lead to the development of climax vegetation. As the level of soil is raised much above the water level by progressive accumulation of humus and soil particles, the habitat becomes more dry and certainly well aerated. In such a habitat, well adapted self-maintaining and self-reproducing, nearly stable and uniform plant community consisting mostly of woody trees develops in the form of mesophytic forest.

In the climax forest, all types of plants are met with. Herbs, shrubs, mosses and shade loving plants represent their own communities. Trees are dominant and they have control over the entire vegetation. Bacteria, fungi, and other micro-organisms are more frequently found in the climax vegetation. They react upon the habitat and make the soil rich in the organic materials. At the climax stage, a complete harmony develops between plant community and habitat. It is now clear that whole sere is a continuously but gradually changing complex in which the changes are forced by biotic, topographic or climatic factors. It is very slow process that cannot be observed in nature. It may require thousands of years to reach the climax stage. One can, however, observe the sequence of hydrosere as he moves in the lake or pond from the deepest region towards the shallower margin.

XEROSERE (A SUCCESSION ON BARE ROCK SURFACE)

Like hydrosere, a xerosere also completes in a series of several orderly steps; each seral stage being characterised by peculiar type of plant community and reaction.

Xeric succession commonly occurs on bare rock surfaces (Lithosere) resulting from glaciation or from erosion by wind and water. The rocky habitat shows many extreme xeric conditions. As the rock is directly exposed to sun, the temperature of rock surface goes very high. There is no water and nutrient holding device on the exposed smooth surface of rock. In such a xeric habitat only those plants can survive which can resist the extreme drought. The various stages of succession are described below:

1. Crustose Lichen Stage

The pioneer colonisers on the bare area are crustose lichens which occur on the rock surface in the form of membranous crusts. Important crustose lichens are *Rhizocarpus, Lacidea,* etc. These plants grow only when water becomes available in the habitat. In dry periods, the pioneers, though they appear to be desiccated, remain alive. Their spongy nature enables them absorb excess amount of water and minerals. These lichens migrate through their spores and soridia and their migration is facilitated by wind and water.

The lichens secrete carbonic acid in excess. That acid is formed when excess CO_2 liberated in respiration combines with water. The reaction completes in the following way :

$$CO_2 + H_2O \rightarrow H_2CO_3 \text{ (carbonic acid)}$$

The carbonic acid reacts with the rocky materials and loosens the rock particles. The corroded rock particles together with decaying lichens make the first thin layer of soil on the rock surface. Nitrogenous compounds formed from the atmospheric gases during the lightning are brought to the soil by rains. Now the habitat becomes less fit for the existing plants and consequently primary colonisers begin to disappear from there.

2. Foliose Lichens Stage

After accumulation of little soil and humus, the rock surface, previously occupied by crustose lichens, now becomes covered with xeric foliose and fructicose lichens, *e.g., Dermatocarpon, Parmelia, Umbilicaria,* etc. These lichens have delicate leaf like thalli which cover the rocks and overshadow the pre-existing crustose lichens. When the supply of light is cut off, the crustose lichens begin to die. Foliose lichens absorb and accumulate water and minerals and check evaporation of surface water. They also secrete carbonic acid which further pulverizes or loosens the rocks into small particles. The water retaining capacity of the habitat increases with the further accumulation of soil particles and humus. Gradually the conditions become less favourable for the existing foliose and fructicose lichens.

3. Moss Stage

When the habitat is changed, the existing foliose lichens start disappearing and in that area xerophytic mosses grow and become dominant. These plants usually grow in the crevices and depressions of the rocks where enough rock particles, humus and moisture are accumulated. Like lichens, mosses are also adapted to survive in extreme drought. The xerophytic mosses develop rhizoids which penetrate deep into the rocky soil. They cover the previous lichens and successfully compete with them for water and mineral nutrients. The decaying older parts of mosses form a thick mat over the rock surface. As this mat becomes more thick, it increases the water holding capacity of soil. Thus the habitat becomes relatively more wet. Now the next seral community may replace the moss community.

4. Herbs Stage

When the soil increases in thickness, the herbaceous vegetation, which consists mainly of annual and perennial herbs, develops very quickly. Increased moisture content of the soil favours the growth of herbs. The roots of these plants penetrate down almost to the level of unpulverized rock where they secrete acids and accelerate the process of rock disintegration. In this way, the thickness of soil on the rock surface increases to a considerable extent. Decaying leaves, stems, roots and other parts of the plants become deposited on the soil surface in the form of humus. This further increases the water holding capacity of the soil. These reactions make the habitat more suited for woody plants than to the existing herbs.

5. Shrub Stage

With the change of habitat, herbaceous vegetation also shows the sign of degeneration and xerophytic shrubs gradually occupy the area. Roots of shrubs also reach the surface of unpulverized rocks and corrode sufficient quantity of rock particles which make the soil more massive. Decaying leaves, twigs and roots of these shrubs also enrich the soil with humus. These important reactions bring about such conditions in the habitat as are most suited to the trees rather than to shrubs and herbs.

6. Xerophytic Trees Stage

Now the xerophytic trees invade the area which has been occupied previously by shrubs. The first trees growing in such areas are dwarf and widely spaced. With the increase in the water holding capacity of the soil these trees gradually disappear and subsequently the mesophytic types will develop.

7. Forest Stage

The mesophytic trees are deeply rooted and their roots are profusely branched, hence they can absorb sufficient quantity of water and nutrients. These mesophytic trees grow densely and become dominant because their seedlings are shade tolerants. In the shade of mesophytic trees, some shade loving herbs and shrubs which are well adapted to humid atmosphere and moist soil also appear and they form their own communities. After very long interval when complete harmony develops between the plant communities and their environments the climax stage is established which remains unchanged unless some major environmental changes disturb it.

Successions on the bare areas do not always advance exactly in the sequences described above. Suppose, on a bare rock sand and gravels are deposited and succession starts on it, then the possibilities are there that some beginning stages may not develop at all and herbs, shrubs or even trees may appear as initial colonizers on such a habitat.

Succession on the sand-dunes starts with drought resistant grasses. The grass vegetation is followed by shrubs and trees. Primary succession on open land may lead to a grassland vegetation rather than a forest climax.

SOME EXAMPLES OF SECONDARY SUCCESSION OR SUBSERES

Subsere develops usually in settled areas which are geatly modified by the activities of man. It has fewer stages and it reaches the climax stage more quickly than the primary ones.

This type of succession may develop on secondary bare habitat, such as burnt lands, lumbered lands, disturbed agricultural lands, etc. The secondary bare area develops after the complete destruction of previous vegetation growing over it.

Secondary succession on burnt or lumbered bare areas may start with annual grasses and other herbs. This pioneer stage in due course of time will be replaced by perennial shrubs and trees. If such an area remains undisturbed by biotic agencies, climax woodland will be established.

From the study of succession, it becomes clear that the vegetation is not merely a haphazard covering of land by plants but it represents an orderly process of colonisation governed by interaction of plants and environments. It is mainly due to succession that the landscapes are changing continuously, some very quickly while others very slowly.

The knowledge of principles of succession is helpful in the management of forests, agricultural lands, mountain ranges and wild life areas.

QUESTIONS

1. What is plant succession? Write an essay on plant succession.
2. What is sere? Describe the various successional stages of hydrosere.
3. Describe the various development stages of succession in hydrosere.
   (Univ. of Pune, B.Sc II, Bot. I, 2010)
4. Write short notes on the following :
   sere, causes of succession, climax, xerosere, secondary succession.
5. Describe the causes, trends and basic types of community succession.
6. Describe the various stages of xerosere or lithosere or succession on rocks.
   (Univ. of Pune, B.Sc II, Bot. I, 2010)
7. Discuss the concept of climax community.
8. Discuss the monoclimax and polyclimax concepts.
9. Define a biotic community. Describe the various features that characterise a community.
   (HNB. Garhwal Univ. B.Sc II, 2010)

8

Plant Adaptations

The living organisms react with their environments and they bear full impression of the environments in which they grow. In order to withstand adverse conditions of the environment and utilize to their maximum benefit the nutrients and other conditions prevailing therein, the organisms develop certain morphological, anatomical, physiological and reproductive features. Any feature of an organism or its part which enables it to exist under conditions of its habitat is called *adaptation.* Every organism develops certain adaptations and so does the population or a community. The completion of life cycle of an organism or stabilization of a community results through a series of adaptations which have survival value. Adaptations of survival value comprise such features as prevent destruction of vital vegetative tissues and help in large production and efficient dissemination of reproductive bodies. Warming (1895) had realised for the first time the influence of controlling or limiting factors upon the vegetation in ecology. He classified plants into several ecological groups on the basis of their requirements of water and also on the basis of nature of substratum on which they grow.

Warming classified plants on the basis of nature of substratum (soil) into the following groups :

(1) Plants of acidic soil (**Oxylophytes**)

(2) Plants of saline soil (**Halophytes**)

(3) Plants growing on the sand (**Psammophytes**)

(4) Plants growing on the surface of rocks (**Lithophytes**)

(5) Plants growing in the crevices of rocks (**Chasmophytes**).

Epiphytes are not included in the above classification because of the fact that they do not have permanent connection with the soil.

Warming's second classification (1909) of the plants is based on their water relations. The supply of water to the plants and regulation of transpiration are the factors that evoke great differences in plant forms and plant life. On the basis of their water requirement and nature of soils, the plants have been classified as follows :

1. **Hydrophytes :** Plants growing in or near water.

2. **Xerophytes :** Plants adapted to survive under the condition of very poor supply of available water in the habitats. Xerophilous plants are further classified on the basis of their habitats as follows :

(*i*) Oxylophytes (on acid soils)

(*ii*) Halophytes (on saline soils)

(*iii*) Lithophytes (on rocks)

(*iv*) Psammophytes (on sand and gravels)

(*v*) Chersophytes (on waste land)

(*vi*) Eremophytes (on deserts and steppes)

(*vii*) Psychrophytes (on cold soils)

(*viii*) Psilophytes (savannah)

(*ix*) Sclerophytes (Forest and bushland)

3. **Mesophytes :** Plants growing in an environment which is neither very dry nor very wet.

The detailed description of only some important ecological groups is given here.

HYDROPHYTES

(*Greek, Hudor* = water and *Phyton* = Plant; water plant)

Plants which grow in wet places or in water either partly or wholly submerged are called hydrophytes or aquatic plants. Examples are *Utricularia, Vallisneria, Hydrilla, Chara, Nitella,* Lotus, *Ceratophyllum, Trapa, Pistia, Eichhornia* (water hyacinth), *Wolffia, Lemna, etc.* Aquatic environment provides a matrix for plant growth in which temperature fluctuation is at minimum and the nutrients occur mostly in dissolved state but light and oxygen become deficient with the increase in depth of water bodies. Zonation of aquatic vegetation with increasing depth is a device for maximum utilization of light energy. The aquatic environment is subject to water movements ranging from small vertical circulation to strong currents. Streams have a unidirectional movement and in seas the movement is reversible. The currents of water often abrade the inhabiting flora and varied modifications are encountered to withstand this abrasive action. Since water makes up a large proportion of the bodies of plants and animals (70 to 90% water in protoplasm), it affects all life processes directly. In plants, the rate and magnitude of the photosynthesis, respiration, absorption of nutrients, growth and other metabolic processes are influenced by the amount of available water. Low relative humidity increases water loss through transpiration and affects plant growth. Conversely, plants in the regions with high moisture show reduced transpiration. Some aquatic groups of higher plants probably originated from mesophytes. In the course of evolution several changes in the physiology, morphology and behaviour, all related to the aquatic mode of life, took place and by these evolutionary changes the mesophytic plants have become adapted to aquatic mode of life.

CLASSIFICATION OF HYDROPHYTES

According to their relation to water and air, the hydrophytes are grouped into the following categories :

(*a*) Submerged hydrophytes

(*b*) Floating hydrophytes

(*c*) Amphibious hydrophytes.

(*a*) **Submerged hydrophytes.** Plants which grow below the water surface and are not in contact with atmosphere are called submerged hydrophytes. Such plants may be free-floating (Fig. 8.1) or rooted (Fig. 8.2). Example *Vallisneria, Hydrilla, Potamogeton, Najas, Ceratophyllum, Myriophyllum, Utricularia, Chara, Nitella* and a number of aquatic microbes.

(*b*) **Floating hydrophytes.** Plants that float on the surface or slightly below the surface of water are called floating hydrophytes. These plants are in contact with both water and air. They may or may not be rooted in the soil. On this ground, the floating plants have been divided into two groups.

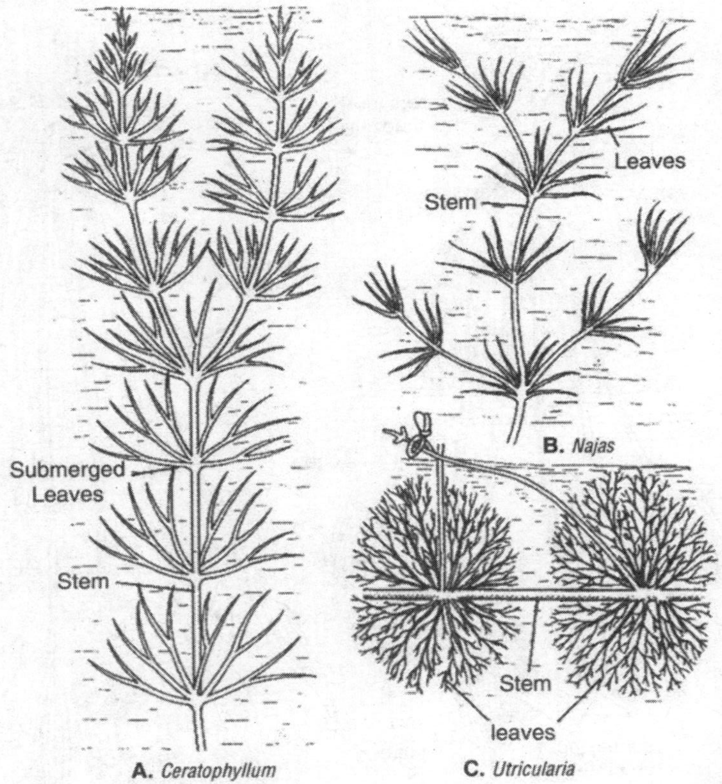

Fig. 8.1. Submerged floating hydrophytes.

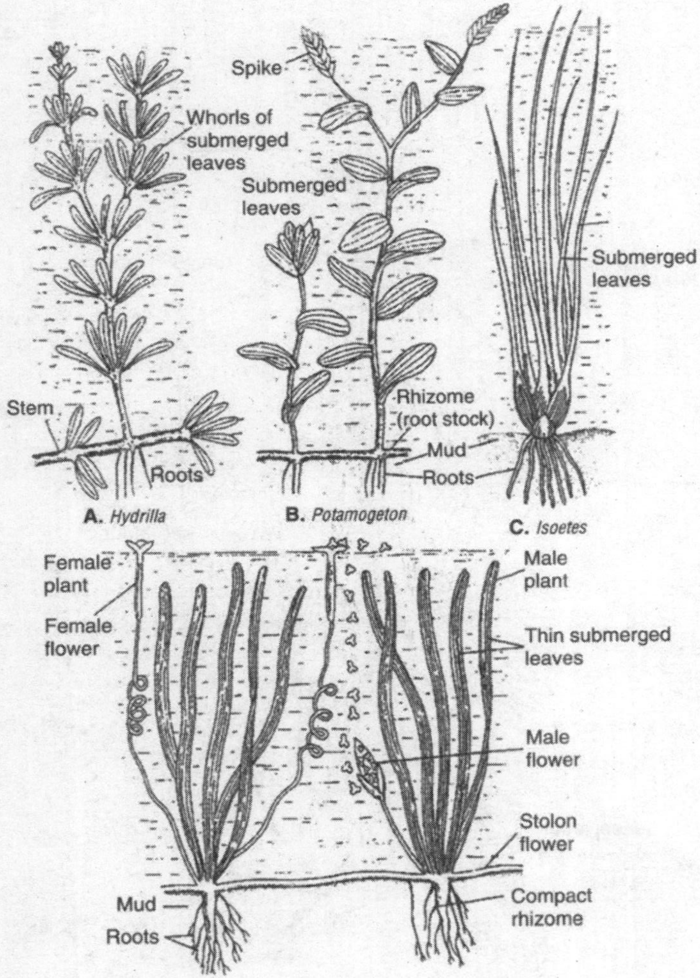

Fig. 8.2. Rooted submerged hydrophytes

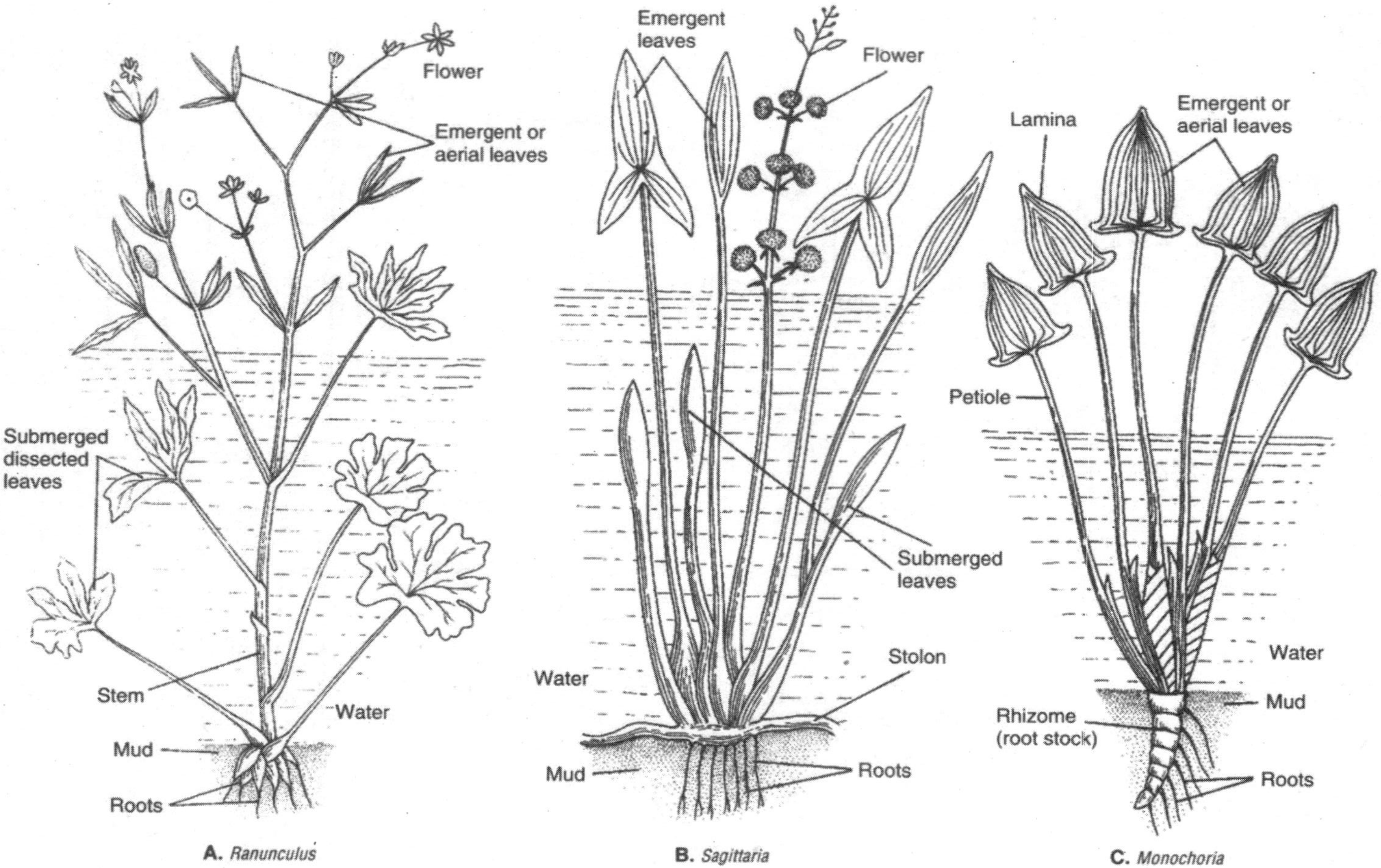

Fig. 8.3. Rooted emergent hydrophytes with heterophylly

(*i*) **Free floating hydrophytes.** These plants float freely on the surface of water but are not rooted in the mud. Examples—*Wolffia arhiza* and *Wolffia microscopica* (a rootless minutes duck weed). *Trapa bispinosa, Lymnanthemum, Eichhornia crassipes* (water hyacinth, verna—Jalkumbhi), *Salvinia* (a fern), *Azolla* (a water fern) (Fig. 8.6).

(*ii*) **Floating but rooted hydrophytes.** Some submerged plants are rooted in muddy substrata of ponds, rivers and lakes but their leaves and flowering shoots float on or above the surface of water. They are grouped as floating but rooted hydrophytes. Examples—*Nymphaea, Nelumbium speciosum* (Lotus), *Victoria regia* (water lily), *Ceratopteris thalictroides* (a hydrophytic fern of family Parkariaceae), etc. (Fig. 8.5).

(*c*) **Amphibious hydrophytes.** These plants are adapted to both aquatic and terrestrial modes of life. Amphibious plants grow either in shallow water or on the muddy substratum. Amphibious plants which grow in saline marshy places are termed as 'halophytes. Roots and some parts of stems and leaves in these plants may be submerged in water or buried in mud but some foliage, branches and flowering shoots spring well above the surface of water or they may spread over the land (Fig. 8.3). The aerial parts of these amphibious plants show mesophytic or sometimes xerophytic features, while the submerged parts develop true hydrophytic characters. Some varieties of rice plants, (*Oryza sativa*), *Marsilea, Sagittaria, Alisma, Jussiaea. Neptunia, Commelina, Polygonum, Ranunculus aquatilis, Phragmites, Enhydra fluctuans,* etc. are familiar examples of this group of hydrophytes.

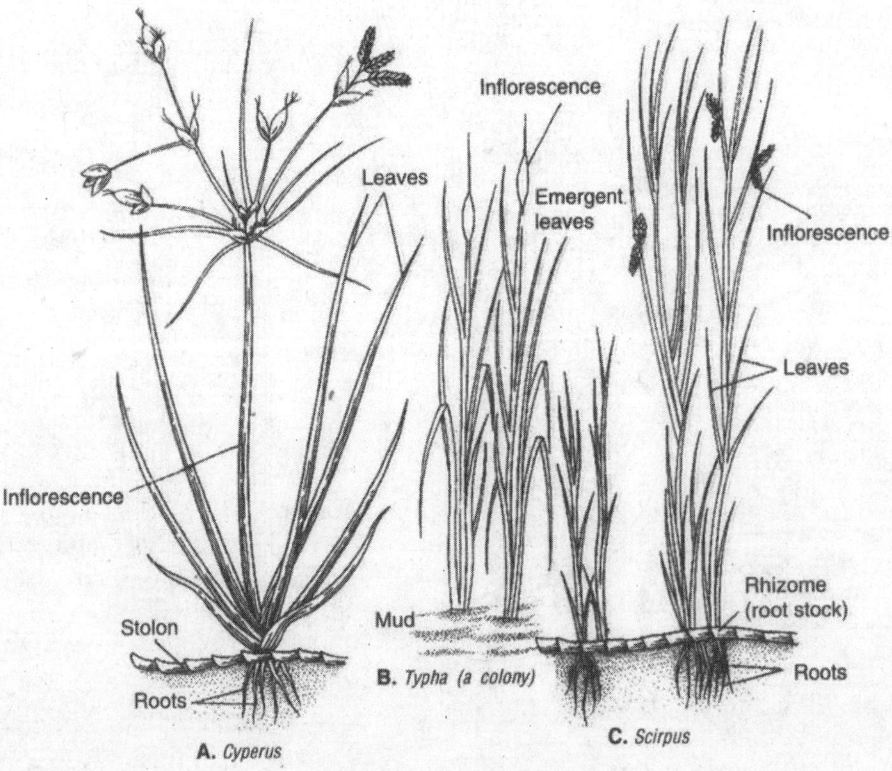

Fig. 8.4. Marsh plants.

In some amphibious plants the shoots are completely exposed to air as in land plants but the roots are buried in water lodged soil or mud. They are called *marsh plants*. The common examples of marsh plants are *Cyperus, Typha, Scirpus, Rumex,* etc. (Fig. 8.4).

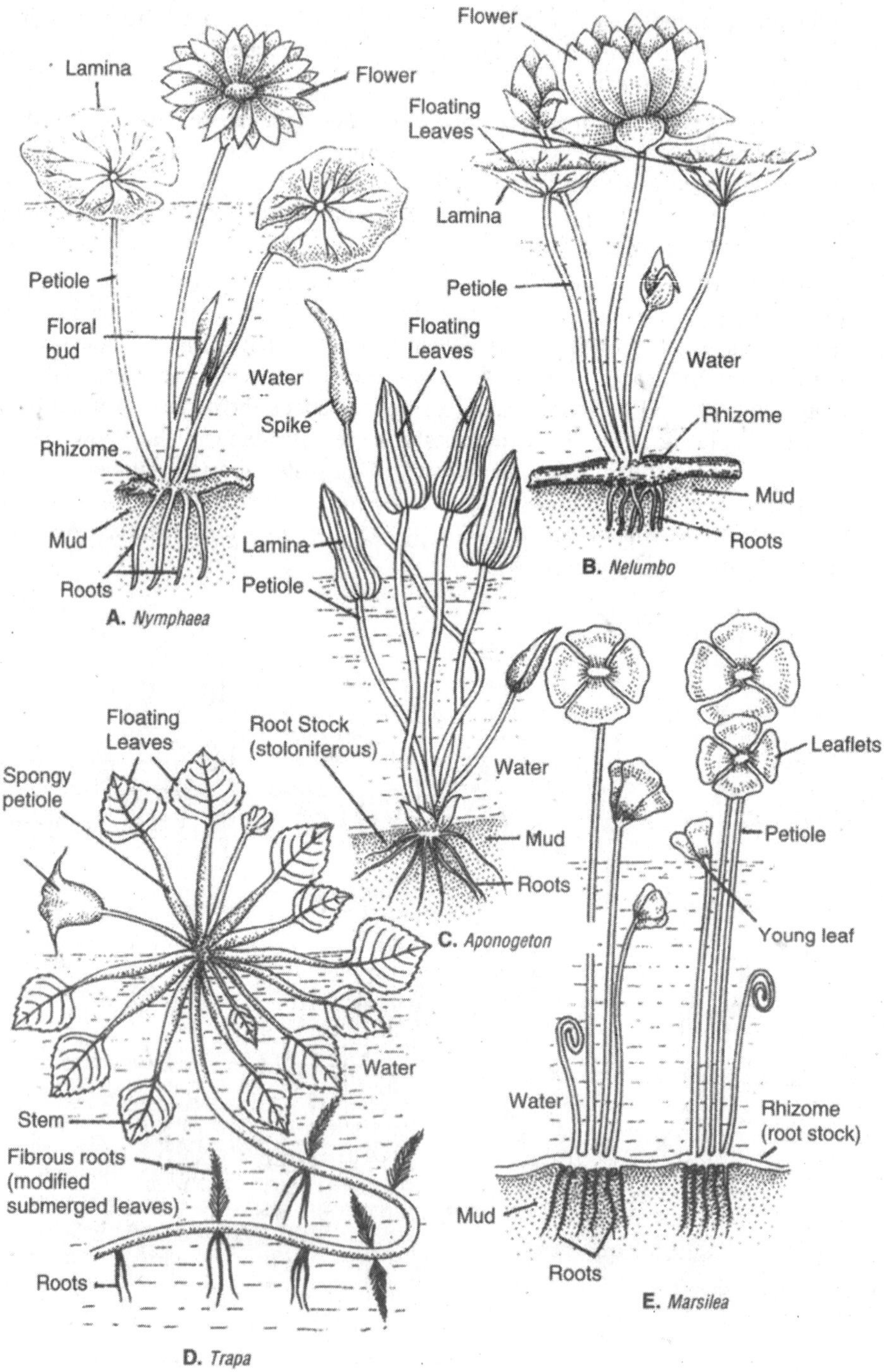

Fig. 8.5. Rooted hydrophytes with floating leaves.

Factors Affecting the Plants in the Aquatic Environment

These are listed below :

1. Temperature of water
2. Osmotic concentration of water
3. Toxicity of water

The osmotic concentration and toxicity are dependent upon the amount and nature of chemical substances dissolved in water. The physiology of aquatic plants is greatly affected by the change in osmotic concentration of water. The aquatic plants are subjected to less extremes of temperature because water is bad conductor of heat (*i.e.*, it takes long time in its heating and cooling). Hydrophytes are less affected as the transpiration from the plant tissue is completely out of question.

HYDROPHYTIC ADAPTATIONS

As the aquatic environment is uniform throughout, the hydrophytes develop very few adaptive features. Important features of these plants are described in the following heads :

A. Morphological

(*i*) **Roots.** Root systems in hydrophytes are poorly developed which may or may not be branched in submerged hydrophytes. Roots are meaningless as the entire surface of the plant body which is in direct contact with water acts as absorptive surface and absorbs water and minerals. This may probably be the reason why roots in hydrophytes are reduced or absent. Roots of floating hydrophytes show very poor development of root hairs. Roots in floating plants do not possess true root caps but very often they develop root pockets or root sheaths which protect their tips from injuries (Fig. 8.5). Exact functions of these root pockets, however, are not fully understood. Some rooted hydrophytes like *Hydrilla* (Fig. 8.6), *Vallisneria spiralis, Elodia canadensis,* though they derive their nourishments from water by their body surfaces, are partly dependent on their roots for minerals from the soil. Roots are totally absent in some plants, *e.g., Ceratophyllum, Salvinia, Azolla, Utricularia,* etc. In *Jussiaea repens* two types of roots develop when the plants grow on the surface of water, some of them are normal while the others are floating roots which are negatively geotropic having spongy structures (Fig. 8.7). The floating roots keep the plants afloat.

(*ii*) **Stem.** In aquatic plants, stem is very delicate and green or yellow in colour (Fig. 8.2 A, B). In some cases it may be modified into rhizome or runner, etc. (Fig. 8.2 D, 8.4).

(*iii*) **Leaves.** (*a*) In floating plants leaves are generally peltate, long, circular, light or dark green in colour, thin and very smooth. Their upper surfaces are exposed in the air but lower ones are generally in touch with water. In lotus plant petioles of leaves show indefinite power of growth and they keep the laminae of leaves always on the surface of water.

(*b*) *Heterophylly.* Some aquatic plants develop two different types of leaves in them. This phenomenon is termed as heterophylly. Examples are *Sagittaria sagittaefolia, Ranunculus aquatilis, Limnophila heterophylla, Salvinia, Azolla* etc. In this phenomenon, generally the submerged leaves are linear ribbon shaped or highly dissected and the leaves that are found floating on or above the surface of water are broad circular or slightly lobed (Fig. 8.3 A, B, C). The occurrence of heterophylly is associated probably with the following characteristic physiological behaviours of these aquatic plants :

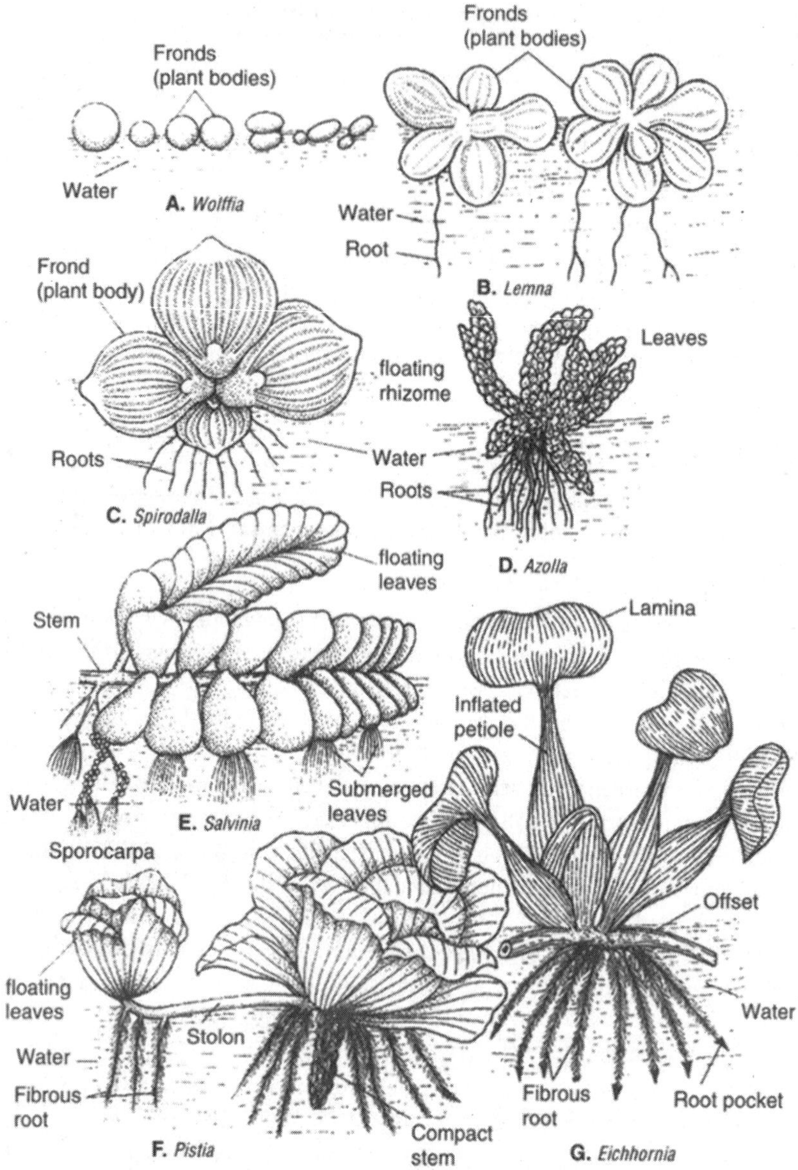

Fig. 8.6. Free floating hydrophytes.

1. Quantitative reduction in transpiration.
2. The broad leaves on the surface overshadow the submerged dissected leaves of the same plant and thus they reduce the intensity of light falling on the submerged leaves. The submerged leaves require light of very low intensity.
3. Plants show very little response to drought because the necessity of excess water during drought period is compensated by submerged leaves which act as water absorbing organs.
4. Variation in the life-forms and habitats.
5. Broad leaves found on the surface of water transpire actively and regulate the hydrostatic pressure in the plant body.

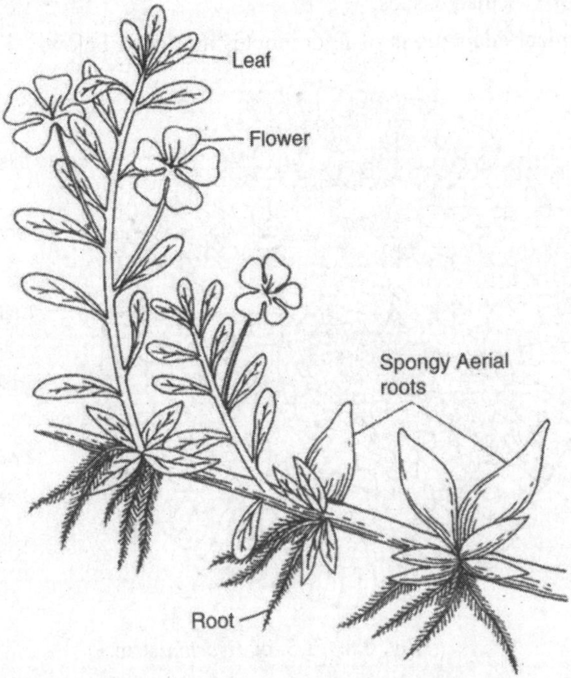

Fig. 8.7. *Jussiaea repens* showing spongy respiratory roots.

(*c*) leaves of free floating hydrophytes are smooth, shining and frequently coated with wax. The wax coating protects the leaves from chemical and physical injuries and also prevents the water clogging of stomata.

(*d*) In floating plants of water hyacinth, *Trapa* etc., the petioles become characteristically swollen and develop spongyness which provides buoyancy to these plants (Fig. 8.6 F).

(*e*) Leaves in submerged hydrophytes are generally small and narrow. In some cases, *e.g.*, *Myriophyllum, Utricularia, Ceratophyllum,* etc., they may be finely dissected (Fig. 8.2). The small slender terete segments of dissected leaves offer little resistance against the water currents. In this way, plants are subjected to little mechanical stress and strain of water.

(*f*) In the Amphibious plants, the leaves that are exposed to air show typical mesophytic features. They are more tough than the leaves of other groups of hydrophytes.

(*iv*) Pollination and dispersal of fruits and seeds are accomplished by the agency of water. Seeds and fruits are light in weight and thus they can easily float on the surface of water.

(*v*) Vegetative reproduction is common method of propagation in hydrophytes. It is accomplished either through fragmentation of ordinary shoots or by winter buds. In algae, reproduction is accomplished by zoospores and other specialized motile or non-motile spores.

B. Anatomical Modifications

The anatomical modifications in hydrophytes aim mainly at :

1. Reduction in protecting structures,
2. Increase in the aeration,
3. Reduction of supporting or mechanical tissues, and

4. Reduction of vascular tissues.

Various anatomical adaptations of hydrophytes are listed below.

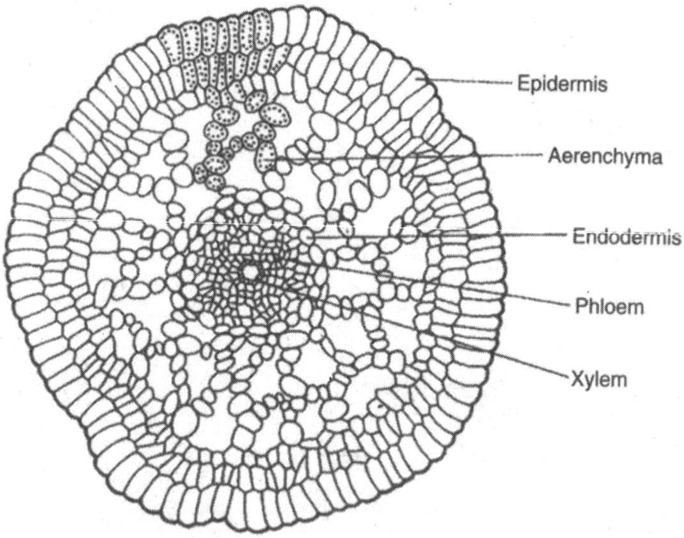

Fig. 8.8. T.S. of *Hydrilla* stem.

1. **Reduction in protecting structures.** (*a*) Cuticle is totally absent in the submerged parts of the plants. It may be present in the form of very fine film on the surfaces of parts which are exposed to atmosphere.

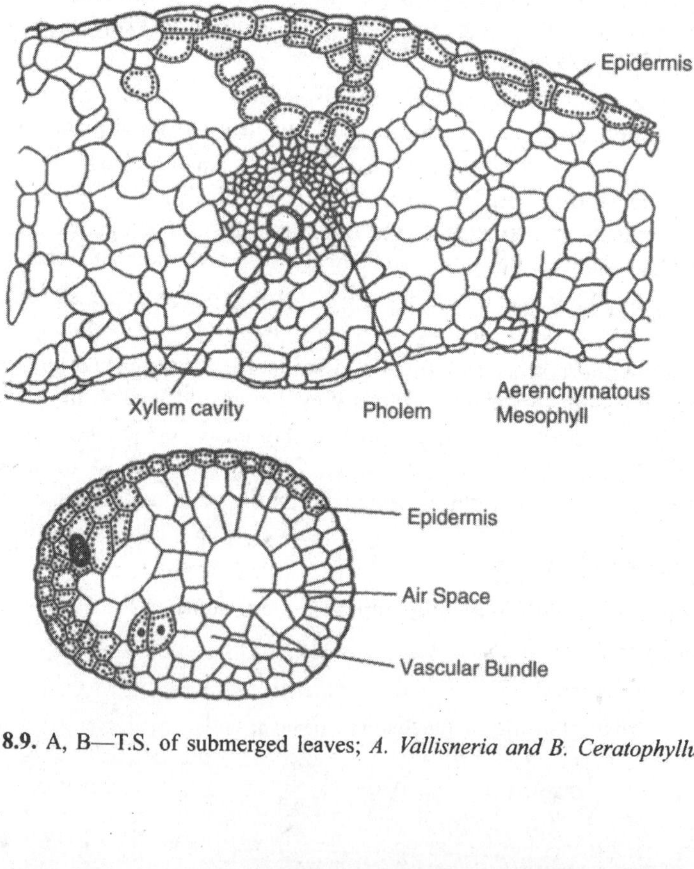

Fig. 8.9. A, B—T.S. of submerged leaves; *A. Vallisneria and B. Ceratophyllum.*

(*b*) Epidermis in hydrophytes is not a protecting layer but it absorbs water, minerals, and gases directly from the aquatic environment. Extremely thin cellulose walls of epidermal cells facilitate the absorption process.

(*c*) Epidermal cells contain chloroplasts, thus they can function as photosynthetic tissue, especially where the leaves and stems are very thin, *e.g. Hydrilla* (Fig. 8.8).

(*d*) Hypodermis in hydrophytes is poorly developed. Its cells are extremely thin walled.

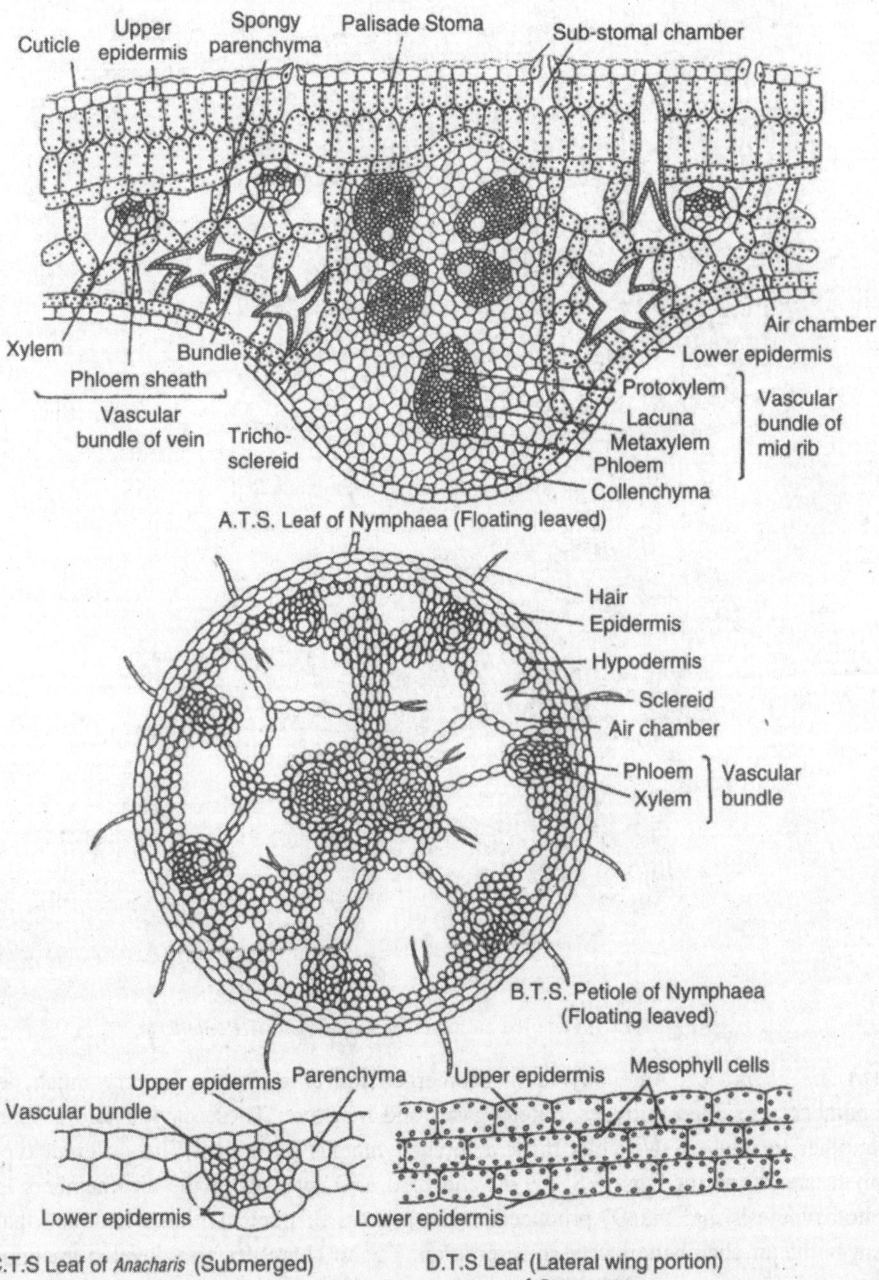

Fig. 8.10. Anatomical features of some hydrophytic leaves.

2. **Increase in the aeration.** (*a*) Stomata are totally absent in submerged parts of the plants (Figs. 8.9, 8.10 C & D). In some exceptional cases, vestigial and functionless stomata have been noticed. In these cases exchange of gases takes place directly through cell walls. In the floating leaves, stomata develop in very limited number and are confined only to the upper surface (Fig. 8.10 A). In amphibious plants stomata may be scattered on all the aerial parts and they develop comparatively in larger number per unit area than those on the floating leaves (Fig. 8.11).

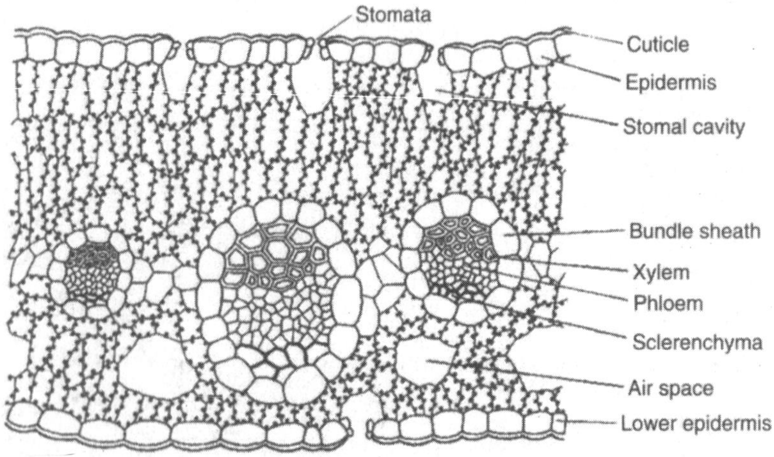

Fig. 8.11. V.S. of an amphibious leaf.

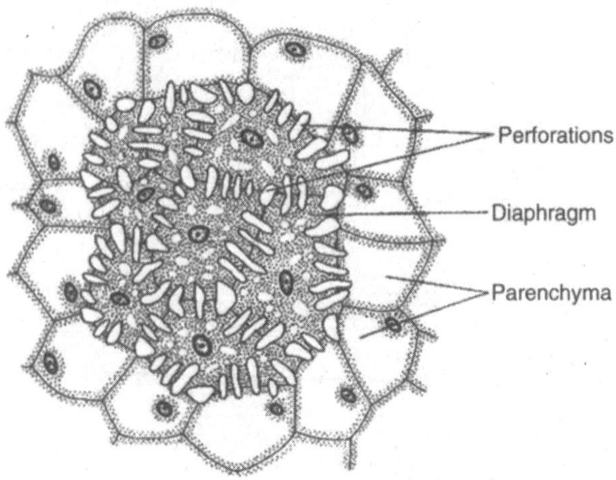

Fig. 8.12. Detailed structure of diaphragm of *Pontederia*.

(*b*) *Air chambers.* Aerenchyma in submerged leaves and stem is very much developed. Air chambers are filled with respiratory gases and moisture. These cavities are separated from one another by one or two cells thick chlorenchymatous partitions. The different types of air chambers are shown in Figs. 8.8, 8.9 A, and 8.10. CO_2 present in the air chambers is used in the photosynthesis and the O_2 produced in the process of photosynthesis and also that already present in the air chambers is used in respiration. The air chambers also develop finely perforated cross septa which are called diaphragms (Fig. 8.12). The diaphragms afford better aeration and perhaps check floating. The aerenchyma provides buoyancy and mechanical support to aquatic plants. Air chambers are abundantly found in the fruits of hydrophytes rendering them buoyant and thus facilitating their dispersal by water.

Development of air chambers in the plants is governed by habitat. This point is clear from the anatomy of *Jussiaea suffructicosa.* In this case, air chambers develop normally if plants are growing in water but they seldom develop if the plants are growing on the land.

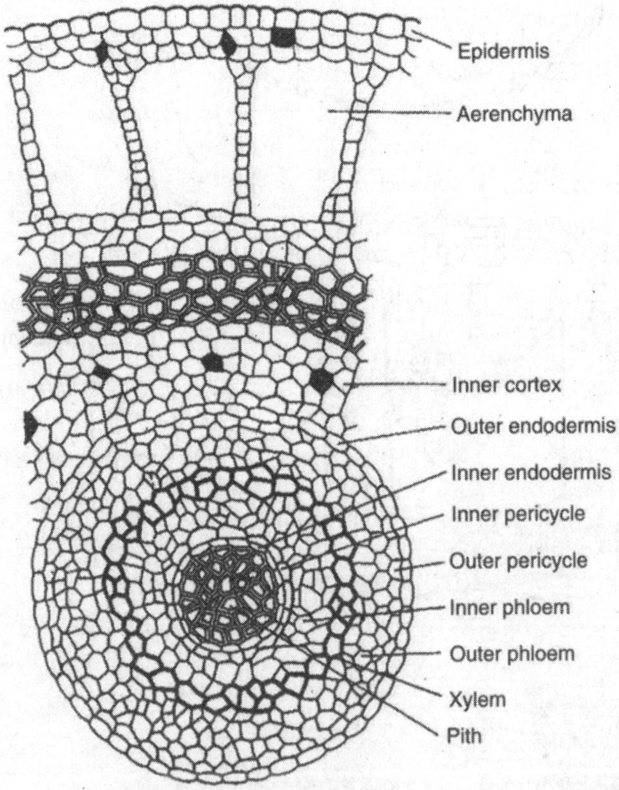

Fig. 8.13. T.S. of *Marsilea rhizome* (An amphibious hydrophyte).

3. **Reduction of supporting or mechanical tissues.** (*a*) Mechanical tissues are absent or poorly developed in the floating and submerged parts of plants because buoyant nature of water saves them from physical injuries. The thick walled sclerenchymatous tissue is totally absent in submerged and floating hydrophytes. They may, however, develop in the cortex of amphibious plants, particularly in the aerial or terrestrial parts (Figs. 8.13, 8.14 B, D). Generally elongated and loosely arranged spongy cells are found in the plant body. These thin-walled cells, when turgid, provide mechanical support to the plants (Figs. 8.15, 8.16 and 8.17).

(*a*) The reduction of absorbing tissue (roots act chiefly as anchors and root hairs are lacking).

(*b*) In water lily and some other plants, special type of star shaped lignified cells, called asterosclereids, develop which give mechanical support to the plants (Fig. 8.10 A,B).

4. **Reduction of vascular tissues.** Conducting tissue is very poorly developed. As the absorption of water and nutrients takes place through the entire surface of submerged parts, there is little need of vascular tissues in these plants. In the vascular tissues, xylem shows greatest reduction. In some cases, it consists of only a few tracheids while in some, xylem elements are

not at all developed (Fig. 8.18). Some aquatic plants, however, show a lacuna in the centre in the place of xylem. Such spaces resemble typical air chambers (Fig. 8.8).

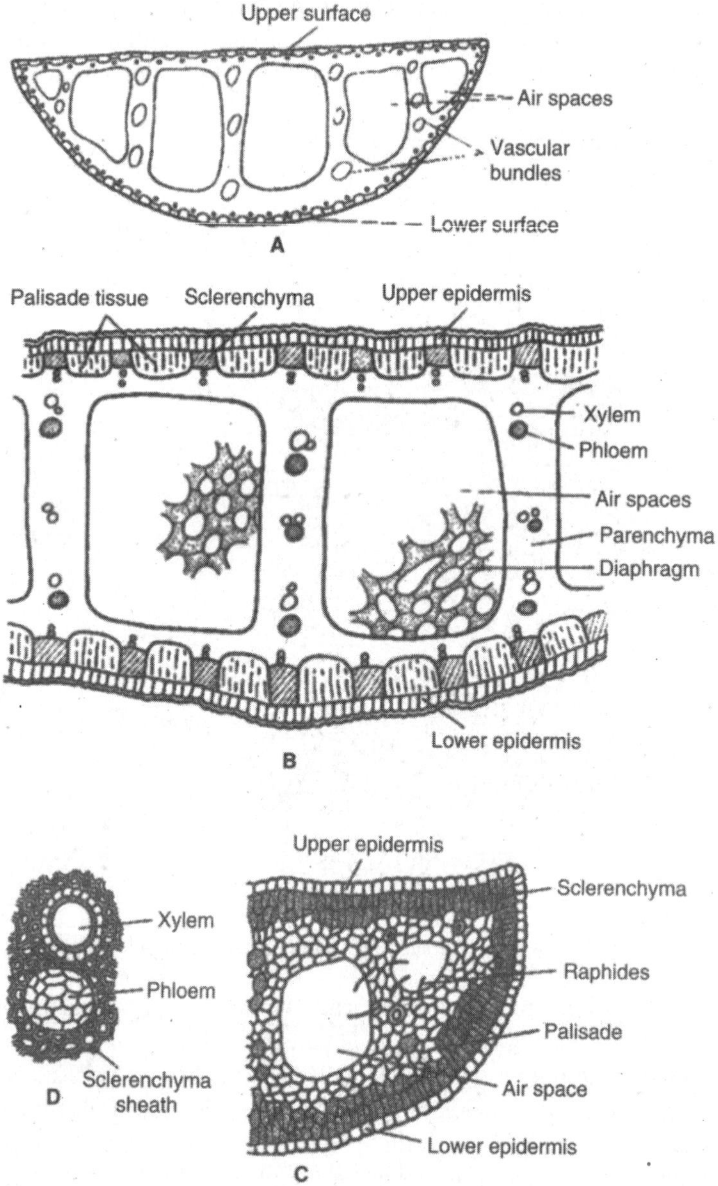

Fig. 8.14. Anatomy of leaf of *Typha,* a marshy hydrophyte. A : T.S. of leaf (diagrammatic), B : detail of a part of A; C, detail of corner of leaf, D : single vascular bundle.

Phloem tissue is also poorly defined in most of the aquatic plants but in some cases it may develop fairly well. Sieve tubes of aquatic plants are smaller than those of mesophytes. Phloem parenchyma is extensively developed. *Endodermis* may or may not be clearly defined.

The *Vascular bundles* are generally aggregated towards the centre. *Secondary growth* in thickness does not take place in the aquatic stem and roots.

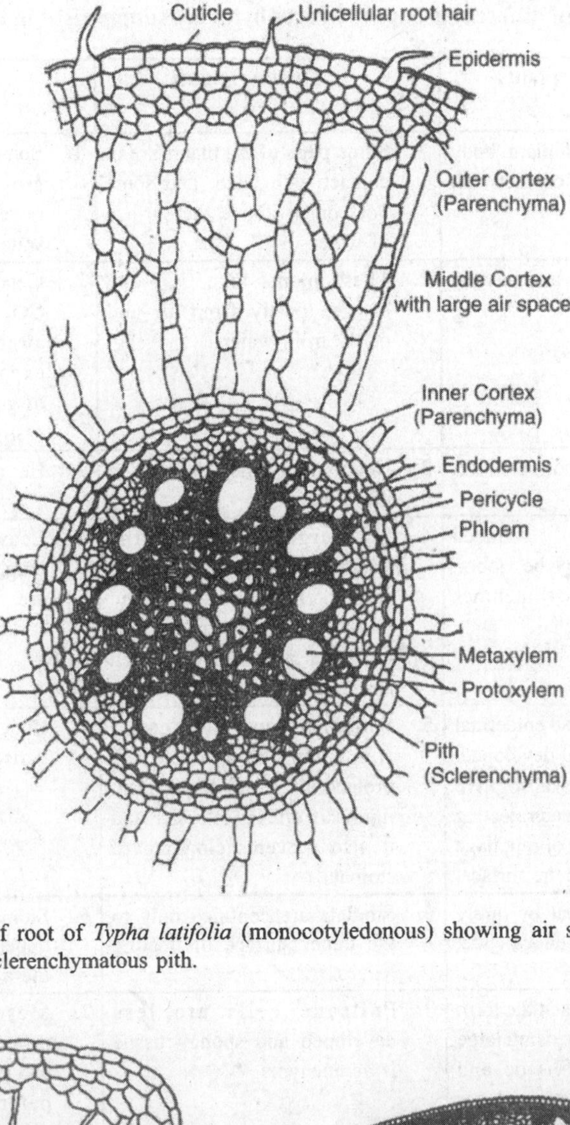

Fig. 8.15. T.S. of root of *Typha latifolia* (monocotyledonous) showing air spaces in cortical region and sclerenchymatous pith.

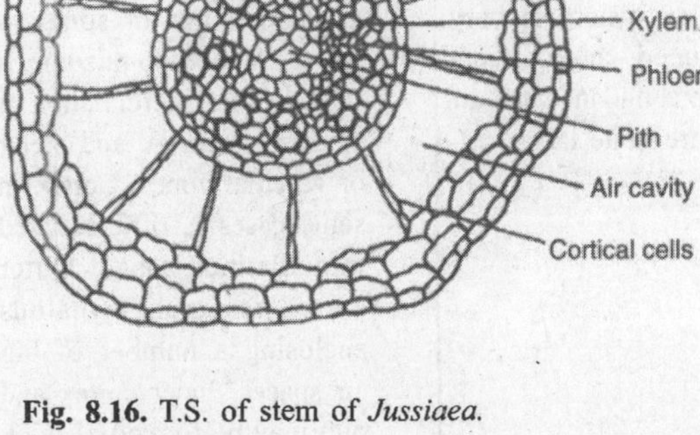

Fig. 8.16. T.S. of stem of *Jussiaea.*

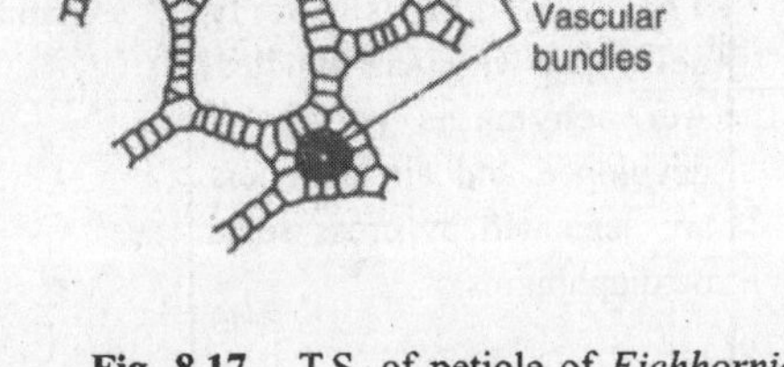

Fig. 8.17. T.S. of petiole of *Eichhornia.*

Distinctive features of different groups of hydrophytes are summarised in the following chart.

	Submerged plants 1	Floating plants 2	Amphibious plants 3
1.	Entire surface of plant body remains in direct contact with water.	1. Some parts of the plants are in contact with water and some float on or above the surface of water.	1. Some parts of the plant body grow in water while some others above the surface of water or even on the land.
2.	Leaves take CO_2 and oxygen form water.	2. These plants take CO_2 and oxygen partly from air and partly from water.	2. Generally the leaves absorb CO_2 and oxygen from the air because they are well exposed and the exchange of gases can take place very easily through stomata.
3.	Heterophylly is not so common.	3. Heterophylly may develop in these plants.	3. Heterophylly is very common.
4.	Leaves are greatly reduced in size. They may be ribbon shaped, thin and sometimes very finely dissected.	4. Submerged leaves in the floating plants are thin, ribbon shaped, entire or dissected but those floating on the surface of water are well expanded and entire.	4. Leaves are large, entire, tough and laminae or leaflets are found much above the surface of water.
5.	Cuticle, suberin and epidermal hairs are not at all developed. Epidermis acts as absorptive surface rather than a protecting layer. A thin layer of mucilage may be present on the surface.	5. Surface of plant body is coated with a thin film of wax which protects the plant tissue from injurious effects of water and it also prevents clogging of stomata.	5. Cuticle develops on the aerial parts of the plants.
6.	Stomata are absent or rarely present in very reduced state.	6. Stomata are confined only to the upper surface of floating leaves.	6. Stomata are confined to both upper and lower surfaces of the aerial leaves.
7.	Mesophyll tissue is in the form of aerenchyma enclosing large air chambers. Palisade and spongy parenchyma are not very well differentiated.	7. Palisade cells are less developed and spongy tissue is aerenchyma.	7. Mesophyll shows clear differentiation into spongy and palisade tissues. Spongy parenchyma is very well developed with large air passages and diaphragms.
8.	Stems are very slender and much reduced. They creep on the substratum either in the form of rhizomes or runners. The stem shows poorly developed vascular bundles. Aerenchyma is very well developed and air chambers are separated by cross septa or diaphragms.	8. Stem is well developed and the fibrovascular system is very much reduced, aerenchyma develops in abundance and air chambers are quite large.	8. Stem is extensively developed but in some, it may be reduced to rhizome. It shows clear differentiation of epidermis, cortex and stelar or vascular zones. Cortex in some cases is differentiated into distinct zones. Outer cortex is aerenchymatous enclosing a number of big air spaces. Inner cortex and pith may be formed of thick walled cells.

9. Roots in submerged plants are greatly reduced, unbranched and without root hairs. They act simply as an anchorage organs. The root system shows poor internal organisation.	9. In some, root is absent while in others, it may be fully developed. Root hairs in free floating plants lie at right angles to the axis of main root. Some roots may be modified into floats, e.g., in Jussiaea. Roots show poor differentiation of internal tissues. It functions as absorbing as well as anchorage organ.	9. Roots in amphibious plants are well developed. They show proper differentiation of all internal tissues. Root hairs develop on the epidermis.

PHYSIOLOGICAL ADAPTATIONS IN HYDROPHYTES

The aquatic plants exhibit a low compensation point and low osmotic concentration of cell sap. Osmotic concentration of cell sap is equal or slightly higher than that of water. Nutrients are absorbed by the submerged plants through the general plant surface. The gases are exchanged from the water through the surface cells. The gases produced during photosynthesis

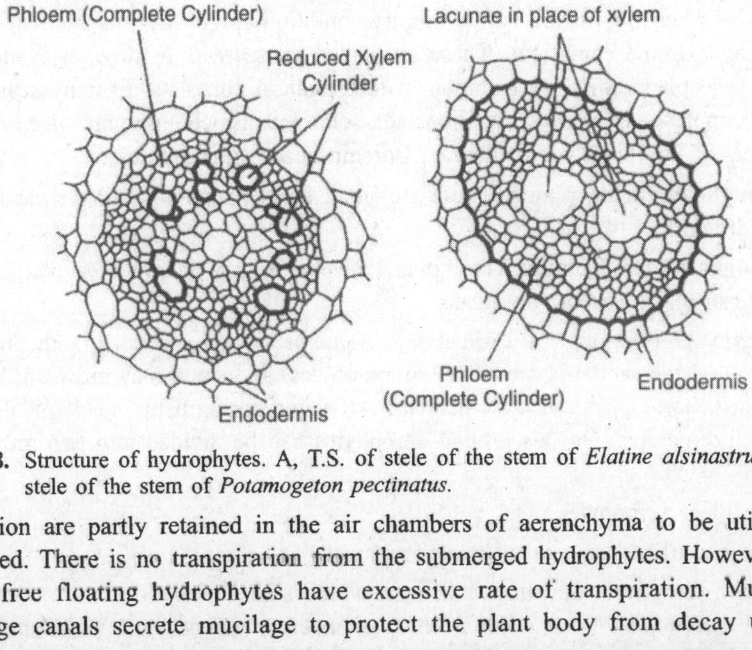

Fig. 8.18. Structure of hydrophytes. A, T.S. of stele of the stem of *Elatine alsinastrum;* B, T.S. of stele of the stem of *Potamogeton pectinatus.*

and respiration are partly retained in the air chambers of aerenchyma to be utilized as and when required. There is no transpiration from the submerged hydrophytes. However, emergent plants and free floating hydrophytes have excessive rate of transpiration. Mucilage cells and mucilage canals secrete mucilage to protect the plant body from decay under water.

XEROPHYTES

Plants which grow in dry habitats or xeric conditions are called xerophytes. Places where available water is not present in adequate quantity are termed xeric habitats. Xeric habitats may be of following types :

1. Habitats physically dry (where water retaining capacity of the soil is very low and the climate is dry, *e.g.*, desert, rock surface, waste land, etc.).

2. Habitats physiologically dry (places where water is present in excess amount but it is not such as can be absorbed by the plants easily. Such habitats may be either too salty or too acidic, too hot or too cold).

3. Habitats dry physically as well as physiologically, *e.g.*, slopes of mountains.

Xerophytes are characteristic plants of desert and semidesert regions, yet they can grow in mesophytic conditions where available water is in sufficient quantity. These plants can withstand extreme dry conditions, low humidity and high temperature. When growing under unfavourable conditions, these plants develop special structural and physiological characteristics which aim mainly at the following objectives :

(*i*) to absorb as much water as they can get from the surroundings;

(*ii*) to retain water in their organs for very long time;

(*iii*) to reduce the transpiration rate to minimum; and

(*iv*) to check high consumption of water.

Xerophytes are categorised into several groups according to their drought resisting power. These groups are as follows :

1. **Drought escaping plants.** These xerophytes are shortlived. During critical dry periods, they survive in the form of seeds and fruits which have hard and resistant seedcoats and pericarps respectively. At the advent of favourable conditions (which are of very short duration), the seeds germinate into new small sized plants which complete their life cycles within a few weeks time. The seeds become mature before the dry condition approaches. In this way, plants remain unaffected by extreme conditions. These are called *ephemerals* or drought evaders or drought escapers. These plants are very common in the semiarid zones where rainy season is of short duration. Examples—*Astragalus* (Papilionatae), some inconspicuous compositae (*e.g.*, *Artemesia*) and members of families Zygophyllaceae, Boraginaceae, some grasses etc.

2. **Drought enduring plants.** These are small sized plants which have capacity to endure or tolerate drought.

3. **Drought resistant plants.** These plants develop certain adaptive features in them through which they can resist extreme droughts.

Xerophytes grow on a variety of habitats. Some grow on rocky soils (Lithophytes), some in deserts, some on the sand and gravels (Psammophytes) and some may grow on the waste lands (Eremophytes). Some plants of xeric habitat have water storing fishy organs, while some do not develop such structures. On this ground xerophytes can be divided into two groups which are as follows :

(1) Succulent xerophytes.

(2) Non-succulents, also called true xerophytes.

Succulent xerophytes are those plants in which some organs become swollen and fleshy due to active accumulation of water in them or in other words, the bulk of the plant body is composed of water storing tissues. Water stored in these tissues is consumed during the period of extreme drought when the soil becomes depleted of available water.

XEROPHYTIC ADAPTATIONS

Plants growing in the dry habitats develop certain structural devices in them. These structural modifications in xerophytic plants may be of two types.

(*i*) **Xeromorphic characters.** Xerophytic characters that are genetically fixed and inherited are referred to as xeromorphic. They will appear in the xerophytes irrespective of conditions whether they are growing in deserts or in humid regions. Halophytic mangroves and many other evergreen trees, although growing in moist conditions always develop xeromorphic characters.

(*ii*) **Xeroplastic characters.** These features are induced by drought and are always associated with dry conditions. They are never inherited. These characters may disappear from plants if all the favourable conditions are made available to them. Important xerophytic features are summarized under the following heads :

(1) Morphological (external) adaptations;

(2) Anatomical (internal) adaptations;

(3) Physiological adaptations.

1. External Morphology of Xerophytes

(A) Roots. Xerophytes have well developed root systems which may be profusely branched. It is extensive and more elaborate than shoot system. Many desert plants develop superficial root system where the supply of water is restricted to surface layer of the earth. The roots of perennial xerophytes grow very deep in the earth and reach the layers where water is available in plenty. Root hairs are densely developed near the growing tips of the rootlets. These enable the roots to absorb sufficient quantity of water.

(B) Stem. Some of the important characteristics of xerophytic stems are listed below :

(*i*) Stems of some xerophytes become very hard and woody. It may be either aerial or subterranean.

(*ii*) They are covered with thick coating of wax and silica as in *Equisetum.* Some may be covered with dense hairs as is *Calotropis.*

(*iii*) In some xerophytes, stems may be modified into thorns, *e.g., Duranta, Ulex,* etc. (Fig. 8.19).

(*iv*) In stem succulents, main stem itself becomes bulbous and fleshy and it seems as if leaves in these plants are arising directly from the top of the roots. Example—*Kleinia articulata.*

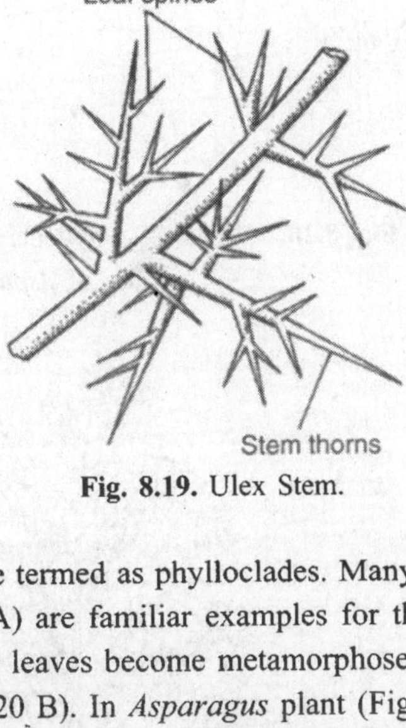

Fig. 8.19. Ulex Stem.

(*v*) Stems in some extreme xerophytes are modified into leaf-like flattened, green and fleshy structures which are termed as phylloclades. Many cacti (Fig. 8.23A, B) and cocoloba (*Muehlenbeckia*) (Fig. 8.20 A) are familiar examples for this. In *Ruscus* plants, the branches developing in the axils of scaly leaves become metamorphosed into leaf-like structures, the phylloclades or cladophylls (Fig. 8.20 B). In *Asparagus* plant (Fig. 8.20 C) also a number of axillary branches become modified into small needle-like green structures which look exactly like leaves. They are called cladodes. A number of species of *Euphorbia* also develop succulence and become green. In these plants, leaves are greatly reduced, so the main function of leaves, the photosynthesis, is taken up by these green phylloclades or cladodes which are modified stems.

(C) Leaves (*i*) In some xerophytes the leaves, if present, are greatly caducous, *i.e.,* they fall early in the season, but in the majority of the plants leaves are generally reduced to scales, as in *Casuarina* (Fig. 8.21), *Ruscus* (Fig. 8.20 B), *Asparagus* (Fig. 8.20 C), etc.

(*ii*) Some evergreen xerophytes have needle-shaped leaves, *e.g., Pinus* (Fig. 8.22 A, B).

(*iii*) In leaf succulents, the leaves swell remarkably and become very fleshy owing to storage of excess amount of water and latex in them. Plants with succulent leaves generally develop very reduced stems. Examples of leaf succulents are *Sedum acre, Aloe spinossissima* (Gheekwar) (Fig. 8.23 C), *Mesembryanthemum, Kleinia ficoides* and several members of family Chaenopodiaceae.

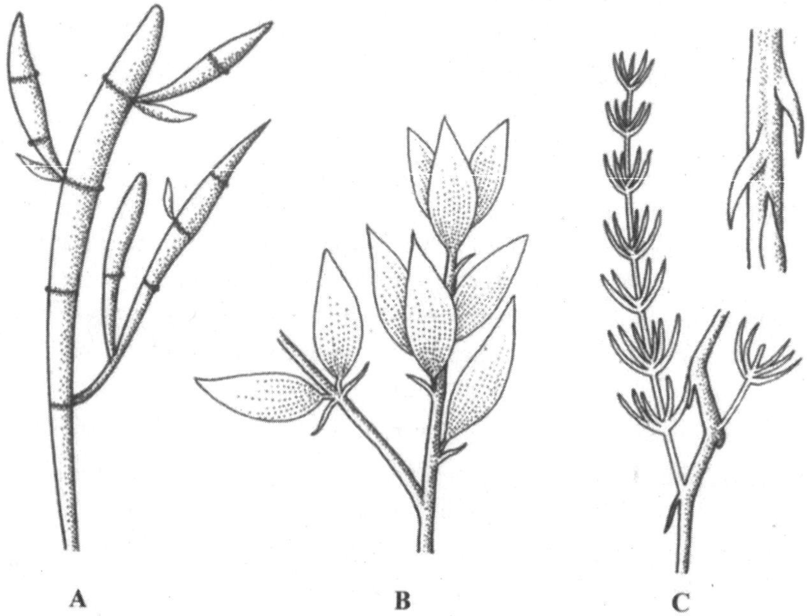

Fig. 8.20. A, B & C—Phylloclade and cladodes. A—Phylloclade of Cocoloba. B—Cladodes of *Ruscus*. C—Cladodes of *Asparagus*.

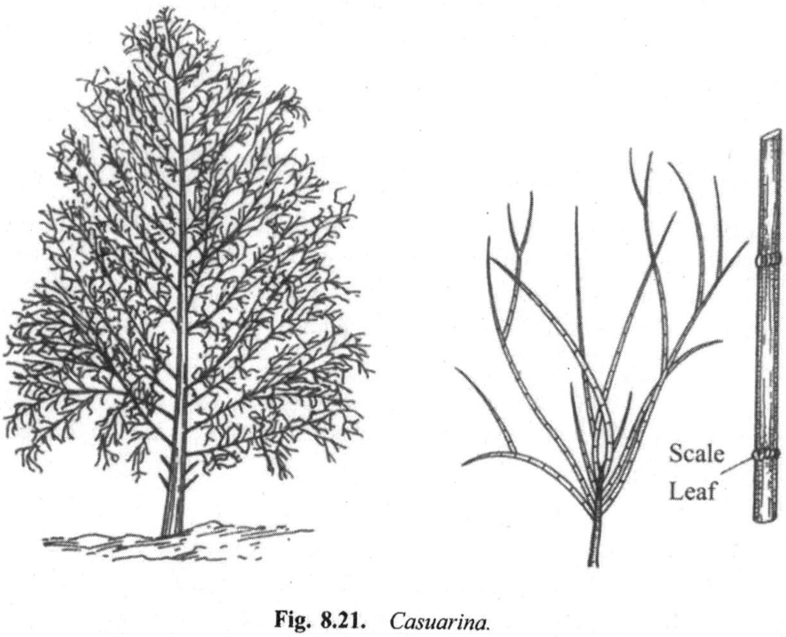

Fig. 8.21. *Casuarina.*

(*iv*) In majority of xerophytes, leaves are generally much reduced and are provided with thick cuticle and dense coating of wax or silica. Sometimes they may be reduced to spines, as for example, in *Ulex, Opuntia, Euphorbia splendens* (Fig. 8.23 A, B), *Capparis* (Fig. 8.24 B) and *Acacia* (Fig. 8.24D).

(*v*) Generally, the leaves of xerophytic species possess reduced leaf blades or pinnae and have very dense network of veins. In Australian species of *Acacia* (Babool) the pinnae are shed from the rachis and the green petiole swells and becomes flattened taking the shape of leaf. This modified petiole is termed as phyllode (Fig. 8.25). The phyllode greatly reduces the water loss, stores excess amount of water and performs photosynthesis.

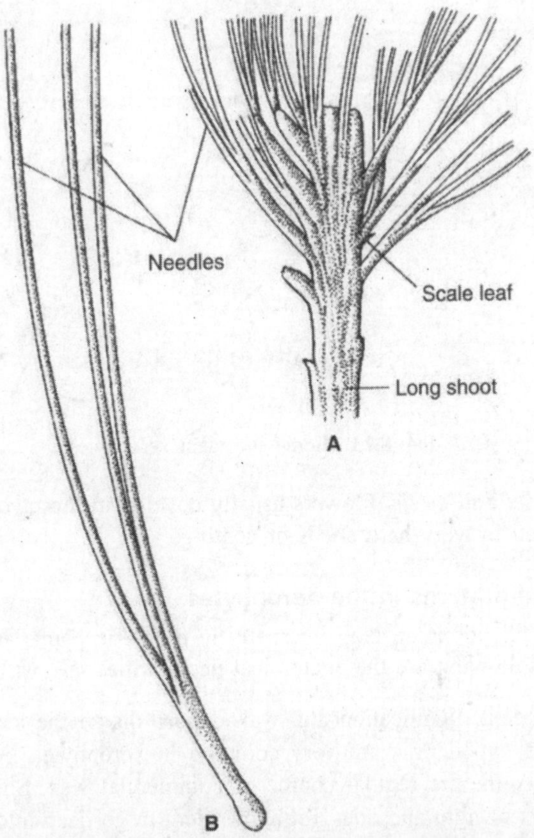

Fig. 8.22. A, B—Long and dwarf shoots of *Pinus roxburghi* showing needle-like leaves and scales.

(*vi*) ***Trichophylly.*** In some xerophytes especially those growing well exposed to strong wind, the under surfaces of the leaves are covered with thick hairs which protect the stomatal guard cells and also check the transpiration. Those xerophytes which have hairy covering on the leaves and stems are known as *trichophyllous plants. Zizyphus* (Fig. 8.24 C), *Nerium, Calotropis procera* (Fig. 8.24 A) are important examples.

(*vii*) ***Rolling of leaves.*** Leaves in some extreme xerophytic grasses have capacity for rolling or folding. In these cases stomata are scattered only on the upper or ventral surface and as the leaves roll upwardly, stomata are effectively shut away from the outside atmosphere. This is effective modification in these plants for reducing the water loss. Sundune grass is an important example for this (Fig. 8.27).

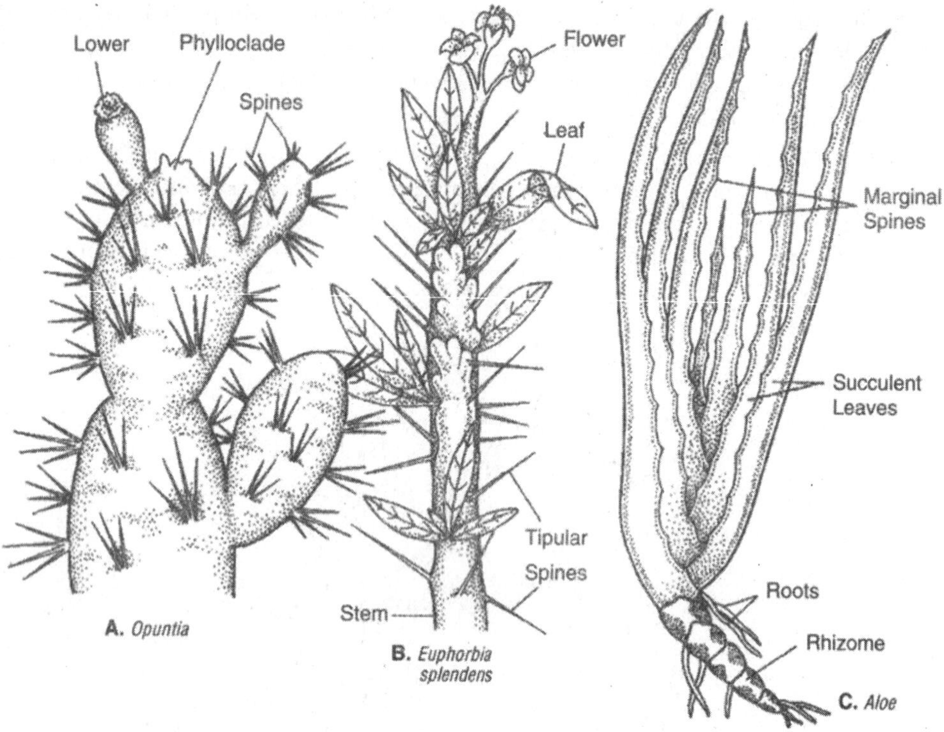

Fig. 8.23 Some succulent xerophytes.

(D) Flowers, fruits and seeds. Flowers usually develop in the favourable conditions. Fruits and seeds are protected by very hard shells or coatings.

2. Anatomical Modifications in the Xerophytes

A number of modifications develop internally in the xeric plants and all aim principally at water economy. The following are the anatomical peculiarities met within xerophytes :

(*i*) Heavy cutinisation, lignification and wax deposition on the surface of epidermis (Fig. 8.26) and even in the hypodermis are very common in xerophytes. Some plants secrete wax in small quantity but some are regular source of commercial wax. Shining smooth surface of cuticle reflects the rays of light and does not allow them to go deep into the plant tissues. Thus, it checks the heavy loss of water.

(*ii*) **Epidermis.** Cells are small and compact. It is single layered, but multiple epidermis is not uncommom. In *Nerium* leaf, epidermis is two or three layered (Fig. 8.26). In stems, the epidermal cells are radially elongated. Wax, tannin, resin, cellulose, etc., deposited on the surface of epidermis form screen against high intensity of light. This further reduces the evaporation of water from the surface of plant body. Certain grasses with rolling leaves have specialized epidermis (Figs. 8.27, 8.28). In these, some of the epidermal cells that are found in the depressions become more enlarged than those found in the ridges. These enlarged cells are thin walled and are called **bulliform cells** or **motor cells** or **hinge cells**. These are found usually on the upper surface of leaves between two parallel running vascular bundles. The highly specialized motor cells facilitate the rolling of leaves by becoming flaccid during dry periods. In moist conditions these cells regain their normal turgidity which causes unrolling of the leaf margins. Bulliform cells are of common occurrence in the leaf epidermis of sugarcane, bamboo, *Typha* and a number of other grasses.

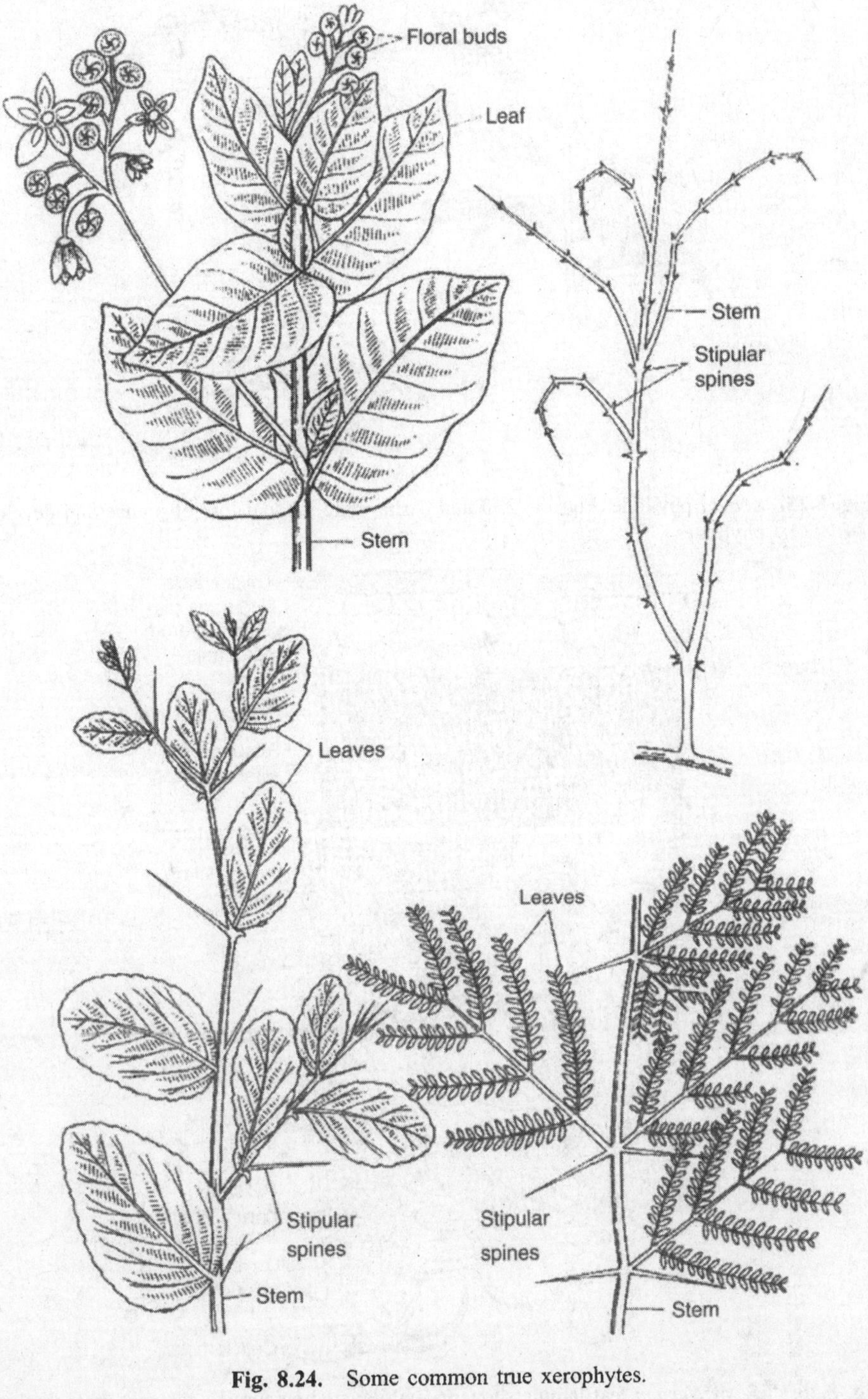

Fig. 8.24. Some common true xerophytes.

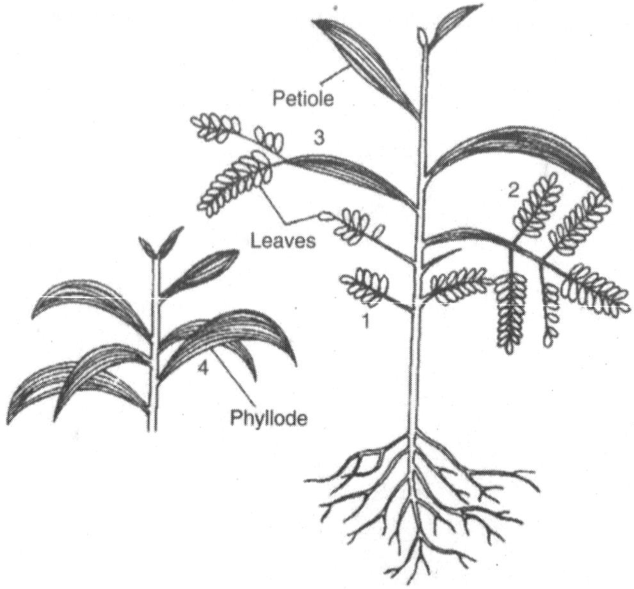

Fig. 8.25. *Acacia* phyllode. Figs. 1, 2, 3 and 4 show the gradual loss of pinnae and development of phyllodes.

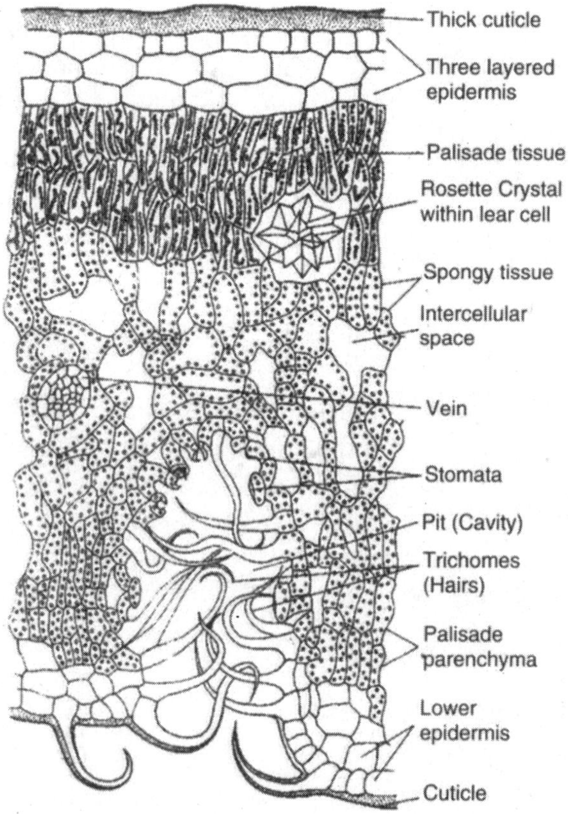

Fig. 8.26. T.S. of *Nerium* leaf lamina showing multilayered epidermis, compact mesophyll and hairy stomatal pit.

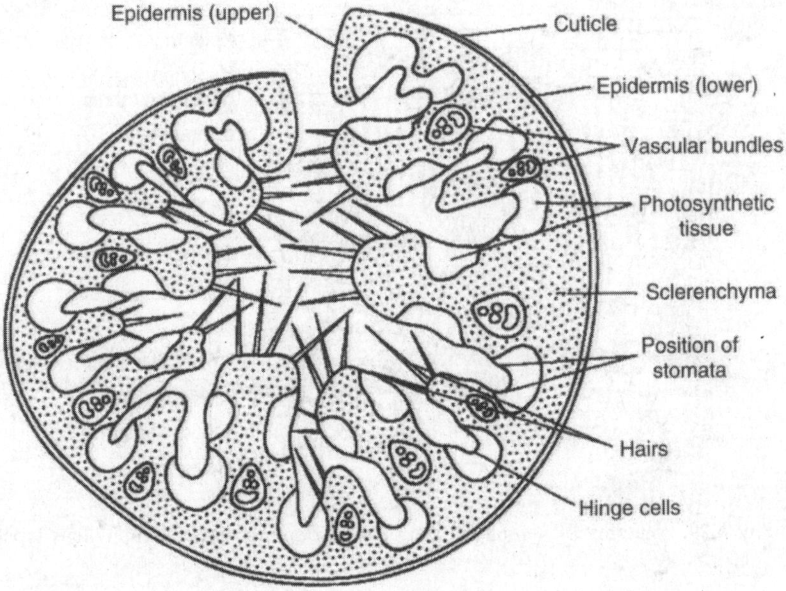

Fig. 8.27. Structure of xerophytic leaf. T.S. (diagrammatic) of *Ammophila arenaria* leaf showing protected stomata.

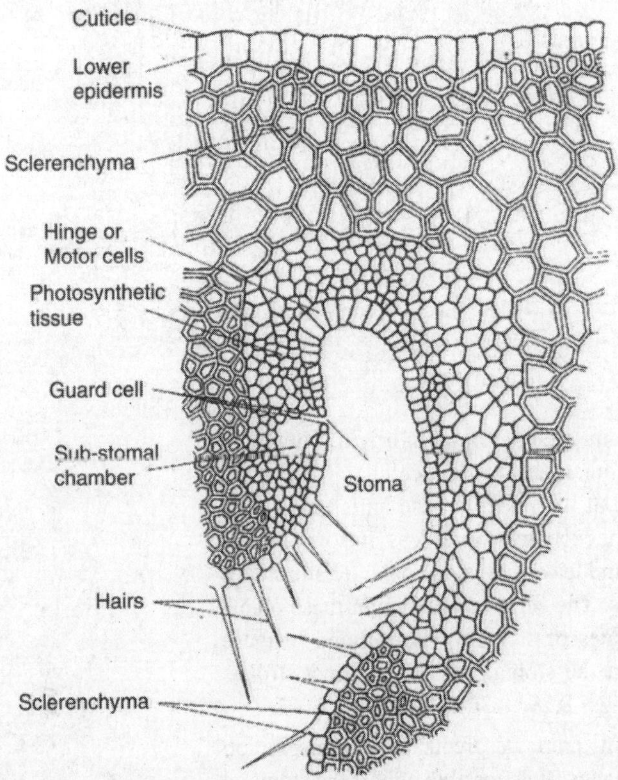

Fig. 8.28. Structure of xerophytic leaf. Part of T.S. of leaf of *Ammophila arenaria* between two ridges (detail).

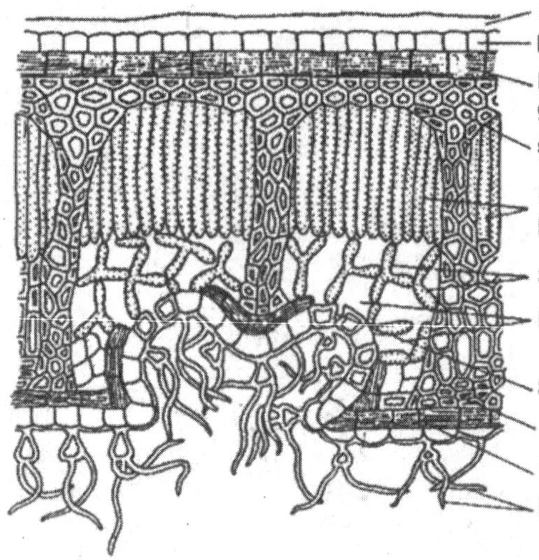

Fig. 8.29. Anatomy of xerophytic leaf. T.S. of *Banksia* leaf (sclerophyllous type).

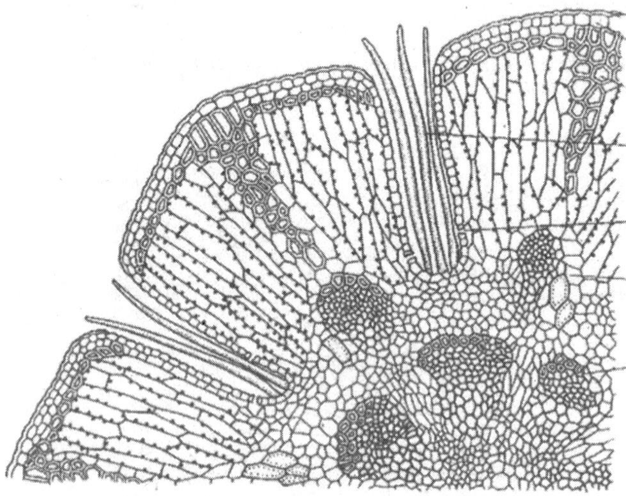

Fig. 8.30. T.S. of *Casuarina* stem.

(*iii*) **Hairs.** Hairs are epidermal in origin. They may be simple or compound, uni- or multicellular. Compound hairs are branched at the nodes. These hairs protect the stomata and prevent excessive water loss. In some plants, surfaces of stems and leaves develop characteristic ridges and furrows or pits. The furrows and pits in these plants are the common sites of stomata. Hairs found in these depressions protect the stomata from the direct strokes of strong wind (Figs. 8.29, 8.30).

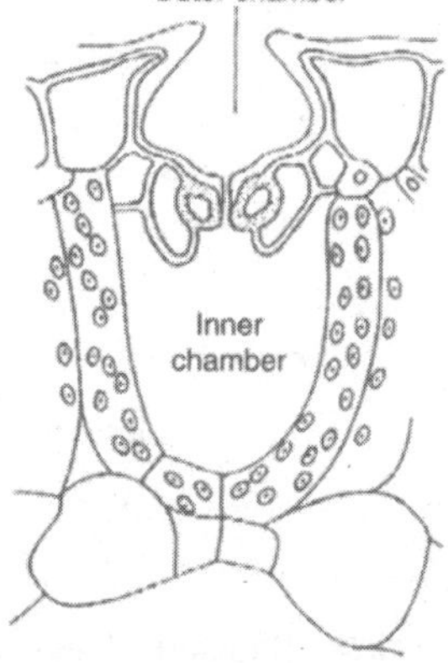

Fig. 8.31. A magnified sunken stoma.

(*iv*) **Stomata.** In xerophytes, reduction of transpiration is of utmost importance. It is possible only if the stomatal number per unit area is reduced or if the stomata are elaborately modified in their structures. In xerophytes, number of stomata per unit area of leaf is greater than in mesophytes. They are generally of sunken type. In some cases, they may be found in the furrows or pits. Subsidiary cells of sunken stoma may be of such shapes and arrangement that they form an outer chamber that is connected by narrow opening or the stoma. Such type of specialised stomata are very common in conifers, *Cycas, Equisetum,* etc. (Fig. 8.31). Walls of the guard cells and subsidiary cells are heavily cutinised and lignified in many xeric plants. These devices have little value in directly reducing transpiration when stomata are open. When the plants are wilting and stomata are closed then only lignified or cuticularized walls of guard cells have protecting properties and under such circumstances only cuticular transpiration is possible which is of little significance. In dorsiventral leaves stomata are generally found on the lower surface, but in rolling leaves they are scattered mostly on the upper surface. In the rolled leaves, stomata are protected from the direct contact of outside wind. This is very important rather secured device for lowering the rate of transpiration in xerophytic grasses.

(*v*) **Hypodermis.** In xerophytes, just below the epidermis, one or several layers of thick walled compactly grouped cells may develop that form the hypodermis. The cells may be much like those of epidermis and may either be derived from epidermis or from the cortex (in case of stem) or from the mesophyll (in case of leaf). The hypodermal cells may sometimes be filled with tannin and mucilage.

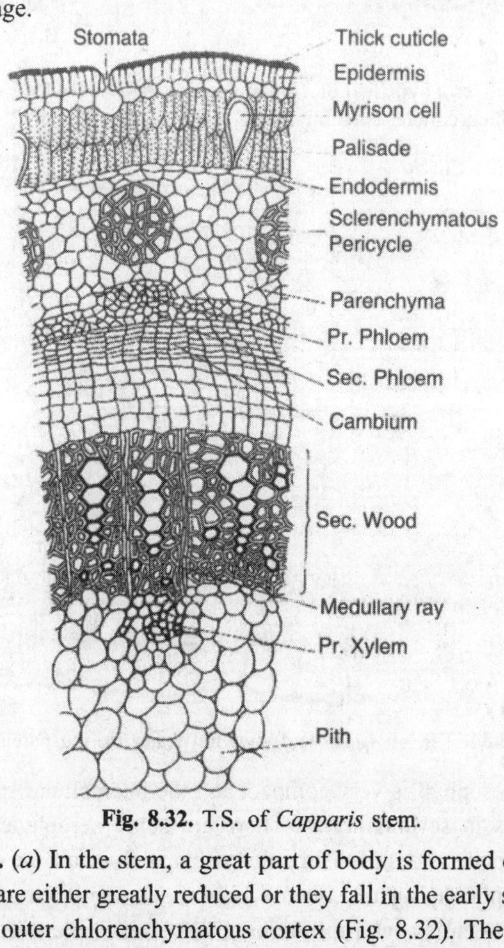

Fig. 8.32. T.S. of *Capparis* stem.

(*vi*) **Ground tissue.** (*a*) In the stem, a great part of body is formed of sclerenchyma. In those cases, where the leaves are either greatly reduced or they fall in the early season, the photosynthetic activity is taken up by outer chlorenchymatous cortex (Fig. 8.32). The chlorenchymatous tissue is connected with the outside atmosphere through stomata. The gaseous exchange takes place in regular manner in the green part of stem.

(*b*) In succulent stems and leaves, ground tissues are filled with thin walled parenchymatous cells which store excess quantity of water, mucilage, latex, etc. This makes thé stems swollen and fleshy (Figs. 8.33, 8.34).

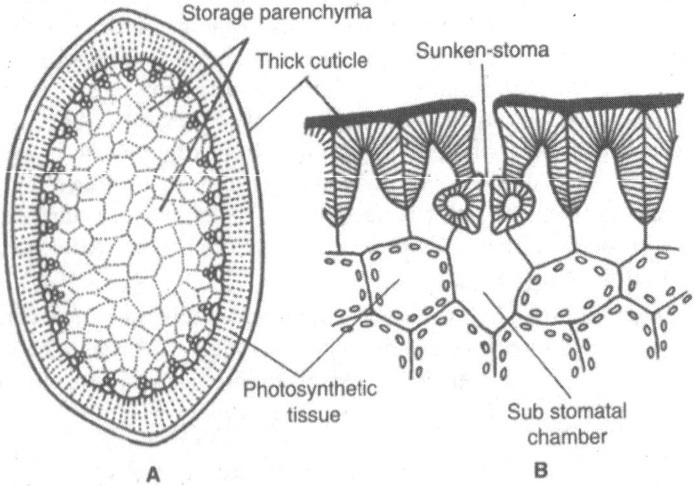

Fig. 8.33. Anatomy of xerophytic leaf of *Aloe* (monocot). A—T.S. of leaf (diagrammatic); B—a portion showing thick cuticle, epidermal cells and sunken stomata.

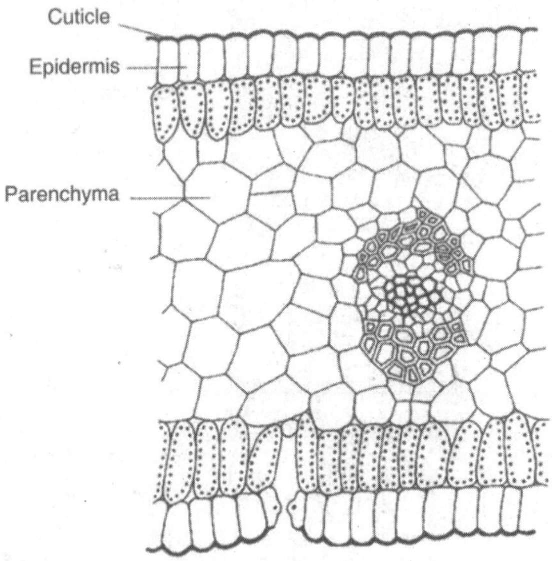

Fig. 8.34. T.S. of *Agave* leaf showing water storing parenchyma.

(*c*) In the leaves, mesophyll is very compact and the intercellular spaces are greatly reduced. Palisade tissue develops in several layers. There are some xerophytes in which mesophyll is surrounded by thick hypodermal sheath of sclerenchyma from all the sides except from below. This sheath forms a diaphragm against intense light. Such xerophytes in which sclerenchyma is extensively developed are called **sclerophyllous** plants. In succulent leaves, spongy parenchyma develops extensively which stores water (Figs. 8.33, 8.34). In *Pinus*, the spongy cells of mesophylls are star shaped (Fig. 8.36).

(*d*) Intercellular spaces are greatly reduced. Cells in the body are generally very small, thick walled and compactly grouped. They may be spherical, rounded or cuboid in shape. Such cells are very common in xerophytes. (Fig. 8.35).

(*vii*) **Conducting tissues.** Conducting tissues, *i.e.*, xylem and phloem, develop very well in the xerophytic body.

3. PHYSIOLOGICAL ADAPTATIONS

It was long assumed that the structural adaptations in the body of xerophytes were useful in reducing the transpiration but now a number of experiments related with the physiology of these plants reveal some facts which are contrary to the early assumptions. Works of Maximov support that except succulents, true xerophytes show very high rate of transpiration. Under similar conditions, the rate of transpiration per unit area in xerophytes is much higher than that in mesophyte. Stomatal frequency per unit area of leaf surface in xerophytes is also greater than that in the mesophytic leaf.

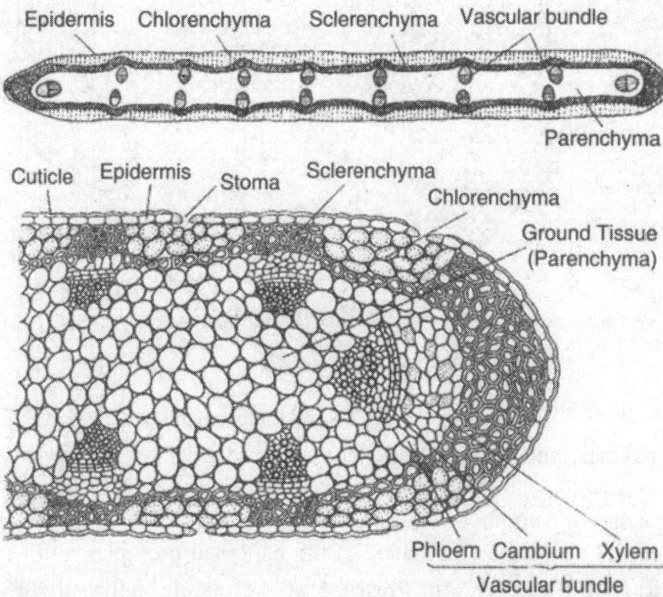

Fig. 8.35. *Homalocladium platycladum* (*Muehlenbeckia platycladus*) phylloclade (dicot); A—transection of phylloclade (diagrammatic); B—transection of phylloclade showing detail structure. The stomata, the cambium, the vascular bundles, the sclerenchyma and the chlorophyllous cells are clearly visible.

(1) Succulents are well known to contain polysaccharides, pentosans and a number of acids by virtue of which they are able to resist drought. The structural modifications in these succulent xerophytes are directly governed by their physiology. How does the succulence develop? Metabolic reaction which induces development of succulence is the conversion of polysaccharides into pentosans. Pentosans have water binding property. These pentosans together with nitrogenous compounds of the cytoplasm cause accumulation of excess amount of water in the cells and consequently the succulence develops.

(2) Another experimented fact in the physiology of succulent plants is that their stomata open during night hours and remain closed during the day. This unusual feature is associated with metabolic activity of these plants. In dark, these plants respire and produce acids. The heavy accumulation of acids in the guard cells increases osmotic concentration which, in turn, causes inward flow of water in the guard cells. When guard cells become turgid the stomata open. In

the sunlight, acids dissociate to produce carbon dioxide which is used up in the photosynthesis and as a result of this osmotic concentration of cell sap decreases which ultimately causes closure of stomata.

(3) In xerophytes, the chemical compounds of cell sap are actively converted into wall forming compounds that are finally incorporated into the cell walls. Conversion of polysaccharides into anhydrous forms as cellulose, formation of suberin, etc., are some examples.

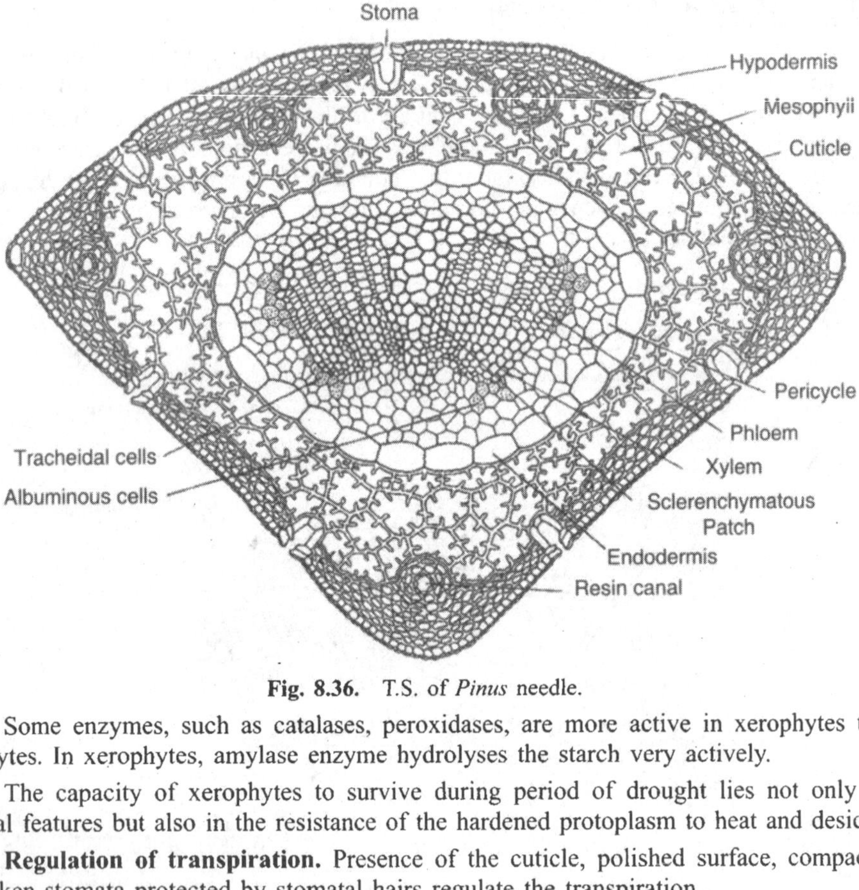

Fig. 8.36. T.S. of *Pinus* needle.

(4) Some enzymes, such as catalases, peroxidases, are more active in xerophytes than in mesophytes. In xerophytes, amylase enzyme hydrolyses the starch very actively.

(5) The capacity of xerophytes to survive during period of drought lies not only in the structural features but also in the resistance of the hardened protoplasm to heat and desiccation.

(6) **Regulation of transpiration.** Presence of the cuticle, polished surface, compact cells and sunken stomata protected by stomatal hairs regulate the transpiration.

(7) **High osmotic pressure of cell sap.** The xerophytes have very high osmotic pressure which increases the turgidity. The turgidity of cell sap exerts tension force on the cell walls. In this way, wilting of cell is prevented. High osmotic pressure of cell sap also affects the absorption of water.

MESOPHYTES

Mesophytes are common land plants which grow in situations that are neither too wet nor too dry. These plants can neither grow in water or water-logged soils nor can they survive in dry places. In other words, mesophytes are the plants of those regions where climates and soils are favourable. Vegetations of forests, meadows and cultivated fields belong to this category. The simplest mesophytic community comprises the grasses and herbs, richer communities have herbs and bushes, and the richest ones have trees (rainforests in tropics).

Mesophytes can be classified into two main community groups :

(1) Communities of grasses and herbs.

(2) Communities of woody plants.

1. Communities of Grasses and Herbs

These include annual or perennial grasses and herbs. The grasslands occur in area of approximately 25 to 75 cm rainfall per annum. They occur over large interior areas in many countries of the world such as U.S.A., Canada, Australia, Southern Russia, Africa, India. The different types of grasslands and herb communities are listed below.

(*i*) **Arctic and alpine mat-grasslands and mat-herbage.** Such communities are restricted to polar regions and mountain tops. The plants are small sized soft shrubs, and the under-shrubs are totally absent. Mosses may be intermingled but lichens do not appear. This groups is subdivided into two :

(*a*) Mat grassland (Gramineae)

(*b*) Mat herbage (Dicotyledonous herbs such as *Saxifraga, Delphinium, Potentilla, Ranunculus,* etc.).

(*ii*) **Meadow.** This forms a connecting link between mesophytes and hydrophytes as they grow in soils where moisture is 60—83%. Plants are tall perennial herbs with long stems. Soil is invisible due to overcrowding of plants. Plants are mostly rhizomatous. The leaves show mesophytic features, *i.e.,* they are thin, broad, flat and glabrous. Members of families Gramineae, Ranunculaceae, Papilionatae and Compositae are found in abundance.

(*iii*) **Pasture on cultivated land.** Vegetation is shorter and more open in pasture than in meadow. It is disturbed very often by grazing. The vegetation usually includes grasses, dicot herbs and some mosses.

2. Communities of Woody Plants (Bushland and Forests)

These are classified and described in the following ways :

(*i*) **Mesophytic bushlands.** Such a mesophytic community occurs where temperature and other conditions are not favourable for the growth of forest but they are too much favourable for mat herbage vegetation. In many places, xerophytic and mesophytic bushlands merge with each other. *Salix, Arabis, Lathyrus, Vicea,* etc., are the important plants of bushlands.

(*ii*) **Deciduous forests.** These forests are found in the areas where rainfall is high enough (about 75—150 cm per year) and evenly distributed and the temperature is moderate. Such forests are characterised by trees which become leafless for certain periods of the year. The foliage persists for about five to eight months. This phenomenon of repeated foliation and defoliation of trees is prominent in temperate and cold regions (where there is long winter) and in tropics as well where the summer is of long duration. Leaves are dorsiventral and they exhibit many shapes and structures. The trees are profusely branched. Mycorrhizae are present on the roots. Epiphytic mosses and lichens grow in abundance on the surface of the trees. The majority of the plants are pollinated by wind. The soil is very rich in microflora. The deciduous forests are named after dominant trees of those particular communities, as for example, *Quercus-Oak* forest, *Betula-Birch* forest, *Fagus-Beach* forest and so on.

Tropophytes (changing plants), an interesting group of tropical plants can be included in this group of mesophytes. Generally in tropical regions, the climate remains, more or less, uniform throughout the year but in some tropics there is alternation between damp and dry cold climates. Plants growing in the tropics of disuniform climate develop some structural modifications through which they can endure the regular cycle of favourable and unfavourable seasons in one way or the other. In other words, tropophytes behave as mesophyte during rainy season and as xerophytes during dry cold season. The shedding of leaves may occur in the beginning of winter season or in the summer. Important adaptive features of these plants are :

(*a*) better protection of winter buds

(*b*) thick bark covering on the stem

(*c*) formation of underground stem which protects the perennating buds from extreme drought and cold. Conifers are the familiar examples of tropophytes.

Fig. 8.37. Forest vegetation of the world

(*iii*) **Evergreen forests.** These forests are found in the tropical and subtropical regions extending into the cold temperate zones of southern hemisphere. Plants in these forests are evergreen (*i.e.*, they retain their leaves for more than one year until new foliage appears). Very few species in these forests may show leaffall. Evergreen forests are of three types :

(*a*) **Antarctic forests.** These forests cover mountains of New Zealand and a number of other countries in the world where annual temperature ranges from 5°C to 70°C and the rainfall is abundant throughout the year. Important plants found in these forests are conifers, Myrtaceae, Hymenophyllaceae. Mosses and Liverworts may also be present.

(*b*) **Subtropical forests.** These forests are found in the regions of fairly high rainfall but where temperature differences between winter and summer are less marked. Winter generally goes without rains. The plants are about 30 metres in height. These forests include *Oaks, Magnolias, Tamarindus* and mosses. Subtropical forests occur in eastern part of U.S.A., South Brazil, South Africa, East Australia, Southern China, and Japan.

(*c*) **Tropical Rain forests or Tropical Evergreen forests.** Tropical rain forests are found in low lying regions near the equator with annual rainfall of 180 cm or more. This type of forest is most dense and is undisturbed by biotic agencies and is therefore, called "primeval forest". The rain forests represent the climax vegetation of the whole world. Climate of such forest is characterized by :

(1) high humidity (air saturated with 95% humidity)

(2) high temperature

(3) daily rains

(4) no distinct dry season

(5) soil very rich in humus, black in colour, and porous.

The plants show luxuriant growth and they are found in several storeys. Humboldt very appropriately commented "forest is piled upon forest", *i.e.,* highest trees form top layer about 40--50 metres up, beneath which is the storey of short trees, then storey of low palms and trees, ferns, then, storey of scattered herbs and shrubs (4 to 5 metres in height). On the surface of ground may be found *Selaginella,* mosses etc. (Fig. 8.38).

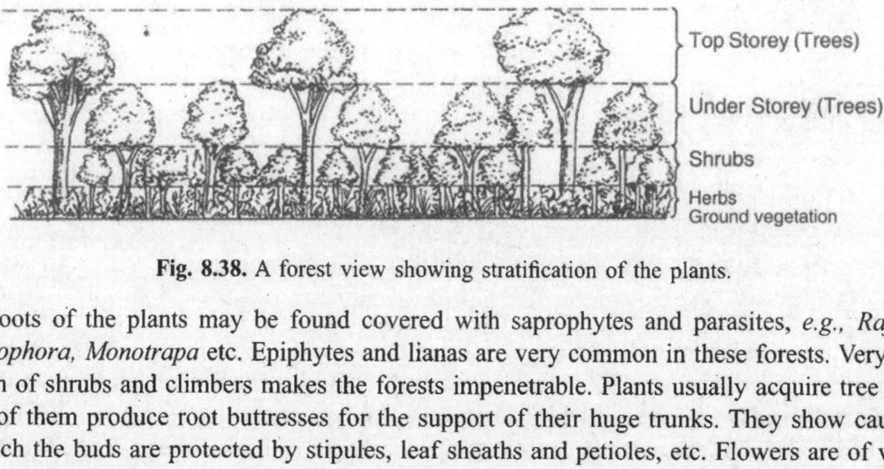

Fig. 8.38. A forest view showing stratification of the plants.

Roots of the plants may be found covered with saprophytes and parasites, *e.g., Rafflesia, Balanophora, Monotrapa* etc. Epiphytes and lianas are very common in these forests. Very dense growth of shrubs and climbers makes the forests impenetrable. Plants usually acquire tree forms. Most of them produce root buttresses for the support of their huge trunks. They show cauliflory in which the buds are protected by stipules, leaf sheaths and petioles, etc. Flowers are of various colours and they develop high over the heads. Plants do not show periodicity for foliation and flowering. Each species has its own flowering and foliation time. The leaves exhibit almost all shapes and are usually directed upwardly to drain off excess water. The leaf surface is cutinised and impregnated with silica which protects them from violent rains. Leaves may be provided with channelled nerves and dripped tips (*i.e.,* they have long and narrow apices). Trees develop thick barks.

Plants belonging to families Leguminoceae, Lauraceae, Myrtaceae, Moraceae, etc. are very commonly found in tropical rain forests.

Vegetational succession in the tropical rain forest takes place in the following sequence :

The pioneer colonisers are deciduous plants that are replaced gradually by semideciduous vegetation that persists for very short period of time after which semi-evergreen plants make their appearance. The semi-evergreen vegetation becomes intermingled with some evergreen plants which finally become dominant. In this way the climax forests develop. This sequence is possible only if the biotic factors are not allowed to affect the vegetation to a major degree. Tropical rain forests are found in central and southern America, central Africa, Pacific Islands, Malaya and in many other equatorial countries of the world. In India, these forests are found in south-eastern Himalayas, tracts of Assam, and western slopes of Nilgiri.

The tropical rain forests are of great economic values to the human beings. These yield timbers of high quality.

QUESTIONS

1. What are hydrophytes? Describe the various morphological and anatomical adaptations met within hydrophytes. **(Nagpur Univ. B.Sc. Final, Bot. II, 2010)**
2. Describe the various features of different groups of hydrophytes. Illustrate your answer with suitable examples.
3. What are xerophytes? Describe the various morphological and physiological adaptations of xerophytes.
4. Describe the main ecological adaptation of plants growing in ponds.
5. Write short notes on the following :

   Succulents, Amphibious plants, Heterophylly, Sclerophylly, Bulliform cells, Trichophylly, Diaphragm.
6. Write short notes on morphological adaptations in xerophytes. **(Nagpur Univ. B.Sc. Final, Bot. II, 2010)**
7. Enlist different plant groups based on ecological adaptations. **(Univ. of Pune, B.Sc. II, Bot. I, 2010)**
8. Write notes on adaptive features of Mesophytes. **(Univ. of Pune, B.Sc. II, Bot. I, 2010)**

9

Epiphytes

Epiphytes (*Epi* = *above*, *phyton* = plant; *i.e.*, plant growing upon plant) are those autotrophic plants which grow on the surface of some other supporting plants and are not permanently rooted in the soil. Orchids, Bo tree (Pipal tree), some algae, lichens and mosses are some of the familiar examples. These plants absorb sufficient moisture from the atmosphere and mineral nutrients from the decaying bark of the supporting plants upon which they are situated. As they are autotrophic in nutrition, they manufacture their own food (carbohydrates) from water and CO_2 in presence of sunlight. These plants differ from parasites because they do not derive nutrients and water from the living parts of supporting plants, and also they differ from lianas (woody stem climbers) because epiphytes, in the real sense of term, are not permanently rooted in the soil. Epiphytes are also called *Aerophytes* or air plants.

DISTRIBUTION

Some epiphytes grow on the surface of submerged aquatic plants, while others may be aerial. Some are found growing on the surface of tree trunks, some on the horizontal forks of the trees and some may grow even on the surface of leaves (*i.e.*, epiphyllous epiphytes). Some epiphytes show specificity in selection of their supporting plants. *Tortula pagorum*, an epiphytic moss, is peculiar in the sense that it grows on the tree trunks within the urban limits. This moss grows well in the city atmosphere presumably because it requires high temperature and smoky air for its normal growth. Both these factors are available to the plants in the city area. Some epiphytic species may often grow on rocks, and some may grow rarely even on the poles and horizontal telephone wires.

Epiphytic vegetation is very rich in moist and cold regions but poor in dry and cold areas. In north-western Himalayas the epiphytic species are much less in number as compared to those present in the eastern Himalayas. In warm and wet regions, members of the families Bromeliaceae and Orchidaceae are found in abundance. In tropical rain forests, epiphytic species found at the tops of trees are xerophytic in nature but those occurring at lower levels are hygrophilous (moisture and shade loving).

IMPORTANT FEATURES

Since the epiphytes depend directly for their water supply on rains, atmospheric moisture, snow and dew, they develop certain structural adaptations for water storage and for reducing excessive water loss. The important features are listed below :

1. Morphological Features

(*i*) **Root system.** In the epiphytic vascular plants, the root system is extensively developed. In these cases, the roots may be of the following three types (Fig. 9.1) :

(*a*) **Normal absorbing roots.** Which absorb water, minerals, and organic nutrients from the moist crevices of decaying barks of supporting plants.

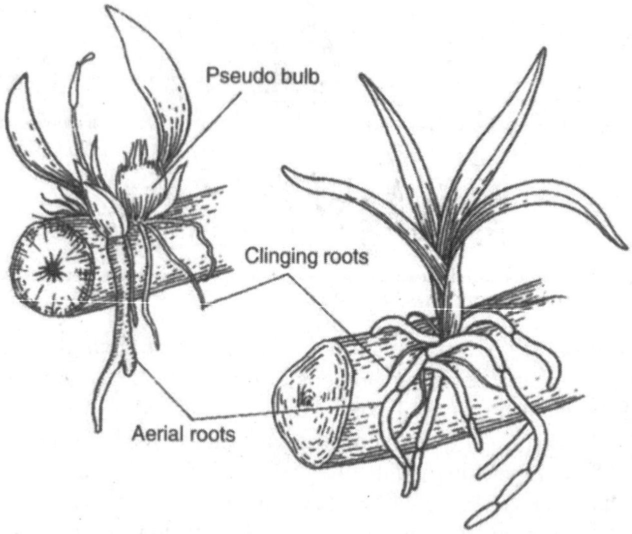

Fig. 9.1. Different types of roots of epiphytes.

(*b*) **Clinging roots.** These roots fix the epiphytes on the surface of the supporting object firmly and also absorb nutrients from the humus and dust that are accumulated on the surface of bark.

(*c*) **Aerial roots.** These are spongy and green roots which hang downwardly in the atmosphere and absorb moisture from the air. These roots can photosynthesize in light because of the presence of green colour in them.

In some epiphytes, the roots collect on their surface good amount of dust that holds water which will finally be absorbed by the roots.

(*ii*) **Stem.** Stem in epiphytic vascular plants may or may not be extensively developed. Some epiphytes develop succulence in their stems and become pseudobulbous or tuberous (Fig. 9.1).

(*iii*) **Leaves.** The majority of epiphytes show considerable reduction in leaf number. Some orchids develop only a single leaf in a growing season. Leaves in some may be fleshy and leathery. In *Dischidia nummularia, Platycerium* and *Asplenium nidus* leaves are modified into the pitchers.

Dischidia nummularia, an epiphytic species of family Asclepiadaceae, growing very commonly in Sunderban shows peculiar types of leaf pitchers. The pitchers have openings through which the adventitious roots enter inside. The roots branch copiously into a number of very delicate rootlets which spread on the entire inner surface of pitcher and form a network (Fig. 9.2). The inner surface of pitcher is coated with wax. Pitcher collects and accumulates rain water, humus and minerals that are absorbed by the root network. Sometimes ants and insects enter the cavity of the pitcher through hole where they may be killed and digested. The dead remains of animals serve as nitrogen source for the plants. Myrmecophily which is a sort of symbiotic association between ants and plants is of common occurrence in the epiphytic vegetation. In the family Bromeliaceae, some species develop spoon-like leaves in rosettes. These leaves collect and store rain water which is finally absorbed by the epidermal hairs present on the concave surface of the leaves.

(*iv*) **Fruits, Seeds and their dispersal.** The fruits and seeds are usually dispersed by wind, insects and birds. When the seeds reach the suitable surface and get favourable environment, they germinate over there and give rise to new independent epiphytes.

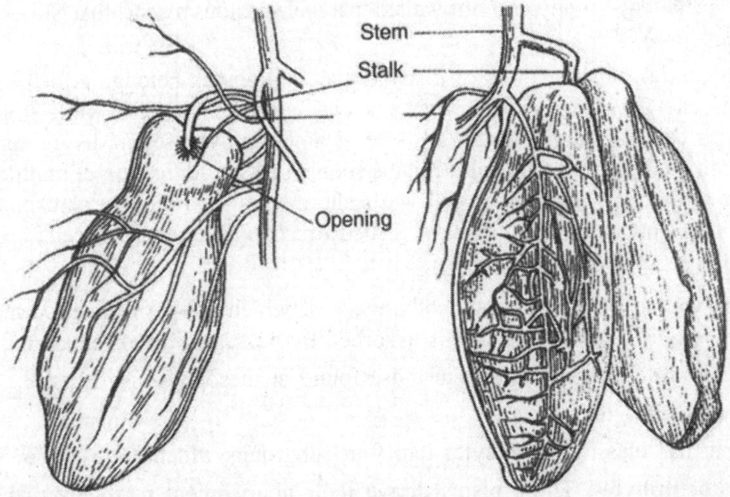

Fig. 9.2. Leaf pitchers of *Dischidia nummularia*. Right pitcher is cut open longitudinally to show network of roots on its inner surface.

2. Anatomical Features

Important anatomical peculiarities in epiphytes are as follows :

(*i*) Presence of thick cuticle and sunken stomata : These two structures greatly reduce the loss of water from the plants. Generally the surface cell of water absorbing organs (roots and some leaves) are not cuticularized.

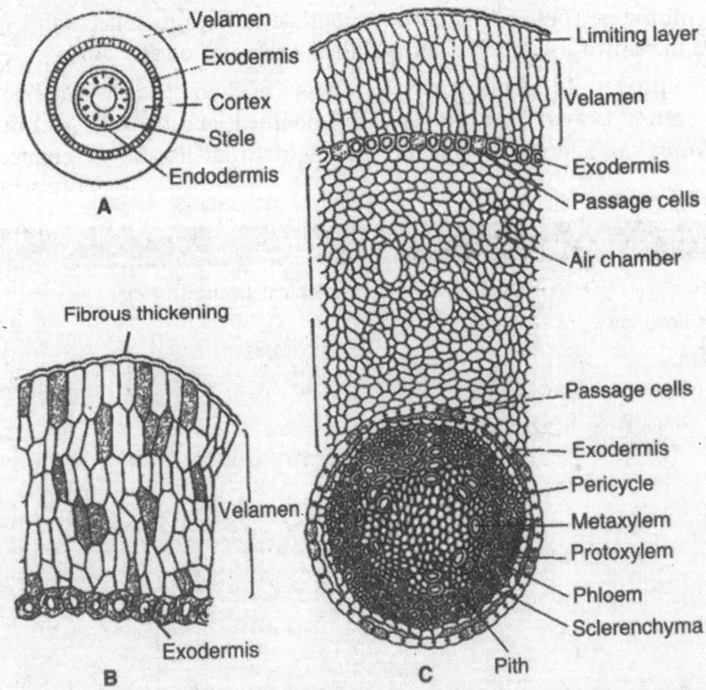

Fig. 9.3. Anatomy of Orchid root (Monocot, Orchidaceae). A. T.S. of root (diagrammatic); B. velamen and exodermis enlarged; C. T.S. of *Dendrobium* (orchid) root showing detailed structure.

(*ii*) In succulent epiphytes, thin-walled parenchymatous tissue that stores water, develops extensively.

(*iii*) The aerial hanging roots of many tropical epiphytes belonging to the families Araceae and Orchidaceae develop on their surface a characteristic greenish white thin-walled massive tissue. That is called velamen (Fig. 9.3). The velamen is hygroscopic tissue that rapidly absorbs moisture from the saturated atmosphere like a sponge. It is modification of multilayered epidermis. Its cells are empty (*i.e.*, dead) and cell walls show spiral or reticulate thickenings. Inner to the velamen there is present a peculiar layer called *exodermis.* Exodermal cells are of two types :

(*a*) Lignified and thick walled cells.

(*b*) Thin-walled cells or passage cells, walls of which are permeable to water. The velamen absorbs and retains moisture till that is absorbed by passage cells of exodermis.

(*iv*) Other structures are similar to those found in mesophytes.

Types of Epiphytes

Schimper has classified epiphytes into four subgroups which are as follows :

(1) **Protoepiphytes.** These plants derive their nourishment partly from the surface of the supporting plants and partly from the atmosphere. They do not develop any adaptive feature in them except perhaps, aerial roots with velamen. Examples : *Peperomia, Dischidia* and some ferns belong to this group.

(2) **Hemiepiphytes.** These plants grow on the supporting plants in the beginning like true epiphytes but later on they establish connection with the soil by their roots. Epiphytic Fig trees, some root climbing Aroids, *Scindapsus officinalis,* etc. are important plants of this group.

Some stem climbing plants grow in the soil but their stems die from below upward and terminal portions live independently like hemiepiphytes. Such plants are termed as Pseudoepiphytes.

(3) **Nest epiphytes.** These plants have appropriate devices to collect large quantity of water and humus for their own use. Orchids are familiar examples of this group.

(4) **Tank epiphytes.** These plants develop fibrous anchoring roots which do not take part in the water absorption. Leaves, that are variously modified, absorb water and manufacture food. *Nidularium, Tillandsia,* and other epiphytic species of Bromeliaceae are common plants of this group.

QUESTIONS

1. What are epiphytes? Describe the various morphological features of epiphytes.
2. Write short notes on :
   (*i*) Velamen
   (*ii*) Root system of epiphytes.
3. What are various types of epiphytes ?

10

Halophytes and Mangrove Vegetation

Some plants grow and complete their life cycle in the habitats with a high salt content. They are known as salt plants or halophytes. According to Stocker (1933), the critical level of salinity for plants is 0.5% of the dry weight. Though the fact that only a small group of higher plants can grow in the saline habitats was recognized many hundred years ago, yet the name "halophyte" was assigned to such plants by Pallas in the early nineteenth century. Such plants are commonly found near sea-shores where mesophytes and fresh water hydrophytes cannot thrive well. Although these plants grow in the areas which are well saturated with water yet they cannot avail of the water because of high concentration of salts in the soils. Thus, the halophytes are plants of physically wet but physiologically dry habitats. Plants cope with the problems of salinity in various ways, some of them avoid salinity, some evade salinity or resist it, and a few others tolerate salinity. Salinity avoidance is usually accomplished by limiting germination, growth and reproduction to specific seasons of the year as well as by growing roots into non-saline layers and limiting salt uptake. Plants evade or resist salinity by accumulating salts in their cells as well as by secretion of excess salts. In the salt tolerants, the protoplasm functions normally and endures a high salt concentration without apparent damage.

Plants occupying only local non-salty ecological niches in an overall saline environment or those which appear in such habitats only for short periods, *i.e.*, during rainy season are called psuedohalophytes or false halophytes (Yoave Waisel, 1972). Saline habitats are not restricted to sea coasts only but they may also be found in many dry places far away from the sea coasts. In India, certain places in Rajasthan and in many other deserted places, the soils are very salty because of presence of sodium chloride, calcium sulphate, sodium bicarbonate, potassium chloride, etc. In these dry and salty habitats generally succulent xerophytes, *e.g., Chaenopodium album, Suaeda fructicosa, Haloxylon salicorneum, Salsola foestida, Tamarix articulata* grow very successfully and form microedaphic formations.

CLASSIFICATION OF HALOPHYTES

Stocker (1933) proposed first classification of saline habitats which is as follows :

1. Aquatic-haline
2. Terrestro-haline
   (*a*) hygrohaline
   (*b*) mesohaline
   (*c*) xerohaline
3. Aero-haline

(*a*) Habitats affected by salt spray (maritime)

(*b*) Habitats affected by salt dust (salt desert)

On the basis of contact between salt and plant in different habitats halophytes can be grouped into the following categories :

1. In terrestrial saline habitat, the contact occurs between the plant roots and soil—*terrestrial halophytes.*
2. In aquatic-haline habitat contact occurs :

   (*a*) between salt and plant roots—*emerged halophytes* or *hygrohalophytes,* and

   (*b*) between salt and entire plant body—*submerged halophytes* or *hydrohalophytes.*
3. In aero-haline habitat, the contact occurs:

   (*a*) between aerial organs of plants and air borne salt droplets (in coastal regions), and

   (*b*) between aerial organs and salt droplets (in dust deserts)—*Aerohalophytes.*

Iversen (1936) classified the haline habitats on the basis of their salt contents. Different habitats and plants found therein are given below:

Habitat	Salt content	Plants
1. Oligohaline	0.01—0.1% NaCl	Oligohalophytes
2. Mesohaline	0.1—1.0% NaCl	Mesohalophytes
3. Polyhaline	1% NaCl and up	Euhalophytes

Besides above-mentioned three types of halophytes, there are some other halophytes which grow in habitats with wider salinity ranges. Plants occurring in both oligohaline and mesohaline habitats are called **oligo-mesohalophytes** and those existing in all the three types of habitats are called **euryhalophytes.**

According to Chapman (1942), halophytes have been classified into the following categories :

(*i*) **Miohalophytes**—Plants growing in the habitats of low salinity (below 0.5% NaCl).

(*ii*) **Euhalophytes**—Plants of highly saline habitats. They have been further sub-divided into the following groups :

(*a*) **Mesohalophytes**—Plants of habitats with salinity range of 0.5 to 1%.

(*b*) **Mesoeuhalophytes**—Plants of habitats with salinity range of 5% and higher.

(*c*) **Eneuhalophytes**—Plants of habitats with salinity range of 1% and up.

Van Eijk (1939) classified halophytes into the following two main categories on the basis of their distribution and their responses to saline habitats.

(*i*) **Salt enduring halophytes** which show optimum development in non-saline habitats but can tolerate salts.

(*ii*) **Salt resistant halophytes** which show optimum development in saline habitats.

Tsopa (1939) classified halophytes into the following four groups on the basis of their response to salinity.

1. **Obligatory halophytes**—Plants requiring salinity throughout their life.

2. **Preferential halophytes**—Plants show optimum growth in saline habitats, despite their appearance in non-saline habitats.

3. **Supporting halophytes**—Non-aggressive plants which are capable of growing in saline habitats.

4. **Accidental halophytes**—Plants which grow in marsh saline habitats only accidentally.

Steiner (1935) classified salt marsh plants into the following three types:

(*a*) **Succulent halophytes**—Plants which can tolerate high concentration of chloride in their cell sap due to increased succulence (as for example, *Salicornia herbacea).*

(*b*) **Non-succulent halophytes**—Plants resisting salts by desalinization of their tissues and secreting excess salts through salt glands (*e.g., Spartina alterniflora*).

(*c*) **Accumulating type**—Plants without any special mechanism of salt removal. Salt concentration in such plants goes on increasing until the death of plants (*e.g., Juncus gerardii, Suaeda fructicosa*).

Coastal angiospermic halophytes have been divided into the following groups :

(*i*) Submerged marine halophytes—*hydrohalophytes,*

(*ii*) Low coast plants—*hygrohalophytes,*

(*a*) *Swamp halophytes,* the **mangrove**, and

(*b*) *Marsh halophytes,*

(*iii*) High coast plant —*aerohalophytes.*

Mangrove is a West Indian name given to formation of trees and shrubs inhabiting the coasts and river estuaries of tropical or subtropical seas. Plants occurring in mangrove vegetation belong to several families. Common mangrove plants are *Rhizophora mucronata, Kandelia rheedii, Ceriops roxburghiana, Bruguiera gymnorhiza,* and many other plants belonging to the family Rhizophoraceae, *Avicennia officinalis* of the family verbinaceae, *Acanthus ilicifolius,* (Acanthaceae), *Ardesia humiles* and *Aegiceras majus* of family Myrisinaceae, *Cassula species* (a very common plant of rice field in northern India belonging to the family compositae), *Cometes surathensis* (Caryophyllaceae) which is commonly found in western Asia, *Avicennia, Sechium edule* (Aroideae)—an Indomalayan plant, *Cryptaeorine ciliata* (Aroideae), a common plant growing in the muddy banks of Ganga, *Poa, Festuca, Melacanna, and Oryza patnai* (rice) of family Gramineae, *Scirpodendron* of Cyperaceae and a number of other plants belonging to the families Palmae, Parkeriaceae Chaenopodiaceae, Plumbaginanceae, Ficoideae, Euphorbiaceae.

Mangrove plants are very commonly found on some saline soils of Indogangetic plains (Bay of Bengal, Sunderbun and Assam), in western India near the sea coasts of Mumbai and Kerala, in the banks of **Gaumati** and Godavari in south India,, particularly in the regions where rivers meet the ocean, and in Andaman and Nicobar Islands.

Mangrove vegetation of Gangetic estuary, particularly of Sunderbun region, according to Prain, is localized in the following three geographical zones :

1. Southern coastal strip and south-western part,
2. Central zone, and
3. North-eastern part.

The Central zone is characterized by climax forest of *Heritiera. Ceriops* is one of the dominant plant species on the higher grounds near the sea and at certain places it may form pure forest in Sunderbun.

Mangrove vegetation in the western sea coast of India is represented by the following plant communities.

(*i*) *Avicennia alba* and *Avicennia officinalis* community.

(*ii*) *Acanthus ilicifolius* and *Avicennia alba* community.

(*iii*) *Suaeda fructicosa* populations,

(*iv*) Pure community of *Salvadora* species,

(*v*) *Sesuvium portulacastrum* population, and

(*vi*) *Aeluropus repens* population.

Godavari delta in Andhra Pradesh shows characteristic mangrove vegetation of *Avicennia alba, Avicennia marina, Rhizophora, Bruguiera, Ceriops, Sonneratia, Acanthus ilicifolius, Myriostachya weightiana, Clerodendron inerme,* etc.

The ecological conditions which are essential for the development of mangrove vegetation or halophytes are :

(*a*) shallow water with thick mud,

(*b*) water logged saline soil or sandy or loose soil or heavy clays containing large amount of organic matter,

(*c*) high rainfall, and

(*d*) high humidity in the atmosphere and cloudy weather.

IMPORTANT CHARACTERS OF HALOPHYTES

As the water in the habitat is not such as can easily be absorbed by the plants, the halophytes develop in them almost all important xerophytic devices for water economy.

1. **Habit.** A great majority of halophytes in the tropical and subtropical regions are shrubs, but a few of them are herbaceous, for example, *Acanthus ilicifolius*. In temperate zones, halophytic vegetation is purely herbaceous.

The shrubs are generally dome-shaped in appearance because of their cymose branching.

2. **External morphology.** (*a*) **Roots** (*i*) Halophytes develop many shallow normal roots. In halophytes, in addition to normal roots, many stilt or prop roots develop from the aerial branches of stem for efficient anchorage in muddy or loose sandy soil. These roots grow downward and enter the deep and tough strata of the soil. In some plants, *e.g., Rhizophora mucronata,* the stilt roots may be strong and extensively developed, but in others they may be poorly developed (*Rhizophora conjugata*). In some plants, the stilt roots may not at all develop (Fig. 10.1).

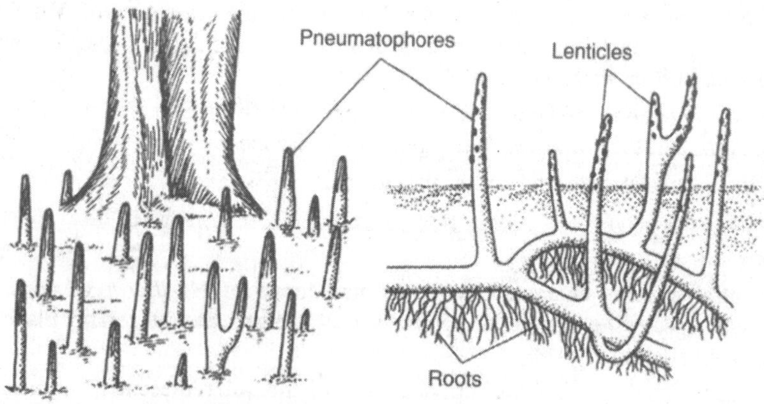

Fig. 10.1. Pneumatophores of mangrove plant.

(*ii*) Sometimes, a large number of adventitious root buttresses develop from the basal parts of tree trunks. These root buttresses provide sufficient support to the plants.

(*iii*) The soil in coastal region is poorly aerated and it contains very small percentage of oxygen because of water logging. Under such conditions, the roots of halophytes do not get sufficient aeration. In order to compensate this lack of soil aeration, the hydrohalophytes develop special type of negatively geotropic roots, called *pneumatophores* or breathing roots (Fig. 10.2). The pneumatophores usually develop from the underground roots and project in the air well above the surface of mud and water. They appear as peg-like structures. The tips of these respiratory roots may be pointed. They possess numerous lenticels or pneumathodes on their surface and prominent aerenchyma enclosing large air cavities internally. The gaseous exchange takes place in these roots through the lenticels. The aerenchyma helps in the conduction of air down to the subterranean or submerged roots. In some plants, *e.g., Bruguiera,* the horizontal roots grow above the surface of mud and then again bend downwardly and enter deep into the mud. In this way, they form knee-like structures. The aerial surface bears a number of pores which facilitate the exchange of gases (Fig. 10.3). Pneumatophores do not develop in some species of *Rhizophora.* In those cases, the upper aerial parts of descending stilt roots probably take up the respiratory activity.

Fig. 10.2. Mangrove forests (Sunderbans)

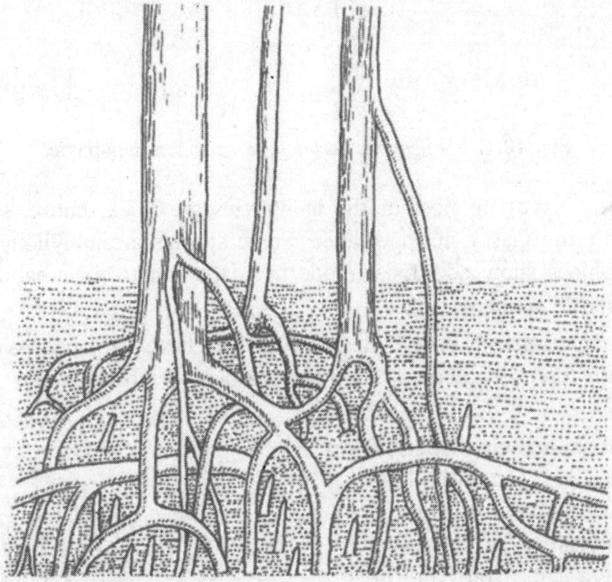

Fig. 10.3. Supporting or stilt roots of mangrove plants developing from the trunk.

(*b*) **Stem.** Stems in several halophytes develop succulence. *Salicornia herbacia* (Fig. 10.4) and *Suaeda maritima* may be quoted as familiar examples for it. According to Arnold (1955) the succulence depends on the ratio of absorbed to free ions in the plant cells rather than absolute amounts of sodium chloride or sulphate present. Succulence is induced only after the accumulation of free ions in an organ increases above a critical level. According to Pokrovskaya (1954, '57) salinity inhibits the cell division and stimulates cell elongation. Such effects cause decrease in the cell number and increase in cell size, so typical of succulents. Repp *et al.* (1959) are of the opinion that succulence is directly correlated with salt tolerance of plants and the degree of their development can serve as an indicator of the ability of plants to survive in highly saline habitats. The temperate halophytes are herbaceous, but the tropical ones are mostly bushy and show dense cymose branching. Submerged marine angiosperms are among the very few species of halophytes that do not become succulent.

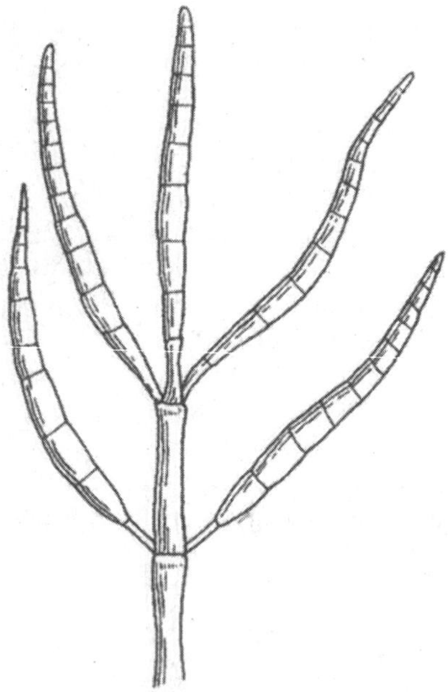

Fig. 10.4. *Salicornia herbacia,* a succulent halophyte.

(*c*) **Leaves.** The leaves in most of the halophytes are thick, entire, succulent, generally small-sized, and are often glassy in appearance. Some species are aphyllous. Stems and leaves of coastal aerohalophytes show additional mode of adaptation to their habitats. Their surfaces are densely covered with trichomes.

Leaves of submerged marine halophytes are thin and have very poorly developed vascular system and frequently green epidermis. They are adapted to absorb water and nutrients from the medium directly.

(*d*) **Fruits, Seeds and their dispersal.** The fruits and seeds are generally light in weight. Fruit walls have a number of air chambers and the fruits, seeds, and seedlings which can float on the water surface for pretty long time are dispersed to distant places by water current. Mangrove vegetations of tropical sea-shores from Australia to East Africa include approximately the same species of plants. Similarly, the mangroves of West Asia show considerable resemblances with those of East Asia and East Africa. It is due, in part, to the fact that medium and temperature remain uniform throughout and partly due to the efficient means of dispersal or migration of plants. A littoral species of *Spinifex* (*S. quarrosus*), a member of Gramineae commonly growing in the sandy saline sea shores in Andhra shows peculiar type of fruit

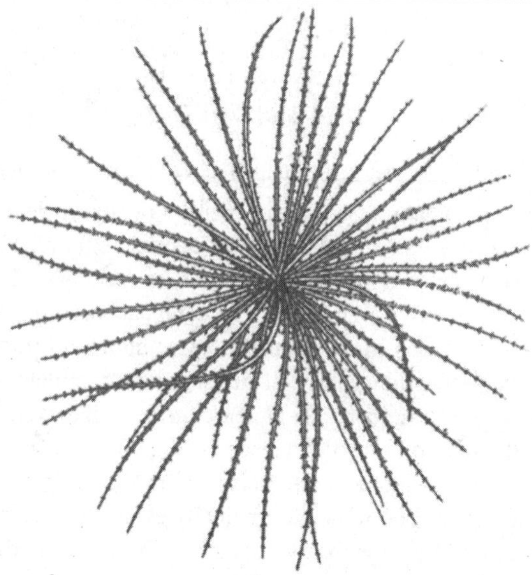

Fig. 10.5. Inflorescence of *Spinifex quarrosus.*

dispersal. In this plant, female inflorescence is spherical in shape and consists of many spikelets (Fig. 10.5). A number of stiff bractiolar bristles of the inflorescence help in the dancing and somersaulting of the inflorescence. When the seeds mature the globular and hairy inflorescence becomes bodily detached from the creeping plant and trails on the sandy substratum dropping its seeds at places. Finally the inflorescence with rest of fruits becomes buried in the mud.

(*e*) **Viviparous mode of seed germination.** Halophytes or mangrove plants growing in the tidal marshes are met with the phenomenon of 'vivipary' which is defined as the germination of seeds while the fruits are still attached to mother plants (Fig. 10.6).

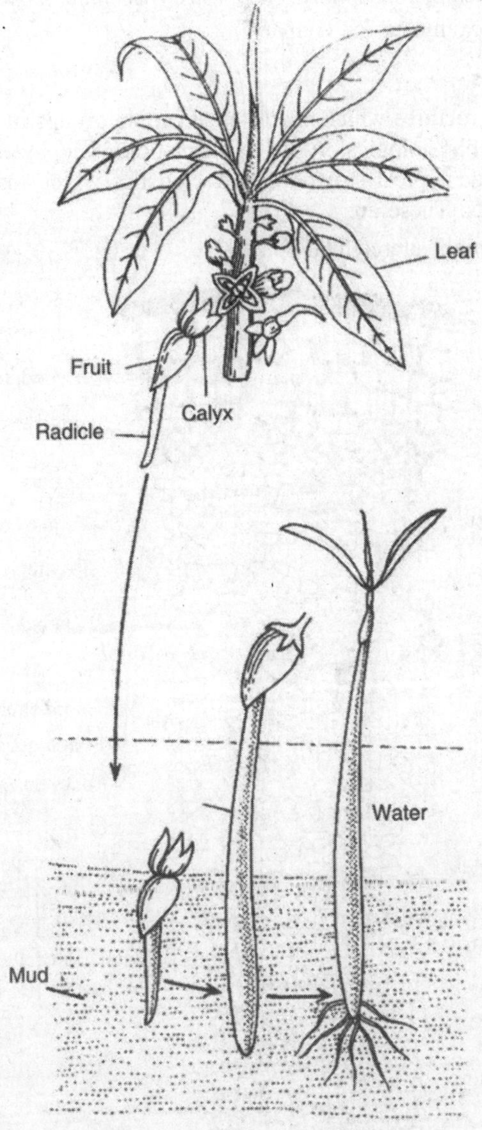

Fig. 10.6. Vivipary in *Rhizophora.*

In *Rhizophora* plants, when the embryo reaches advanced stage of development the massive club-shaped hypocotyl and terminal radicle pointing downwardly emerge out of the fruit. When the hypocotyl attains a length of several centimetres (about 50-80 cms), the seedling falls vertically down. Thus, the radicle and a part of hypocotyl become fixed in the mud and the remaining upper part of hypocotyl along with other embryonal parts, such as plumule and cotyledons remains above the surface of mud or water. Within a few hours the radicle develops a tuft of roots and plumule also starts growing rapidly. Sometimes, seedlings fall in deep water and they float on the surface of water vertically with hypocotyl pointing downwards. On reaching shallow areas, the radicle becomes fixed in the soft mud and the plant starts growing rapidly. It is noticed that high degree of salinity in the soil or water checks the germination of seeds. So the viviparous germination is a very significant adaptation in these plants to avoid the retarding effects of salinity on seed germination. Species of *Rhizophora, Aegiceras, Avicennia, Cassula, Ranansatia vivipara* are some of the common examples for vivipary.

3. Anatomical Features

The appearance and structures which characterise certain groups of plants sum up to a great extent their ecological and physiological means of adaptation. Halophytes are no exception to this rule because of specific and typical structural characteristics which make them distinguishable from other groups of plants. These are :

1. Large cells and small intercellular spaces,

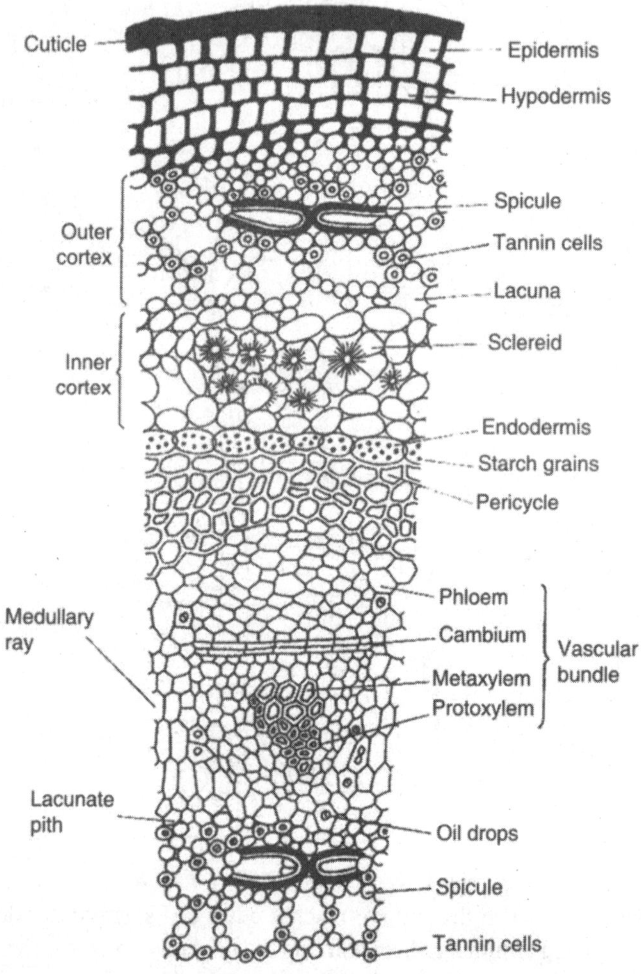

Fig. 10.7. *Rhizophora mucronata.* Transverse section of young stem.

2. High elasticity of the cell walls,
3. Extensive development of water storing tissues,
4. Smaller relative surface area (surface/volume ratio),
5. Small and fewer stomata, and
6. Low chlorophyll content.

Anatomy of halophytes reveals a number of xerophytic features in them. These are as follows :

(*i*) Presence of thick cuticle on the aerial parts of the plant body. The epidermis of xerosucculents and coastal halophytes is characterised by a cover of waxy layers in addition to thick cuticle (Uphof, 1941) (Fig. 10.7).

(*ii*) Leaves may be dorsiventral or isobilateral. They develop protected stomata which are not deeply sunken. Epidermal cells are thin-walled. The palisade consists of several layers of narrow cells with intercalated tannin and oil cells (Fig. 10.8).

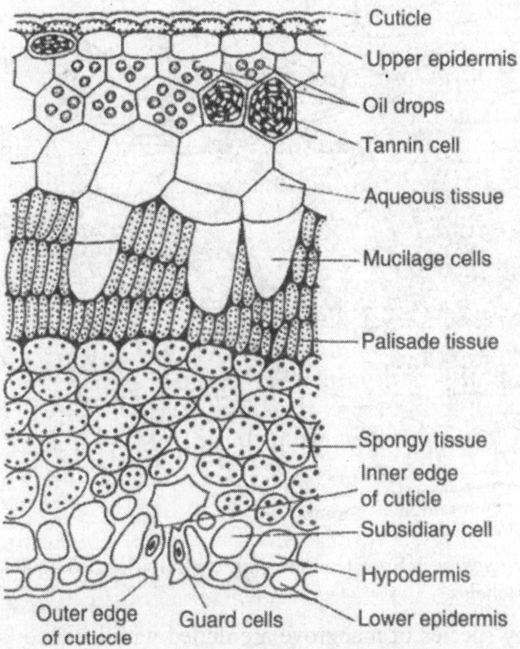

Fig. 10.8. *Rhizophora mucronata.* V.S. of leaf showing detailed structure.

(*iii*) Stems in the succulent plants possess thin-walled water storing parenchyma cells in them. Mucilage cells may be found in abundance. Epidermal cells of various mangrove species contain large quantities of tannins and oil droplets. Cortex is fleshy, several cells thick and in old stems it may become lacunar. Salinity causes extensive lignification of stele (Fig. 10.7).

(*iv*) The leaves and stems of coastal halophytes are abundantly covered with various types of simple and branched trichomes giving the plants a greyish appearance (Fig. 10.9). The trichomes may exert a protective function in plants by :

(*a*) affecting water economy,

(*b*) affecting the temperature of the leaves, and

(*c*) preventing sea water droplets from reaching the live tissues of leaves.

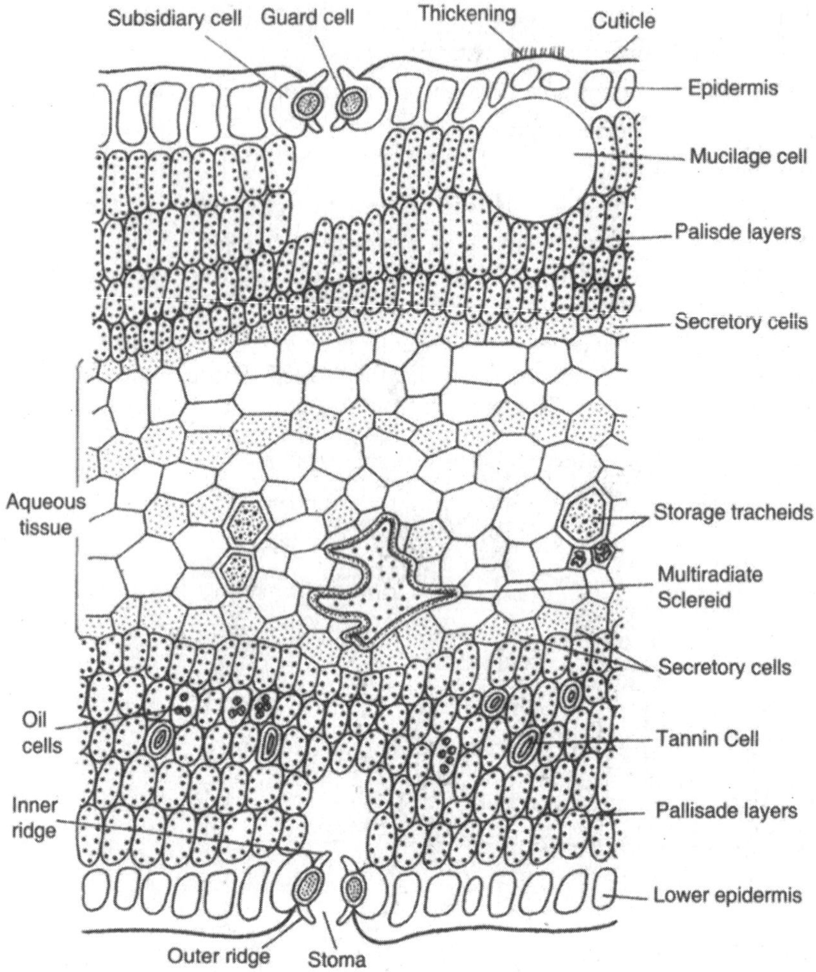

Fig. 10.9. *Sonneratia apetala.* V.S. of leaf showing centrally located aqueous tissue with sclereids and storage tracheids.

(*v*) Leaves of many species of mangrove are dotted with local cork formation "cork warts". Leaves of *Sonneratia* and *Aegiceras* and *Nitraria* (a desert shrub), *Suaeda monoica* contain well developed aqueous tissue (Figs. 10.8, 10.9, 10.10) (Mullan, 1932). Salt secreting glands may be found in some halophytes.

(*vi*) The stilt roots of mangrove plants show normal features with periderm on the surface, aerenchymatous cortex containing sclereids, normal endodermis, secretory pericycle, radially arranged xylem and phloem and extensively developed pith (Fig. 10.11). Pneumatophores develop a number of lenticels on their surface. The cortex is spongy and consists of extensively developed aerenchyma enclosing large air chambers. Highly developed air chambers are continuous with the stomata of leaves and with the cortex and primary phloem of the stems. Pneumatophores show variations in their internal structures. Generally, they show conjoint, collateral vascular bundles with endarch xylem at maturity. Very young pneumatophores, however, show root features, *i.e.,* exarch xylem and radial arrangement of vascular tissues (Fig. 10.11). Generally the negatively geotropic breathing roots show features of stem and not of roots (Fig. 10.12).

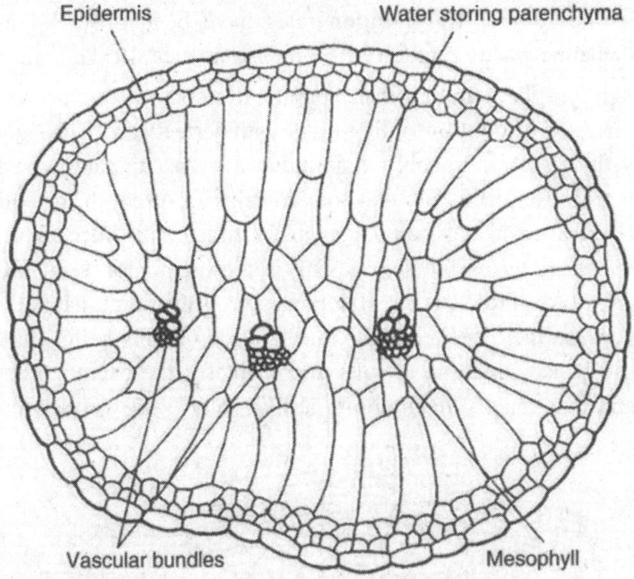

Fig. 10.10. T.S. of succulent leaf of *Suaeda monoica* showing water storing parenchyma.

4. Physiological Adaptations in Halophytes

Morphology and anatomy of the halophytes clearly show xeromorphic features in them. Now, these plants are growing in an environment where water is available to the plants in abundance then why xeromorphy develops in halophytes? Previously physiological drought was believed to be the main cause of developments of xeromorphy in halophytes, but recent physiological experiments on these plants have proved that xeromorphism in these plants is, apparently, an example of purposeless adaptation. Physiological experiments make it clear that the halophytes do not experience difficulties, whatsoever, in absorbing too saline water. This point is concluded taking into consideration the following reasonable facts:

(*i*) they show high rates of transpiration,

(*ii*) they show exudation of sap that contains dissolved salts, and

(*iii*) they develop many shallow absorbing roots.

Saline conditions are not essentially "dry" for all plant species. Under saline

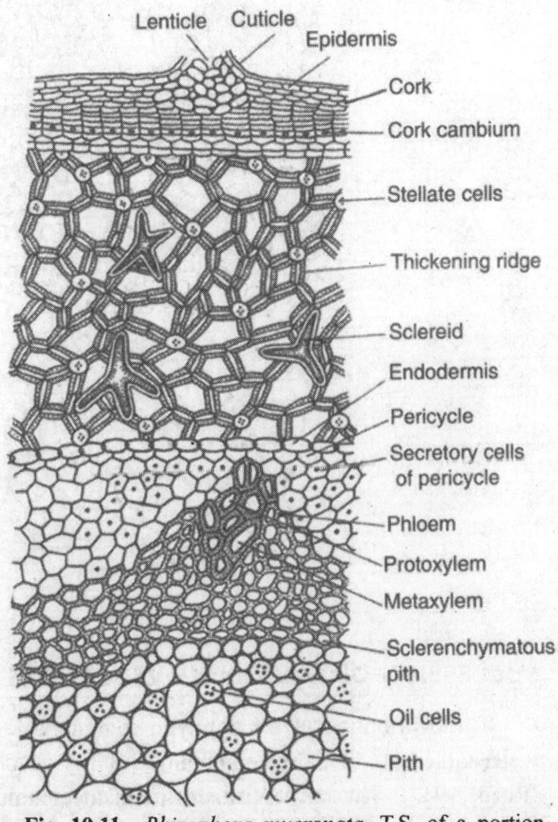

Fig. 10.11. *Rhizophora mucronata.* T.S. of a portion of subterranean stilt root.

conditions sometimes higher transpiration rates have been observed in halophytes than in neighbouring salt hating plants (Delf, 1911; Braun-Blanquet, 1931).

It should, therefore, be admitted that the halophytes show xeromorphism for enduring high salinity of soil water and also for absorbing water with perfect ease. The significance of succulence is not so clearly understood. Probably, it is induced by accumulation of salts in cytoplasm. It seems reasonable because of the fact that sodium salts if present in the soil water will definitely stimulate succulence even in non-halophytes and characteristic succulence of some plants may disappear if they are grown on the soil lacking in these common salts. Excessive accumulation of sodium does not harm these plants. Halophytes grow in saline habitats not because they are salt loving but because they tolerate high concentration of salts better than other plants of non-saline habitat. Active accumulation of salts also increases the osmotic concentration of cell sap in these plants and thus makes them able to absorb salty water very easily.

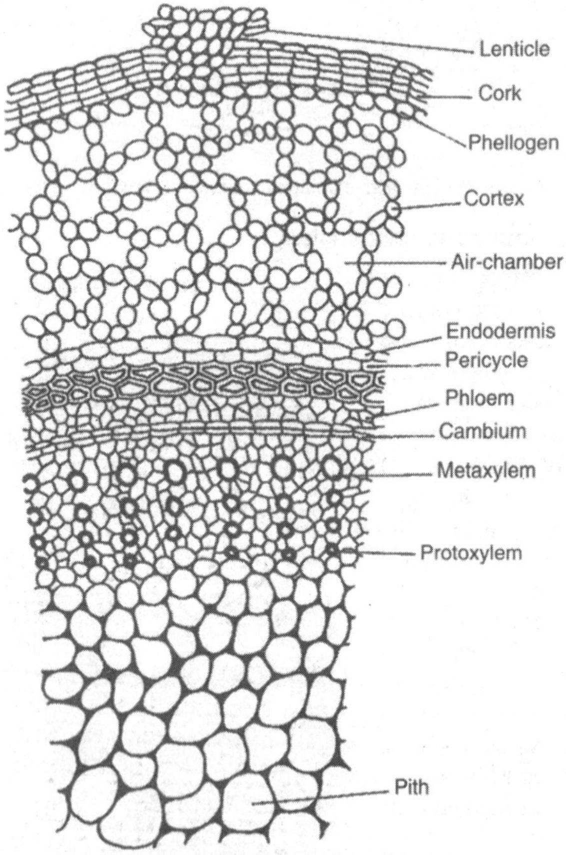

Fig. 10.12. T.S. of a pneumatophore.

SUCCESSION OF MANGROVE VEGETATION IN SEA COAST

The distribution of a halophytic community appears to be limited by salinity and depth of water table, as well as the competitive ability of the members of next community in the halosere (Reed, 1947). The aggressiveness of plant communities in saline habitat is due to changes in the salinity level.

In coastal region, nature of vegetation is greatly affected by the gradual elevation of sea coast. Succession of mangrove formation in coastal regions may take place very slowly in the following sequence :

(1) In deep water generally true mangroves, *e.g.*, the species of *Avicennia* grow.

(2) When the bottom of the sea is slightly raised up, *Avicennia, Rhizophora, Ceriops, Bruguiera,* etc. form mixed mangroves vegetation in the shallow water.

(3) As the ground is exposed, true mangroves disappear and other halophytes, *e.g.*, species of *Aegiceras, Excoecaria,* etc. gradually invade the land within short period. These halophytic communities are interspersed by salt tolerating succulents and the grasses make the soil fit for the cultivation of some crop plants. Several varieties of paddy, such as *Oryza sativa var, achra,* wild species of *Oryza coarctata* grow very well in these areas.

Successsion of angiospermic halophytes varies also in different habitats in accordance with other ecological conditions besides salinity. Thus, in reality there is no general trend for development of various halophytic plant communities around the world and local variations are encountered in each specific site (Yoav Waisel, 1972).

QUESTIONS

1. What are halophytes? Describe the various characteristic features of halophytes. How do they resemble xerophytes and hydrophytes?
2. Write short notes on :

   Mangrove, Vivipary, Pneumatophore, Physiological adaptation of halophytes.
3. Discuss briefly the classification of halophytes and give an account of distribution of mangrove vegetation in India.
4. What are the various anatomical features which characterise halophytes?
5. How is it that mangrove plants show xerophytic structures in spite of abundance of moisture in the habitat?

11

Principles of Phytogeography, Climate, Vegetation and Botanical Zones of India

The aspect of ecology which deals with the present and past geographical distribution, establishment and survival of organisms in their habitats is called **Biogeography**. It is broadly divided into (*i*) Zoogeography and (*ii*) Phytogeography. The study of migration and distribution of animals is known as **Zoogeography** while that of plants is known as Phytogeography. Alexander Von Humboldt is said to be the father of plant geography who wrote the first book on the subject in 1806.

PHYTOGEOGRAPHY OR PLANT GEOGRAPHY

According to Campbell (1926), the main theme of plant geography is to discover the similarities and diversities in the plants and floras of the present and past found in widely separated parts of the earth. Wulff (1943) states that Phytogeography is the study of distribution of plant species in their habitats and elucidation of origin and history of development of floras.

According to Croizat (1952), Phytogeography is the study of migration and evolution of plants in time and space.

Major Divisions of Phytogeography

There are two major divisions of Phytogeography :

(*i*) **Descriptive or Static Phytogeography**

(*ii*) **Interpretive or Dynamic Phytogeography**

Descriptive Phytogeography. This deals with the actual description of floristic or vegetational groups found in different parts of the world. Early plant geographers described floras and attempted to divide earth into floristic and botanical zones.

Interpretive or Dynamic Phytogeography. This deals with the dynamics of migration and evolution of plants and floras. It explains the reasons for varied distribution of plant species in different parts of the world. It is a borderline science involving synthesis and integration of data and concepts from several specialized disciplines like ecology, physiology, genetics, taxonomy, evolution, palaeontology and geology.

Good (1931), Mason (1936), Cain (1944) and some others have pointed out the factors involved in the distribution of plants. Lowerence (1951) has suggested the following thirteen modern principles of phytogeography which are classified into four groups :

I. Principles concerning environment.

1. The distribution of plants is primarily controlled by climatic conditions.

2. There has been variation in climate during geological history in the past which affected migration of plants.

3. The relations between land masses and seas have varied in the past. The large land masses split up to form new land masses or continents which separated and reoriented. Land bridges between continents acted as probable routes for migration of plant and animal species. The land bridges became submerged in sea with the passage of time and the possibility for migration of plants and animals from one continent to another disappeared for ever.

4. Soil conditions on plains and mountains of different land masses show secondary control on distribution of vegetation. Halophytes, psammophytes, calcicols, calcifobs etc. have developed because of edaphic conditions.

5. Biotic factors also play important role in distribution and establishment of plant species.

6. *The environment is holocentric,* i.e., all environmental factors have combined effects on the vegetation of a place (Ale & Pank, 1939).

II. Principles concerning plant responses.

7. *Range of distribution of plants is limited by their tolerances.* Each plant species has a range of climatic and edaphic conditions.

Therefore, tolerance of a large taxon is the sum of tolerances of its constituent species.

8. *Tolerances have a Genetic basis.* The response of plants to environment is governed by their genetic make up. Many of the crops through breeding and genetic changes have been made to grow in wider range of environmental conditions. In nature, hybrid plants have been found to have wider range of tolerances than their parents.

9. *Different ontogentic phases have different tolerances.* Different developmental stages of plants show different degree of tolerances, as for example seeds and mature plants are more tolerants to temperature and moisture variations than their seedlings.

III. Principles concerning the migration of floras and climaxes.

10. *Large scale migrations have taken place.* The fossils and palaeoecological evidences reveal that large scale migrations of plants and animals have taken place during Mesozoic era and Tertiary periods.

11. *Migration resulted from transport and establishment.* In the process of migration plants are dispersed to new habitats through their propagules such as spores, seeds, bulbils etc, and there they are established if environmental conditions are favourable. Plants grow and reproduce there and progeny perpetuates through ecological adjustments.

IV. Principles concerning the perpetuation and evolution of floras and climaxes.

12. *Perpetuation depends first upon migration and secondly upon the ability of species to transmit the favourable variations to the progenies.*

13. *Evolution of floras and climaxes depends upon migration, evolution of species and environmental selections.*

DISTRIBUTION

On the basis of area of the earth surface occupied by the plants, the various taxa are categorised as under :

1. Wides.
2. Endemics.
3. Discontinuous species.

1. Wides

Plants widely distributed over the earth in definite climatic zones and the different continents are referred to as wides. Cosmopolitan is applied for wides but, in fact, no plant is cosmopolitan in real sense of the term. *Taraxacum officinale* and *Chaenopodium album* are the common examples of the wides. Plants of tropical regions are called *Pantropical.* The plants of very cold climate may not only be found in the arctic regions but also in alpine zone of mountains in tropical and subtropical regions. These are called *arctic-alpine* plants.

2. Endemics

A taxon whose distribution is confined to a given area is said to be **endemic** to that area. The taxon may be of any rank, although it is usually at a family level or below, and its range of distribution may be wide, spanning an entire continent, or very narrow covering only a few square metres : The concept of endemism is important because in the past the formulation of biogeographic regions was based on it. The limits of a region are determined by mapping the distributions of taxa; where the outer boundaries of many taxa occur, a line delimiting a biogeographic region is drawn. Major regions are still determined as those that have the most endemics or, stated another way, those that share the fewest taxa with other regions. As regions are further broken down into subdivisions, they will contain fewer unique taxa. This has been criticized because it assumes that species ranges are stable, which they are not. An alternative method of determining biogeographic regions involves calculating degrees of similarity between geographic regions. The concept of endemic distribution of plants was putforth by A.P. de Candolle (1813). Engler (1882) suggested two categories of endemic forms; **Palaeo-endemics** which are survivors of ancient forms and **indigenous** or **native forms** which are confined to a particular locality. According to area of distribution, the species may be **continental endemics** (restricted to a continent), **endemic to a country, provincial, regional** or **local endemics** (restricted to valley, hills, islands, etc.).

Now the endemic species have been grouped into the following categories:

(*i*) **Relics or Palaeoendemics.** They are the survivors of once widely distributed ancestral forms, for example, *Ginkgo biloba* (restricted to China and Japan), *Sequoia sempervirens* (confined to coastal valleys of California, U.S.A.). *Agathis australis, Metasequoia* (confined to Single valley in China). These species are called **Palaeoendemics** or **epibionts.** A great majority of the endemic species belonging to this type have many fossil relatives. They are also called living fossils. Because of little variability the endemics are adapted only to a particular environment and even if they reach new areas, they fail to establish themselves in new. environment.

(*ii*) **Neoendemics.** The other endemics may be modern species which have had not enough time for occupying a large area through migration. They are called ***neoendemics***. There are several such genera which are widely endemic or few species of which are endemic. Neoendemics show good variability and have many biotypes, grow in diverse habitats and have wide tolerance for habitats. Some of the well known endemic genera in Indian flora are *Mecanopsis* (Papaveraceae). *Chloroxylon swietenia* (Flindersiaceae, formerly Rutaceae). *Catenaria* and *Butea* (Papilionaceae), *Caesulia* (Compositae), *Petalidium* (Acanthaceae), etc. *Eletteria repens* (Zingiberaceae), *Piper longum* (Piperaceae), *Piper nigrum* (Piperaceae), *Ficus religiosa* (Moraceae), *Shorea robusta* (Dipterocarpaceae), *Venda caerulea* (Orchidaceae), *Salmalia malabarica* (Bombacaceae), *Eleusine coracana* (Gramineae) are the well known endemic species of Indian flora.

There are some special terms to designate the quality of these endemics, *viz.* **Local endemics** which are found in small land features, **progressive endemics** which tend to spread with time, **retrogressive** endemics in which case the area of distribution is contracting and **microendemics** (*i.e.*, the endemics of lower groups).

Pseudo endemics. These endemics arise due to mutation in existing population at a particular place. These pseudo endemics or mutants may or may not persist for long in the particular area where they originate.

Endemism results from the failure on the part of species to disseminate its seeds, fruits, spores or propagules because of existence of great barriers like mountains, oceans and large deserts. The oceanic islands which are isolated from rest of the world by large expanses of waters abound in endemic species and water barrier checks the migration of those species outside their original habitat.

3. Discontinuous Distribution

When plants occur at two or more distant places of the world which are separated by overlands or oceans hundreds or thousands of kilometres apart. Such a distribution is called **discontinuous** or **disjunct** distribution. Three genera *Nothofagus*, *Jovellona* and *Adansonia* for example are found in parts of South America, South Africa and Australia which are separated by vast oceans. The significant phytogeographical causes for discontinuous distribution are as follows :

(*i*) The species might have evolved at more than one place and they failed to migrate outside their original habitats because of barriers.

(*ii*) The species which were once widely distributed in the past disappeared from certain areas and are now surviving in some distant pockets.

(*iii*) The climate may also be a factor for discontinuity in distribution of species. Plants having specific climatic requirements are found in widely separated areas with similar environmental conditions, as for example, plants of arctic regions are also found in alpine zone of high mountains in tropics and subtropics. *Salix* and *Silen* species show discontinuous distribution in arctic-alpine regions.

THEORIES OF DISCONTINUOUS DISTRIBUTION

1. Theory of Land Bridge. According to this theory, land bridges occuring in between the separated continents are believed to have helped in the migration of various taxa from one continent to the other. Uniform distribution of plants and animals in different parts of the world during Palaeozoic era is believed to have been due to those land bridges. With the passage of time the land bridges became submerged in sea and the connections between the various continents snapped beyond the dispersal capacity of organisms resulting thereby the discontinuity in the distribution.

2. Theory of Continental drift. The theory of continental drift was propounded by Wegner (1912, 1924). According to him the whole land- mass of the world was a single super continent during Palaeozoic era. He named it as **Pangaea**. That super continent was surrounded by sea on all the sides which was named **Panthalassa**. During Mesozoic, Pangaea split up into two large landmasses; **Laurasia** in the north and **Gondwanaland** in south. The two landmasses were separated by *Tethys sea.* Du Toit (1937), however, suggested that Laurasia and Gondwanaland existed from the very beginning. The two large landmasses having characteristic flora and fauna broke up into new landmasses called **continents**. Laurasia gave rise to Eurasia, Greenland and North America and similarly Gondwanaland gave rise to South America, Africa, India, Polynesia, Australia, Antarctica etc. About 135 million years ago *reorientation of continents* began. The continents were drifted apart by the oceans. This is called **Continental Drift**. The occurrence of Dinosaurs and many fossil plants lend support to the existence of Laurasia and Gondwanaland. With the separation of continents the distribution areas of several plant and animal species got separated and gave rise to discontinuous distribution areas.

Factors Affecting Distribution of Species

Several factors are known to affect the geographical distribution of plant species, some of which are as follows :

1. Geological history and distribution,
2. Migration, and
3. Ecological amplitude.

1. **Geological history and distribution.** The place where a species first originated is called its centre of origin. Evolution of species is a slow but continuous process. Some of the species in present day flora are quite old while a great majority of them are recent in origin. The process of species differentiation involves (*i*) hybridization between the related species as well as mutation and (*ii*) the natural selection of the hybrid and mutant populations. In the selection process not all the hybrids and mutants are selected by nature and only the fittest individuals which find the habitat conditions within their ecological amplitudes are selected and the individuals least fit are eliminated. Changing climate has also played important role in the origin of new species. In the course of evolution several old species became extinct, some of which can be found even today as fossils. The fossils provide direct evidence for the existence of various taxa in the past.

Age and Area Hypothesis. This hypothesis was proposed by J. W. Willis (1915) on the basis of his extensive studies of geographical distribution of certain plant species in tropics. On the basis of his findings Willis postulated that the species which evolved earlier occupy greater areas than those which appeared later in the evolutionary sequence. According to this hypothesis, the frequency of a species over an area is directly proportional to its age in scale of evolution and age of species is directly related with the area of its distribution. Thus a small area of distribution of a species will indicate its relative young age. Willis has quoted several examples such as *Impatiens, Primula, Gentiana, Rhododendron* in support of his hypothesis. Genus *Coleus* may be quoted here as an example in support of this hypothesis. There are two species of *Coleus* namely *C. elongatus* and *C. barbatus.* The former species is endemic while the latter is widely distributed. On the basis of areas under distribution of these species Willis considered *C. elongatus* less evolved and derived from *C. barbalins.* Willis has also pointed out that the majority of endemics are found to be members of large and successful genera.

The age and area hypothesis, however, is not universal and it has been criticised by many.

2. **Migration.** The newly evolved species starts migration to new areas and side by side it undergoes further evolutionary changes. The dispersal of germules and propagules is brought about by several agencies like wind, water, glaciers, insects, animals, even man. The dispersal is followed by *ecasis.* Migration may be adversely affected and sometimes even totally stopped by some factors called ***migration barriers.*** Barriers in the dispersal of species may be classified as **ecological or environmental** and **geographical**. The climate, an ecological barrier, plays important role in distribution and establishment of species. Unsuitable climatic condition or change of climate in particular area forces the species to migrate from one place to another and the failure of some species to migration leads them to gradual extinction. Besides climate, there are geographical barriers, as for example, high mountains, vast oceans or deserts. The fresh water plants, for example, cannot be dispersed across oceans if their propagules are suitable only for fresh water dispersal and similarly germules or propagules of land plants from one country cannot reach other country separated by vast oceans and mountains. Species are called **natives** of the place of occurrence if they originated there. Outside the area of its origin, the species is referred to as **exotic**. Exotic species reach new area through migration. If any species is introduced intentionally in new area by man then it is called **introduced species**.

3. **Ecological amplitudes and distribution.** Environmental conditions not only influence the life and development of plants but also determine the presence or absence, vigour or weakness and relative success or failure of various plants in a particular habitat. Each plant species of a

community has a definite range of tolerance towards physical and biological environment of the habitat. This is referred to as **ecological amplitude.**

The presence of species at a particular place, no doubt, indicates that the environmental conditions of that habitat are within its ecological amplitude but the absence of a species from one place does not necessarily indicate that the environment is not suitable for that species.

The ecological amplitude is governed by genetic set up of the species concerned and thus different species have different ecological amplitudes which may sometimes overlap only in certain respects. Further, some species may occur at different geographical regions as and when the conditions fall within their ecological amplitudes. As for example, some plants of temperate region, say conifers, may also be found in alpine zone of high mountains in tropical and subtropical regions. The other consideration in ecological amplitude as a factor in plant distribution is its change with time. In sexually reproducing plants the hybridization between related species results in offsprings with new genetic composition. With the change of environment the plant species also make adjustments with new environment by shifts in their ecological amplitudes facilitated by changes in the genotype. Within a species there may occur several genetically different groups of individuals (populations) which are adjusted to particular set of ecological conditions. These populations are called ecotypes or ecological races or ecological populations. In *Euphorbia thymifolia*, for example, there are two major populations—one is calcium loving or *calcicole* and the other type is calcium hating or *calcifuge*. Similarly ecological races of *Xanthium strumarium* and *Ageratum conyzoides* differ in the photoperiodic requirements. The existence of ecotypes within the species widens the area of its geographical distribution.

CLIMATE OF INDIA

Atmospheric and meteorological influences, principally moisture, temperature, wind, atmospheric pressure and evaporation of a region collectively form the climate. Climate has marked influences on the growth, distribution and development of vegetation types in India. India lies north of equator between latitudes 6° and 38° N and longitude 68° and 97° 25' E. The Himalaya mountain is present in its north. India is surrounded on its south, east, and west by Indian ocean, Bay of Bengal and Arabian Sea respectively. The country as a whole can be divided into three main parts : (*i*) The mountain wall, (*ii*) The plains of northern India, and (*iii*) The plateau of Peninsular India.

Climate of India is not homogeneous throughout but it is highly variable. The country stands in the tropical and subtropical belts and the climate is greatly modified by oceans and typography of several mountain ranges, as Himalayan ranges, mountain ranges of Meghalaya and Arunanchal in North east and Western Ghats with straight and Eastern Ghat in South. The climate in South and East is more typically tropical rainy, near temperate at high altitudes in the north (Kashmir), and semiarid and arid (dry hot) in the north-western part. The temperature is usually high throughout the year in most parts of the country except at high altitude in extreme north where snowfall occurs during winter and temperature goes below 0°C for short period.

The amount of rainfall varies greatly in different periods of the year and in different parts of the country; some parts receiving above 400 cm annual rainfall and some receiving as little as 20 cm annual rainfall or even no rains at all in some years. The distribution of annual rainfall in India is shown in Fig. 11.1.

Sir John Elliot was the first man to prepare a detailed climate map of India and later Kendew (1952), C.W.B. Normand (1937), Stamp (1946) and Waheed Khan (1959) also described the seasons and the climates of India. The Indian Meteorological Department (IMP) has adopted the following 4 seasons :

1. Cold season (from mid December to February).

2. Hot season (from March to mid of June).
3. Rainy season (from mid June to mid September).
4. Season of retreating monsoons (from mid of September to mid December).

1. Cold season. The season starts in December and persists upto February. In this season the atmospheric temperature goes very down, chilly winds blow from west or north-west to east or south-east direction, air is dry and fresh and the sky is clear. The cyclonic rains are prevalent at the top of Peninsula, foothills of Himalayas and the margins of the plains in the Punjab and extending up to U.P. and southward into Rajasthan, Maharashtra and M.P. The winter rain, although light, greatly influences the growth of vegetation because runoff is very little, evaporation rate is very low and a large amount of it is available to the root zone.

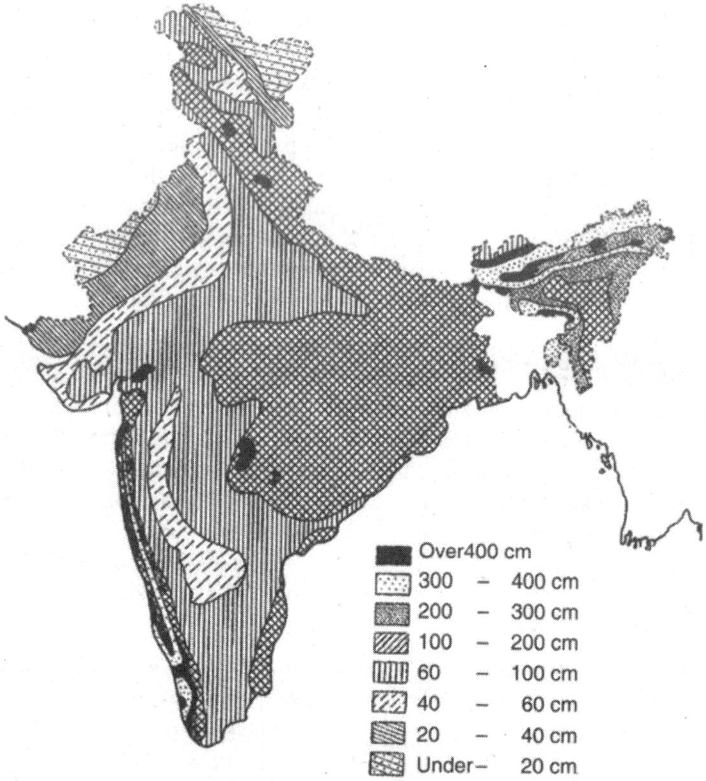

Fig. 11.1. Rainfall map of India.

2. **Hot season.** In the beginning of March, temperature starts rising and humidity decreases. The onset of summer season is indicated by leaf-fall and almost simultaneous appearance of new leaves and blooming in many plants as Mango, Amaltas (*Cassia fistula*), Gulmohar (*Delonix regia*) and so on. The temperature continues to rise till middle of June when mercury level may touch 45° to 49°C in upper Gangetic plains. The high temperature causes decrease in atmospheric pressure in most parts of the country. This results in hot winds which blow from south and south-east to north and north-west. These winds carry clouds with them.

3. Rainy season. This season starts with the bursting of the advancing monsoon winds during the middle of June or beginning of July. The monsoon winds arising in the Indian Ocean and passing over the Bay of Bengal enter the Gangetic plains and produce maximum rains in the north-eastern parts of the country and heavy rains in the plains of Bengal, Bihar, U.P., Orissa, Punjab, part of M.P. Kashmir, etc. The amount of rainfall decreases towards west. The annual rainfall in these regions ranges between 100 and 200 cm. Parts of Punjab, Rajasthan, Gujarat and Maharashtra receive only scanty rainfalls and a part of Rajasthan west of Aravalli ranges shows arid desert conditions with almost no rains in some years. The monsoon winds arising over Arabian sea produce heavy rains in South India and western coasts and lesser rains in the interior. The rain is not continuous throughout the season but it continues upto middle or end of September interrupted by bright or cloudy rainless days. The mountain ranges play significant role in the distribution of rainfall. Aravalli ranges are largely responsible for the arid climate in western Rajasthan.

4. **Season of retreating monsoons.** Towards the end of September the monsoon winds start retreating and produce only occasional showers. The sky becomes clear. The temperature decreases gradually and in the mid December cold sets in. Due to sudden variations in temperature local cyclones develop in the region of Bay of Bengal and the east coast of Tamil Nadu and cause isolated rains in certain areas.

CLIMATIC REGIONS OF INDIA

On the basis of rainfall climatic conditions in different physiography determine the geographic and not temperature called climate regions. India can be roughly divided into the following four climatic zones; wet zone, Intermediate zone, Dry zone and arid zone.

1. **Wet zone.** The average annual rainfall in this zone is more than 200 cm. Wet zone comprises western ghats including western coastal part of Mumbai, Karnataka, Kerala, Assam, eastern half of Bengal, Tarai region of U.P., Orissa, Chhattisgarh and North Bihar. Natural vegetation includes normally evergreen and semi-evergreen tropical forests in the west coasts and Assam but dry deciduous forests (mostly *Shorea robusta*) in other parts. Paddy is the main agricultural crop.

2. **Intermediate zone.** The mean annual rainfall in this zone is above 100 cm but less than 200 cm. The zone comprises south-west Bengal, Bihar, Orissa, Part of M.P., east U.P., north-eastern Andhra Pradesh, eastern slopes of Western ghats and east Tamil Nadu. Natural vegetation includes dry deciduous forests dry deciduous thorn forests and scrubs. Rice is the major agricultural crop.

3. **Dry zone.** This zone receives annual rainfall between 50 and 100 cm and includes western U.P., N-E. Punjab, Delhi, S-E. Mumbai, Eastern part of Rajasthan, and western half of Andhra Pradesh. Natural vegetation includes dry deciduous forests, shrubs and thorn scrubs. Wheat and millets are the major agricultural crops.

4. **Arid zone.** This zone receives rainfall less than 50 cm in a year and includes south-western parts of Rajasthan and Punjab, N-E. Mumbai, and N-W. Gujarat. Natural vegetation includes thorny scrubs and short-lived herbs.

VEGETATION OF INDIA

Vegetation is the total plant cover of a region or area or place. Geographically, India is a tropical country but whole of the Gangetic plain lies outside the tropics. It has strong monsoonic climate and differs from other tropical regions of the world. As pointed out earlier, there are four climatic regions in India which are characterised by different types of natural vegetation. Several factors are responsible for the occurrence of a variety of vegetational types in this country . Besides the climatic factors, soils upon which vegetation is found greatly determine the nature of vegetation. Vegetation in this subcontinent has been influenced by biotic factors as well as human culture for a long time. The plant formations determined by specific and continuous operation of man himself are called anthropogenic formations.

Depending upon the climate, the vegetation of India may be divided into (*i*) Tropical, (*ii*) Subtropical, (*iii*) Temperate and (*iv*) Alpine

Some of important criteria in the classification of vegetation are physiognomy, structure, function, composition, dynamics (successional phenomenon), habitat relations and history. Fosberg (1961) found that the structure and function are the most satisfactory criteria for classification of vegetation. Champion (1936) gave the first complete classification of vegetation of India. Since then many workers have modified Champion's classification and some of them have given their own classification. Fosberg (1961) recognised the following three major structural groups in the vegetation of India.

1. Closed vegetation.
2. Open vegetation
3. Sparse vegetation.

Indian vegetation presents a varied assemblage of plant communities and includes several exotic elements, especially from China, Japan, Malaya, Africa, Tibet and Siberia. The following recognised plant formation types cover whole of India:

1. Forests,
2. Grasslands,
3. Vegetation of ponds and lakes, and
4. Vegetation of sea-shore.

FORESTS

Champion (1936) recognised 13 major forest types in India. Champion and Seth (1968) recognised sixteen forest types which are listed below. (Fig 11.2)

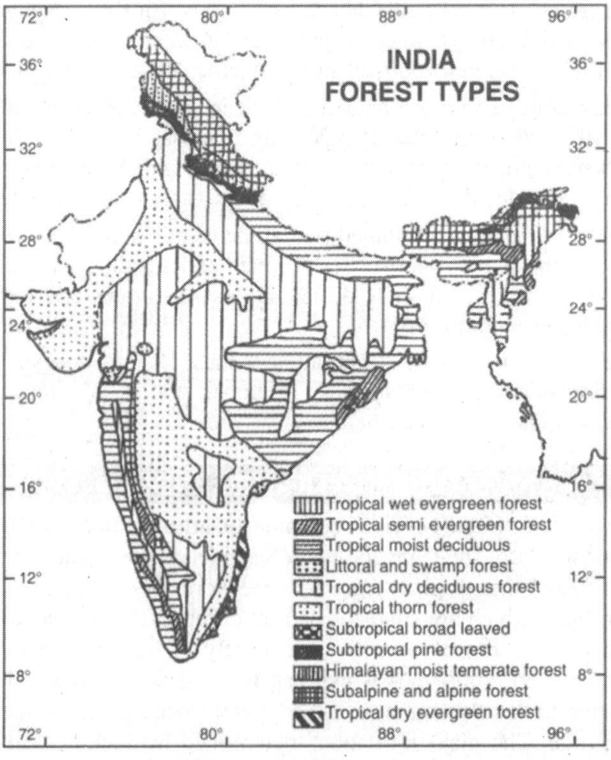

Fig. 11.2. Outline map of India showing distribution of different forest types (*After Champion and Seth, 1968*).

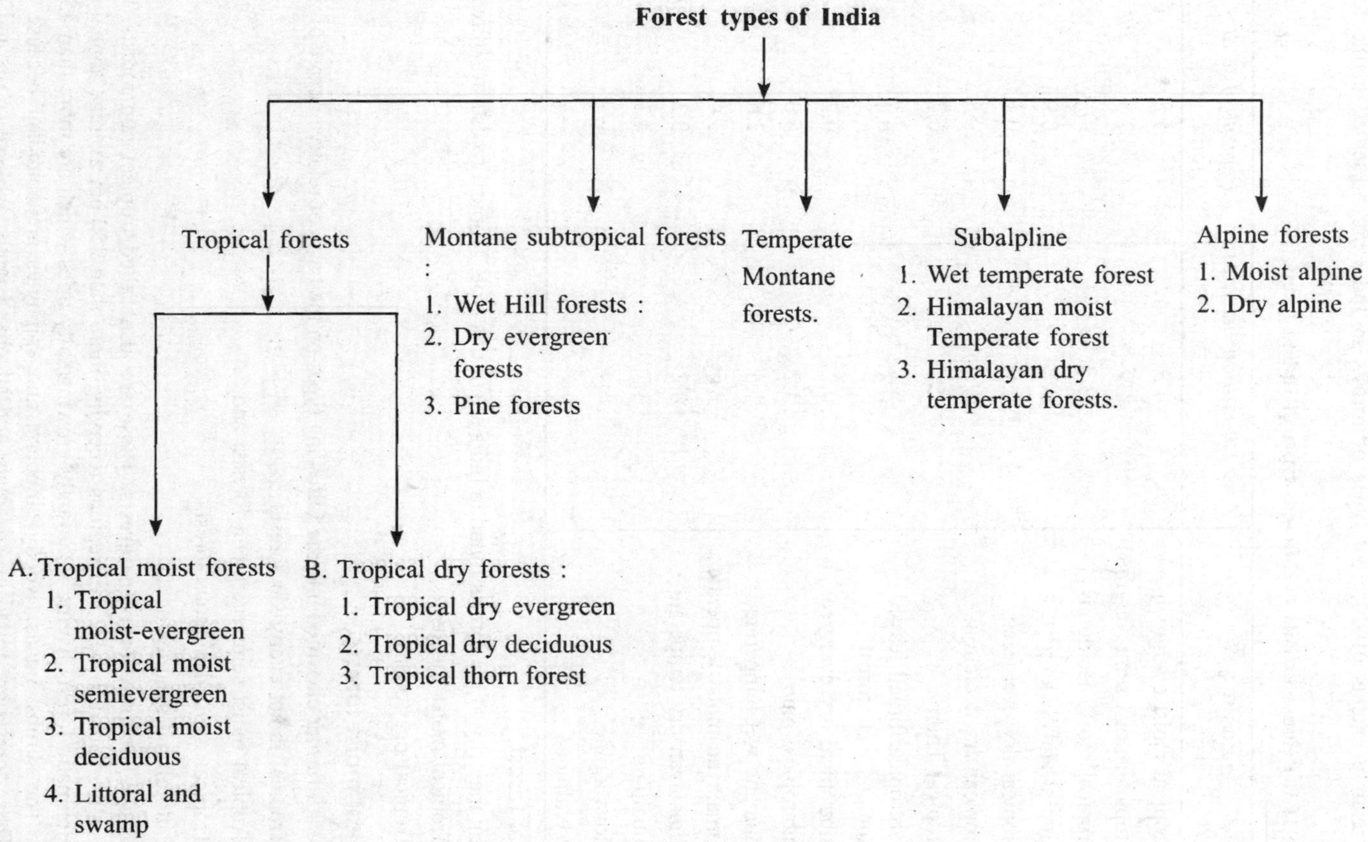
Forest types of India
Tropical forests
A. Tropical moist forests
1. Tropical moist-evergreen
2. Tropical moist semievergreen
3. Tropical moist deciduous
4. Littoral and swamp
B. Tropical dry forests :
1. Tropical dry evergreen
2. Tropical dry deciduous
3. Tropical thorn forest
Montane subtropical forests :
1. Wet Hill forests :
2. Dry evergreen forests
3. Pine forests
Temperate Montane forests.
Subalpline
1. Wet temperate forest
2. Himalayan moist Temperate forest
3. Himalayan dry temperate forests.
Alpine forests
1. Moist alpine
2. Dry alpine

Hanson (1962) defines forest as "a stand of trees growing close together with associated plants of various kinds". The following types of forests are found in India which cover nearly 17 per cent of the total area of Indian territory.

The relative extents of different types of forests in India are presented in the following Table 11.1

Table 11.1. Relative extents of different types of Indian forest.

Forest type	*Area (in million hectare)*	*Percent of total forest area*
Tropical moist evergreen	4.5	5.8
Tropical moist semievergreen	1.9	2.5
Tropical moist deciduous	23.3	30.3
Littoral and Swamp	0.7	0.9
Tropical dry evergreen	0.1	0.1
Tropical dry deciduous	29.4	38.2
Tropical Thorn	05.2	6.7
Subtropical broad leaved montane wet forest	0.3	0.4
Subtropical dry evergreen	0.2	0.2
Subtropical pine	3.7	5.0
Montane wet temperate	1.6	2.6
Himalayan moist temperate	2.6	3.4
Himalayan dry temperate	0.2	0.2
Subalpine	3.3	4.3
Moist alpine	—	—
Dry alpine	—	—

I. Tropical forest

A great majority of the forests found in India are of this type. Tropical forests are of two types:

(A) Tropical moist forests.

(B) Tropical dry forests.

A. Tropical moist forests

These are further classified into the following types on the basis of relative degree of wetness:

(*i*) Tropical moist evergreen forests,

(*ii*) Tropical moist semi-evergreen forests, and

(*iii*) Tropical moist deciduous forests.

(*iv*) Littoral and swamp forests.

(*i*) **Tropical moist evergreen forests.** These are also called tropical rain forests. In India such forests are found in very wet regions receiving more than 250 cm average annual rainfall. These are climatic forests having luxuriantly growing lofty trees which are more than 45 metres in height. The shrubs, lianas (woody climbers) and epiphytes are abundant because of high rainfall. These forests are found in Andaman and Nicobar Islands, Western coasts and parts of Karnataka (N. Canara), Annamalai hills (Koorj), Assam and Bengal. The detail account of this is given in the description of mesophytes.

(*ii*) **Tropical moist semi-evergreen forests.** These forests are found along the western coasts, eastern Orissa and upper Assam where annual rainfall is between 200 and 250 cm. They are characterised by giant and luxuriantly growing intermixed deciduous and evergreen species of trees and shrubs. The important plants in these forests are the species of *Terminalia, Bambusa, Ixora, Dipterocarpus, Garcinia, Sterculia, Mallotus, Calamus, Albizzia, Elettaria, Pothos, Vitis, Shorea, Cinnamomum, Bauhinia, Albizzia*, etc. Orchids, ferns, some grasses and several other herbs are also common.

(*iii*) **Tropical moist deciduous forests.** These cover an extensive area of the country receiving sufficiently high rainfall (100 to 200 cm) spread over most of the year. The dry periods are of short duration. Many plants of such forests show leaf-fall in hot summer.

The forests are found along the wet western side of the Deccan plateau, *i.e.* Mumbai, N-E. Andhra, Gangetic plains and in some Himalayan tracts extending from Punjab in west to Assam valley in the east. The forests of Southern India are dominated by Teak (*Tectona grandis*), *Terminalia paniculata, T. bellerica, Grewia tilliaefolia, Dalbergia latifolia, Lagerstroemia, Adina cordifolia*, etc. are the other common species in forests of South India. In north, they are dominated by shal (*Shorea robusta*). Some other common associates of shal are *Terminalia tomentosa, Dellenia species, Eugenia species, Boswellia species and Mallotus philippensis.* These forests produce some of the most important timbers of India. Grasses become important both in seral stages and in the areas under fire.

(*iv*) **Littoral and Swamp Forests.** Littoral and Swampy forests include the following types :

(1) Beach forests

(2) Tidal forests or Mangrove forests

(3) Fresh water swamp forests.

Beach Forests. The beach forests are found all along the sea beaches and river deltas. The soil is sandy having large amount of lime and salts but poor in nitrogen and other mineral nutrients. Ground water is brackish, water table is only a few metres deep and rainfall varies from 75 cm to 500 cm depending upon the area. The temperature is moderate. The common plants of these forests are *Casuarina equisetifolia, Borassus, Phoenix, Manilkara littoralis, Callophyllum littoralis, Pandanus, Thespesia, Barringtonia, Pongamia, Cocos nucifera, Spinifex littoreus* and a number of twiners and climbers.

Tidal or Mangrove forests. Tidal forests grow near the estuaries or the deltas of rivers, swampy margins of Islands and along sea coasts. The soil is formed of silt, silt-loam or silt-clay and sand. The plants are typical halophytes which are characterised by presence of prop roots with well developed knees for support and pneumatophores and viviparous germination of seeds. Tidal forests one distinguished into the following four types with overlapping constituent species :

(*i*) Tree mangrove forests

(*ii*) Low mangrove forests,

(*iii*) Salt water forests and

(*iv*) Brackish water forests.

Tree Mangrove forests. These forests occur on both east and west sea coasts. The best development occurs in Sundarbans. The forest floor is flooded with salt water daily. Plants may attain a height 10-15 metres and form an almost closed evergreen forests. The common trees of these forests are *Rhizophora mucronata, R. conjugata, Avicennia alba, Bruguiera conjugata, B. parviflora, B. caryophylloides, Kandelia candel, Xylocarpus molluccensis, X. granatuns, Ceriops tagal, Avicennia officinalis, Excoecaria agallocha, Sonneratia acida, Lumnitzera racemosa, L. littorea, Aegiceras carniculatum* and two most frequently occurring palms; *Nipafruticans* and *Phoenix paludosa.* In addition to these trees *Acanthus ilicifolius* and other shrubs also grow at places.

Low Mangrove Forests. These forests grow on soft tidal mud near estuaries, which is flooded by salt water. Forest is dense but the trees with leathery leaves attain maximum height of 3-6 m. The vegetation consists of a few species which show gregarious growth habit. Important tree species are *Ceriops decandra, Avicennia alba, Aegialitis rotundifolia* and *Excoecaria agallocha.* Besides, a common shrub *Acanthus elicifolius* and some grasses also occur at places. Low mangrove forests are more developed on east sea coast than on west coast.

Salt water Mangrove Forests. These forests occur beyond tree mangrove forests in big river deltas where the ground is flooded with tidal water. Silt deposition and salt content in soil are low. Tree height is upto 20 m or so but girth is not large. Forests are dense. Pneumatophores are common. The common plants are *Heritiera minor, Excoecaria agallocha, Ceriops decandra, Xylocarpus molluccensis, Bruguiera conjugata, Avicennia officinalis* and *Nipa* at places.

Brackish Water Mangrove Forests. They grow near the river deltas where forest floor is flooded with water at least for some times daily. Water is brackish (salty) but during rains it is nearly fresh. Tree height may reach 30 m or so. Forest is dense. Common species of the forests are *Heritiera minor, Xylocarpus molluccensis, Bruguiera conjugata, Avicennia officinalis, Sonneratia caseolaris, S. acida, Excoecaria agallocha, Ceriops decandra Cynometra ramiflora, Amoora cuculata, Pandanus,* and two palms; *Nipa* and *Phoenix paludosa.*

Fresh Water Swamp Forests. These forests grow in low lying areas where rain or swollen river water is collected for some time. Water table is near the surface. Important plants include *Salix tetrasperma, Acer, Putranjiva, Holoptelia, Cephalanthus, Barringtonia, Olea, Phoebe, Ficus, Murraya, Adhatoda, Canna* and a variety of grasses.

B. Tropical dry forests.

These are classified into the following types:

(*i*) Tropical dry evergreen forests,

(*ii*) Tropical dry deciduous forests, and

(*iii*) Tropical thorn forests.

(*i*) **Tropical dry evergreen forests.** These forests are found in the areas where rainfall is in plenty but dry season is comparatively longer. The trees are dense, evergreen and short (about 10 to 15 metres high). These forests are found in eastern part of Tamil Nadu, in east and west coasts. The common plant species are much the same as in Tropical moist evergreen forests. Species of *Maba, Calotropis, Pabatta, Feronia, Canthium, Zizyphus, Randia* etc. are most common. Bamboos are absent but grasses are common.

(*ii*) **Tropical dry deciduous forests**. These forests are distributed in the areas where annual rainfall is usually low, ranging between 70 and 100 cm, such as, Punjab, U.P., Bihar, Orissa, M.P. and large part of Indian peninsula. The largest area of the country's forest land is occupied by Tropical dry deciduous forests. The dry season is long and most of the trees remain leafless during that season. The forest trees are not dense, 10 to 15 m in height, and undergrowth is abundant. In north, the forests are dominated by shal and in south by teak (*Tectona grandis*). The common constituents of these forests in South are *Dalbergia, Terminalia, Dillenia, Acacia, Pterospermum, Diospyros, Anogeissus, Boswellia, Chloroxylon, Bauhinia, Hardwickia, Gymnosporia, Zizyphus, Moringa, Dendrocalamus*, and so on. The other species of trees and shrubs of Sal dominated forests of northern region are *Terminalia, Semicarpus, Buchnania, Carissa, Modhuca, Acacia, Sterculia, Launea, Salmalia Adina, Bauhinia, Aegle, Grewia, Phyllanthus,* etc.

(*iii*) **Tropical thorn scrubs.** These forests occur in the areas where annual rainfall is between 20 to 70 cm, dry season is hot and very long. They are found in South Punjab, most of Rajasthan and part of Gujarat. The vegetation in these region occurs only along the rivers. The land away from the rivers and devoid of irrigation is mostly sandy and devoid of trees. The vegetation is of open type consisting of small trees (8 to 10 m high) and thorny or spiny shrubs of stunted growth. The forests remain leafless for most part of the year and are sometimes called thorn scrub or scrub jungles. There is luxuriant growth of ephemeral herbs and grasses during the rainy season. Towards the desert region the vegetation diminishes and in arid parts there is almost no vegetation. The species of *Acacia, Cassia, Calotropis, Randia, Albizzia, Zizyphus, Erythroxylon, Euphorbia, Cordia, Prosopis. Salvadora, Aegle, Gymnosporia, Atriplex, Grewia, Asparagus, Berberis, Butea, Kochia, Leptadenia, Capparis, Adhatoda,* etc characterise the plant formations of semiarid regions of India.

Champion (1938) named the natural vegetation of desert as *tropical thorny forest*. Bharucha (1955) divided the Rajasthan desert into the following vegetational zones :

(*i*) Area of shifting sand dunes at and around Jaisalmer and Bikaner.

(*ii*) Area of established sand dunes near Jodhpur.

(*iii*) Sand stone rocks covered by xerophytic plants like *Euphorbia nerifolia.*

(*iv*) Area of halophytic vegetation.

(*v*) Sandy-loam soil vegetation.

II. Subtropical montane forests

These forests are found in the region of fairly high rainfall but where temperature differences between winter and summer are less marked. Winter generally goes without rains. They are found upto the altitude of about 1500 metre in south and upto 1800 metre in the north. In composition, subtropical forests are almost intermediate between tropical forests and temperate forests and a sharp demarcation can seldom be made between tropical and subtropical or subtropical and temperate forests. These forests have been grouped into the following three types :

(*i*) Wet hill broad leaved forests,

(*ii*) Dry evergreen forests, and

(*iii*) Pine forests.

(*i*) **Wet hill broad leaved forests.** They are found in Mahabaleshwar, Coorg, Karnataka, parts of Assam, Panchmarhi and other parts of M.P. The important plants found in the wet hill forests of south are the species of *Eugenia, Randia, Terminalia, Eleganus, Murraya, Gymnosporia, Atylosia, Ficus, Pterocarpus, Lantana,* etc while those of the north are *Castonopsis, Calamus, Alnus, Quercus, Betula, Schima phoebe, Cedrella, Garcinia, Populus* etc.

(*ii*) **Dry evergreen forests.** They occupy the foot-hill areas of Himalayas. The common constituents of vegetation are *Acacia modesta, Olea cuspidata,* etc.

(*iii*) **Pine forests.** They are found mostly in western and central Himalayas and in Assam hills. The forests are dominated by species of *Pinus* (*Pinus khasya* and *P. roxburghii*). Species of *Quercus, Berberis, Carissa, Bauhinia* may also occur rarely in pine forests.

III. Temperate Montane forests

These forests occur in the Himalayas at the altitude from 1800 to 3800 metres where humidity and temperature are comparatively low. Montane forests have been classified into the following three types on the basis of moisture regime :

(*i*) Montane Wet temperate forest,

(*ii*) Himalayan Moist temperate forest, and

(*iii*) Himalayan Dry temperate forest

(*i*) **Montane Wet temperate forests.** These are found in Himalayas extending from Nepal to Assam at the altitude from 1800 to 3000 m, as well as in some parts of South India (Nilgiris). The forests in south are evergreen and are called **sholas**. The forests are dense with closed canopy and the trees may be 15 to 20 m high. Epiphytes are in abundance. Important plants constituting the vegetation in Eastern Himalayas are species of conifers, *Hopea, Balanocarpus, Elaeocarpus,*

Artocarpus, Pterocarpus, Myristica, Hardwickia, Salmelia, Dioscoria. The members of family Compositae, Rubiaceae, Acanthaceae and Leguminosae form the undergrowth.

(*ii*) **Himalayan Moist temperate forests.** These forests develop in the areas of lesser rainfall. The trees are high, sometimes upto 45 metres tall. The dominant elements of vegetation are oak and conifers. Undergrowth is shruby and consists of deciduous species of *Barberis, Spiraea, Cotaneaster*, etc.

(*iii*) **Himalayan Dry temperate forests.** These forests dominated by Rhododendrons, oaks and conifers from a narrow belt at the altitude from 3000 to 4000 m in the western Himalayas extending from a part of Uttranchal through Himanchal Pradesh and Punjab to Kashmir. The other commonly found species belong to genera *Daphne, Desmodium, Indigofera, Artemisia, Cannabis, Plectranthus, Fraxinus,* several epiphytic mosses, Lichens, etc.

IV. Sub-alpine Forests

The sub-alpine forests are found throughout Himalayas from Ladakh in the west to Arunachal in the east at the altitude from 2800 m to 3800 m. Annual rainfall is less than 65 cm. but snowfall occurs for several weeks in a year. Strong winds and below 0°C temperature prevail for greater part of the year. Trees are like those of temperate zone. Epiphytic mosses and lichens are in abundance. Champion (1939) has recognised the following two types of forests in sub-alpine zone :

(*a*) **Silver Fir-Birch forests** which are found on glacial moraines. *Abies spectabilis, Abies densa, Pinus wallichiana, Betula utilis, Quercus semecarpifolia, Pyrus spp, Rhododendrons, Juniperus recurva, J. wallichiana, Berberis, Salix fruticulosa* are common plants of these forests.

(*b*) **Birch-Rhododendron forests** which grow on rocky substrata. The common trees are *Betula utilis, Quercus semecarpifolia,* many species of *Rhododendron Pyrus* spp. *Acer* spp, *Salix, Juniperus* spp etc.

V. Alpine forests

Alpine vegetation has been classified into the following three types:

(*a*) Alpine forests,

(*b*) Moist Alpine scrubs, and

(*c*) Dry Alpine scrubs.

(*a*) **Alpine forests.** Plants growing at the altitude from 2900 to 6000 m are called alpine plants. In India, alpine flora occurs in Himalayas between 4500 and 6000 metres. At lower level, alpine forests consist of dwarf trees with or without conifers and at higher level scrubs and only scattered xerophytic shrubs are left to merge with alpine meadows. The common plants of alpine forests are *Abies, Pinus, Juniperus, Betula,* shrubby Rhododendrons, *Quercus, Pyrus, Salix* etc.

(*b*) **Moist Alpine scrubs.** This type of vegetation is distributed extensively throughout the Himalayas above 3000 metres. It is most often dense and composed of evergreen dwarf *Rhododendron* species, some birch and other deciduous trees. Mosses and ferns cover the ground with varying amounts of alpine shrubs, flowering herbs and ferns. Alpine pastures include mostly mesophytic herbs with very little grasses.

(*c*) **Dry Alpine scrubs.** These are open xerophytic formations spread in U.P., Himachal Pradesh, Punjab and Kashmir. Species belonging to *Artemisia, Potentilla, Kochia, Juniperus* predominate in the vegetation which develops generally on lime stone rock.

2. GRASSLAND VEGETATION

According to Ecological Society of America (1952), "grassland is a community dominated by grasses or grass-like plants". Approximately 24 percent of total vegetation cover is grassland. Grasslands occupy about 45.0 million sq. km. area throughout the world in the tropics, temperate and alpine regions (Shantz, 1954). Natural grasslands occur in temperate zones with annual rainfall 25 to 80 cm, while in tropics they may be found in areas receiving rainfall upto 150 cm. The favourable conditions for development of a stable grassland are frequent rainfall and sufficient warmth during the growing season. Grasslands do not form a prominent feature of vegetation in tropical part of India because of the following reasons:

(*i*) In the moist lowlands, grasses face very tough competition from trees and shrubs.

(*ii*) In the drier parts, the conditions (dry period, high temperature and evaporation), are so severe that they do not permit the extensive development of grasslands. Indian grasslands are not climax formations but they have developed secondarily under the influence of two factors, namely edaphic and biotic. In most cases, grasslands are maintained in their present seral stage due to biotic influences. Fire alone or fire coupled with grazing and browsing may be very potent factors causing the development of grasslands in forest areas. If a grassland exists in forest territory for a number of years, the forest growth cannot eliminate it even if the fire factor is absent. Once the land is covered with grasses, tree seedlings find it difficult to gain a foothold. There the grassland may persist as a very stable preclimax. With the fire and Grazing which destroy seedlings of trees as soon as they enter the grassy stands, preclimax of grasses persists for indefinite period. Once a forest has been destroyed and grasses have started secondary succession (secondary seres) over there, succession may be arrested in seral stage.

In India, there are no examples of typical subtropical climax grasslands and typical tropical Savannah type vegetation is also absent. In the tropical zone, the grassland is common enough as a secondary seral stage and may be stable preclimax under the influence of fire and grazing.

The grasses are successful invaders and are capable of colonisation in various habitats due to their following adaptabilities :

(1) They can grow in both hydrophytic and xerophytic habitats
(2) Many of the grasses show xeromorphic adaptations
(3) Their life span is from annual to perennial
(4) Habit varies from trailing to erect
(5) Size varies from small to dendroid
(6) Meristematic activity both apical and inter calary.
(7) Superficial rooting habit
(8) High reproductive capacity and capability for both sexual and vegetative reproduction
(9) Wind pollination
(10) They produce light seeds which are easily and quickly dispersed by wind and animals
(11) Position of buds below the soil
(12) Elasticity of internodes to bend
(13) Nodes have ability to produce roots and shoots
(14) Grasses can withstand trampling, Grazing and fire
(15) Grasses have wide ecological amplitutes
(16) Grasses tolerate poor growth conditions
(17) They can act as poineer coloniser in Primary and Secondary successions

There are several grassy flats and grassy plains of seral nature in Nilgiri, Khasi hills and Naga hills: At elevations higher than 800 metres, the rainfall is heavy, the soil is dark in colour and rich in humus, and the vegetation consists of grasses which grow forming tufts at lower altitude and a turf at higher altitude. In swampy places above 1800 metre elevations, species belonging to Cyperaceae are in abundance. In the regions where forests are destroyed by fire, vegetation consisting of tall grasses with scattered trees (*i.e.*, **Savannah**) develops.

Grasses are of two types : **bunch grasses** which form clumps and **Sod grasses** which develop from rhizomes. Besides grasses, many nongraminaceous herbs, shrubs and trees also occur in grasslands. Depending upon the size of grasses and the abundance of nongraminaceous flora, grasslands have been assigned different names. These are as follows :

(*i*) **Grassy Plains.** These are with short grasses.

(*ii*) **Prairies.** They contain tall grasses and shrubs

(*iii*) **Pampas or Velds.** They are similar to Prairies but with more scattered bushes and a few trees.

(*iv*) **Savannah or Savanna.** A grassland having tall grasses (1 - 4.5 m tall) and scattered trees (3-10 m tall). Savannahs are found in Africa, South America, Australia, India, and Burma. Savannahs are named after the dominant trees as for example *Acacia* savannah, *Phoenix* Savannah, *Eucalyptus* savannah and so on.

Grasslands of the world are categorised under the following types on the basis of physiognomy and habitat (Table 11.2)

Table 11.2. Grassland types of the world

Grassland type	*Distribution*	*Approximate area (million sq. k.m.)*
Marsh grass swamps	Tropical and Temperate swamps	2.0
Mountain grass meadows	High mountains in tropics	1.5
Shortgrass (Prairies, Pampas, Velds)	North and South America, Africa, Soviet Russia	3.0
Tallgrass Prairies, Pampas and Velds	North and South America, Africa	5.50
Desert grass Savannahs	Thorn bush and steppe	6.0
Tall grass Savannahs	South America, Africa, India, Australia	14.0
High grass Savannahs	Africa, South Asia	11.0

The tropical grasslands of India may be classified into the following types :

(*i*) **Xerophilous grasslands.** These are found in dry regions of North-West India under semidesert condition. The common species of xerophilous grasslands are *Andropogon contortus, Cenchrus ciliaris, C. barbatus,* etc.

(*ii*) **Mesophilous grasslands.** They are also called Savannahs. They are extensive grassy flats or grassy plains typically occurring in moist deciduous forests of U.P. The dominant species of such grasslands may be *Saccharum munja, S. narenga,* etc.

(*iii*) **Hygrophilous grasslands.** These are called wet savannahs adapted to very wet soil.

All these types being biotically controlled are inferior to temperate grasslands and suffer considerably from monsoon nature of climate and lack of proper legume mixture.

Based on their floral characteristics Indian grasslands have been divided into the following eight major types by R.O. Whyte (1957). (Fig. 11.3). The geographical distribution and environments of these grassland types are given in Table : 11.3.

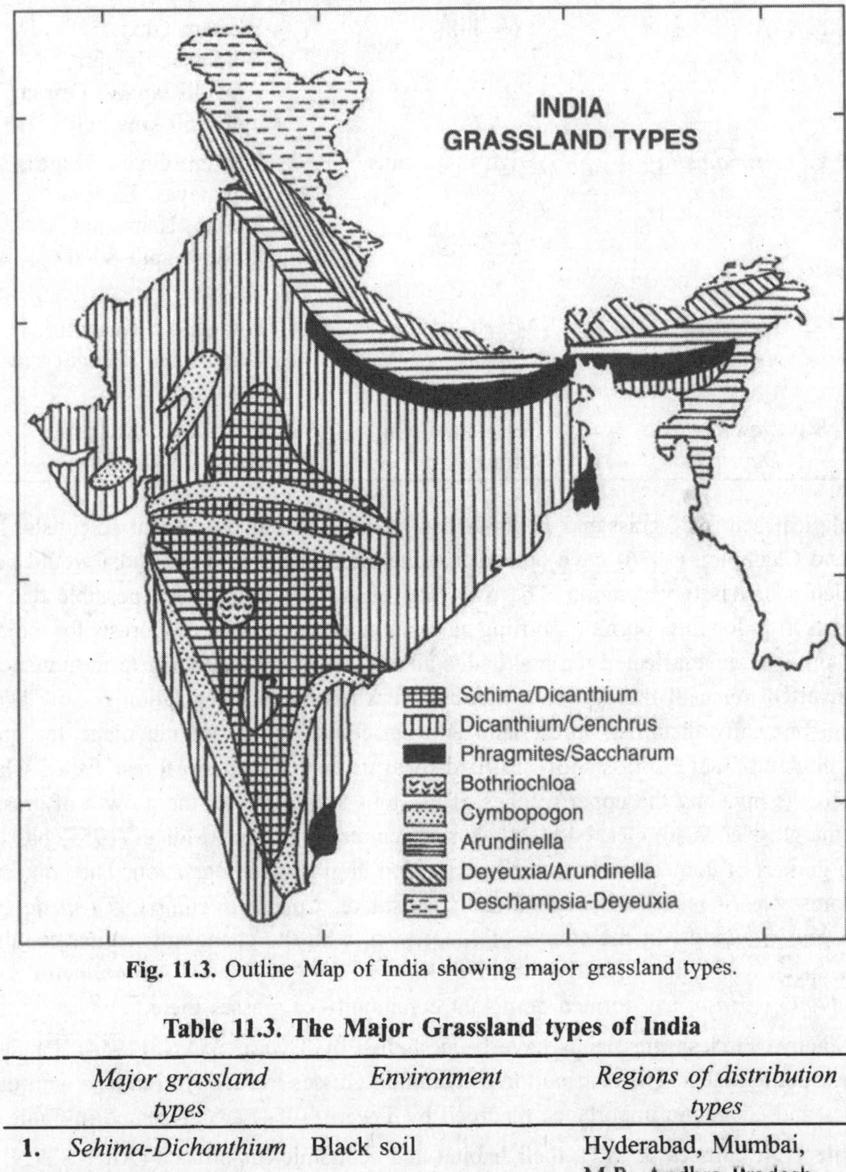

Fig. 11.3. Outline Map of India showing major grassland types.

Table 11.3. The Major Grassland types of India

	Major grassland types	*Environment*	*Regions of distribution types*
1.	*Sehima-Dichanthium*	Black soil	Hyderabad, Mumbai, M.P., Andhra Pradesh, Tamil Nadu, S-E. U.P.
2.	*Dichanthium-Cenchrus*	Sandy loam soil	Punjab, Delhi, Rajasthan, Gujarat, Cutch, Eastern U.P., Bihar.
3.	*Phragmites-Saccharum*	Marshy areas	Terai of U.P., Uttaranchal, Bihar, Bengal, Assam, Sunderban and other deltas.
4.	*Bothriochloa*	High rainfall and low lying areas	Lonavala tract of Maharashtra.

5.	*Cymbopogon*	Low hills	Western Ghats, Vindhyas, Satpura, Aravalli ranges, Orissa upto 160 kms belt.
6.	*Arundinella*	High mountains	Western Ghats, Nilgiris, Himalayas, Eastern Punjab, Hamachal Pradesh upto 3000 metres.
7.	*Deyeuxia-Arundinella*	Mixed temperate climate	Himalayas, Kashmir, Uttaranchal, Bengal and Assam.
8.	*Deschampsia-Deyeuxia*	Temperate-alpine Climate	Himalayas, Kashmir above 2500 metre.

Ecoligical status of grassland in India has been discussed by several scientists. N.L. Bor (1942) and Champion (1936) have opined that the climax vegetation of India would be forests, or grasslands or desert vegetation. The existence of grasslands has been possible due to biotic activities such as lopping, burning, shifting cultivation and grazing of the forests for the centuries. Human activities have affected the grasslands allover the world. The long term human activities have converted much of the land into agroclimax where climax vegetation is now agricultural crop. The biotically disturbed forest land and agriclimax areas provide niche for grasses on Nilgiris in South India. Grasslands or Maldeows are maintained by forest factor which kills trees, seedings invading the upper reaches of hill tops and promotes the growth of grassland. In Punjab, the grasslands are cleaned mostly for agricultural purposes. Mohan (1955) has listed the common grasses of Punjab and discussed succession in grassland vegetation. The poineer species in the forest clearings were the common grasses like, *Conchrus ciliaris, Themeda anathera, Chrysopogon montanus*. In the course of succession, with the change of habitat conditions the previous species disappeared and gradually *Dichanthium annulatum, Hateropogon contortats, Bothriochloua pertusa*, etc. formed dominant community of grasses there.

In Madhya Pradesh grasslands have been studied by Tiwari, S.D.N. (1954), Pandeya, S.C. (1955) and some others. The distribution of common grasses in Madhya Pradesh with their habit prefrences and economic importance recorded by Tiwari (1954) are presented in Table 11.4.

Table 11.4. Grasses of M.P., their habitat and economic importance (After S.D.N. Tiwari, 1954)

Species	Habitat	Economic value
Alloteropsis cimicina	Common in wasteland, on sandy loam soil	Fodder
Amphiopsis glabra	Common in rocky places	Fodder
Andropogon ascinodis	Commonly occurs in forest banks in hilly areas	Fodder
Aplluda aristata	Occurs sporadically in hill slopes, on bunds of fields in Mandla.	Fodder
Apocopsis pallida	Common on the periphery of forest	Fodder

A. Wightii	Occurs in wet places on poor soil, along rocky nullahs.	Fodder
Aristida funiculata	Sandy soils in Dhaba range.	
Arundinella bengalensis	Sandy loam soils in marshes near nullahs.	
Brachiaria ramosa	Rare	Fodder
Cenchrus ciliaris	Rare	Fodder
Cimbopogon martini	Commonly occurs in Mandla & Bastan	Fields common Russa oil
Chrysopogon	found in heavily grazed grasslands	Fodder

Pandey (1955) has worked out the Grassland associations of Saugar District of Madhya Pradesh and classified the grassland into 8 types based on apparent floristic similarty, limits of the area occupied, time duration and correlated them with the soil types (Table 11.5)

Table 11.5: Correlation of grassland types with soil types at Saugar (M.P.)

	Soil type	
Grassland Associations	**High incidence**	**Low incidence**
Themeda - Caudata	Cheromozen soil	Drifted soil not on lime bed
Eulalia – Cymbopogon	Drifted soil on lime bed	Chermozem soil
Sehima – Tropogon	Thin Coarse, limerich soil	Drifted and freshly weather soil
Bothriochloa – Dichanthium	Drifted soil not on lime bed	Chernozem soil
Heteropogon contortus	Latiarite soil	Freshly weathered soil
Aristida – Melanocenchrus	Freshly weathered soil	—
Coix – Ischemum	Drifted soil not on lime bed drfted	—
Cynodon-Bothriochloa and Dichanthium	Drifted soil not on lime bed	—

Seth (1955) recognised 13 types of grasslands in Uttar Pradesh (including Uttaranchal) on the basis of moisture relations of soils and dominant species (Table 11.6).

Rowntree (1954) has described grasslands of Assam in two associes:

(*i*) *Imperata – Saccharum – Themeda* on higher well drained lands throughout Assam.

(*ii*) *Alpinia – Phragmites – Saccharum* on clay alluvium of rivers & flood plains with *Salmalia & Albizzia* as common tree associates.

Rangnathan (1938) considered upland grassy areas of Nilgiris as climatic climax with species of *Andropogon pertusus, Themeda imberbis, Cymbopogon polyneuros* and *Eragrostis nigra* as common grasses. Meher Homji (1972) has discussed the phytogeographical aspects of shola grasslands and considers frost as a controlling factor for perpetutation of these grasslands.

Table 11.6. Grasslands of Uttar Pradesh and Uttaranchal (Seth, 1955)

Grasslands of U.P. (Uttaranchal Included)

- Xerophilous grasslands
  - Physically dry soils
    - Textually dry soils in moist regions
      - (*i*) *Aristida-Neyraudia associes*
      - (*ii*) *Chrysopogon-Euloliopsis* associes.
    - Soils of dry & arid regions
      - (*i*) Dry Scrub and ravine grasslands
      - (*ii*) *Themeda-Sehima* associes
  - Physiologically dry soils
    - In moist region
      - (*i*) *Saccharum - Vetiveria associes* (on clay water logged soils)
    - In direr regions
      - (*i*) *Sporobolus – chloris associes* (on alkaline soils)
- Mesophilous grasslands
  - Himalayan meadows
    - Alpine meadows
  - Moist Savannahs
    - Savannahs on high alluvium
      - (*i*) *Narenga* associes
      - (*ii*) *Saccharum – Themeda-Vetiveria* associes
    - Savannahs on low alluvium
      - (*i*) *Erianthus munja* associes
      - (*ii*) *Erianthus ravennae* associes
- Hydrophilous grasslands.
  - (*i*) *Erianthus ravennae* assories
  - (*ii*) Reed swamp: *Phragmites-Arundo* associes

VEGETATION OF PONDS AND LAKES

The aquatic vegetation is affected by supply of O_2, CO_2, mineral salts, nature of substratum, depth of water, temperature and light intensity. Some of the common aquatic plants which are found in the ponds and lakes of India have been discussed in "Hydrophytes" under the chapter "Plant adaptations".

BOTANICAL ZONES OR PHYTOGEOGRAPHIC REGIONS OF INDIA

Vegetation of any place is modified by the environmental factors; climate, geology and biotic factors. The great area of Indian subcontinent has wide range of climate and corresponding diversity in the vegetation. Important contributors who attempted to analyse the vegetation of India are Hooker (1907) and Chatterjee (1939). Hooker (1907) analysed vegetation of India (before partition) into the following Botanical provinces :

(1) Eastern Himalayas, (2) Western Himalayas,
(3) Indus plain, (4) Gangetic plain,
(5) Malabar, (6) Deccan,
(7) Assam, and (8) Maldives.

Chatterjee (1939) modified the Hooker scheme and suggested the following Botanical provinces in undivided India, excluding Malaya and Ceylon. These provinces are as follows :

(1) Western Himalayas, (2) Eastern Himalayas,
(3) Central Himalayas (now in Nepal), (4) Gangetic plains,
(5) Indus plain, (6) Deccan,
(7) Malabar, (8) Assam,
(9) Upper Burma, and (10) Lower Burma.

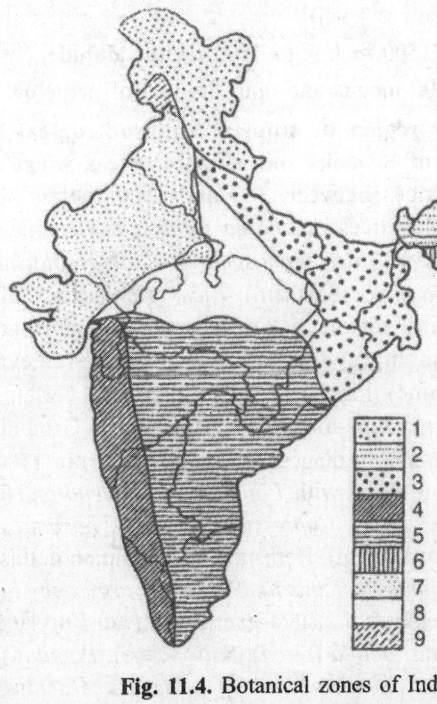

Fig. 11.4. Botanical zones of India.

1. Western Himalayas, 2. Eastern Himalayas, 3. Gangetic plain, 4. Western Coasts of Malabar, 5. Deccan, 6. Assam, 7. Indus plain, 8. Central India, 9. Bay Islands of Andaman and Nicobar.

Recently, India (after partition) has been divided into the following botanical provinces by D. Chatterjee (1962) Fig. 11.4

(1) Western Himalayas,

(2) Eastern Himalayas,

(3) Indus plain,

(4) Gangetic plain,

(5) Central India,

(6) Deccan,

(7) Western coasts of Malabar,

(8) Assam, and

(9) Bay Islands of Andaman and Nicobar.

1. Western Himalayas

The northern part of our country is bounded by highest ranges of Himalayas and is one of the important botanical regions of the world with climate and vegetation ranging from truly tropical near the low altitudes to temperate arctic types at the high altitudes. The northern mountain division can phytogeographically be divided into western, central and eastern zones.

Western Himalayas consist of north Kashmir, south Kashmir, a part of Punjab, H.P., Garhwal and Kumaon. This zone is wet in outer southern ranges and slightly dry in inner northern zone. The average annual rainfall in this region is from 100 to 200 cm. Snowfall occurs in this region during winter season. The region may be divided into three subzones (Fig. 11.5):

(*i*) *Submontane zone* or *lower region* or tropical and subtropical belts (upto about 1500 metres altitude from the sea level).

(*ii*) *Temperate zone* (from 1500 metres to 3500 metres altitude),

(*iii*) *Alpine zone* (above 3500 metres and upto the line of perpetual snow).

(*i*) **Submontane or lower region or tropical and subtropical belts.** It includes outer Himalayas, particularly region of Siwaliks and adjoining areas where annual average rainfall is over 100 cm. This zone ranges between 300 and 1500 metres above sea level. In this zone, forests dominated by timber trees of *Shorea robusta* are common. Other important tree species are *Salmalia malabaricum, Butea monosperma, Acacia catachu* and *Zizyphus species.* In the swampy areas, *Dalbergia sisso* (Shisham), *Ficus glomerata, Eugenia jambolana* are of common occurrence. In west dry regions sal trees are replaced by xeric plants particularly *Zizyphus, Carissa, Acacia,* and thorny Euphorbias. At higher elevation, around 1000 to 1500 metre altitude, cheer (pine) forests are also found at certain places. The common species of pine are *Pinus longifolia* and *Pinus roxburghii.* Ground vegetation is scanty.

(*ii*) **Temperate zone.** It commonly ranges at the altitudes from 1500 to 3500 metres above the sea level. Oaks are dominant along with *Populus, Rhododendron, Betula* and *Pyrus. Pinus excelsa, Cedrus deodara, Picea, Abies, Cupressus* and *Taxus baccata* are found in the heavy rainfall region (between 1600 and 1800 m). Herbs are also common in this region. Common herbs are *Ranunculus hirtila, Polygonum, Pedicularia, Potentilla argyrophylla, Primula, Delphinium, Clematis,* crucifers and many members of asteraceae. In cultivated drylands of Punjab, wheat and barley are main crops. In Kashmir, *Betula* (birch), *Salix* (cane), *Populus* (poplar) are of common occurrence. Besides these, *Quercus semicarpifolia, Q. dilatata, Aesculus indica* (chestnut) and

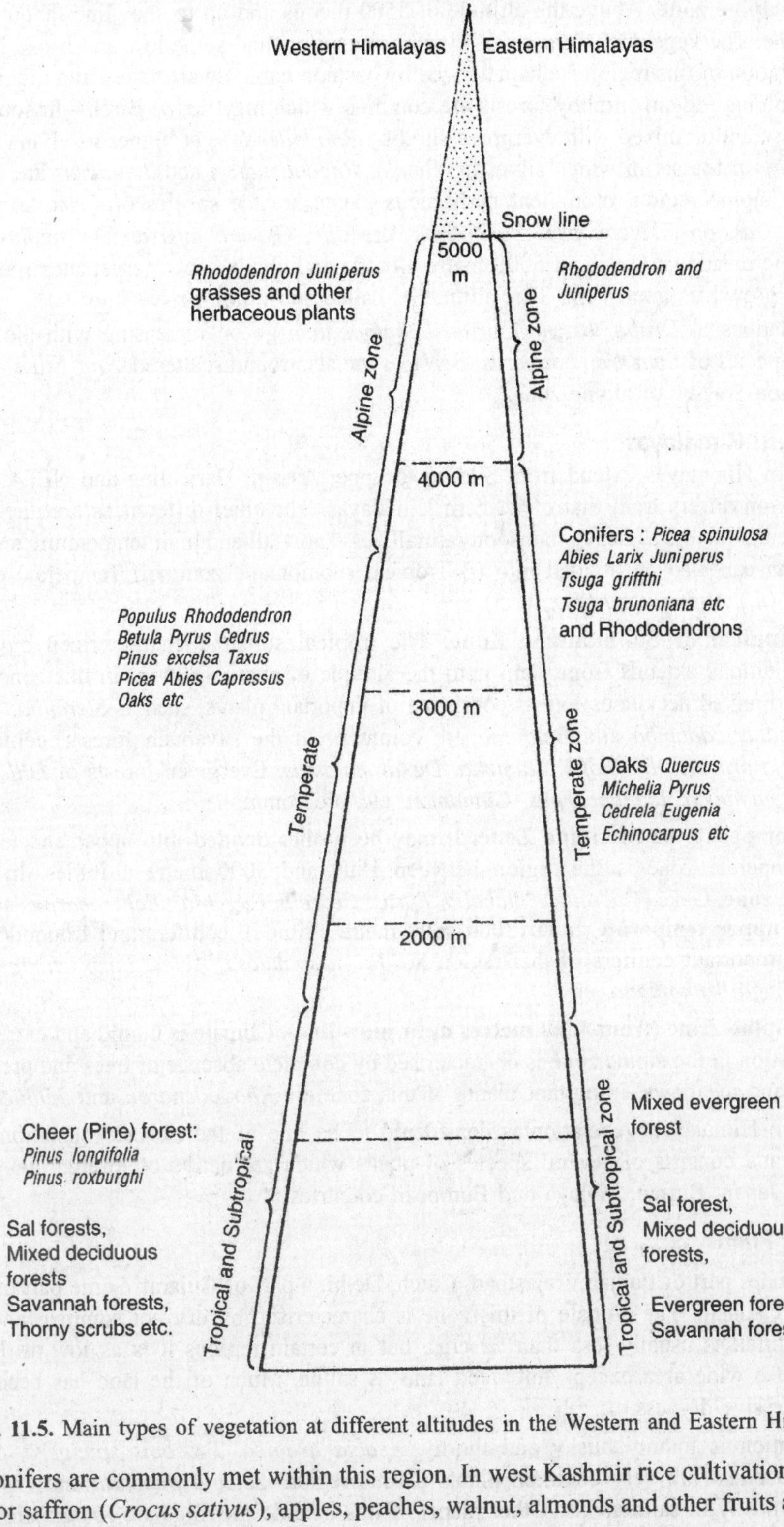

Fig. 11.5. Main types of vegetation at different altitudes in the Western and Eastern Himalayas.

many conifers are commonly met within this region. In west Kashmir rice cultivation is common. Keshar or saffron (*Crocus sativus*), apples, peaches, walnut, almonds and other fruits are important economic plants of Kashmir region.

(*iii*) **Alpine zone.** Above the altitude of 3500 metres and up to snowline (about 5000 m) is alpine zone. The vegetation consists of evergreen conifers and some low and broad leaved trees. The vegetation of this region is characterized by cushion habit, dwarf nature and gregarious habit. In lower alpine region, shrubby forests are common which may be (*a*) Birch—fir forest which is fairly dense and is mixed with evergreen shrubby *Rhododendron* at higher level and (*b*) Birch—*Rhododendron* forests in which silver fir, *Betula, Rhododendron* and *Juniperus* are common. In the upper alpine region, prominent herbaceous plants are the species of *Primula, Polygonum, Gentiana, Cassiope, Meconopsis, Saxifraga, Potentilla, Geranium, Aster, Astragalus* etc. which form alpine meadows. At about 5000 metre altitude and above snow perpetuates round the year and plant growth is almost nil. This altitude is called **snow line or ice line.**

Populations of *Draba, Braya, Cortia, Leontopodium* go on increasing with the increase in altitude. Species of *Ephedra, Juniperus, Berberis* are also found scattered. *Poa, Stipa* and *Fectuca* are common grasses of alpine zone.

2. Eastern Himalayas

Eastern Himalayas extend from Sikkim to upper Assam, Darjeeling and NEFA. Vegetation of this region differs from that of western Himalayas. The chief differences are due to changed environmental factors as heavy monsoon rainfall, less snowfall and high temperature and humidity. This region can also be divided into (*i*) Tropical submontane zone (*ii*) Temperate or Montane zone, and (*iii*) Alpine zone (Fig. 11.5).

(*i*) **Tropical or Submontane Zone.** The tropical subzone characterized by warm and humid conditions extends from plain upto the altitude of about 1800 m. In this zone mostly sal forests, and mixed deciduous forests consisting of important plants, such as *Sterculia, Terminalia, Anthocephalus cadamba* and *Bauhinia* are common. In the savannah forests, common plants are *Albezzia procera, Bischofia, Salmelia, Dendrocalamus.* Evergreen forests of *Dillenia indica, Michelia champaca, Echinocarpus, Cinnamon*, etc. are common.

(*ii*) **Temperate or Montane Zone.** It may be further divided into upper and lower zones. Lower temperate zone is the region between 1800 and 3000 metre altitudes. In the **lower temperate zone,** Oaks (*Quercus*), *Michelia, Pyrus, Cedrela, Eugenia, Echinocarpus* are common plants. In **upper temperate zone** (3000-4000 metre altitude), conifers and Rhododendrons are common. Important conifers of this region are *Picea spinulosa, Abies, Larix, Juniperus, Tsuga griffithi, Tsuga brunoniana,* etc.

(*iii*) **Alpine Zone (from 4000 metres upto snow line).** Climate is humid and extremely cold. The vegetation in the alpine zone is characterised by complete absence of trees and predominance of shrubs and meadows. Important plants of this zone are *Rhododendron* and *Juniperus.*

Eastern Himalayan vegetation is considered to be one of the richest vegetational units in the world and consists of several species of plants which are native of foreign countries, such as, China, Japan, Burma, Malaya and European countries.

3. Indus Plains

It includes part of Punjab, Rajasthan, Cutch, Delhi, a part of Gujarat. Some part of this plain is now in Pakistan. The climate of this zone is characterised by dry hot summer, and dry cold winter. Rainfall is usually less than 70 cms, but in certain regions it is as low as 10-15 cms. The soil of a wide area except cultivated land, is saline. Much of the land has become desert due to excessive dryness.

Vegetation is mainly bushy and thorny. *Acacia arabica, Prosopis spicigera, Salvadora, Capparis decidua* are very common plants of this region. *Salsola phoetida* and Lunakh grass are found mostly in saline soils. Other plants of this botanic province are *Anageissus, Eugenia, Mango, Dalbergia sisso, Albizzia lebbek, Zizyphus nummularia,* etc.

Historical evidences indicate that the area was covered by dense forest some 2000 years ago, but gradual destruction of vegetation cover either by biotic agencies or by any other agency led to the development of desert in this plain. *Saccharum munja, Cenchrus ciliaris, Prosopis spicigera, Acacia leucophloea, A. senegal* are the important plant species which are grown for checking the spread of desert.

4. Gangetic Plains

This is one of the richest vegetational zones in India. This zone covers flat land of a part of Delhi, whole of U.P., Bihar, West Bengal and also a part of Orissa. Rainfall in this zone is from 50 cm to 150 cm.

A great part of the land is under cultivation. The common crop plants are wheat, barley maize, *Sorghum* (jowar), Bajra, urad, Moong (*Phaseolus mungo*), *Cajanus cajan,* til (*Sesamum indicum*), sugarcane, Pea (*Pisum sp.*), gram (*Cicer arietinum*), potato, *Brassica*, rice.

In western part of U.P. annual rainfall is from 50 cm to 110 cm. Dry deciduous and shrubby forests are common in this part. Important plants of south-western part of U.P. are *Capparis, Saccharum munja, Acacia arabica.* In the north-western part of U.P. near Himalayas foothills *Dalbergia sisso, Acacia arabica* are most common plants.

In eastern gangetic plain, the conditions are cold and wet (annual rainfall, 150 cm in West Bengal). In this part evergreen forests are common. In central part, the annual rainfall is about 100 cm to 150 cm. The vegetation consists mainly of deciduous trees. Sal trees are dominant. Other common trees are *Terminalia tomentosa, T. belerica, Acacia species, Bauhinia, Diospiros* (Biri Ka patta or tendu) *Eugenia sp.*, neem trees (margosa), *Madhuca indica* (Mahua), *Cordia myxa* (Lasora), *Tamarindus*, Mango (*Mangifera indica*). *Ficus* etc.

In Bihar and Orissa hills, *Rubus*, *Potentilla*, *Fragaria (Rosaceae), Pyrus* etc. are common. Mangrove vegetation is common in tidal regions in West Bengal near Sunderban, and Orissa. *Rhizophora mucronata, R. conjugata, Sonneratia, Ceriops roxburghiana* and *Acanthus ilicifolius, Kandelia rheedii, Bruguiera gymnorhiza* are common mangrove plants in those regions.

5. Central India

Central India covers Madhya Pradesh, part of Orissa (now Odisha), Gujarat and Vindhya. The areas are hilly. The average rainfall per annum may be 100-170 cm. Some places are at the altitudes of 500-700 m from the sea level. Biotic disturbances are very common in this botanical province which have led to the development of the thorny vegetation in open areas. In this region teak *(Tectona grandis) and sal (*shorea robusta*) forests are* very common. Other trees are *Terminalia tomentosa, Bauhinia,* Mango, *Phyllanthus, Ficus glomerata*, etc. Among common shrubs are *Mimosa rubricaulis, Desmodium, Acacia sp., Zizyphus rotundifolia* and other. Entire forest vegetation of central India may be divided into (*i*) sal forests (*ii*) mixed deciduous forests (*iii*) thorny forests.

At Sarguja (M.P.) many species have been reported to occur. Some of them are *Pyrus, Berberis asiatica, Rubus, elipticus*, etc.

6. Deccan

This region comprises whole of the southern peninsular India including Satpura and southern part of Godawari river. Average annual rainfall in this region is about 100 cm. It may be divided into the following two subdivisions:

(*i*) Deccan plateau

(*ii*) Coromandel coast.

In Deccan plateau, teak forests containing *Diospiros, Acacia, Prosopis spicigera, Santalum album* (chandan tree) and *Cedrela toona* are common. On rocks, *Capparis*, Euphorbias, *Phyllunthus* are common. Teak, *Pterocarpus, Borassus, Foenix silvestris* are also common in this area. In Chhota Nagpur plateau, important species are *Clematis natans, Barberis, Thallictrum* and also many members of Annonaceae, Rosaceae, Compositae, Araliaceae, Apocynaceae, Lauraceae, Amaranthaceae, Orchidaceae. Some ferns are also common.

In Coromandel coast vegetation consists largely of some halophytic species.

7. Western Coast of Malabar

This is small botanical province covering Cape Comorin to Gujarat and Western Ghat. This is a region of heavy rainfall. In this zone, four types of forests are common :

(*i*) Tropical forests (occur at 700 m altitude),

(*ii*) *Mixed deciduous forests* (found at the altitude upto 1600 m).

(*iii*) *Temperate evergreen forests* (occur above 1200 m altitude), and

(*iv*) *Mangrove vegetation.*

In tropical evergreen forest the trees are tall and they have root buttresses. Important species are *Cedrela toona, Dipterocarpus, Mangifera indica, Sterculia alata, Artocarpus hirsuta*. In the mixed deciduous forests, important plants are *Terminalia tomentosa, Terminalia peniculata, Tectona grandis, Dalbergia, Lagerstroemia lanceolata* and bamboo species, particularly *Dendrocalamus* and *Bamboosa arundinacea*. On the Nilgiri hills sub-tropic and temperate conditions exist. Important plants of Nilgiri vegetation are *Rubus, Rhododendron arboreum, Barberis, Thallictrum, Ranunculus, Fragaria, Potentilla*. Many other herbs along with many grasses are also common.

Temperate forests commonly called as "sholas" contain *Gardenia obtusa, Michelia nilgirica. Eugenia* species are also common. In Malabar, plants belonging to family Dipterocarpaceae, Tiliaceae, Anacardiaceae, Meliaceae, Myrtaceae, Piperaceae, Orchidaceae and many ferns are common. The west coast of Malabar region receives very high rainfall. In the coastal region mangrove plants grow luxuriantly.

8. Assam

This botanical province is very rich in vegetation and covers valley of Brahmaputra, Naga hills and Manipur. This is the region of heaviest rainfall. Cherapunji is one of the rainiest places in the world where annual rainfall often exceeds 1000 cm. Excessive wetness and high temperature in this zone are responsible for the development of dense forests. Broad leaved, tall, evergreen angiosperms and some conifers are very common in the forests. Common plants occurring in this region are *Ficus, Artocarpus, Michelia champaca, Sterculia alata, Morus* species. Besides these, bamboos, canes, climbers, and green bushes are also common. Prominent plants in the northern forests of this zone are *Alnus nepalensis, Betula, Rhododendron arboreum, Magnolia, Michelia and Prunus*. Sal also occurs at Garo hills. Orchids and fern species are very rich in this zone.

9. Bay Islands of Andaman and Nicobar (India)

Islands: These are represented by the Andaman and Nicobar islands in the east and Lakshadweep islands in the west. The Andaman and Nicobar islands are a group of more than 300 islands, which support many characteristic plants and animals. The forests range from tropical evergreen to moist deciduous and even mangroves. The Lakshdweep group of islands comprise 36 major islands, which together from an area of 32 sq km. Many varied marine fauna are present here that include turtles, crabs, molluscs and fishes. Beautiful coral reefs are also present in this part of India.These bay islands represent elevated portions of submarine mountains. Climate is humid in the coastal region. In Andaman, beech forests, evergreen forests, semi-evergreen forests, deciduous forests and mangrove vegetation are of common occurrence. *Rhizophora, Mimusops,*

Calophyllum, etc. are common plants in mangrove vegetation. In the interior evergreen forests tall trees are common. Important species of trees are *Calophyllum, Dipterocarpus, Lagerstroemia* and *Terminalia* etc. Some part is under cultivation. The important crops are paddy and sugarcane.

QUESTIONS

1. What are phytogeographical principles? Discuss in brief the factors which affect distribution of plant species.
2. Write short notes on :
   (*a*) Discontinuous distribution of species
   (*b*) Endemism.
3. Describe the climates of India and their bearing in distribution of plant formations.
4. Describe the various forest communities of India.
5. Give an account on grassland vegetation in India.
6. Describe in brief different phytogeographic regions of India.
7. Describe the vegetation of Eastern or Western Himalayas. **(Nagpur Univ. B.Sc. (F) Bot.-I, 2010)**
8. Write brief notes on:
   (*i*) Tropical dry deciduous forest **(Nagpur Univ. B.Sc. (F) Bot.-II. 2010)**
   (*ii*) Grasslands of India **(H.P. Univ. B.Sc. III, Bot-IX, 2010)**

12

Ecological Indicators

The heredity and environment both are equally important in the expression of phenotypic characters. Heredity performs its action through environment. Species differ in their environmental requirements and establish themselves where conditions are favourable. It is found that certain species of plants, animals and micro-organisms have one or more specific requirements which very much limit their distribution. The occurrence, character and behaviour of a plant are thus indicator of the combined effect of all factors prevailing in a habitat. Since a plant species or plant community acts as a measure of environmental conditions, it is referred to as *biological indicator* or *bioindicator* or *phytoindicator.* In other words, plants which indicate some very specific conditions of environment are called plant indicators.

The knowledge of relationship between plants and ecological factors can be used as an indicator of environment. Many plants are used as indicators of environment. In a plant community some plants are dominant and found in abundance. These plants are important indicators because they bear full impact of habitat. It has been seen, in general, that plant communities are better indicators than individual plants. Individual plants or plant communities are used to determine the types of soil and other conditions of the environment. Sometimes these also indicate past or future conditions of the environment.

The knowledge of plant indicators can be helpful to determine local soil, thus it can be decided which crops should be cultivated in a particular soil and which soil should be left for pasture or other purposes. Plant indicators are also used to determine optimum use of land resources for forest, pasture, and agricultural crops. Many plants also indicate the presence of particular mineral or metal. So the presence of precious metal can be detected by the growth of the specific plant in an area.

CHARACTERISTIC FEATURES OF PLANT INDICATORS

The characteristic features of plant indicators are as follows :

1. On the basis of distribution the indicators may be 'steno' species or 'eury' species. The 'steno' is used to indicate narrow limits of tolerance and 'eury' is used to indicate wide limits of tolerance. A plant may show wide limits of tolerance for certain conditions and narrow limits of tolerance for other conditions. For example, a plant may be indicator of wide limits of tolerance for heat but of narrow limits of tolerance for water. Plants with wide limits of tolerance of heat are called *eurythermal* and those with narrow limits of tolerance for water are called *stenohydric.*

2. Plants of large species are better indicator than the plants of small species.

3. Before relying on a single species or group of species as indicators, there should be abundant field evidence.

4. Numerical relationships between species, population and whole communities often provide more reliable indicators than single species

DIFFERENT TYPES OF PLANT INDICATORS

Different types of plant indicators have different roles in different aspects which are described below :

Plant indicators for agriculture. Many plant indicators decide whether soil is suitable for agriculture or not. The growth of a particular crop plant is seen under different environmental conditions and if growth is satisfactory in a particular soil that soil is considered to be suitable for agriculture. For example, growth of the short grasses indicates that water is less in the soil. A natural growth of tall and short grasses indicates that soil is fertile and is also suitable for agriculture. Dhawan and Nanda (1949, 50) and some other workers have recorded plant indicators for different types of soils as given in Table 12.1.

Table 12.1. Plant indicators for different types of soils.

Plant Indicators	*Characteristic of soil*
Salvadora oleoides	High calcium and boron, good soil suitable for crop plants
Zizyphus nummularia	Good soil for agriculture
Prosopis cineraria	Good soil for crop plants provided irrigation is there
Peganum harmala	Soil is rich in nitrogen and salts and good for agriculture
Butea monosperma	Heavy alkaline soil
Capparis decidua	Alkaline soil
Rumex acetosella	Acid grassland soil
Pinus and Juniperus	Uranium rich soil
Salsola, Sueda spp.	Saline water condition
Andropogon scoparium	Sandy loam type soil
Argemone mexicana	Recently disturbed or flooded soil
Lippia nodiflora and *Rumex* species	Nitrate rich soils

Plant indicators for groundwater : Certain plant communities indicate the depth of ground water. Central Arid Zone Research Institute, Jodhpur has made the use of certain plant communities to indicate the depth of groundwater and salinity level in the groundwater. Chatterjee and Bhaskar (1977) have listed the plant communities as ecological indicators for ground water in Indian deserts (Table 12.2)

Table 12.2

	Plant communities	*Indicated depth of ground-water*	*Indicated total soluble salt ppm in ground-water*
1.	*Euphorbia caducifolia*	12—18 m	1,500—7,000
2.	*Acacia senegal—Anogeissus pendula*	12—18 m	500—1,500
3.	*Salvadora persica —Tamarix sp.*	6m	1,500—3,200
4.	*Salvadora oleoides—Prosopis cineraria*	10—20 m	500—2,000
5.	*Prosopus cineraria—Zizyphus nummularia-Capparis decidua*	6—18 m.	5,000—10,000

6.	*Salvadora oleoides—Capparis decidua*	6—12 m	5,00—10,000
7.	*Salvadora oleoides—Zizyphus nummularia*	18—28 m	5,000—10,000
8.	*Panicum turgidum—Zizyphus complex*	6—18 m	5,000—10,000
9.	*Panicum turgidum— polygonoides Calligonum*	6—18 m	1,000—2,000
10.	*Crotalaria burhia—Leptadenia pyrotechnica*	6—20 m	500—3,200
11.	*Suaeda fruticosa—Alurophus lagopides*	6 m	11,000—12,000
12.	*Capparis decidua*	12—20 m	180—15,000
13.	*Acacia indica—Prosopis cineraria—Salvadora oleoides*	12—20 m	400—1,500

Plant indicators for Over-grazing. Many plants are over grazed which result in modification of grassland. It has been seen that grasses are removed by overgrazing while others are disturbed and forage production is considerably reduced. Some plants which are vigorous and undisturbed, remain viable and become distinct from rest of the plants. Some plants show characteristic indication of overgrazing which can be recognised. The predominance of annual weeds and short-lived impalatable perennials indicate severe grazing. Examples of such plants are *Polygonum, Chenopodium, Lepidium* and *Verbena.* Some plants are less pronounced and show poor or no over-grazing. Examples of these plants are *Opuntia, Grindelia, Vernonia* etc.

Plant indicators of forest. Some plants indicate the characteristic types of forest and they grow in an area which is not disturbed. *Narenga porphyrocoma* is a grass which binds the soil. In such a soil sal (*Shorea robusta*) can be cultivated. *Viola* species in western Himalyas is a suitable indicator for plantation of *Cedrus deodara* and *Pinus wallichiana.* If we know that a particular forest grows better in certain area of specific soil the productivity can be increased. For example, *Quercus stellata* and *Q. mariandica* grow on upland, lowland or on sterile sandy soils.

Sometimes forests is destroyed due to fire, overgrazing and other environmental factors and the area is left to reach upto climax. In this, subdominant species get favourable chances for growth and survival. This indicates the future plants to come and establish.

Plant indicators for humus. Some plants act as humus indicators. *Monotropa, Neottia* and mushrooms indicate the presence of humus in soil. *Strobilanthes* and *Impatiens* indicate the presence of high humus or litter which prevents regeneration of tree species.

Plant indicators for moisture. Plants which prefer to grow in arid area indicate the poor or very low moisture content in the soil. *Saccharum munja, Acacia nilotica, Calotropis, Agave, Opuntia, Argemone* are such plants. Some plants grow in low soil moisture as *Citrullus colocynthis. Eucalyptus* lowers the water table. *Echinops echinatus, Cassia auriculata* are found in the area of deep water table. *Typha, Phragmites,* and *Vetiveria* grow in water-logged soil. Growth of *Typha, Phragmites, Juncus* and *Carex* indicates the swampy condition. Mangrove vegetation and *Polygonum* are found in water-logged saline soils.

Plant indicators for Soil types. Many plants indicate the characteristic soils. For example, *Casuarina equisetifolia, Ipomoea pes-caprae, Citrullus colocynthis, Calligonum polygonoides, Lycium barbarum* and *Panicum* grow in sandy soil. *Saccharum munja* prefers to grow in sandy loams. *Imperata cylindrica* and *Vetiveria zizanioides* grow on clayey soils. Cotton prefers to grow in black soil.

Plants indicators for soil reaction. Many plants indicate whether the soil is acidic or basic. For example, *Rumex acetosa, Rhododendron, Polytrichum* and *Sphagnum* indicate acidic soils. Many forest trees as *Shorea robusta, Pinus roxburghii* are calcium loving. *Tectona grandis* (teak), *Cupressus torulosa, Ixora parviflora* and *Taxus baccata* are calcicoles. Some mosses *e.g. Tartulla* and *Neckera* grow on lime stones. Halophytes such as *Suaeda fruticosa, Tamarix ariculata, Salicornea, Chenopodium, Salsola foetida* grow in salty soil.

Plant indicators for minerals. Many plants indicate the presence of characteristic minerals in the soils. These plants are called *metallocoles* or *metallophytes.*

The following plants grow in the presence of specific metals :

(*i*) **Diamond.** *Vallozia candida* grows in presence of diamond in Brazil.

(*ii*) **Gold.** *Equisetum arvense, Lonicera confuse, Papaver libonoticum, Alpinia speciosa, Thuja* species indicate the presence of gold minerals in the soil.

(*iii*) **Silver.** *Eriogonium ovalifolium* indicates the presence of silver minerals in soils in U.S.A.

(*iv*) **Mercury.** *Stellaria setacea* grows in Spain in mercury rich soils.

(*v*) **Uranium.** *Astragalus* species grows in USA in uranium rich soils.

(*vi*) **Selenium.** *Astragalus* species, *Neptunia amplexicaulis, Stanleya pinnata, Onopsis condensator,* etc. grow in selenium rich habitat.

(*vii*) **Copper.** *Viscaria alpina* in Norway, *Gymnolea acutiloba* in America, *Gypsophila patrini* in USSR grow on the soil rich in copper.

(*viii*) **Zinc.** *Viola calaminara, V. lutea* in Europe grow on the soil rich in zinc minerals.

(*ix*) **Boron.** *Salsola nitrata, Eurotia cerutoides* grow in boron rich soils.

(*x*) **Cobalt.** *Silene cobalticola* in Congo and *Nyssa sylvatica* grow in America in cobalt rich soils.

(*xi*) **Nickel.** *Lychnis alpina* grows in Sweden in presence of nickel.

(*xii*) **Sulphur.** *Allium, Arabis, Oenothera,* and *Atriplex* grow on soils rich in sulphur minerals.

(*xiii*) **Lithium.** *Lycium Juncus, Thalictrum* grow on soils containing lithium.

(*xiv*) **Iron.** *Damara ovata, Dacrydium caledonicum* grow in Scotland on the soils rich in iron.

(*xv*) **Aluminium.** *Ulex aquifolium* grows in Italy on soils rich in aluminium.

Besides above, the mineral content in a plant tissue can be employed in biogeochemical prospecting. Lyon and Brooks (1969) have found *Olearia rani* to be valuable for the molybdenum. Similarly, silver has been discovered in certain localities in leaves of plants. Sulphate content of leaf can directly be related to SO_2 concentration in air. Farrar (1977) has suggested that high sulphur content in pine needles indicates high concentration of SO_2 in atmosphere. Fluoride content in *Sorghum vulgare* leaves indicates the distance up to which air pollution by a fluoride source can fall out and this distance may be upto 4 km. In some cases higher copper content may be due to high tension copper wires. Mercury concentration in *Festuca rubra* grass may be due to chloroalkali set-up and lead in leaves may increase due to automobile exhaust.

Indicators of fires. Some plants are well adapted to grow in burnt and highly disturbed areas as for example, *Agrostis hiemalis, Epilobium spicatum, Populus tremuloides. Pteris aquilina,* and fungus *Pyronema confluens* grow in areas subjected to fire.

Indicators of petroleum deposits. Some protozoans, as Fusilinds indicate petroleum deposits in the area.

Plant indicators for pollution. The use of vegetation as biological indicator of environmental pollution has a long history. Knowledge of specific resistance to pollutant is of practical significance when plants grow in industrial or thickly populated areas. Species differ in sensitivity to pollutants. In general, plants are more sensitive to pollutants than human. Therefore, plants can be used for the bioindication of environmental pollution. Sensitive species can serve as indicators and resistant species as accumulators which collect large amount of pollutants without damage. Mosses, lichens and some fungi are much sensitive to SO_2 and halides. Even 1% SO_2 concentration is harmful to higher plants. Lichens do not survive in areas exposed to SO_2 for long time.

Many chemicals, fertilizers, pesticides and fossil fuels release toxic substances into the environment that are taken up by the plants from air, water, and soil. Atmospheric pollutants, particularly SO_2 halides (HF, HCl), Ozone and Peroxi-acetyl-nitrate (PAN) produced from automobiles, industrial fumes and strong radiations are dangerous to plants. Harmful substances that reach plant through the air are SO_2, nitrogen oxides, hydrocarbons, dust, and smoke. Plants growing in water are severely affected by toxic chemicals like cyanide, chlorine, hypochlorate, phenols, benzyl derivatives and heavy-metal compounds of sewage.

The effects of different kinds of pollution can be determined by the nature of pollutants, their concentrations and the period of exposure. Under exposure to high concentration, plants suffer acute injury with externally visible symptoms, such as chlorosis, discolouration, necrosis and death of entire plant. Besides morphological changes, biochemical, physiological and fine structural changes also occur in plants.

Pollution damage can be recognised by the accumulation of toxic material in the plant, changes in pH, reduced or increased activity of certain enzymes, increase in compounds with SH groups and phenols, lowered ascorbic acid level in the leaves, depression of photosynthesis, stimulation of respiration, low dry matter production, changes in permeability, disturbances in water balance, reduced fertility under prolonged exposure. The disturbances in metabolism develop due to chronic injury with irreversible consequences. Plants show reduced productivity and yield and quality is also lowered. Besides above, the structure of wood is changed, branches dry out and gradually the trees die. The symptoms of pollution affected plants are varied and unspecific. A particular pollutant affects different plants in very different ways and a particular symptom can be produced by a variety of substances. The influence of external factors (pollutants) on plants depends on the species, state of development and the organ or tissue involved. Morphological alteration of a plant and floristic composition of a plant community are commonly used to indicate changes in the environment. According to Van Haut and Stratmann (1970), visible plant symptoms are most commonly used to indicate the responses of plants to pollutants. Jacobson and Hill (1970) have studied the effects of common pollutants on plants. It is possible that any part of plant body, if it responds specifically or characteristically to any pollutant, can be used for its indication. Goldstein (1974) emphasized that the number and kind of biological indicators can be subdivided in order of decreasing biological complexity, such as organism, organ , tissue, cell, cell-free preparation and enzymatic studies. M.U. Beg (1980) from Industrial Toxicology Research, Centre, Lucknow has reported the responses of pollutants as a biological indicator taking several parameters into consideration.

Attempts have been made to use certain structures and functions of plants, such as seed germination, growth of plant, development of lateral branches, expansion and colour changes in leaf, flower and fruit formation, decolouration of flower, loss of physiological control, mineral composition, chemical constituents of cells, enzymatic activity and pollen germination as indicators of pollution stresses. Important aspects in response to pollution are summarised in Table 12.3.

Seed germination has been used by many workers to monitor pollution responses. Several growth parameters such as percentage of germination, seedling survival, seedling height, cotyledonary expansion and fresh and dry weight have been taken as criteria to assess plant response to a specific pollutant. *Phaseolus vulgaris* has been grown in smoke-free and smoke-affected regions by Sorauer (1899). The toxic effect of thiosulphate has been indicated as germinator inhibition in many plants. Houstan and Dochinger (1977) have evaluated germination inhibition in relation to pollution by sulphur dioxide and ozone. The effects of lead, cadmium, NO and CO_2 have been studied on many plants. Besides seed germination, pollen germination in *Nicotiana sylvestris* has been used to indicate pollution.

Some plant species are good indicators of pollution. *Polygonum*, *Rheum*, *Vicia*, *Phaseolus*, and *Capsella* have been observed as pollution indicators. According to Brandt (1974), a large number of plant species are capable of indicating specific contaminants. Generally, the plants response to pollutants is characteristic rather than specific. Efforts have been made to develop certain plant strains which can specifically be used as indicator for a particular pollutant. Stunting of corn, sweet potato and rye has been reported due to high toxicity. Reduction in root length, shoot length, numbers of tillers, leaves, ears and grains in wheat have been reported under condition of cement dust pollution. Similarly plant height, number of leaves and bolls per plant are reduced in cotton exposed to particulate pollution. Inhibition of lateral growth of forest trees is caused by lime stone dust. Pine trees do not flourish in SO_2 polluted areas. It has been noticed that leaf is the most sensitive organ to pollution. The pollution indicator value of leaf has been exploited by many workers in response to a variety of conditions. Leaf injury is a characteristic symptom to various pollutants. The characteristic symptoms on leaf include pigmentation, chlorosis, yellowing, necrosis etc. The leaves of dicotyledons generally exhibit spotted markings between the veins while monocotyledons usually show necrotic streaks between parallel veins. Injury may also occur along the leaf margin and tip. Symptoms produced by ozone, oxides of nitrogen and chlorine are almost similar. Reduced expansion of cotyledonary leaves in response to pollution has been observed in several cases.

Recently epidermal morphology has been studied as indictator of different pollutants especially SO_2. Cuticular and epidermal damage can be used to indicate air pollution. Dry weight of leaf, decrease in leaf thickness, cell size, loss of leaves and early senescence may be due to smoke and SO_2 pollution. Yunus and Ahmad (1980) have observed that leaves in the polluted area of cement factory showed higher stomatal and trichome densities, smaller epidermal cells and trichomes as compared to leaves obtained from unpolluted atmosphere.

Table 12.3. Effects of Pollution on plants and their characteristic features

Pollutant	*Plant indicator*	*Characteristic features*
Ozone	*Salvia, Dahlia, Pinus.*	Red or brown spots or streaks on the upper surface of the leaves, in presence of more pollution leaf margins are curled, wilting of apical part of pine needle occurs.
SO_2	*Ficus, Xenia, Pinus*	Light spots on the margin or near the veins, killing of tissues.
Hydrogen fluoride	*Gladiolus, Pinus*	Tissues of apical part and margin of leaves are destroyed.
Peroxi-acetyl nitrate	*Petunia, Salvia Chrysanthemum, Primrose*	Destruction of chlorophyll of leaves, and killing of cells of lower surface.

BIOCHEMICAL AND PHYSIOLOGICAL CHANGES

Chemical composition of leaf has widely been used to indicate environmental conditions. Among the biochemical estimations, the most important parameter is pigment analysis. Chlorophylls *a* and *b* have been measured as index for response to different types of pollution. In *Cassia and Cynodon,* 5% reduction of chlorophyll has been observed while in *Saccharum* the pigment is least affected. Chemical estimation like proteins, aminoacids, soluble sugars, sucrose, starch, reducing sugars, vitamin C, riboflavin, thiamine and carbohydrate are used to indicate foliar sensitivity to air pollution.

Physiological activities as opening of stomata, and rate of photosynthesis can also be used as indicators of pollution. Photosynthesis as a parameter has been used for mixed exposure of SO_2, NO_2 and dust.

Enzymatic parameters are also used to indicate the presence of particular pollutant. Peroxidase was found to be most sensitive indicator of pollutants in the absence of visible injury. Kellar (1974) and Jager (1975) have reported a differential response of enzymes in areas affected by fluoride, automobile pollution and SO_2. Thus on the basis of enzyme activity, the susceptible species of plants can be identified. Many workers have reported that enzymatic activity has been related to air pollution. Other common enzymatic parameters used are ribulose diphosphate carboxylase, glutamate- pyruvate transaminase, glutamate-oxaloacetate transaminase and peroxidase for SO_2 pollution.

QUESTIONS

1. What are plant indicators ? How can indicators be used for different purposes?
2. Write short notes on :
   (*a*) Biological indicators;
   (*b*) Pollution indicators;
   (*c*) Plant indicators for agriculture.
3. Describe different kinds of plant indicators with suitable examples.

13

Environmental Pollution

Today millions of people are without basic needs of cloth, shelter, health, education and employment. This is not due to overpopulation alone but also due to environmental consequences. The loss of forests, fertility of soil, productivity and energy crisis have created many problems. The pollution created by industries, technology etc and over-consumption by the affluent society lead to the rapid depletion of basic natural resources. Many human problems are also due to mismanagement of environment which is created by man himself.

Environmental implications due to pollution are in various aspects. These produce serious problem for human beings to maintain its existence, protection, survival and for the improvement of general standard. The basic needs of human beings have disturbed natural resources and finally led to a situation which has threatened to be disastrous. In the recent years everyone has started thinking over the problem of over-population and its consequences, which is primarily concerned with the environmental pollution and every effort should be made to focus public attention to save mankind from self destruction and steps should be taken at national and international levels so that the consequences may not become worse.

The ecological state of biosphere is becoming more and more disbalanced day by day due to technical and industrial advancements as well as population explosion. Vast changes are taking place in the environment due to interaction between human society and environment itself. Man is exploiting the natural resources for its own interest and many such instances are there as clearly indicate that man has disturbed the natural balance for the sake of small benefits and has changed the environment of many places to such an extent that they are not fit for inhabitation by living beings.

The environmental science is concerned with the study of all the systems of air, land, water, energy and life that surround us. Environmental problems are so diverse and diffused that virtually every activity of civilization interacts with the environment. The addition of extraneous materials or energy in a particular environment in concentrations greater than the normal renders the environment partially or wholly unfavourable for human life. This is referred to as environmental pollution. *"Environmental pollution is the unfavourable alteration of our surrounding, wholly or largely as byproducts of man's action through direct or indirect effects of changes in energy patterns, radiation levels, chemical and physical constitutions and abundance of organisms". These changes may affect man directly or through his supplies, of water and agricultural and other biological products, his physical objects or possessions, or his opportunities for recreation and appreciation in nature.*

—from U.S. Products Science Advisory Committee, Environmental Pollution Panel (1965)

Pollution and *contamination* are two terms sometimes used interchangeably. Contamination is the presence of harmful substances or organisms that may cause diseases or discomfort to human beings. Polluted material need not necessarily be contaminated. Dependent as he is on air, water and food from environment, man is the main culprit in polluting these natural resources to the point of no return.

Pollution is defined as the addition of extraneous materials to water, air or land which adversely affect the natural quality of the environment. In some cases, it may involve the removal, rather than addition, of constituents from the environment. A **pollutant** is a substance which may alter environmental constituents or cause a pollution. A pollutant can also be defined as constituent in the wrong amount at the wrong place or at the wrong time. For example, nitrogen, phosphorus and potassium are extensively used in agriculture to increase crop yields but sometimes they cause pollution of lakes and rivers by promoting algal growth.

The natural sources of pollution are, no doubt, important on a global scale, man generated pollutants may be more important in urban and industrial areas where the adverse effects of pollution are most severe. There is accumulating evidence that many types of pollutants can be distributed over the whole earth in relatively short period of time. Radioactive fallout from atmospheric nuclear explosion test is detectable throughout the world within a few days `or weeks; not even the polar regions are immune from fallout. Comfort giving automobile's are polluting the atmosphere with oxides of C, N, and other noxious gases. Coal, diesel oil and other fossil fuels are emitting suffocating SO_2 and choking our lives. Synthetic chemicals like plastics are adding to the problem of solid waste disposal; while detergents that cannot be decomposed by micro-organisms are making the natural rivers and streams polluted bubble boxes. Some other effects of environmental pollution are problems of health, soil erosion, sanitation, water supply, energy crisis, population and depletion of natural resources.

The law of conservation of mass or material equally applies to the pollutants and while one cannot destroy them, they may be changed from one state to another or from one compound to a another. The presence of small amount of pollutants may make profound influence on human health even when the level of air pollutants is so low that they cannot be detected except with special instruments. Some pollutants may harm living creatures exposed for long periods of time.

Pollutants are divided into two categories :

(*i*) **Biodegradable pollutants.** These pollutants are natural organic compounds which are degraded by biological or microbial action *e.g.*, sewage.

(*ii*) **Non-biodegradable pollutants.** These are not acted upon by microbes but are oxidised and dissociated automatically. They are further divided into two classes :

(*a*) **Wastes** *e.g.*, glass, plastics, phenolics, aluminium cans, etc.

(*b*) **Poisons** *e.g.*, radioactive substances, pesticides, heavy metals like mercury, lead, cadmium etc.

As regards the nature of pollutants, the problem of pollution can be divided into the following categories :

1. Pollution caused by solid wastes
2. Pollution caused by liquid wastes
3. Pollution caused by gaseous wastes
4. Pollution caused by wastes without weights.

1. **Pollution caused by solid wastes.** Solid wastes are useless and unwanted substances that are discarded by human society. Solid wastes include the heterogenous mass of urban wastes as well as the more homogenous accumulation of agricultural, industrial and mining wastes.

The proportions of different constituents of solid wastes vary from season to season, place to place, extent of industrial and commercial activities and so on.

1. **Urban solid wastes** are those which are collected and disposed of by municipal bodies. These include the following :

(*i*) **Garbage** such as wastes from kitchen, slaughter houses, canning and freezing industries.

(*ii*) **Rubbish** including the **combustible** wastes such as leaves, grasses, plants from the gardens, clothes, paper etc and **noncombustible** wastes such as bottles, crockery and plastic materials, glass, metals etc, generated from households, commercial establishments and markets.

(*iii*) Ashes.

(*iv*) Wastes formed due to demolition and construction processes, as for example, bricks, stones, plaster, plastic materials, furnitures etc.

(*v*) Dead animals.

(*vi*) Sludge, settled solid components of sewage wastes.

2. **Agricultural Solid Wastes** such as manure, crop residues, pesticides, insecticides, wastes of farm animals and so on.

3. **Mining Wastes** such as wastes from coal mines, mines of metal ores, radioactive substances etc.

4. **Industrial Solid Wastes (ISW).** These are generated by various industrial units such as chemical plants, paint industries, cement plants, metallurgical plants, power plants etc. Industrial solid wastes can be broadly classified into the following two groups :

(*i*) Non-hazardous wastes.

(*ii*) Hazardous wastes.

NON-HAZARDOUS INDUSTRIAL SOLID WASTES.

The major industries in urban areas that generate substantial amount of non-hazardous solid wastes are fruit and food processing plants, cotton mills, paper mills, sugar mills, textile mills etc. The wastes generated from these industries are biodegradable.

Non-hazardous non-biodegradable solid wastes are coal ash or fly ash generated by thermal power plants, blast furnace slags, wastes of steel melting units, muds etc.

HAZARDOUS INDUSTRIAL SOLID WASTES.

Some of solid wastes generated by industries are inflammable, corrosive, explosive, chemically highly reactive and toxic. The major industries that produce hazardous wastes are metal, chemical, drug or pharmaceutical, leather, pulp and paper, electroplating, refining, pesticide, dye, rubber etc. It is estimated that currently the industrial sector generates about 100 million tonnes of non-hazardous solid wastes and two million tonnes of hazardous wastes a year.

The surveys conducted by Environment Protection Training and Research Institute from 1971 to 1995 revealed that the urban population generated 375 g. of solid wastes per head per day in 1971 and in 1995 it was estimated to be 490 g. per head per day (Table 13.1).

Table 13.1. Urban wastes generated annually in Indian Cities.

Waste generated	Year of survey			
	1971	1981	1991	1995
Daily per capita generation in gm.	375	430	460	490

The problem of domestic solid waste whether it is garbage, litter or rubbish cannot be underestimated. In advanced countries it is estimated that at an average a city dweller produces more than half a ton of garbage per year.

2. **Pollution caused by liquid wastes.** Major portion of water on the surface of earth is not in a form that can be used for domestic purposes by man since it is saline. Man obtains fresh water from the well known hydrologic cycle. Surface runoff gathers minerals and organic impurities as it moves down to the sea. Under normal conditions river takes care of many

polluting substances that enter its body. Green plants and algae take up CO_2 from water and in presence of sunlight synthesize carbohydrates and O_2 is produced by splitting of water molecules. Animals take up O_2 and give up CO_2 and other compounds which are used by the plants. This is ecological balance in a natural stream. If some organic matters, that can be food materials for bacteria, enter water course then bacteria oxidise these materials and in that process take up O_2 from water. If the process of re-oxygenation is slower than the process of de-oxygenation then river will be devoid of life sustaining dissolved oxygen and aquatic animals and plants will die and under anaerobic condition foul smelling hydrogen sulphide and other products are formed. The river can be termed dead in the sense that it cannot sustain normal aquatic life.

Sewage. The important source of organic pollutants is sewage which contains faecal matter, urine, and kitchen washings and some soil washing. Sewage contains large number of bacteria, both pathogenic and harmless. The strength of organic waste materials of sewage is measured in terms of demand for dissolved oxygen required in oxidation of organic matter by micro-organisms. This value is expressed in terms of mg of O_2 per litre of waste. Since biological reaction is dependent on time and temperature, this O_2 demand (called biochemical O_2 demand or B.O.D.) is given for 5 days at 20°C. If the value of B.O.D. is below 1500 mg per litre, the sewage is termed *weak waste*, if it is below 4000 mg per litre it is *medium* an above this value it is termed *strong waste*.

Domestic sewage in small quantities rarely gives trouble. However, if liquid industrial wastes enter the river along with acid or alkali and poisonous substances like cyanides, etc., the aquatic life in the river is affected and self-purification system of water is impaired. Surface run-off from agricultural fields can carry nitrogen and phosphate fertilizers that enter the aquatic envrionment and later undergo decomposition adding to the organic loading of the stream. Pesticides, and herbicides which enter waters may kill some organisms or accumulate in the fishes which when consumed by man, pass on the chemicals giving rise to cumulative poisoning.

3. Pollution from gaseous wastes. The gaseous wastes—most dangerous to people—are the ones that threaten the life of animals and plants. Carbon monoxide, SO_2, NO_2 (nitrogen dioxide), ozone and the so called "smog gases" made up of complex mixtures of hydrocarbons are common gaseous pollutants. These are the gases generally found in the atmosphere of industrial cities.

Carbon monoxide which is a product of incomplete combustion is deadly poisonous at high concentration. It has high affinity for haemoglobin in the blood and prevents that from transporting oxygen from the lungs to the tissue of the body. Fortunately, most of the times the amount of CO in the open air is too low to do much damage to human health.

Hydrogen sulphide is another deadly gas when it occurs in large concentration which is rare.

Sulphur dioxide Of the common gaseous pollutants, (Sulphur-di-oxide SO_2) is regarded as one of the most dangerous gases to human health. It attacks the respiratory tracts and interferes in the breathing mechanism. Concentrations above 1 ppm (part per million) can begin to affect people.

Nitrogen dioxide gas in sufficient quantities in air may attack lungs and cause eye irritation. Some hydrocarbons and nitrogen dioxide under the influence of sunlight produce complex substances affecting the eyes and mucous membrane.

Smog is the result of fog and the photochemical oxidation products of hydrocarbons and other organic compounds released from automobile exhaust.

4. Pollution caused by wastes without weight. The fourth kind of pollution caused by "wastes without weight" can also be called pollution by energy waste. Certain types of pollution cannot be seen and these are included in this group. Wastes without weight include the following:

(*a*) Radioactive substances and lethal radiations

(*b*) Heat, and

(*c*) Noise

(*a*) **Radioactive substances.** The amount of radioactivity in this atomic age has increased in the environment many folds as compared to Thirties and Forties. According to one authority, man may be already having radioactive cesium in his muscles, radioactive strontium in his bones and radioactive iodine in his thyroid indicating the extent of radioactive pollution. Although every precaution is taken in the functioning and maintenance of nuclear reactors, it has been shown that minute, yet measurable amount of radioactive waste material escapes into the environment. From the minining operation of uranium to the use and final disposal of wastes from the reactors, radioactive waste materials continuously escape out. Besides emission from nuclear installations, there is worldwide fallout of radioactive substances from atomic explosions. This kind of radioactive pollution caused alarm in the late Fifties and Sixties leading to a moratorium on surface and aerial tests explosions by the super powers. China and France incidentally continue to make occasional tests.

The underground storage of radioactive material in concrete or steel containers still does not ensure complete safety. Radioactivity cannot be destroyed and hence it has cumulative effect. The minute radioactive atoms, molecules or ions in water to which they may have leaked out pass on to aquatic organisms and since they are part of food chain, the radioactivity goes on accumulating in the higher organisms. In a survey of Columbia river into which cooling water of plutonium producing reactors was dumped, it was estimated that compared to the radioactivity of river water the microscopic plants and animals had one thousand times more radioactivity, fly larvae had 35,000 times more radioactivity, while the eggs of ducks which fed on the larvae etc., had 40,000 times and some birds had almost 75,000 times more radioactivity than that of water.

(*b*) **Heat.** A large quantity of waste heat energy by way of hot liquid streams or hot gases released by industries, automobiles etc., dissipates in to atmosphere and water and enhances the temperature.

(*c*) **Noise.** Noise is unwanted sound. It has come to be regarded as an important pollutant of the environment. The sources of noise for the general public are the machine in the industry, traffic noise due to trucks and cars as well as due to the indiscriminate use of transistors, radios and public address systems. Processions, public broadcasting of films, music of high pitch on festive occasions are the noises that pollute the environment. Noise produced from the aircraft, especially the supersonic jets, is harmful.

DIFFERENT TYPES OF POLLUTION

WATER POLLUTION

Ground Water Pollution

More and more people throughout the world, including India, are now dependant on ground water for drinking, agricultural and industrial use. In the recent past, ground water was considered generally pure and safe for drinking. However, recent studies have revealed that ground water sources are also facing major threats of pollution.

Ground water is being polluted by percolation of contaminated surface water through the layers of the earth. Release of raw sewage in unlined soak-pits and release of toxic effluents by the industries into surface water bodies, are the main causes of ground water pollution Indiscriminate and overuse of fertilisers, chemicals and pesticides have also caused ground water pollution through the seepage of irrigation water into ground water reserves.

The hazards of ground water pollution depends on several factors such as:

- Concentration or toxicity of the pollutant
- The level of ground water - if the level is higher chances of contamination are more
- Conditions of ground water recharge

Water is essential for all forms of life and none can survive on this earth without water. The surface of earth measures 50,000 billion hectare of which about 70% is covered by water and the rest is land. The total volume of water on the earth is 1011 million cubic kilometres of which about 97% *i.e.*, 986 million cubic kilometres is contained in oceans and an additional 3 million cubic kilometres of salty water is buried underground and the remaining 2.5% (22 million cubic kms) account for the total fresh water, frozen water of glaciers and polar ice caps. Of the 113 river basins in India, 14 are major, 44 are medium and 55 are minor (major basins are larger than 20,000 square km., minor basins are smaller than 2000 sq km). Three major rivers, namely the Ganga, the Brahmaputra, and the Indus, are snowfed rivers originating from the Himalayas and the other major rivers originate either in the central plains or in peninsular India.

Chemically, water contains two parts hydrogen and one part oxygen. Water for human consumption should be clean, colourless, odourless, well aerated, cool, soft, palatable, free from dissolved toxic substances and suspended particles. But water is rarely found in pure state. Man has polluted the bodies of water to an alarming state. In recent years the pollution of water has become one of the most significant environmental problems in the world. There is a great concern for rapidly deteriorating quality of water. The causes of water pollution are many but urbanisation, industrialization and increasing pollution are more prominent among them. Water is said to be polluted when its quality or composition is changed either naturally or as a result of human activities and it becomes unsuitable for drinking and less suitable for domestic, agricultural, industrial, recreational and other uses.

The dissolved or suspended substances which deteriorate the quality of water and make it unfit for human consumption are called **water pollutants**. In other words, water pollutants are those physical, chemical or biological factors which are harmful to aquatic life and to those who consume water. Water pollutants include several chemicals in dissolved or suspended state, some physical factors such as heat, radiations, and some biopollutants such as aquatic microorganisms, particularly pathogens (Table 13.2).

Table 13.2 Major water pollutants, examples and sources

Category	Examples	Sources
1. Affecting health		
Infectious agents	Bacteria, viruses and parasites	Sewage, human and animal excreta
Organic chemicals	Pesticides, plastics, detergents, oil	Agricultural, industrial and domestic wastes
Inorganic chemicals	Acids, caustics, salts, metals	Industrial and domestic effluents
Radioactive materials	Uranium, thorium, radon, etc.	Mining, power plants, natural sources
2. Affecting ecosystem		
Plant nutrients	Nitrates, phosphates, etc	Chemical fertilisers, sewage, manure
Sediments	Silt, soil	Soil erosion
Thermal	Heat	Industries, power plants
Oxygen demanding	Agricultural wastes, manure	Sewage, agricultural runoff

Most of the wastes generated by human society are disposed of in the bodies of water such as rivers, lakes and oceans. Some of the wastes which are discharged in air or on land may also ultimately enter the bodies of water. When human population is concentrated too much along the water ways they are unable to handle huge quantities of domestic wastes released into them and consequently they become polluted with unhappy results. Growth of big cities and industrial setups at the banks of rivers have posed serious problems. Huge discharge of industrial wastes and sewage into the rivers, lakes, ponds etc. at times, polluted water to such an extent that dead fishes were seen floating on the surface of water. Today there is a growing concern about the pollution of our lakes and rivers. The results of water pollution are quite apparent whether they are measured in terms of living resources, or hazards to human health or reduction of immunities.

Marine Pollution

Marine pollution is the matter of International concern from the point of view of conservation of living resources. All coastal nations dispose of millions of gallons of untreated sewage, millions of tonnes of garbage, unlimited amount of low level radioactive wastes etc. into the seas. There can be no doubt in it that there has been a serious decline in environmental quality and production in estuarine and shallow coastal waters as a result of pollution. Oceans are considered the ultimate waste disposal sinks. Though, the oceans have enormous capacity to dilute, degrade and disperse large amounts of wastes, there is a limit to it. In addition to the marine environment, areas along the coasts, such as, estuaries, reefs, wetlands, mangroves, etc are adversely affected due to enormous dumping of pollutants into the ocean. This problem is further aggravated due to the fact that about 40% of the world's population lives near the sea.

The oceans cover major part of the earth's surface (71%) and play an important role in maintaining the different geochemical cycles. The unceremonious dumping of pollutants in to the marine water would thus cause severe imbalances in the earth's natural cycles.

The main sources of marine pollution are:

- Muncipal wastes and sewage
- Industrial effluents
- Runoff agricultural wastes
- Oil spills from tankers
- Offshore drilling and mining
- Submarine nuclear testing
- Dumping of radioactive wastes

The consequences of marine pollution are as follows:

- The polllutants adversely affect the productive ocean regions, thus causing huge losses of fish populations and coral reefs. This results in economic losses amounting to billions of dollars per year.
- Eutrophication, due to the influx of organic pollutants, results in the formation of **red tides**. These are blooms (massive growth) of red algae, which inhibit the movement of ships and also kill marine fauna.
- Dumping of huge amounts of toxic wastes in a short duration of time, creates areas of oxygen-depleted zones in the coastal waters. In these zones, most of the aquatic lives die or migrate elsewhere.
- Discarded garbage, sewage, plastic refuse, etc that are dumped in the oceans sometimes accumulate in the beaches. This spoils the aesthtetic beauty of the region and results in loss of tourism.

Water Pollutants and their Effects

Most of the rivers and fresh water streams in India are badly polluted by industrial wastes or effluents. The major sources of pollution of some Indian rivers are listed in Table 13.3.

Table 13.3. Indian rivers and sources of their pollution

Name of river	Sources of pollution
1. Kali	Sugar mills: distilleries; paint, soap, rayon, silk, yarn, tin and glycerine industries.
2. Yamuna	D.D.T. factory, sewage, Indraprastha Power Station, Delhi.
3. Ganga	Jute, chemical, metal and surgical industries; tanneries, textile mills and great bulk of domestic sewage of highly organic nature.
4. Gomti	Paper and pulp mills sewage.
5. Dajora	Synthetic rubber factories.
6. Damodar	Fertilizers, fly ash from steel mills, suspended coal particles from washeries, and thermal power station.
7. Hoogly	Power stations; paper pulp, jute textiles, chemical mills, paint, varnishes, metal, steel, hydrogenated vegetable oil, rayon, soap, match, shellac, and polyethene industries and sewage.
8. Sone Bhadra	Cement, pulp and paper mills.
9. Cooum, Adyar and Buckinghum canal (Chennai)	Domestic sewage, automobile workshops.
10. Kaveri	Sewage, tanneries, distilleries, paper and rayon mills.
11. Godavari	Paper, mills.
12. Siwan	Paper, sulphur, cement, and sugar mills.
13. Kulu	Chemical factories, rayon mills and tanneries.
14. Suwao	Sugar industries.

Contamination of water with industrial wastes is most dangerous. The sewage of big cities is often drained into rivers. This sewage promotes the growth of phytoplanktons. The excessive growth depletes the oxygen of water. This reduction of oxygen and the presence of poisonous wastes affect the fish population. Besides these, rivers, lakes and ponds are also used directly by people for bathing or washing. This contaminates the water with the germs of various diseases-like cholera, dysentery and hepatitis.

The effluents produce physical, chemical and biological changes in water. Some pollutants produce only temporary effects in water whereas others have long standing effects. There are several types of physical and chemical effects produced by pollutants. These are: (1) addition of poisonous substances; (2) addition of suspended particles; (3) addition of non-toxic salts; (4) water deoxygenation, and (5) Heating of water. Some of the important effects of water pollutants are discussed here as under :

1. **Sewage and other oxygen demanding wastes.** These are largely organic materials that can be oxidised by micro-organisms to CO_2 and water. When the amount of sewage discharged is relatively small, the river will not become badly polluted and the biological degradation will

soon remove most of the wastes. However, medium and strong sewage as well as other O_2 demanding wastes from industry and agriculture can lead to depletion of dissolved O_2 in water. Septic conditions are said to prevail when the dissolved O_2 is very low.

Sewage contains human excreta, dung and urine of animals, some dissolved proteins, carbohydrates, fats and a variety of inorganic wastes such as nitrites, nitrates, phosphates, chlorides, carbonates, sulphates and mineral elements like sodium, potassium, calcium, magnesium, aluminium, nickle, chromium etc. Through drainage system the sewage is discharged into the rivers, canals, lakes and sea which makes the bodies of water polluted. Pollution of water promotes vigorous growth of algae and other microbes which results in the development of yellow or grey scum on the surface of water. Sometimes water smells stingy and appears to be coloured. This is referred to as water bloom. In natural state, the organic wastes are degraded or decomposed by micro-organisms. Sometimes, the rate of decomposition of pollutants is much slower than their addition leading to depletion of oxygen which makes the pollution problem serious and complicated. In developed countries and in a great majority of developing nations including India, different types of sewage disposal systems are employed for the removal of organic and inorganics wastes and other harmful substances from sewage before that is discharged into the body of water.

Sewage and other O_2 demanding wastes are classified as water pollutants because of the following reasons :

(*i*) their degradation leads to O_2 depletion which affects or even kills fishes and other aquatic life;

(*ii*) They produce foul odour and undesired colours ;

(*iii*) They may lead to scum and sludge that render water unfit for recreational use.

2. **Biopollutants.** Microorganisms such as algae, fungi, bacteria, viruses, protozoa, etc. often reach to water bodies through surface runoff, domestic wastes and sewage. These microbes produce several undesirable and harmful effects in water. Many of them cause diseases in human beings and equatic animals through contacts. Waste water from municipalities, sanatoria, tanning and slaughtering plants and boats discharged into the rivers, streams, lakes, etc. may be a potential source of infective bacteria and other microbes which cause diseases in man and other animals. Cholera, typhoid and many skin diseases are transmissible though polluted water. Recent Jaundice epidemic in Delhi and Kanpur was considered to be due to the excessive pollution of water of Jamuna and Ganga respectively, which are used for drinking purpose.

3. **Plant nutrients.** Surface runoff from agricultural fields carries nitrogenous and phosphate nutrients that increase the growth of aquatic plants and later undergo decomposition adding to organic loading of the streams. Excess algal growth has been of particular concern since algae lead to depletion of O_2 in water, create problems for municipalities and industries and make water unfit for recreational uses.

4. **Exotic organic chemicals.** These include surfactants, detergents, pesticides, various industrial products, oils and decomposition products of other organic compounds. Good amount of oil along with waste products of industry reaches the sea and rivers. Oil discharged from coastal industries and from ships is most likely to be responsible for the tainting of fishes and shell fishes. Sea food with oily taints is highly objectionable and the suspected presence of tainted consignment can damage the market generally.

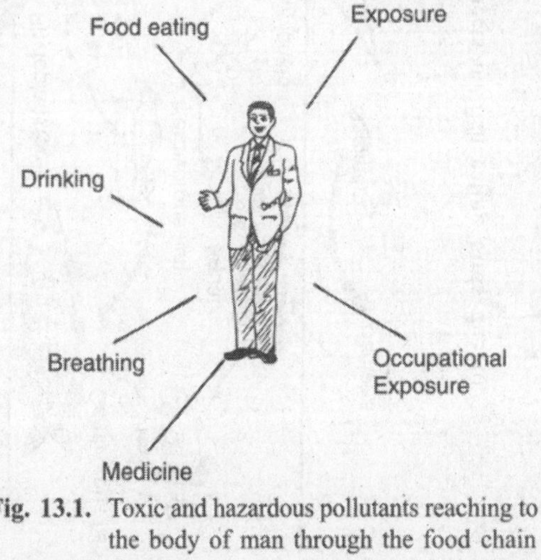

Fig. 13.1. Toxic and hazardous pollutants reaching to the body of man through the food chain

The most important groups of compounds which are both toxic and persistent are the chlorinated hydrocarbons including organochlorine pesticides, such as DDT, Dieldrin, BHC, Polychlorinated biphenyls (PCBs) and benzene products. The pesticides as shown in Fig.13.2 are dispersed in the environment through air and water. They become concentrated in successive transfers from plants to herbivores to carnivores. This is termed biomagnification. Agricultural pesticides poison or kill millions of fishes and other aquatic life. DDT may upset behaviour pattern in fishes and survival of young molluscs and crustacea. PCBs compounds seem to be substantially less toxic than DDT but they certainly accumulate in marine animals in the similar manner. There is suggestion

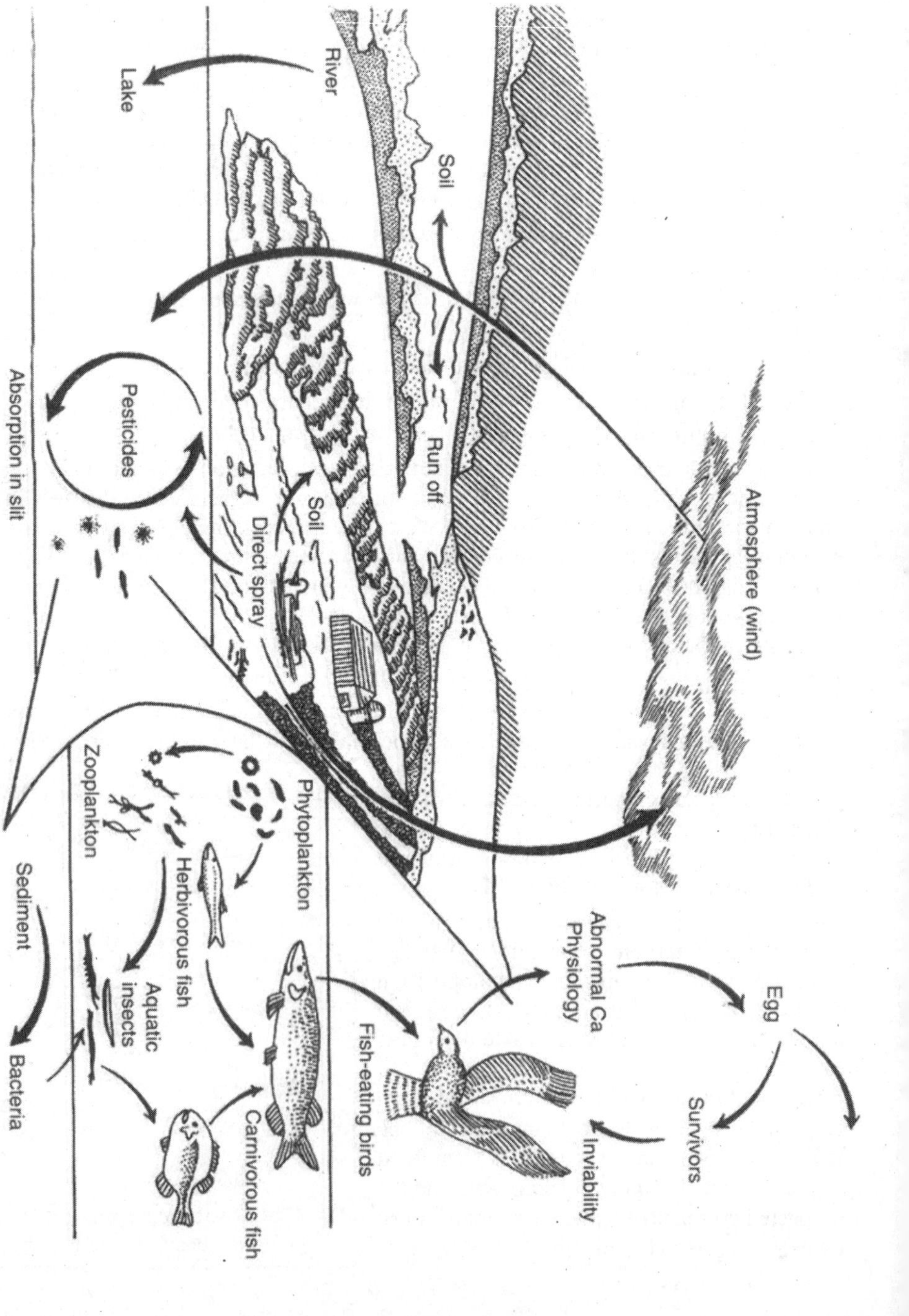

Fig. 13.2. Cyclic distribution of pesticides in nature and their effects on fish eating birds due to biomagnification through aquatic food chain.

that they were responsible for sea birds' deaths at the U.K. coast in 1969 but it could not be substantiated. Some plastic intermediates and the by-products appear to be widespread in the sea. Very insufficient information is available to assess their toxicity which may well accumulate in marine animals and plants.

Biomagnification. At various places, water becomes accumulated with some toxic compounds such as arsenic, lead, cadmium, mercury and some pesticides etc. These reach to underground water by percolation. When the people consume water, these compounds remain stored in the body tissue. Many plants and some animals also store such toxic compounds in their biomass. Human beings consume such toxic compounds through plants, animals and polluted water and store in their body tissues (Fig. 13.1). This phenomenon of accumulation of biotoxins is called **bioaccumulation**. As these compounds do not metabolise but they get magnified and become concentrated. Thus, the amount of toxic compounds is increased through biological food chain; the process is called **Biomagnification** or **Biological magnification**. During deficiency of food in the body, the reserve food is utilized along with these toxic compounds, which are passed into blood stream causing toxic effects.

Figure 13.2 shows the various steps of food chain in which the concentration of DDT increases from 0.02 ppm in the pond water to 5 ppm in phytoplanktons (free floating plants) to 40-300 ppm in plant eating fishes and ultimately reaches to 2000 ppm in fish eating birds.

5. **Inorganic minerals and chemical compounds.** Inorganic chemicals of many types find their way into waters through municipal and industrial wastes and the urban runoff. These pollutants can kill and injure fishes and other aquatic life and they can interfere with the suitability of water for drinking and industrial purposes. Mercury, lead, cadmium and copper are the important metals that cause most concern. Other metals reaching the rivers and sea in substantial quantities are zinc and chromium. Cyanides, thiocyanates, chromates, acids, alkalies, organic solvents and several other industrial wastes are causing serious concern to general public.

Aresenic. Arsenic is absorbed through skin and lung. It causes cancer of skin and lung. Chronic exposure to arsenic causes "Black-foot" disease which is prevalent in the parts of West Bengal and Bangladesh (Fig. 13.3).

Chromium. It is toxic to aquatic life.

Mercury occurs naturally in the sea as the result of weathering of mercury bearing rocks. It is also present in fossil fuels, coal and oils and reaches the sea by serial transport. It is highly persistent and said to be converted into a highly toxic monomethyl mercury (CH_3Hg) and dimethyl mercury $(CH_3)_2Hg$ which produce nervous disorders in marine animals at an ordinarily low level of dietary intake. Normal level of Hg in fish probably ranges between 0.02 and 0.2 ppm. In situations exposed to industrial discharges containing Hg wastes in Japan, Sweden and North America, Hg levels above 1.0 ppm have been found. Consumption of Hg contaminated fishes may be hazardous to man. High concentration of mercury is injurious to human life. Human beings feeding on such poisoned animals develop a crippling deformity

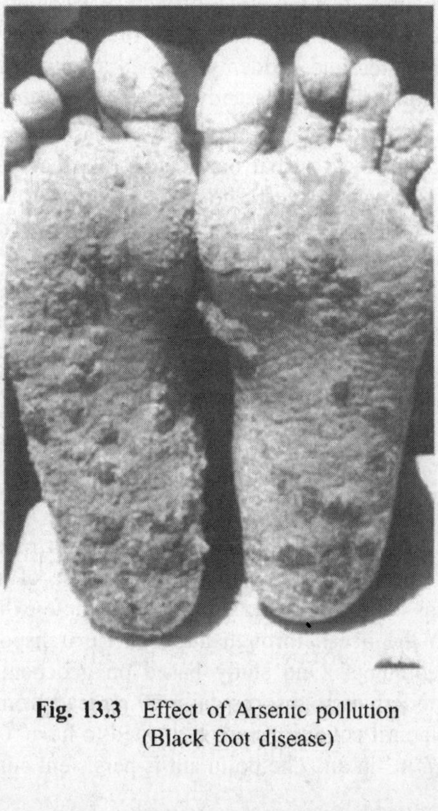

Fig. 13.3 Effect of Arsenic pollution (Black foot disease)

called *minamuta disease*. Mercury causes chromosome damage and inhibits chromosomal disjunction during gamete formation and brings about genetic changes (Ramel,1974). It is now known that anaerobic bacteria in bottom mud can convert Hg into monomethyl mercury (CH_3Hg).

Minamata Disease. Minamata disease is caused by organic mercury poisoning. In 1956, an epidemic broke out in minamata, a small town on the coast of Kyushu which is the southern most Island of Japan. The disease took a heavy death toll. The cause of disease was mercury waste dumped into Minamata Bay by Chisso Corporation. The effluent of chissofactory contaminated the Mianamata Bay with mercury. The mud samples collected from the Bay near factory outlet contained about 200 ppm mercury. Fish and shellfish samples were also found to be contaminated with high methyl mercury levels ranging from10–40 ppm. Autopsies of Manimata Victims revealed about 70 ppm in liver 140 ppm in kidney and 24 ppm in brain which is very sensitive to methyl mercury. Even hair samples of residents of the Minamata town with no outward symptoms of disease showed 100–150 ppm methyl mercury, normal level ranging from 8–90 ppm. Upto early 1960s, it was widely assumed that elemental mercury discharged into the biosphere was not hazardous as it did not react easily with other substances and it was nearly insoluble in water. Swedish scientists for the first time noted this fact that mercury dumped in the bay was in elemental or inorganic form whereas the fishes inhabiting in the waters contained toxic form, the methyl mercury. In 1972, it was demonstrated beyond doubt that aquatic micro organisms living in the mud of natural water ways could transform insoluble elemental mercury into highly soluble and highly toxic form, the methyl mercury. That soluble methyl mercury was absorbed by tissues of fishes and from fishes that reached to man and other predators that consumed contaminated fishes.

The symptoms of minamata disease are muscular pain, madness convulsions, arching of body like a bow, loss of speech and vision, paralysis, crippling of arms and legs. At the last stage of disease, the patient loses imotional control, passes into coma and finally dies.

Soon after 1956 epidemic outbreak, Chisso's own physician Dr. Hajime Hosokawa had quietly begun to test the effect of factory waste by feeding that to laboratory cats and, to this surprises, he found the same results. On 7th October, 1959 he found that cats suffered convulsions, profusely salivated and suddenly crashed against the laboratory wall and died. Bur Dr. Hosakawa did not publish his results and that secret remained unexplained for some ten years.

In 1966, again an identical case of mercury poisoning broke out in Niigata town in central Japan and for that outbreak Denko company was held guilty of dumping toxic substances into the Angano river. The victims sued the company was held guilty of dumping toxic substances into the Angano river. The victims sued the company and the court decided the case in favour of victims. This success caused fierce resentment in Minamata against Chisso factory and a criminal Law suit was filed in a Japanese Court against the factory. In 1968, the company stopped discharging mercury waste into Minamata Bay. On March 20, 1973 the court gave verdict against Chisso factory and the plaintiffs and dependents of those who had died were paid heavy damage compensation ranging from 60,000 to 80,000 dollars to each and company had paid 200 million dollars. In 1979, 1401 individuals were certified as victims of Minamata disease and out of this number 353 had died. This is continuation of the account on lead.

Lead. Lead is the natural nonessential element generally found in all the components of ecosystem. It is one of the important pollutants of water, air and biosphere. Lead is accumulating in the marine environment as a result of drainage from rocks bearing this metal, industrial and domestic sources particularly the use of anti-knock motor fuels containing lead compounds. It has been calculated that motor vehicle exhausts contribute upto 2×10^5 tonnes of lead annually to the ocean through atmospheric transport. Much of this accumulates in the surface layer of sediments. One study based on geochemical relationships and material balances suggest that modern man absorbs daily 20 μg lead from food, 1 μg from water and upto 10μg from urban air. Natural conditions are estimated to have .01 ppm lead in food, .0005 ppm in water and 5×10^{-4} $\mu g/m^3$ in air. The pollutant is persistent cumulative and may show biological amplification. It has mutagenic effects (Hickey *et al.* 1971) and it may cause congenital deformities. Lead disturbs metabolism through its effect on enzymes. Symptoms of lead poisoning include loss of appetite, appearance of bluish lines round the gums, anaemia, etc. The rapid increase in the lead level in air and aquatic environment in recent years must be viewed with concern.

Copper and **zinc** have long been known to accumulate to very high levels in molluscs where they may colour the flesh and render them unmarketable but marine fishes do not exhibit this characteristic to the same degree. No toxic effects on consumers have yet been recorded.

Besides above metal elements, molybdenum, barium, manganese, cobalt, are other metallic pollutants which may show toxic effects on human health.

Cadmium is also accumulated by some marine animals and is only slowly excreted. It is highly toxic but there is lack of information as to its biological effects. It was the cause of *etaietai* disease in Japan. This metal leads to the formation of kidney stone. Liver and pancreas are also effected.

The information regarding chromium is also very sparse.

Liquid wastes or effluents from industries. The liquid effluents of industries containing a variety of poisonous chemicals are discharged into the bodies of water. They not only change the pH of water but also adversely affect the aquatic plant and animal life and sometimes cause large scale killing of fishes and other aquatic animals.

Oil spills. Petroleum extracted from the area of continental shelf is transported from one country to another through sea. After unloading, the tankers are washed in sea. This causes oil slicks or spills in the sea especially near the ports and shore lines. Because of accidents during transportation and extraction process, oil is spread hundreds of kilometres on the water surface which causes serious problems. A line of oil spilled into water can spread and cover an area of four thousand square metres of water surface. Three fourth of the oil spilled into the water is subjected to emulsification and the rest is evaporated (Fig. 13.4). The effects of oil pollution are as follows :

Fig. 13.4. Oil spill on the surface of sea water

(*i*) oil film on the surface of water prevents oxygenation of water.

(*ii*) it inhibits plankton growth and photosynthetic activity of aquatic plants.

(*iii*) it causes death of aquatic animals due to reduced oxygen availability in water, poisoning of food and toxic effect of oil.

(*iv*) oil spilled over water surface may catch fire and cause damage to aquatic life.

(*v*) detergents used to clean oil spill are equally harmful to aquatic life.

Effects of Water Pollution

Water pollution is the cause of many dangerous diseases. In India, still more than 50% people suffer due to disease caused by drinking of unsafe water. The various effects of water pollution are discussed here as follows :

1. Physicochemical Effects

Many pollutants produce undesirable colours, taste and odour in water and make it unpleasant and unfit for drinking and domestic uses. These changes may be in O_2 contents, temperature, and pH which affect the physicochemical nature of water.

The addition of organic substances results in depletion of oxygen and increase in CO_2 due to their decomposition by aerobic bacteria. The addition of nutrients through various sources enhances the algal and other biological growths which help in decomposition and lead to depletion of O_2. This is called entrophication. The decomposition of organic materials in absence of O_2 produces unpleasant odour and unaesthetic conditions due to release of several obnoxious gases as ammonia, methane, hydrogen sulphide etc. Algal photosynthesis produces increase in pH of water by consuming CO_2 . CO_2 dissolved in water produces carbonates.

2. Biological Effects

Excess pollutants affect the aquatic flora and fauna. Most of the fresh water algae are highly sensitive to pollutants and their elimination modifies the predatory relationship by breaking down the food chain which results in change of plant and animal communities, Biodiversity decreases in presence of a few tolerant species in polluted aquatic medium. Excess of nutrients in water promotes algal growth and formation of water blooms by blue-green and green algae. Many of the blue-green algae are not consumed by fishes and other aquatic animals and some of them produce toxic secretions which disturb the aquatic ecosystems.

3. Toxic Effects

Some pollutants for example, heavy metals, biocides, cyanides, and other organic and inorganic compounds are harmful to aquatic organisms. Although these substances have low permissible limits, their presence beyond those limits can make water unfit for aquatic life, human beings, as well as for other uses. These chemicals prove to be toxic to aquatic organisms and many of them, especially non-biodegradable ones accumulate in the body of organisms and cause long-term effects.

4. Pathogenic Effects (Vector-borne Diseases)

Vectors are certain organisms that act as the carriers of many pathogens, for example mosquitoes. They themselves are unaffected by the pathogens, but transmit the pathogens to the human bodies, causing a number of deadly diseases. The vectors breed in unsanitary and stagnant water bodies. Some common diseases spread by vectors are malaria, filaria, encephalitis, etc. (Table 13.4).

Few wastes, for example sewage, contain several pathogenic fungi, bacteria and viruses. *Clostridium perfringens, Streptococcus faecalis* cause various types of food poisoning. Besides, several diseases are water-borne which spread by sewage contaminated water. *e.g.*, cholera, typhoid, paratyphoid, dysentery, colitis, jaundice etc (Table 13.4).

Table 13.4 Some diseases caused by pathogens

	Disease	Symptoms	Prevention
Bacteria	**Cholera** Caused by *Vibrio cholerae.* Epidemic disease caused by contaminated food and water. Spreads through germs through housefly.	• Vomiting and thirst • Acute diarrhoea • Pain and fatigue • Suppression of urine and dehydration	• Boiled and saline water for drinking • Onion and mint juice should be given • Antibiotics and immunisation
	Tuberculosis Caused by *Mycobacterium tuberculi.* Acute bacterial infections that primarily attacks the lungs, but may also affect the kidneys, bones, lymph nodes and brain.	• Coughing, chest pain, shortness of breath, loss of appetite, weight loss, fever, and fatigue • Tubercles are formed in the affected area	• Complete rest • Balanced diet • Improved hygienic condition • BCG vaccine, antibiotics
	Typhoid Caused by *Salmonella typhi.* Spreads through contaminated food and water.	• Prolonged continuous fever, headache and uneasiness • Small white eruptions appear on the body	• Complete rest • Milk and fruits, light diet • Use of antibiotics • Vaccination
	Diarrohoea Caused by number of bacteria like *Escherichia coli.*	• Frequent loose motions • Dehydration	• Antibiotics
Virus	**Polio** Caused by *Poliomyelitis* virus.	• Infection can result in permanent paralysis of limbs	• Oral polio vaccine
	Jaundice Viral hepatitis A, B, E	• Affects the liver functions. • Extreme infection can cause liver cirrhosis.	• Vaccination • Boiled drinking water
Fungi	**Ring worm and Itch** Caused by mycelium of the fungus *Aspergillus* below the epidermis of skin.	• Reddish spherical spots on the skin • Spots produce itching • Increases with moisture	• Bathe with antiseptic water • Local application of antifungal creams
Protozoa	**Dysentery** Caused by *Entamoeba histolytica.* Acute or chronic disease of the large intestine of humans. Spreads through contaminated food and water.	• Frequent watery stool, sometimes containing blood and mucus • Dehydration	• Food articles should be covered, protected from dust and flies • Use of suitable drugs
	Malaria Caused by *Plasmodium vivax.* Transmitted by the bite of female *Anopheles* mosquitoes.	• Chills, shaking and periodic high fever	• Use of mosquito net or repellents • Control of mosquito breeding

Control of Water Pollution

Controlling water pollution is a multi-disciplinary approach that relates to maintaining the quality of water for human, industrial and agricultural use. This can be achieved by the following steps:

- Treatement of effluents before discharge, which involves the removal of particulates, dissolved organic contaminants and dissolved inorganic contaminants.
- Installation of waste-water treatment plants should be made mandatory for every industry (Fig. 13.5).
- Proper use of fertilisers, chemicals and pesticides, and preventing their runoff into water bodies.
- Tree plantation is very effective to check excessive runoff of polluted agricultural water.
- Adopting **Integrated Pest Management** (IPM) to minimise the use of pesticides for the control of pests. It involves the use of chemical, biological and cultural methods to control weeds, insects and plant pathogenrs simultaneously.
- Proper disposal of municipal sewage so as to avoid contamination of ground water reserves.
- Marine polution should also be checked, especially in sensitive coastal ecosystems. This can be done by avoiding oil spills, curbing developmental activities on the coastalines and avoiding the dumping of toxic wastes.
- The government has undertaken several projects to clean the rivers, the first of which was the Ganga Action Plan.

Fig. 13.5 Water Treatment Plant.

PROBLEM OF WATER SUPPLY AND SEWAGE DISPOSAL

Supply of Potable Water

The main sources of drinking water supply in the cities are rivers, lakes and streams. Water from such sources is purified or made pollutant-free and germ-free before that is supplied for drinking and other domestic purposes. To make the raw water clean and pollutant-free, the following three steps are followed :

(*i*) Sedimentation
(*ii*) Filtration
(*iii*) Chlorination

(*i*) **Sedimentation.** In this process, alum, aluminium sulphate or iron sulphate are mixed with raw water drawn from lakes or rivers into the mixing tank which form jelly-like floccules with dissolved and suspended substances. Water mixed with flocculants is allowed to flow into the flocculation tank where the floccules along with suspended soil particles, other extraneous materials and microbes settle at the bottom.

(*ii*) **Filtration.** After precipitation of floccules clean water is allowed to pass through special types of filter to remove the micro-organisms from it. For this purpose, water is allowed to percolate down through several alternating superimposed layers of sand and gravels.

(*iii*) **Chlorination.** Water after filteration process is subjected to chlorine treatment. In this process chlorine gas is passed through water which being a strong oxidant causes quick degradation of organic substances and at the same time kills the remaining bacteria. Water so obtained is then supplied to public for drinking and other domestic purposes.

Sewage Disposal

Sewage treatment aims chiefly at the removal of solid wastes and their degradation and conversion into simple inorganic substances through microbial activities.

The following methods are employed for the disposal of sewage:

1. Soaking pits
2. Septic tank
3. Municipal sewage disposal plants.

1. **Soaking pits.** In this process a big perforated underground tank made up of concrete and cement is used (Fig. 13.6). The sewage is discharged into the tank through a pipe. Sewage water from the tank comes out through the holes and percolates into the soil. The solid wastes are decomposed by micro-organisms inside the tank.

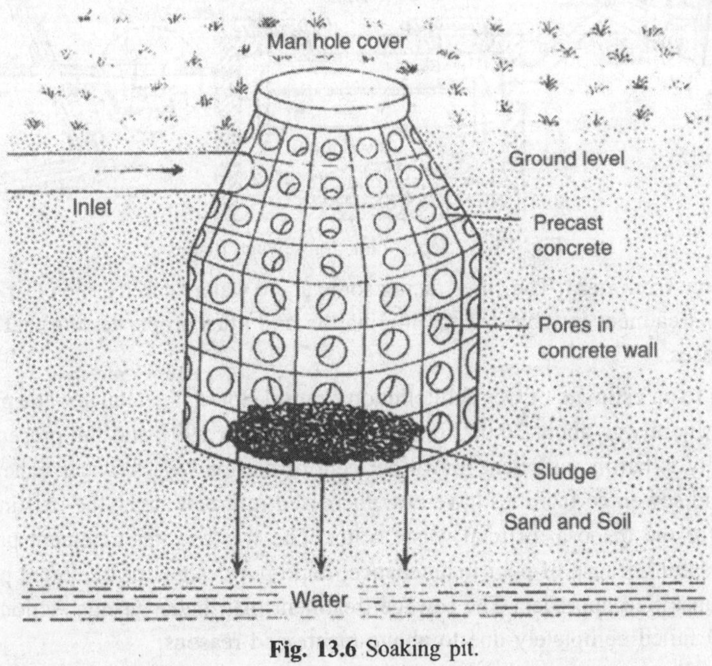

Fig. 13.6 Soaking pit.

2. **Septic tank.** In this method the sewage from house is discharged into underground septic tanks through pipes. The solid fraction of the sewage settles at the bottom of the septic tank and fraction flows out into the distribution pipes fitted in the upper part of the tank and finally drained into the field (Fig. 13.7). The solid fraction of sewage collected at the bottom is decomposed quickly by microbes.

3. **Municipal sewage disposal plants.** Treatment and disposal of sewage in big cities involve the following three steps :

(*i*) **Primary treatment.** For primary treatment the sewage is carried into the big open tanks through pipes. The solid fraction of the sewage settles at the bottom of the tanks which is drained through pipe system into the aerobic digester tank and is decomposed. The watery fraction of sewage from the primary settling tanks is drained into the secondary settling tank and mixed with aluminium sulphate or iron sulphate which forms jelly-like floccules. The floccules along with micro-organisms and suspended solid particles settle at the bottom of the tank in the form of sludge which is then drained through pipes into the aerobic digester tanks. (Fig 13.8).

(*ii*) **Secondary treatment.** The watery fraction of sewage which contains bacteria and other microbes as well as dissolved organic wastes is collected into the secondary settling tanks and air current under pressure is passed through the fraction to promote microbial decomposition of dissolved organic wastes. After sometime the fraction is passed through sand filters to remove microbes. The clean water is then allowed to flow into the rivers and oceans

The solid wastes and sludge carried to the digester tank are attacked and decomposed by aerobic bacteria. The decomposition of wastes results in the formation of NH_3, methane, hydrogen sulphide gases which are collected for various industry purposes.

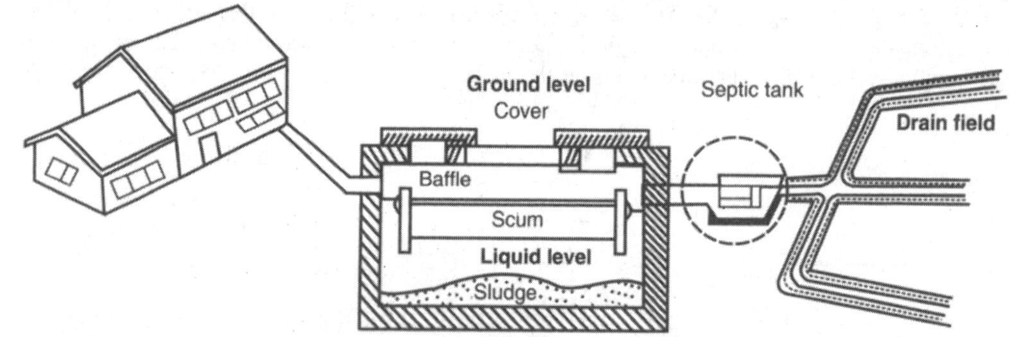

Fig. 13.7. Septic tank for sewage disposal.

(*iii*) **Tertiary treatment.** The cities facing some acute water scarcity, the clear water obtained after secondary treatment is subjected to chlorination and after proper testing that is supplies for domestic purposes.

According to an estimate of Central Pollution Control Board, the total sewage production from urban areas in India was about 30,000 billion litres a day in 1997 and the present total sewage treatment facility is hardly sufficient for 10 per cent of the total waste water generation. Though the drainage and sewerage facilities have now increased in urban areas, the existing facilities are not sufficient for the disposal of total waste water. The waste water treatment programmes are not fully successful because of poor maintenance, improper designs of treatment plants and non-technical and unskilled approach. The sewage treatment plan under Ganga Action Plan between 1980 and 1990 failed completely due to above-mentioned reasons.

Due to poor facilities of sewage and waste water treatment most of the pollutants find their way into groundwater, rivers, and other water bodies. In some pockets of India, the villagers are still dependent for drinking water on natural water reservoirs and face many problems as detailed below:

1. Drinking water is loaded with pollutants.
2. Water contains pathogens of cholera, typhoid, and a number of skin diseases.

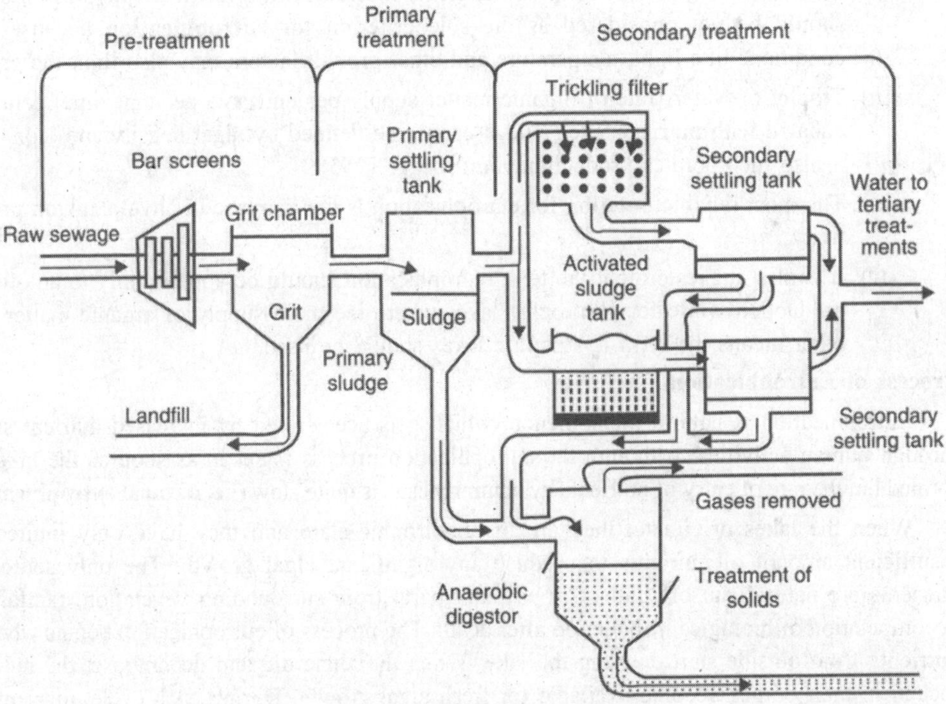

Fig. 13.8. Municipal sewagé disposal plant.

3. In some localities water is highly saline and contains fluorides or other toxic elements.

In some urban areas too, the supply of pure drinking water has become a big problem. According to an estimate of World Bank (1998), about 60 percent deaths in urban areas were due to such water related diseases as cholera, dysentery, gastroenteritis, hepatities etc.

Eutrophication

The ever-increasing human pressure on water bodies due to demographic growth, modern technology and agriculture has caused several water pollution problems. One of the most sever and commonest problems is due to enrichment of waters by plant nutrients which leads to biological growth and renders water unfit for diverse uses.

Additional nutrients in the form of nitrogen and phosphorous compounds from fertilisers, sewage, detergents and animal wastes increase the rate of growth of aquatic plants and algae. The excessive growth of algae and other aquatic plants due to added nutrient is called **eutrophication** This results in high biological productivity in certain aquatic plants, which is manifested in the form of **blooms**. This makes the water oxygen deficient due to degradation of organic matter in bodies of water which adversely affects other organisms. Algae and larger aquatic plants may interfere with use of water by clogging water intake pipes, changing the taste and odour of water and causing the buildup of organic matter on the bottom. As this organic matter decays, oxygen level reduces, ultimately fishes and some other aquatic species may die.

Weber (1907) while studying the North German peat bogs noticed that the upper layers contained more nutrients in upper layer of lake in comparison of lower ones. He used the term **eutrophic** (rich in nutrients) and **oligotrophic** (poor in nutrients) to distinguish between those two layers. The use of these terms in limnology was made for the first time by Naumann (1919). According to the present day concept of eutrophication,

(*i*) enrichment of water with plant nutrients increases the growth of phytoplankton but should be not considered as the sole criterion for **eutrophication** because other conditions like light, temperature and other growth factors may also limit the growth,

(*ii*) Trophy of water (rate of organic matter supply per unit area per unit time) cannot be equated with nutrient levels and it cannot be defined by algal density and biomass as it also incorporates production (Findenegg, 1955).

(*iii*) The most reliable criterion for eutrophication is the increase in phytoplankton productivity.

(*iv*) It is also suggested that the term eutrophication should be applied only to autotrophic production while for allotrophic lakes where the main supply of organic matter is by other means the term **dystrophic** lakes should be used.

Process of Eutrophication.

Eutrophication is natural phenomenon, which gets accelerated by increased nutrient supply through human activities. Although the eutrophication process is set in as soon as the lakes are formed but the rate of entry of nutrients by natural means is quite slow **(i.e. natural eutrophication)**.

When the lakes origimate, they are in oligotrophic state and they have only limited and insufficient amount of nutrients to produce any significant algal growth. The only sources of nutrients are natural run off, fall of dried plant parts from surrounding vegetation, rainfall and decomposition of biological production after death. The process of eutrophication begins when the nutrients from outside start entering the lake. When the algae die and decompose, the nutrients locked in their bodies become available for fresh algal growth. During each cycle, nutrients are progressively increased in lakes and after sometime, the cycling of nutrients does not maintain a balance between addition and decomposition with the result that an ever increasing organic matter in the lake ultimately gets deposited at the bottom. This leads to the formation of swamps bogs, marshes and finally water body disappears. This is why the process of eutrophication is referred to as aging of lakes. It is thus evident that with the advancement of eutrophication more and more nutrients are added to the body of water and ultimately nutrient cycle is unable to maintain equilibrium between addition and decomposition. The speed of eutrophication depends on the rate of supply of nutrients as well as on some other factors like climate etc. Generally, speed of eutrophication is high in hot climate which promotes nutrient utilization and algal growth in comparison of the rate in cold and temperate climate. The rate of eutrophication cation slows down with time due to reduced light penetration as are salt increased turbidity and consequent fall in primary production.

Effects of Eutrophication.

When there is a departure from balance between photosynthesis (*P*) and respiration (*R*) it indicates pollution. At equillibrium ($P = R$) there is no change in chemical and biological composition of water; a condition found in unpolluted water, with no supply of nutrients from outside. When photosynthesis exceeds respiration activity it indicates eutrophication of water bodies. It is characterised by progressive increase of algae leading to organic overloading. In deep lakes excussive production at the surface of lakes ($P >> R$) is balanced by saprophytic conditions at the bottom (R >> P) when respiration exceeds photosynthesis, the dissolved O_2

becomes exhausted forcing reduction of several oxidised chemicals like NO_3^-, SO_4^{-2} and CO_2 into N_2, NH_4^+, H_2S and CH_4 which produce dirty odour and are harmful to several aquatic species. Poole *et al.* (1978) has reported 11 mg per litre as lethal concentration 50% (LC 50) for H_2S for some aquatic organisms.

The eutrophication induces many physical and chemical changes in waters which bring about changes in flora and fauna. Many desirable species including fishes are replaced by undesirable ones. There is an algal succession and blue green algae become dominant, many of them such as *Microcystis, Anabaena, Oscillatoria* produce blooms. Algae like *Chlorella, Scenedesmus* may also form blooms. *Spirogyra, Cladophora, Zygnema* and many other filamentous green algal may form floating mat on the surface of waters. These algal blooms and thick mat reduce the intensity of light below the surface.

Eutrophication leads to change in the characteristics of bottom sediments. Accumulation of organic matter affects the benthic community.

Algal blooms affect the recreational value of water bodies. The death and decay of algae produce foul odours and tastes in water. The algal scum checks the oxygen penetration into water and may kill the fishes and other organisms. In the initial stage of algal growth, enough oxygen is produced but when the algal blooms die, the water become deficient in O_2 because oxygen production is reduced and consumption is increased due to decomposition of dead algae by aerobic bacteria. The fall in the level of dissolved O_2 in water may be the cause of mortality of fishes and other aquatic organisms.

The algal blooms cause discolouration of water. The overall effects of eutrophication render the water to be unfit for human consumption and various other purposes. Besides, the treatment cost of water is inflated too.

Water Quality

The evaluation of water quality is done in terms of several parameters such as alkalinity, dissolved oxygen, Biochemical oxygen demand (5 days), number of coliform bacteria, colour, hardness, odour, pH, salinity, temperature, total solids, turbidity, salts–chlorides, fluorides, nitrates, phosphates and sulphates, presence of trace elements like Al, As, Ba, Cd, Cr, Fe, Pb, Mn, Hg, Se, Ag, Sn, Zn and B, pesticides and radioactivity. Among these attributes, the amount of dissolved oxygen, biochemical oxygen demand and total coliform counts are good indicators of the quality of water. These are briefly discussed here as under :

Dissolved oxygen. It is a measure of the ability of water to support a well balanced aquatic life. Sufficient amount of dissolved oxygen in a water body brings about quick microbial degradation of organic wastes. Biochemical oxidation of ammonia to nitrate in natural water requires dissolved oxygen. Insufficient amount of dissolved oxygen in water adversely affects the microbial decomposition and methane is released instead of CO_2, odourous amines result from nitrogen instead of NO_3 and NH_3 and foul smelling H_2S gas is formed from sulphur instead of SO_2.

Biological or Biochemical Oxygen Demand (BOD)

The most common index of water pollution is biochemical oxygen demand (BOD) which refers to the quantity of oxygen required by bacteria to decompose the organic wastes aerobically to CO_2 and water. The BOD test normally measures the amount of oxygen used in the first five days of aerobic microbial decomposition in a certain volume of effluent at 20°C. This is also called BOD_5. Thus 100 ppm BOD means 100 mg of oxygen consumed by one litre of test sample over 5 days at 20°C. Domestic sewage generally has BOD_5 of around 200 milligrams of oxygen per litre and for industrial wastes BOD may be around thousand mg per litre. BOD of

0.17 pound or 77 gm is also called population equivalent, being roughly equal to the requirements for the domestic wastes of one per cent concentration. The capacity of sewage treatment plant is generally measured in terms of population equivalents per day. Contamination of water by sewage is the main cause of water borne diseases *e.g.*, cholera, typhoid, paratyphoid fever, dysentery and infective hepatites.

Total coliform counts. BOD gives a rough measure of the quality of water. It does not accurately indicate the risk of disease. For that purpose more specific parameters are required. One of the most common parameters is the number of coliform intestinal bacteria especially *Escherichia coli* in feces per unit volume of water. Although coliform bacteria are harmless, their presence in large numbers indicates that pathogenic germs might be present in the sample.

Water quality of river water is monitored at 480 stations under different programmes such as MINARS (Monitoring of Indian National Aquatic Resources), GEMS (Global Environmental Monitoring systems), and G.A.P. (Ganga Action Plan). The number of stations under the MINARS programmes initiated in 1979 increased gradually and at present the number of stations is 260.

A number of physical, chemical, biological and bacteriological parameters are being considered under the programme to determine the quality of water but important ones are DO, BOD and T.C. (Total coliform counts).

The different categories of water in response to quality and their respective uses are as follows :

Class A— Drinking water source without conventional bacteria in water.
Dissolved oxygen more than 5 mg/litre, T.C. less than 50/100 ml.

Class B— Water for bathing, swimming and recreational use, D.O. > 4 mg/litre and T.C. < 500/100 ml.

Class C— Drinking water source after conventional treatment.

Class D— Water for wildlife, fisheries etc. D.O. > 4 and T.C. < 500/100 ml.

Class E— Water for irrigation, industrial cooling, no fishing, swimming or drinking. D O. > 3mg/tire.

AIR POLLUTION

The atmosphere is a mixture of gases, predominantly nitrogen (78.09%), oxygen (20.94%) and some other gases (less than 1%). Table 13.5 gives composition of unpolluted atmosphere. These values remain constant upto 50,000 feet altitude.

Table 13.5. Composition of unpolluted atmosphere

Constituents	Volume
N_2	78.09%
O_2	20.94%
Argon	0.93%
CO_2	0.032%
Neon	18 ppm
Helium	5.2 ppm
Methane	1.3 ppm
Krypton	1.0 ppm
Hydrogen	0.5 ppm

Constituents	Volume
N_2O	0.25 ppm
CO	.1 ppm
O_3	.02 ppm
SO_2	.0002 ppm
NO_2	.0001 ppm

Water vapour which is present in the range of 1-3% is an important constituent of atmosphere. The other minor constituents of atmosphere may be dust particles, smoke, H_2S, hydrocarbons, bacteria, spores of plants. *The atmosphere is said to be polluted if the concentration of any constituent increases or decreases drastically to cause harmful effects on human health or his environment.* The addition of foreign materials into the atmosphere may also cause pollution.

The air pollution may be defined as qualitative and quantitative changes in the atmospheric constituents due to addition or contamination of such substances as may be harmful to man and his environment. Air pollution may be categorised into ***indoor*** and ***outdoor pollutions***. Pollution of atmosphere inside buildings, offices and residences is called **indoor air pollution** and the atmospheric pollution in the open space is referred to as **outdoor air pollution.** The different indoor and outdoor air pollutants and their sources are listed in the following Table 13.6.

Table 13.6. Indoor and outdoor air pollutants and their sources

Pollutants	Sources
A. Outdoor pollutants:	
Lead	Automobiles, industrial emission.
Ozone	Photochemical reactions in atmosphere.
Sulphur dioxide	Fuel combustion, smelters.
Organic substances	Petrochemical solvents, natural sources, vaporisation of unburnt fuels.
Spores and pollengrains	Bacteria, fungi and higher plants.
B. Indoor and outdoor pollutants :	
Carbon dioxide	Metabolic activities of living organisms, combustion of fuels.
Carbon monoxide and Nitrogen oxides	Burning of coal and other fuels, Tobacco smoking.
Organic substances	Volatilization, combustion of coal, kerosine and other fuels, pesticides, metabolic activities, adhesives, cosmetics, solvents etc.
Hydrocarbons Particulate matters	Fuel burning, tobacco smoke, condensation of vapours, combustion of coal, kerosine and other fuels.
C. Indoor pollutants :	
Water vapour	Biological activities, combustion, evaporation of water.
Aerosols	Consumer products, varnish and paints.
Allergens	Dust particles, spores and pollengrains.
Ammonia	Metabolic activities, cleaning products.
Asbestos and synthetic fibres	Fire retardants, thermal or electrical insulations.
Formaldehyde	Particle board, insulation, furnishing, tobacco smoke.
Mercury	Spills in dental care facilities, thermometer breakage.

Air forms nearly 80% of man's daily intake by weight. A normal man breaths nearly 2200 times a day inhaling about 16 kg air. The problem of air pollution is not new. Primitive man too introduced foreign substances into the air by way of burning and other activities. Modern man has accelerated the process of pollution to an alarming rate. Many of the long known effects of air pollution are reduction of sunshine, morbidity, mortality from respiratory ailments, dust fall, corrosion of metal and so on. Extensive experimental and epidemiological studies carried out in 20th century, especially during the last two decades have verified these scientifically. Although the adverse effects of air pollution are well known, our existence in a pollution free environment is still far beyond our reach. India's main problem of atmospheric pollution is due to unplanned and disproportionate industrialization and unrestricted urbanization. About 80% of the industries are concentrated in 10 or 12 big cities.

In India, air pollution is caused mostly by automobiles, power plants and oil refineries, The vehicles on the road and oil refinery plants emit several types of atmospheric pollutants such as carbon monoxide, nitrogen oxides, smoke, organic vapour and hydrocarbons as is clear from the following Table 13.7.

Table 13.7. Potential sources of specific emissions from oil refineries

Emissions	Sources
Aldehyde	Catalytic regenerators
Ammonia	Catalytic regenerators
Carbon monoxides (CO)	Catalytic regenerators, compressors, incinerators.
Hydrocarbons (HC)	Loading facilities, turn around, sampling storage tanks, waste water separators, blow down systems, catalyst regenerators, pumps, valves, cooling towers, vacuum jets, condensors, air blowing, high pressure equipment handling, compressor engines, heaters, process boilers.
Oxides of sulphur (SO_2, SO_3)	Boilers, process heaters, catalyst cracking units, catalyst regenerators, treating units, hydrogen sulphide flares, sludge disposal.
Oxides of nitrogen (NO, NO_2)	Process heaters, boilers, compressor engines, catalyst regenerators.
Suspended Particulate Matter (SPM)	Catalyst regenerators, boilers, process heaters, decoking operations, incinerators.
Odour	Treating units, air blowing streams, drains tank vents, barometric condensors, pumps, waste water separators.

CAUSES OF AIR POLLUTION

Atmospheric pollution is increasing due to the following major factors:

1. Over population and increasing urbanization.
2. Increasing traffic
3. Industrialization and energy consumption.

1. Over population and Urbanization

According to current estimates, the world population is increasing at an average by two per cent annually which means the addition of about one million people every five days or five people every couple of seconds. In India population has already crossed 103 crores (1.03 billion). Increase in population would demand more food, water and land, the three items that are limited on the earth. The increasing population has led man to exploit the natural resources, especially land

and water to support life. This has resulted in disturbance in ecological balance, disruption of natural bio-geochemical cycles and fast depletion of natural resources.

Technological advances in farming practices, use of high yielding cultivars, increased application of chemical fertilizers and pesticides have made it possible to boost agricultural production. Pesticides are being used at the rate of 570 gms/acre/year and are on further increase in India. These pesticides have become contaminants of air, water and food.

Growing population has created problems of employment. This leads people to move from villages to urban areas in search of employment, comfort and facilities. As a result, the cities are expanding in area and are being thickly populated. About 40% of the population in India is concentrated in the states of Bihar, Uttar Pradesh, Haryana, Punjab and West Bengal. Thickly populated areas face a variety of problems such as conjestion, large number of vehicles, accumulation of solid and liquid wastes, poor sanitation and management problems.

Particulate matter in city atmosphere comes from cooking fire, industrial activities, automobile exhausts and dirt arising from general transportation. Accumulation of particulate matter in air may eventually cause some changes in climatic patterns of the city.

The disposal of urban sewage and household wastes into the body of water makes that dirty and contaminated. In India about 50 billion litres of waste water from domestic and municipal sources are released into the body of water daily. This is the potent cause of pollution of water.

2. Increasing Traffic

Rapid increase in the number of automobiles is one of the potential factors responsible for air pollution in urban areas. The vehicles on road are mobile sources of air pollution. During the last two decades, there has been a tremendous increase in the number of vehicles. The number of vehicles per 1000 populations was 3 in 1972-73 and 25 in 1990-91. According to a report of Ministry of Surface Transport, 1996 the number of vehicles increased from 0.3 million in 1951 to 30.3 million in 1995. Out of the total vehicles, 32% were confined to 23 metropolitan cities. Data further indicated that Delhi registered highest increase in the number of vehicles which was followed by Chennai, Hyderabad, Bangalore, Mumbai. With the rapid increase in the number of vehicles, the emission of different pollutants has also increased considerably. According to a report of the Ministry of Environment and Forest (1995), the vehicles in major metropolitan cities of India release upto 43% CO, 20% Hydrocarbons (HC), 30-40% NO_x, 5% suspended particulate matter (SPM) and 2% SO_2. The amount of these pollutants is highest in Delhi, followed by Mumbai, Bangalore, Calcutta and Ahmedabad.

The daily pollution load emitted by automobiles in 12 big cities of India is presented in the following Table 13.8 :

Table 13.8. Daily emission of pollutants from vehicles in 12 metropolitan cities of India (According to a report of Central Pollution Control Board, 1995)

Cities	Amount of pollutants (tonnes per day)					
	SPM	SO_2	NO_x	HC	CO_2	Total
Delhi	10.3	9.0	126.5	249.6	651.0	1046.3
Mumbai	5.6	4.0	70.8	108.2	469.9	651.6
Bangalore	2.6	1.8	26.2	78.5	195.4	304.5
Calcutta	3.3	3.7	54.7	43.9	188.2	293.7
Ahmedabad	3.0	2.9	40.0	67.8	179.1	292.7
Pune	2.4	1.2	16.2	73.2	162.2	255.3
Chennai	2.3	2.0	28.2	50.5	143.2	226.3

	SPM	SO_2	NO_x	HC	CO_2	Total
Hyderabad	1.9	1.6	16.8	56.3	126.2	202.8
Jaipur	2.0	1.3	15.3	21.0	51.3	89.0
Kanpur	1.1	1.1	13.4	22.2	48.4	86.2
Lucknow	1.1	1.0	9.7	22.5	49.2	83.5
Nagpur	0.6	0.4	5.1	16.3	35.0	57.4
Total	35.3	29.8	422.9	810.0	2299.2	3597.2

The total estimated pollution load from transport sector increased from 0.15 million tonnes in 1947 to 10.3 million tonnes in 1997. In 1997, emission of CO was highest which was 43%. The total emission of NO_x was 30% and those of Hydrocarbons (HC) 20%, SPM 5% and SO_2 2% (Table 13.9). The emission of pollutants depends on the type of vehicle, engine design and type of gasolene diesel used. Leaded petrol is the main source of toxic lead in atmosphere. Keeping this fact in view, recently a worldwide drive has been initiated to supply leadfree gasolene for vehicles.

Table 13.9. Average Emissions from Automobiles.

Type of emissions	Kgs/1000 litres of gasolene used	Kgs/1000 litres of diesel used
Carbon monoxide	300.00	7.00
Hydrocarbons	25.00	17.00
Oxides of nitrogen	14.00	27.00
Oxides of sulphur	1.00	5.00
Aldehydes	0.50	1.00
Organic acids	0.50	4.00
Benzopyrene	0.06	0.10
Particulate matter	1.50	14.00

Combustion of fuel in aircraft engines produces hydrocarbons, nitrogen oxides, carbon monoxide, small amount of particulate matter, and sulphur oxides. Amounts of pollutants depend upon the fuel used, type of aircraft engine. The various pollutants released into the atmosphere during the take off and landing of aircrafts are listed in the following Table 13.10.

Table 13.10. Emission from aircraft below 1000 metre altitude in kgs per flight (Landing+Take off)

Type of emission	Jet aircraft four engine conventional	Piston-engine aircrafts		
		Fan jet	2-engine	4-engine
Carbon monoxide	17.5	10.3	67.0	163.0
Oxides of nitrogen	11.5	4.6	3.2	7.8
Hydrocarbons	5.0	9.5	12.5	30.0
Aldehydes	2.0	1.1	0.3	0.3
Particulates	17.0	3.7	0.3	0.7

3. Industrial Growth and Energy Consumption

Industries, thermal power plants and oil refineries are the other important sectors which are major sources for air pollution in the country. Industries are producing a variety of toxic substances. Thermal power plants and oil refineries are releasing tremendous amount of fly ash and many toxic gases into the atmosphere. Domestic burning, agricultural refuse, firewood and dry dung burning are also contributing considerable amounts of pollutants to atmosphere.

According to an estimate of World Bank, the major air pollution is contributed by a few industrial sectors. In India, petroleum refineries, textile mills, pulp and paper industries and chemical industries produce 87% of the atmospheric pollutants.

Common air pollutants and their effects. There are two major factors which contribute to the problem of air pollution : increasing human population and increased productivity each increasing by about 2 to 3% every year. The United States alone discharges roughly 125 million tonnes of junk into the atmosphere each year and a conservative estimate for the whole world would be about 520 million tonnes every year.

The pollutants of atmosphere may be in gaseous or vapour state and in particulate form. Chemical vapours found in atmosphere are derived from those substances which have boiling point less than 200°C. The following are the major air pollutants :

Air pollutants

Gases & Chemical vapour

Particulates

(size 20μ – 100mμ in diametre)

1. Solid particles (dust & smoke)
2. Liquid droplets (fumes & mists)

1. **Inorganics :**

   Oxides of carbon (CO, CO_2)

   Oxides of nitrogen (NO, NO_2)

   Oxides of sulphur (SO_2, SO_3)

   Other inorganics (NH_3, HF, H_2S, etc.)

2. **Organics**

   (*i*) Simple hydrocarbons

   (*ii*) *Aliphatics :*

   alkenes (ethene, propene)

   alkanes (ethane, methane)

   alkylene (acetylene)

   (*iii*) Aromatics (benzene, toluene, xylene)

   (*iv*) Oxygenated Hydrocarbons:

   Aldehydes (Formaldehyde, acetaldehyde)

   (Formic acids & acetic acid) Acids

Ketones (Acetone)

The sources of atmospheric pollution are many. P.K. Zutshi of Bhabha Atomic Research Centre (BARC) in 1970 classified the sources of pollution as follows (Table 13.11);

Table 13.11. Classification of Air Pollution Sources, and Emissions

Type 1	Category 2	Examples 3	Important Pollutants 4
Combustion	Fuel burning	Domestic burning, thermal power plants	Sulphur and nitrogen oxides
	Transportation	Cars, trucks, aeroplanes and trains	Carbon monoxide, nitrogen oxides,
	Refuse burning	Open burning dumps	lead, smoke, organic vapours, odours, Fly ash and particulates. etc.
Manufacturing process	Chemical	Petroleum refineries, fertilizers, cement, paper, ceramics, clay products, and glass manufacture	Hydrogen sulphide, SO_2, flourides, odours, organic vapours and dusts.
	Metallurgical plants	Aluminium refineries, steel plants etc.	Metal fumes (Pb and Zn) flourides and particulates.
	Waste recovery	Scrap metals.	Metal fumes, organic vapours, odours, smoke, soot, etc.
Agricultural activities	Crop spraying	Weed and pest control	Organic phosphates, chlorinated hydrocarbons, Pb and Hg
	Field burning	Burning of refuse, firewood and dry dung	Smokes, fly ash, soot, Sulphur dioxide, particulates and organic vapours
Solvent usage	Spray painting;	Furniture and appliances, finishing, dyeing, printing and chemical separation	Hydrocarbon and other organic vapours.
	Solvent extraction, inks.		
	Solvent cleaning	Cleaning, degrading etc.	
Nuclear energy	Fuel fabrication	Gaseous diffusion	Fluorides
	Ore preparation	Crushing, grinding and screening	Uranium and Beryllium dust and other particulates, $argon^{40}$, $iodine^{51}$.
	Nuclear testing device, Spent fuel processing	Bomb explosion, Chemical separation	Radioactive fallout Sr^{90}, Cs^{137}. Cs^{14}, etc.

The main primary air pollutants are oxides of carbon, sulphur and nitrogen, particulate matter and hydrocarbons. Besides these, there are some other pollutants of air. The following table (13.12) highlights the atmospheric pollution levels in four cities of India:

Table 13.12. Atmospheric pollution levels in four cities of India (based on P.K. Yennawar and others, 1970)

Cities	SO_2 average (ppm)	NO_2 average (ppm)	H_2S average (ppm)	O_3 average (ppm)	Suspended dust average (mg/m^3)
Delhi	0.0156	0.011	0.0032	0.0088	924
Mumbai	0.0396	0.0108	0.018	0.008	239.4
Kolkata	0.0222	0.0128	0.0048	0.0148	5266
Kanpur	0.0416	0.04	—	0.048	488

The important sources of air pollutants are motor vehicles (gasoline and diesel), aircrafts, rail and road transportation, coal, fuel oil, natural gas, and wood fuel combustion, industrial processes, solid waste disposal, forest fire, agricultural burning, etc. Exhausts and other particulate matters in the presence of sunlight produce the photochemical **smog** which is no less than plague. The December 1952 smog caused 15030 deaths in London within 5 days. Air pollution is clearly a great danger to man and it can lead to disastrous incidence. But not many are however aware of this fact that the air we breath is contaminated with dust and gases in harmful proportions.

Suspended particulate matters. Solid and liquid aerosols suspended in the atmosphere are referred to as suspended particulate matters. (SPM). They arise from grinding, erosion, spraying, etc.

Particulate matter refers to the varying mixture of particles in different proportions, suspended in the air. Such suspended particles are normally present everywhere, but high concentrations of specific types of particles pose danger to human health. SPMs are emitted both by natural processes and anthropogenic activities. Almost all the industries release particulate matter into the atmosphere. These include materials such as soot, dust, asbestos particles, heavy metals such as copper, lead, zinc, arsenic, etc.

The SPMs, being of very small size (less than 10 micron in diameter) are easily inhaled into the lungs and have chronic affects on respiration. They are also absorbed by the blood streams and may cause multiple health hazards, as many particulates are carcinogenic in nature. The SPMs spread over large areas, alter food chains and affect the ecosystems. They also block sunlight and may cause changes in the climate. Suspended particles in the air sometime cause reduction of visibility, adversely effecting air and rail and road traffic.

Aerosols are the chemicals which are released in to the air with force in the from of a mist or vapour. Aerosols have caused serious environmental pollution in the recent years. Jet aeroplane emission is an important source of aerosols in upper atmosphere. Layers of pollutants which are largely due to high flying commercial jets can be found as high as 43000 feet. Although smoke is popularly used to denote mixtures of particulate matter, fumes, gases and mists, 'dust' refers to solid dispersion aerosol and 'mist' to liquid aerosol. Dust particles greater than 10 μ in size are called settleable dust. These are measured as dust fall in tonnes /sq mile/ month. Smaller particles suspended in air are referred to as suspended dust. The aerosols contain flourocarbons. Fluorocarbons deplete the ozone layer in the stratosphere and thus permit more harmful ultraviolet radiations to reach the earth surface ; ozone acts as preventive covering against ultraviolet light. From the emission of supersonic aeroplanes nitrogen and sulphur oxides are released which also cause thinning of the ozone layer. Carbon tetrachloride (CCl_4) is also known to destroy some of the stratospheric ozone. Pilots have discovered pollution high over the middle of the oceans in the north polar region.

The important effects of air pollutants are as follows:

1. Atmospheric particles can scatter and absorb sunlight, thus reduce the visibility. Reduced visibility is aesthetically undesirable and it is also dangerous for aircraft and motors. In general, cities receive above 15 to 20% less solar radiation than rural areas and the reduction of sunlight can become as high as one-third in the summer and two-third in winter. The reduction of sunlight is largely due to fuel combustion for industrial and household heating purposes.

2. The effects of particulate matter include corrosion of metals, erosion and soiling of buildings, sculptures and painted surfaces and soiling of clothings and draperies, damage of electric equipments, etc.

3. The toxic effects of particulate matter on animals and human beings can be classified as :

(*i*) **Intrinsic toxicity** due to chemical or physical properties. Carbon monoxide in congested areas removes 5 to 10% of blood from circulation. Although body tissues extract only 25% of oxygen from the blood, the heart needs 75 per cent. So there is little margin for safety.

(*ii*) **Interference** with clearance mechanism in the respiratory tracts. Chronic bronchitis and emphysema have also been found to be caused by SO_2. A 24 hour exposure to about 0.2 ppm of SO_2 may cause serious health problems. Lung cancer has been found to be correlated with air pollution. Polycyclic aromatic hydrocarbons (PAH) are found to be related to the pathogenesis of lung cancer.

(*iii*) **Toxicity** due to absorbed toxic substance.

Many toxic particles including metal dusts, asbestos, aromatic hydrocarbons have been discovered in a polluted urban atmosphere. Lead from vehicle exhausts resulting from the use of tetraethyl lead as an anti-knock additive to petrol may build up to dangerous levels in urban areas adjacent to busy road complexes. Lead in high doses kills outrightly. Lower doses in dense traffic areas shortens life span and causes deterioration of nervous system. Retarded children have a higher lead content in their body than the normal ones.

4. **Benzpyrenes.** Their concentrations are extremely small but they play a role in increasing cancer rate in urban areas as compared to rural areas. Peroxyacetyl nitrates (PAN) or photochemical smog may constitute a serious problem where high levels of vehicular emissions occur in cities experiencing bright sunlight and ambient temperatures above 21°C.

5. The small solid particles can serve as carriers for micro-oganisms and other infective agents and thereby spread diseases. Large dust particles are trapped in nose and throat and very tiny particles which stay in the lungs may start an ugly chain of events leading to serious illness and deaths. Air pollution causes coughing, sneezing, thickening of mucus and narrowing or complete closure of glottis due to presence of gases, especially SO_2, nitrogen dioxide and oxidants. The silicon particles may cause 'silicosis' and fibrous particles 'fibrosis'. It is suspected that some pollutants can start lung cancer.

Sulphur dioxide (SO_2). The most important air pollutant is SO_2 which is colourless, nonflammable gas with pungent irritating odour. About one-third of SO_2 present in the atmosphere is believed to be produced by man's activity. Robinson and Robbins (1971) have estimated that SO_2 from man's activity introduces 6.6 million tones of sulphur into the atmosphere annually. SO_2 present in atmosphere comes largely from coal and petroleum combustion. Biologically produced H_2S when oxidised produces SO_2. The following tables (13.13 and 13.14) give an idea about the rough estimates of SO_2 and suspended particulates in air in major cities of India :

Table 13.13. SO_2 and SPM in air major cities of India

City	Mean value of SO_2 (microgram/m^3)	Suspended particulate matter (microgram/m^3)
Ahmedabad	10.66	306.6
Mumbai	47.11	240.8
Calcutta	32.88	340.7
Delhi	41.43	601.1
Hyderabad	5.06	146.2
Jaipur	4.15	446.1
Kanpur	15.97	543.5
Chennai	98.38	100.0
Nagpur	7.71	261.6

Source: NEER1, 1970

Table 13.14. Concentration of SO_2 in different city atmosphere in India

	Concentration of SO_2 ($\mu g/m^3$)	
City	1970	1979
Kolkata	85	33
Mumbai	83	47
Ahmedabad	71	11
New Delhi	41	39
Hyderabad	26	5
Kanpur	25	12
Jaipur	17	4
Chennai	16	8
Nagpur	12	8

Effects of SO_2. 1. SO_2 can damage materials and properties mainly through its conversion into the highly reactive H_2SO_4. It causes discolouration and physical deterioration of building materials and sculptures. Deterioration and fading are also produced in fabrics as cotton, nylon, rayon, leather and paper.

2. It accelerates corrosion of metals especially iron, steel and zinc.

3. SO_2 has been found to affect vegetation adversely even at the concentration below 0.32 ppm. It is suspected that sulphur dioxide pollution in urban and industrial areas of industrialised countries has a major impact on the respiratory condition of the population and also has significant effects on the crops and other vegetations of the surrounding rural areas. High concentration of SO_2 over short period of time can produce leaf injury, such as necrosis in plants or brownish colouration in the tips of pine needles. Lower concentrations over long period lead to chronic leaf injury such as gradual chlorosis.

4. SO_2 and H_2SO_4 both are capable of causing irritation in respiratory tracts of animals and human beings and high concentrations of SO_2 cause severe heart and lung diseases.

Carbon monoxide and carbon dioxide. Carbon monoxide is colourless, tasteless, odourless gas which is deadly poisonous in high concentration. It is one of the important gaseous air pollutants. It originates from incomplete combustion of carbonaceous materials. Man's activities result in the production of nearly 250 million metric tons of CO annually. Aircrafts are responsible for about 2.5% CO and 1.1% hydrocarbons in USA. Space rocket engines produce many dangerous exhaust products. In one study it was found that engine of Apollo Lunar module which took two astronauts to the surface of moon produced NH_3, water, CO, NO, CO_2 and N_2O as major exhaust products and wide variety of different minor exhaust products. The toxic effects of CO on human beings or animals arise from its reversible combination with haemoglobin in the blood. Haemoglobin has great affinity with CO which lessens the oxygen carrying capacity of blood. It also reduces the dissociation of oxyhaemoglobin. Carbon dioxide released in to our atmosphere after combustion of fossil fuels has brought a 10 per cent increase in the level of atmospheric CO_2 over the past century. The regular measurement of CO_2 content in the atmosphere began in 1958. In 1959 the content of CO_2 in atmosphere was found to be 316.1 ppm. (part per million or part per ten lakhs) and in 1994 it was found to be 350.0 ppm. The study on the amount of CO_2 trapped in old ice revealed that prior to industrialization the content of this gas on an average was 280 ppm, but from the start of industrialization in the end of eighteenth century right upto 1959 the CO_2 content increased from 280 ppm to 316 ppm. We are all aware of this fact that the consumption of fossil fuels such as coal, petroleum products, diesel, gas fuels is rapidly increasing and the forest area is decreasing day by day. Consequently, the content of CO_2 in the atmosphere is increasing quickly and in case the present rate of CO_2 increase continues it may touch or even exceed 500 ppm mark soon. A continued increase in excess unabsorbed CO_2 could have a catastrophic warming effect on the atmosphere, melting of polar ice, change in the ecosystems of seas and even floods on an undreamed scale.

Nitrogen oxides. Robinson and Robbins (1971) have estimated that biological production of NO and NO_2 amounts to about 1 billion metric tons annually while combustion processes produce 48 million metric tons NO_2 annually. Man-made nitrogen oxides pollution is insignificant in urban areas. These oxides play role in production of photochemical smog.

The effects of nitrogen oxides are as follows :

1. Nitrogen oxides are known to produce fading of textile dyes, deterioration of cotton and nylon and corrosion of metals due to production of particulate nitrates.
2. NO_2 causes leaf injury and reduction of growth in several sensitive plants.
3. It affects lungs, heart, liver and kidney at higher concentrations (15 to 50 ppm for two hours). Besides, it is considered to be a major factor in causing eye irritation.

ACID RAIN

In industrialised countries of the world large quantity of gaseous pollutants is released into the atmosphere by the factories. The most common among the gaseous pollutants are the oxides of nitrogen and sulphur.

Air borne SO_2 and SO_3 produced naturally or from human activities such as combustion of fossil fuels etc. react with water or water vapour to form H_2SO_3 and H_2SO_4.

$$SO_2 + H_2O \rightarrow H_2SO_3$$
$$SO_3 + H_2O \rightarrow H_2SO_4$$

Nitrogen oxides get dissolved into water vapour to form nitric acid.

$$N_2O_5 + H_2O \rightarrow 2HNO_3$$

These acids are dissolved in water falling down in the form of rain drops. This is called acid rain.

When the pH of rain water or snow is less than 5.7 it is called **ACID RAIN**. Acid rain is man-made ecological event. It is mainly due to H_2SO_4 and HNO_3. The ratio of acids varies depending upon the amounts of oxides of sulphur or nitrogen liberated into the atmosphere. On an average 60 to 70 per cent of the acidity is caused by H_2SO_4, and about 30-40% of the acidity is caused by HNO_3. On the earth surface acid rain increases the acidity of water and thus destroys aquatic vegetation and lowers the oxygen content of water.

Sources of Acid Rains

There are two main sources of acid rains:

1. *Oxides of sulphur.* Sulphur dioxide (SO_2) and sulphur trioxide (SO_3) are released into the atmosphere by the burning of fossil fuels.

2. *Oxides of nitrogen.* Oxides of nitrogen are released into the atmosphere by the automobiles exhaust and other sources.

Acid rain problem has increased tremendously by industrialization. Burning of fossil fuel for energy production has contributed 60 - 70 per cent of the total SO_2 liberated globally. Acid rain problem has become a global problem because these oxides can travel thousands of kilometres and during their journey they may undergo some physical and chemical transformations which may cause acid rains and produce hazardous products.

Formation of Acid Rain

1. ***Action of SO_2 leading to acid rain.*** It is formed in two steps:

(*a*) Conversion of SO_2 into SO_3 by absorption of UV radiation

(*b*) Reaction of secondary pollutant SO_3 with moisture to form sulphuric acid. Sulphuric acid dissolved in rain drops falls on the earth surface as acid rains.

$$SO_3 + H_2O \xrightarrow{\text{UV radiation}} H_2SO_4$$

2. ***Action of NO_2 leading acid rain.*** NO_2 acts to produce acid rain in two steps.

(*a*) Formation of NO_2 and N_2O_5 by absorption of UV-radiation from the sun.

(*b*) Reaction of NO_2 and N_2O_5 with atmospheric moisture to form nitric acid which falls down as acid rain

$$N_2O_5 + H_2O \xrightarrow{\text{UV radiation}} 2HNO_3$$

Factors Determining the Acidity

Rain water gets acidified on its way down to the earth. Its acidity depends on the amounts of oxides of sulphur and nitrates available for dissolution in the water.

The acidified rain water is mixed with already existing water on the earth. The resulting pH after this mixing determines the pH of the aquatic environment. This is influenced by the following factors :

(*i*) The neutralizing ability of the aquatic medium and

(*ii*) The acidity of rain water

Generally, lakes and rivers show a poor neutralizing capacity because of their low mineral contents. Only the rocky areas having $CaCO_3$ exhibit a buffer action and maintain a pH of about 5 - 6. But continued occurrence of acid rain is bound to lower the pH of the aquatic medium.

Consequences of Acid Rain

1. Acid rain declourizes the leaf pigments thus causes chlorosis and necrosis of leaves, which adversely affect agricultural production.
2. It inhibits CO_2 fixation and photorespiration and adversely affects the productivity of crops.

3. It inhibits seed germination and growth of seedlings. Alfalfa, barley, cotton, lettuce and spinach are most susceptible to acid rains.
4. It affects the enzymatic activities.
5. It may cause wilting of plants.
6. Acid rain increases the acidity of water bodies and consequently destroys the various aquatic vegetation and animals. Many fishes die in the lakes and their population are affected.
7. Acid rain causes irritation in the eyes and skin of human beings.
8. Since the acid rain water may dissolve some toxic metallic pollutants, it can directly enhance their toxic action and may pose a serious threat to plant and animal life.
9. Acid rains change the pH of soils and affect the plant growth in various ways.
10. Besides above, it may cause corrosion of many buildings, bridges, monuments, railings, fancings etc. This would result in putting off external surface and weakening of the structures.

Controlling acid rain

There are several measures through which acid rain can be checked:

(*i*) Emission of sulphur dioxide and nitrogen oxides from industries and power plants may be checked by using pollution controlling devices.

(*ii*) Liming or addition of calcium carbonate should be done to neutralise the adverse effects of acidity.

(*iii*) The long-term solution to check acid rain is to reduce the emission of the chemicals that cause acid rain.

Hydrocarbons and other chemical pollutants. Many hydrocarbons are known to be present in the atmosphere as pollutants. Natural sources of hydrocarbons are largely biological. Worldwide methane production mainly from anaerobic decay of organic matter probably amounts to one billion metric tons annually. Some plants produce volatile terpenes. In urban areas, ethylene is known to inhibit the plant growth. Open chain hydrocarbons affect human beings. Aromatic compounds cause irritation and injury in the mucous membrane at 500 ppm. Hydrocarbons are of particular concern because of their involvement in the production of photochemical oxidants which cause irritation and other effects. Southwick (1976) has listed common chemical air pollutants, their main sources and pathological effects on man (Table 13.15).

Table 13.15. Some Chemical Air Pollutants

	Pollutants	Main sources	Pathological effect on man
1.	Aldehydes	Thermal decomposition of fats, oil or glycerol.	Irritate nasal and respiratory tracts.
2.	Ammonia	Chemical processes—dye making, explosives, lacquer, fertilizer.	Causes inflamation of upper respiratory passages.
3.	Arsines	Processes involving metals or acids containing arsenic soldering.	Breakdown of red blood cells, damage kidneys, cause jaundice.
4.	Carbon monoxide	Gasoline motor exhausts; burning of coal.	Reduces oxygen carrying capacity of blood.
5.	Chlorine	Bleaching cotton and floor, many other chemical processes.	Attacks entire respiratory tracts and mucous membranes of eyes, causes pulmonary oedema.

6.	Hydrogen cyanide	Fumigation, blast furnaces, chemical manufacturing, metal plating.	Interferes with nerve cells, produces dry throat, indistinct vision and headache.
7.	Hydrogen fluoride	Petroleum refining, glass itching, aluminium and fertilizer production.	Irritates and corrodes all body passages.
8.	Hydrogen sulphide	Refineries and chemical industries, bituminous fuels.	Smells like rotten eggs, causes nausea, irritates eyes and throat.
9.	Nitrogen oxides	Motor vehicle exhausts, burning of soft coal.	Inhibit ciliary action so that soot and dust penetrate far into the lungs.
10.	Phosgene (carbonyl chloride)		Induces coughing, irritation, and sometimes fatal pulmonary oedema.
11.	Sulphur dioxide	Chemical and dye manufacturing.	Induces coughing, irritation, and sometimes fatal pulmonary oedema.
12.	Suspended particles (ash, soot, smoke)	Coal and oil combustion, incinerators and almost all manufacturing units.	Cause emphysema, eye irritations and possibly cancer.

Air pollution due to odours. Many chemical compounds, perfumes, whiskies and wines, and many other rotten substances emit undesirable smells which pollute the local atmosphere. Sometimes only by odour the quality identification of particular chemical is made and many edible things are evaluated only through odours. In dairy industry also smelling and testing of milk and its products may be used to determine the quality of raw material. Such testing is made only by experts. For food and perfumery industries, correct odour is essential.

Odour is a smell which can be determined by the nose. According of *Karolinska Institute Symposium on Environmental Health*, a product for the activation of the sense of smell, an olfactory experience, is called odour and any chemical compound which can stimulate the olfactory sense is called odour. Odours are described on the basis of both intensity and characters.

Odour is a physiological response of individual person. There are some instruments which can measure the physical or chemical characteristics of odour indirectly but these instruments are not better than human nose to determine the smell.

Other air pollutants. A number of other minor emissions such as vapour resulted from cosmetics, the smoke from tobacco products, aeroallergens (spores and pollengrains) which cause allergy in some persons are also regarded as air pollutants. Industrial processes especially chemical industries produce special air pollutants such as H_2SO_4, HCl, formaldehyde, alcohols and many other chemicals. A particular air pollutant that has attracted special interest in the recent years is asbestos. Its occupational hazards are known since 1943. All the urban dwellers are exposed to asbestos pollution. Pesticides and radiations are other important air pollutants which are very dangerous for human health.

AIR QUALITY AND AIR QUALITY INDEX (AQI)

A number of pollutants are known to occur in air but for the purpose of assessment of the air quality of a particular place, the following five pollutants are considered to be important. These are:

(*i*) Suspended Particulate Matter (SPM)

(*ii*) Nitrogen oxides (NO and NO_2)

(*iii*) Carbon monoxide (CO)

(*iv*) Sulphur oxides (SO_2, SO_3)

(*v*) Hydrocarbons (HC)

Recently lead has also been added to this list in view of its increasing content in the atmosphere and corresponding health effects. The maximum permissible limits or standards of these major air pollutants are presented in the following Table 13.16.

The standards are in two part: (*i*) Primary standards which apply exclusively to human health and (*ii*) secondary standards which apply to property damage and general welfare.

Table 13.16. Standards for different air pollutants

Pollutants	Primary standards		Secondary standards	
	ppm	µg/m³	ppm	µg/m³
Suspended Particulate Matter (SPM)				
Maximum 24 hrs conc. of	—	260	—	150
Sulphur oxides				
Maximum 24 hrs conc. of	0.14	365	0.1	26
Carbon monoxide				
Maximum 8 hrs conc.	9	10,000	same as primary	
Maximum 1 hr conc.	35	40,000	”	”
of Nitrogen oxides				
Annual arithmetic mean	0.05	100	”	”
Hydrocarbons	0.24	160	”	”
Maximum 3 hrs conc.				

The differences observed among the primary permissible limits of these pollutants can be attributed to their toxicities. The combined adverse effects of two or more pollutants should therefore be considered carefully while adopting these standards. A comparison of individual air pollutants against the maximum permissible limits, no doubt, gives some clues on the quality of air, the cumulative combined effect of all the pollutants provides real information on the quality of air, whether good or bad.

Air Quality Index (AQI) is a tool to reduce the vast amount of data to the simplest form rating all the essential information. The Air Quality Index (AQI) is a useful tool as it helps in rating the overall air quality based on the observed values of different air pollutants. Such an index is always associated with specific descriptor category. It may be noted that AQI is a complementary tool and should not replace precise scientific informations.

At present a variety of Air Quality indices are adopted by different organization. Here AQI developed by Thomas and co-workers (1971) at the Oak Ridge National Laboratory, USA is discussed. It is called ORAQI (Oak Ridge Air Quality Index).

This is based on 24 hr average concentrations of five major air pollutants (CO, NO_2, SPM, SO_2, Hydrocarbons). AQI has been modified for three pollutants (SO_2, NO_2 and SPM) for the same 24 hrs standards as follows:

$$\text{ORAQI} : 39.02\ \Sigma^3\ (X_i \ /\ X_s)\ 0.967$$

where X_i and X_s denote the pollutant concentration and the sum total of respective permissible limits of the major criteria of air pollutants.

For 5 critical Air pollutants ORAQI is represented mathematically by the following equation :

$$\text{ORAQI} = \left\{5.7\sum_{I=1}^{I=5}(C_i / S_i)\right\}137$$

where C_i = concentration of pollutant

S_i = sum of the standards for 5 criteria of air pollutants. ORAQI system use namograph, a scale representing concentrations for various pollutants and their relation to the established standard value. Thus the pollutants concentration in a fraction of the appropriate standard (C_i/S_i) coefficient (5.7) and the exponent 1.37 in the expression are used to scale the ORAQI in a such way that non-polluted background level attains Index value of 10 and the polluted air attains an index value between 10 and 100. ORAQI categories are as follows (Table 13.17).

Table 13.17. Index values for different categories of polluted cities

Categories	Index
1. Non-polluted	0—10
2. Excellent	10—20
3. Good	20—39
4. Fair	40—59
5. Poor	60—79
6. Bad	80—99
7. Dangerous	100 and above

Central Pollution Control Board (CPCB) in 1994 used SPM, SO_2, NO_2 as Air Quality data for calculating Air Quality Index for important cities of India. Because there is no clear cut demarcation between industrial and non-industrial areas, the annual average values for residential areas were used in calculation of AQI.

There are 14 cities which are in bad category. Seven cities are in good category, namely Tuticorin, Coimbatore, Daman, Silvassa, Vasco, Kozhikode and Guwahati. The rest of the cities come in either fair or poor category.

Central Pollution Control Board has published data for the purpose of analysing the Air Quality Index for the period from 1987 to 1996 of the following cities :

1. *Top ten polluted cities.* On the basis of 1991 census data, top ten polluted cities are : Mumbai, Kolkata, Delhi, Chennai, Hyderabad, Bangalore, Ahmedabad, Pune, Kanpur and Nagpur,

2. *Next other smaller cities.* These are Chandigarh, Dehradun, Faridabad, Howrah, Vadodara, Mysore, Cochin, Bhopal, Jaipur, Shimla, Agra.

The most polluted cities in terms of suspended particulate matter (SPM) are Dehradun, Gajroula, Surat, Kanpur, Delhi and Kolkata. In most of the cities the average value of suspended particulate matter (SPM) in open area may be double or tripple to that of residential areas. Among top 10 polluted cities, the worst in terms of SPM are Kanpur, Delhi, Kolkata, where the average concentrations are nearer to the levels of maximum permissible limit.

In terms of NO_x (nitrogen oxides) some top polluted cities are Howrah, Kota, Dhanbad, Jharia, Gajroula and Mumbai.

The Central Pollution Control Board has suggested the following classification of Indian cities on the basis of Air Quality Index (Table 13.18.)

Table 13.18. Classification of Indian cities on the basis of Air quality index

Dangerous	Bad	Poor	Fair	Good
Howrah	Faridabad	Vapi	Paonta Sahib	Tuticorin
Kolkata	Satna	Haldia	Chennai	Coimbatore
Delhi	Ahmedabad	Patna	Kottayam	Daman
Kanpur	Mumbai	Chandipur	Aurangabad	Silvassa
Lucknow	Indore	Talcher	Ponda	Vasco
Vadodara	Nagda	Yamuna Nagar	Mysore	Kozhikode
Jalandhar	Jabalpur	Bhopal	Nasik	Guwahati
Dhanbad	Rourkela	Chandigarh	Shimla	
Ghaziabad	Jaipur	Angul	Hyderabad	
Alwar	Varanasi		Nagpur	
Ankleshwar	Pondicherry		Cochin	
Jharia	Raipur			
Surat	Pune			
Rajkot				
Patiala				
Ludhiana				
Agra				
Sindri				
Kota				
Korba				

DISEASES DUE TO AIR POLLUTION

Air pollution has caused the most serious threat to human health. The common air pollutants and their diseases are:

1. **Sulphur dioxide:** Causes respiratory diseases like cough, damage to lungs, asthma, bronchitis.
2. **Carbon monoxide:** Heart and brain damage, headache and giddiness, fatigue.
3. **Oxides of nitrogen:** Respiratory tract diseases, irritation in the eyes, allergies.
4. **Ozone:** Affects the eyes, damages the lungs, affects the heart, soreness.
5. **Hydrogen sulphide:** Extremely toxic even at low concentrations, affects the nervous system.
6. **Gaseous hydrocarbons:** Affects the lungs and is carcinogenic (causes cancer).
7. **Tobacco smoke:** Main source of indoor pollution, it is carcinogenic.
8. **Particulate matter:** These are the solid tiny particles that remain suspended in the air. Their effect on human health depends on their nature and type. Some common particulate pollutants are:
   - **Lead:** Extremely toxic and affects the kidneys, liver and nervous system. It enters the blood stream and also affects pregnant women, foetuses and newborn babies.
   - **Mercury:** Higher concentration of mercury is injurious to human life; it causes chromosome damage, loss of memory.

- **Cadmium:** It was the cause of **itai-itai** disease in Japan; It leads to the formation of kidney stone. Liver and pancreas are also affected.
- **Arsenic:** It causes diarrhoea, it is absorbed through the skin and lungs and causes skin and lung cancer. Chronic exposure to arsenic is the cause of **black-foot** disease, which is common in parts of West Bengal and Bangladesh.
- **Copper:** Carcinogenic.
- **Chromium:** Specially toxic to aquatic life.
- Some other metallic pollutants like molybdenum, barium, manganese, cobalt also affect human health.
- **Spores and pollens:** These remain suspended in the air for long time and cause respiratory diseases, asthma and allergies.

Respiratory Diseases

Air pollution mostly affects the lungs and other organs of the respiratory system. Some of the common respiratory diseases are:

Bronchospasm: The pollutants mainly cause irritation in the respiratory tract. Sometimes the muscles of the bronchioles get inflamed and obstructed due to excessive irritation. This is known as bronchospasm.

Bronchitis: This is caused due to swelling of the lining of the bronchi. It causes severe coughing and mucus production.

Emphysema: Here the walls of the individual alveoli breakdown and the alveoli join together to form a sac like structure. This reduces the efficiency of the lungs.

Pulmonary Fibrosis: This is caused due to dust and other solid suspended particles. As dust is not eliminated from the alveoli, some fibrous tissues develop around it in due course of time. This also decreases the efficiency of lungs.

Asthma: It is an allergic disease, which is caused due to foreign bodies inside the respiratory tract. In this bronchospasm occurs and the airways are narrowed. It is marked by difficulty in breathing.

Pneumoconiosis: This is a respiratory tract disease caused due to inhalation of dust particles, which induces severe coughing and congestion.

CONTROL OF AIR POLLUTION

Air pollution can be better controlled by (*i*) reducing the emission of pollutants and (*ii*) by using non-polluting materials. Some of the major steps for reducing air pollution are as follows:

1. Controlling particulate pollutants by using scrubbers, filters, cyclone separators and electrostatic precipitators in the industries (Fig. 9.4)
2. Change from high-sulphur coal to low-sulphur coal, especially for the thermal power plants, to check the emission of SO_2.
3. Locating industries far from human settlements
4. To check automobile pollution by the use of unleaded petrol, better engines with low emissions and proper maintenance of vehicles
5. Installation of catalytic converters in the vehicles
6. Conducting regular pollution control checks on the vehicles
7. Adopting less polluting fuels like Compressed Natural Gas (**CNG**) and Liquefied Petroleum Gas (**LPG**)

8. Adopting more non-conventional energy sources that are non-polluting, such as solar energy, wind energy, etc.
9. Using mass transport systems and switching over to non-polluting modes of transport like bicycles
10. To check deforestation and encourage plantation especially around the polluting industries

The Taj Mahal was being adversely affected by sulphur dioxide smoke and suspended particulate matter released by many industrial units located around it, at Agra. As a result, the 'White Wonder' was loosing its beauty due to corrosion of the marble and was developing a yellowish tinge. Noted environmental activist lawyer, Sri S.C. Mehta, filled a Public Interest Litigation in 1984, expressing concern over the effect of the polluting industries on the Taj Mahal. The Honourable Supreme Court, in 1996, ordered the industries to be shut down, or to adopt strict pollution control measures like use of natural gas.

AGRICULTURAL POLLUTION

Improved agronomic practices, artificial fertilizers, farm chemicals and improved methods of preserving and transporting food are playing roles in agricultural pollution. Although air and water pollutions have affected agriculture adversely, agricultural methods themselves have contributed to environmental pollution. Some of the important agricultural pollutants are as follows :

Farm animal wastes. Over the century farm animal wastes, such as faecal matter, urine, wastes from slaughter houses, dead animal wastes, have been regarded as important sources of soil fertility. Nevertheless, the presence of farm animal wastes pose serious problems of foul odour and water pollution. Sometimes public health problems arise from their role in transmitting diseases. Animal wastes pose serious problems when they enter the water supplies.

Soil erosion. Eroded materials affect the fertility of the soil. Strong winds carry soil particles into the air and water and thus produce air and water pollution. When they are carried to the streams by storms or by runoff water they settle at the bottom and destroy irrigation systems. Sediments can clog gills of fishes. It interferes with the industrial uses of water and leads to extra-expenses for purification of water. It can reduce the dissolve O_2 level and thereby it adversely affects aquatic life. It affects the recreational values of lakes, reservoirs and streams. Control of soil erosion consequently can control the environmental pollution.

Plant residues. Plant residues from crop fields and orchards contribute much to environmental pollution. They carry plant diseases and pests. When they are burnt they emit smoke and hydrocarbons. Since both the situations correspond to an unfavourable alteration of environment, agricultural burning is important pollution problem. Burning of plant residues gives—

8.3% CO,

8.5% particulate matter,

5.3% hydrocarbons,

1.5% nitrogen oxides.

Much of these burnings is for sanitation purposes since plant debris can carry diseases or pests for succeeding crops. Agricultural burnings and the resulting air pollution are likely to continue until better methods of controlling the plant diseases are available.

Agricultural chemicals. In recent years there has been increased use of agricultural chemicals, notably pesticides and fertilizers. **Pesticides** is a general term applied to the substances that kill the disease pests. These include insecticides, fungicides, nematicides, rodenticides and herbicides or weedicides. Thus, agricultural chemicals have no doubt increased crop yields but they have contributed air and water pollution to a significant level. They have many adverse effects on fishes, wild life and sometimes human beings. Excessive application of fertilizers may create problems since fertilizers can be transported to ground water by leaching or to waterways by natural drainage and surface runoff. Nitrates in drinking water can cause methaemoglobinemia in babies (blue bodies), because in infant's stomach the NO_3 is converted into NO_2 which when reaches the blood reacts with haemoglobin to form methaemoglobin. Phosphorus is often implicated in algal blooms in waterways.

A. Insecticides. Farmers and gardeners use pesticides to avoid the damage of crops by pests. Insecticides have come in use from the last two decades. Mitter discovered DDT (Dichlorodiphenyl trichloroethane) which is the first chlorinated hydrocarbon to be used as pesticide. Later on other pesticides were discovered. Insecticides are classified as follows :

(*i*) stomach poisons (*ii*) contact poison, and (*iii*) fumigants.

(*i*) **Stomach poisons** are arsenic and fluorine compounds; lead arsenate, inorganic fluorine compounds, such as cryollite, sodium fluoride, mercury compounds (HgCl, $HgCl_2$), boron compounds (borax and boric acid), chlorinated hydrocarbons, Organophosphates, carbamates, and certain botanical compounds as pyrethrum (powdered flowers of *Chrysanthemum*), nicotine. Organophosphorus compounds are most toxic insecticides that affect insects and mammals. Many of them (parathion, thimet, phosdrin, paraxon) are supertoxic and are human poisons giving many symptoms as nausea, vomiting, diarrhoea, sweating, salivation and muscular tremors.

(*ii*) **Contact poison insecticides** are chlorinated hydrocarbons and their metabolites, *e.g.*, DDD, TDE , BHC , DDT which have primary effects on central nervous system. In the environment DDT is degraded into different forms. Chlorinated hydrocarbons are known to be present in the body of fatty human beings, the main source being food supply. Studies have shown that daily content of DDT in meal is about .04 to .05 mg or more. Chlorinated Hydrocarbons persist in the environment for long time. They not only affect large number of insect species but also they are responsible for undesirable side effects.

(*iii*) **Fumigants.** There are gases which are used to kill pests of stored products or nursery stock, *e.g*, methyl bromide, carbon disulphide, carbon tetrachloride, nicotine, naphthelene, phosphine and other compounds.

B. Fungicides. These are used to kill fungal pathogens of plants. Sulphur, organic mercuric compounds, formaldehyde, copper sulphate and some others are commonly used.

C. Herbicides. These are chemicals which kill herbs. They are of two types:

(*i*) **Non-selective herbicides** which kill all the plant. The examples are sodium arsenite, sodium chlorate, H_2SO_4.

(*ii*) **Selective herbicides** which kill only certain types of plants. Most commonly used herbicides are 2, 4-D and 2, 4, 5-T and their esters and growth regulating substances.

Extensive use of herbicides like 'picloram' and cacodylic acid in South Vietnam by the U.S army has changed the entire ecology of about one-third land area. They caused defoliation of forest trees and destroyed crops and fauna to a great extent. It is horrible act to play with ecosystem and upset the balance of nature by polluting or poisoning the environment just on political ideology. 2,4,5-T and 2,4-D are potential herbicides. Spraying of these herbicides completely destroyed the vegetation and fauna in estuarian areas of Vietnam. Application of defoliants completely destroyed the forest trees which provided thick cover to Vietnam army. The destruction of climax forest has caused problems of increased erosion, leaching of nutrients and migration of animals from the area.

Herbicides and insecticides are very potent pollutants and they greatly affect the structure and function of ecosystems. Their indiscriminate use may be suicidal for our own race.

D. Rodenticides. They are used to kill rodents, rats and mice. Some have direct immediate action, *e.g.* strychnine, sodium fluoroacetate, P-thallium and ANTV and others are **blood coagulants.**

The demand for pesticides as a plant protection measure has rapidly increased in India since the introduction of high yielding varieties of crops. A large number of pesticides approved for use under the Insecticide Act, 1968 have been found to be effective against various pests either through direct contact or through systemic action. Some are beneficial as fumigants for protection of food grains during storage. According to a report of the International Development Research Centre. Ottawa, Canada, about 10,000 people die of pesticide poisoning every year in the developing countries. The alarming rise in the number to such deaths is attributed to increasing use of toxic pesticides. There are about 40,000 such chemicals. A number of such toxic chemicals banned in Western countries are dumped in the Third World countries.

In India, the use of pesticides has risen from 2,000 tonnes a year in the fifties to over 80,000 tonnes in 1985-86 and the areas of application cover both agriculture and public health. The cropped area under plant protection coverage increased from 6.4 million hectares in 1960-61 to over 80 million hectares. The demand for pesticides in 1989-90 was estimated to be about 1.20 lakh tonnes. BHC, DDT, and malathion which account for more than 50 per cent of total production are the cheapest and most popular with small farmers. DDT is mostly used by public health department.

Despite restrictions and regulations on pesticide use, India accounts for one-third of pesticide poisoning cases in Third World. Farm labourers employed for spraying operations are the worst affected as they get exposed for longer period by working continuously during the week. Cases of blindess, cancer, stunted growth of farm workers' children, deformities, diseases of liver and nervous system due to pesticide poisoning have been indentified in the cotton growing districts of two states in India. Studies conducted by the Industrial Toxicological Research Centre (ITRC), Lucknow in collaboration with the Neurology Department of S.J.M. Medical University, Lucknow revealed that one-fifth of the farm workers have damaged eye-sight following muscular degeneration.

There were 106 deaths in Kerala from consumption of wheat flour and sugar that got contaminated with folidol due to leakage during shipment from Mumbai to Cochin, which caused myosis, salivation, pain in abdomen, diarrhoea and unconsciousness. Earlier, Harijans in the Chikmagalur district of Karnataka were affected with severe paralysis and other fatal diseases after eating crabs from pesticides treated paddy fields.

Organic pesticides are more effective and relatively less hazardous than inorganic ones. Chlorinated compounds are quite stable and persist in the environment for long period. Their accumulation in the tissue lipoids causes acute toxicity affecting central nervous system leading to hypersensitivity, convulsions, paralysis and death through respiratory arrest. Some pesticides of this group are DDT, BHC, aldrin, dieldrin, lindane, methoxychlor, endrin, etc.

Ethylene dichloride, ethylene dibromide, methyl bromide (halogenated compounds) and aluminium phosphide (inorganic) are commonly used for fumigating cereals and pulses. There is a risk of bromide residue in stored grains. It accumulates in human and animal tissues ; liver, kidney, heart, spleen and causes degenerative lesions.

People engaged in agriculture are exposed to the risk of pesticides and other chemicals. Due to continuous exposure to such pollutants through air, soil and water, the people face many complications ranging from neurological diseases, gastric disease, respiratory diseases and even cancer (Table 13.19).

Table 13.19. Toxic pollutants and their effects

Pollutant	Source	Effect on health
DDT	Insecticide	Affects central nervous system, carcinogenic
Aldrin	Insecticide	Convulsions, damages the kidneys

Dioxin	Herbicide	Carcinogenic, causes acne
Chlorinated organic compounds	Chlorination of water	Carcinogenic
Nitrates and nitrites	Septic tanks, heavily fertilised crops and sewage treatment plants	Reduces oxygen carrying capacity of blood, fatal for infants
Polychlorinated biphenyls (PCBs)	Electric appliances, plastic containers, ingredients in soap, cream, paint, etc.	Vomiting, fatigue, abdominal pain, gastric disorder, temporary blindness
Vinyl chloride (PVC)	Used in plastic	Affects liver, bones and circulatory system
Benzene	Detergents, mouldings, fibres, insecticides	Anaemia, leukaemia and bone marrow disorder
Nistrosamine	Chemical reaction of nitrites with amines present in food	Carcinogenic

RADIATION POLLUTION

Radiation refers to the energy produced by fast moving particles or waves of energy. Low levels of radiations occur naturally from several sources and this process has been continuing ever since the universe came into existence. There are certain radioactive elements that occur as natural minerals in the Earth's crust which continuously emit some degree of radiation. However, high levels of radiation are extremely dangerous to all organisms. Radiation pollution refers to the increase in radiation that is injurious to life.

The process of radiation or radioactivity involves radioisotopes of certain elements that undergo radioactive decay. During the process of decay, the radioisotope changes from one isotope to another and emits certain types of radiation. Isotopes are atoms of an element that have the same atomic number but different in atomic mass. For example, the two isotopes of uranium are ${}^{235}U_{92}$ and ${}^{238}U_{92}$. An important parameter in the study of radioactivity is the **half-life period.** It is defined as the time required for one half of a isotope to decay into another form. The half-life varies from element to element, some of which are given below:

- Uranium-235 - 700 million years
- Carbon-14 - 5570 years
- Radon-222 - 3.8 days
- Polonium-218 - 3 minutes

The isotopes with short half-life periods are present only for a brief time in the environment, whereas those with a long half-life remains in the environment for a long period.

There are three kinds of nuclear radiation: **alpha particles, beta particles** and **gamma rays.** Alpha particles are positively charged, beta particles are negatively charged and gamma rays are very high energy electromagnetic radiations. Radioactivity is the spontaneous disintegration of atomic nuclei and the emission of subatomic particles called alpha particles and beta particles, or of electromagnetic rays called X-rays and gamma rays. The alpha particles can travel only upto 5 to 8 cm in air and are least penetrating. Beta particles are nearly 100 times more penetrating and the gamma rays are much more penetrating than beta particles. Alfa radiation is relatively harmless unless inhaled or ingested, beta radiation is intermediate in toxicity and gamma rays are most toxic and dangerous.

SOURCES OF RADIATION POLLUTION

Ever since the earth came into existence, all life forms have been exposed to some degree of low-level radiation (natural background radiation). Scientists consider that this exposure has played a key role in the evolution of life. However, in the present scientific and industrial world, the sources of radiation have increased due to multiple human activities. Thus the sources of radiation can be broadly classified into two major groups:

Natural sources

Our surrounding environment, such as the air, the rocks and soil, our house, even the food we eat and the water we drink are all sources of radiation. Some amount of radiation is also present within the human body; for example, polonium and radium are present in the bones while radioactive carbon and potassium are found in the muscles. The natural sources can be further categorised into the following:

Solar radiation: The radiation from the sun comprises both ultraviolet and infrared radiations.

Cosmic radiation: Cosmic rays constantly rain down in the Earth's atmosphere from outer space. These primarily contain protons, few electrons, helium ions, etc which continuously interact with the Earth's atmosphere.

Terrestrial radiation: Terrestrial sources are the most important natural source to which the Earth's organisms are exposed. It results from the cumulative effect from a number of materials around us, such as water, air, rock, soil, food, etc. The sources of terrestrial radiation are the radioactive minerals that are present in the Earth's crust.

Man-made sources

Due to the advancements in science and technology, humans have been using radiation technology for energy requirements and for health benefits. The important man-made sources of radiation pollution are as follows:

- Radioactive wastes generated during the mining and processing of radioactive minerals.
- Nuclear fuels used in nuclear reactors for power generation.
- Nuclear fall out due to the testing of nuclear weapons.
- Medical use such as X-rays, radiotracers for detecting and curing cancers, etc.

Sources of radiation pollution

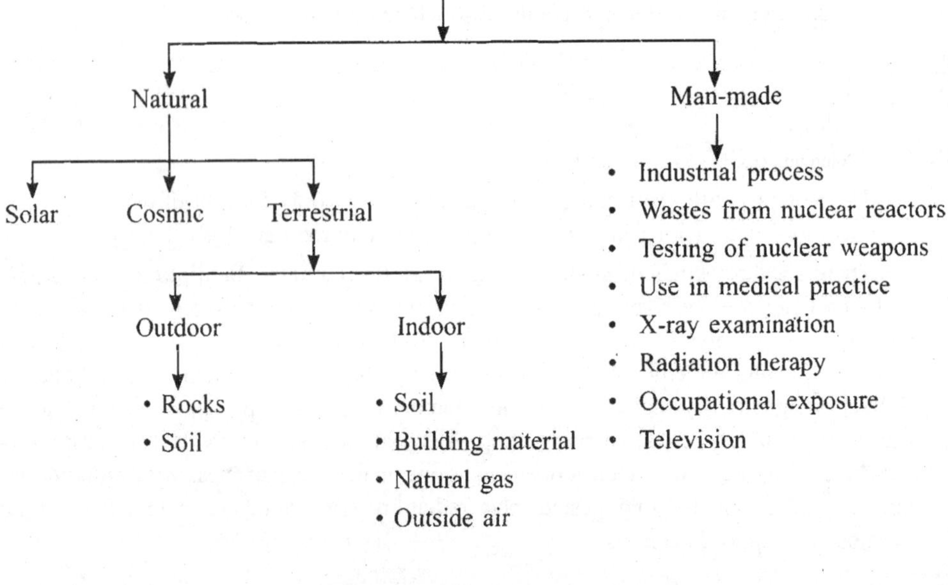

Consequences of Radiation Pollution

People who live or work in the vicinity of radioactive substances, such as mine workers, medical personnel, technologists working in nuclear power plants, etc. are more prone to the radiation hazards. Most serious hazards are caused due to accidents in the nuclear reactors. The severity of the injury depends on the type of radiation, the absorbed dose, the rate at which the dose was absorbed and the radio-sensitivity of the tissues involved.

The damages to the organisms are of two main types:

- **Somatic damage:** The body of the organism is directly affected causing burns, mental retardation, reduced fertility, blindness, cancer, etc.
- **Genetic damage:** In this case, the genes, chromosomes and the DNA of the organisms are affected and transmitted to the next generations.

NON-IONIZING RADIATION POLLUTION

Non-ionizing radiation includes radio-frequency radiation or electromagnetic fields (EMFs), from sources such as power lines, radar, communication networks, cellular phones and microwave ovens. (Fig. 13.9). Hign doses of such radiation are known to be harmful, causing burns, cataracts, temporary sterility, etc. Since the 1980s, with the proliferation of such devices, the possible effects of long-term exposure to low levels of non-ionizing radiation are a matter of great concern.

Fig. 13.9. Microwaves cause non-ionising radiation pollution

P.K. Zutshi of Health Division, Bhabha Atomic Research Central (BARC) has thrown sufficient light on the problems related to radioactive pollution in a paper entitled "Environmental Pollution Studies at BARC" published in *Nuclear India*, Vol. 12.

Fallout of nuclear explosions. Nuclear explosion has both immediate and long-term effects. It can reduce big cities like New York or London to ashes in no time, yet there is no practical method of protecting the population from it. It can char woods and even ignite them within 16 km from the site of explosion. The temperature at the site of explosion is so high that metals and minerals not only melt but are vapourised also. During atomic explosions the radioactive substances are rapidly injected into the upper layer of the atmosphere where they condense due to cooling from gaseous state to liquid and then to fine dust. Radioactive dust particles are suspended in the air at an altitude of about 10 - 15 km and form together with water vapour, a radioactive cloud which moves in the direction of wind. The radioactive cloud settles down slowly on the surface of earth over several years. Most of the radioactive materials produced in explosion are short-lived but some, such as strontium^{90} Cesium^{137}, Co^{60} , remain radioactive for many years. About 5% of the radioactive matter injected into atmosphere after nuclear explosions consists of radioactive Sr which is easily absorbed by the living organisms. On reaching to the earth, it poisons water, air and soil and is absorbed by grasses and vegetables and thus gets into the milk and finally into human body. There, it accumulates in the bones and other tissues and causes diseases. Low level radioactive liquid wastes, radioactive gases and dusts are released from nuclear installations. The liquid wastes are released into the waters and the gases are disposed through high stacks. Besides stack emission from nuclear installations, there is Worldwide fallout of radioactivity from atomic bomb tests. BARC has been assessing exposure of the population in India to radioactivity arising from nuclear tests explosions. It is also studying the atmospheric air circulation pattern by which radioactive pollutants are dispersed on a global scale. Level of radioactivity after test explosions is reported to be insignificant for short period immediately after explosions but it increases appreciably after some time and subsides within a few weeks. Radioactive fallouts have been observed from the upper atmosphere where the debris of tests explosions stays for a long time.

Hoda, M.M. (1977) in his article on *"Nuclear Energy and Environments"* highlighted some of the problems relating to hazards of radioactive materials. Much has been said about the nuclear weapons, although there is a great possibility that it may not be used again, the greatest danger to humanity is now posed by the so called peaceful uses of atomic energy but not so many of us realise this fact. While man can and does create radioactive elements, there is nothing to reduce their activity, once he has created them. No physical interference, no chemical reaction but only the passage of time can reduce the intensity of reaction once it has been set going. The fission of Uranium^{235} results in a mixture of a number of long-lived and short-lived isotopes. The disposal of radioactive wastes presents enormous problem. It was thought at one time that radioactive wastes could be safely dumped into the deepest part of the ocean on assumption that no life could subsist at such depth, but within a short time after dumping these materiels in water the great bulk could be found in the living organisms. No international agreement has been reached so far on disposal of nuclear wastes. Nuclear wastes with high level of radioactivity continue to be dumped into the sea while large quantities of so called 'intermediate" and *"low level"* wastes are discharged into the rivers or directly into the ground. An Atomic Energy Commission report observes that liquid wastes find their way slowly into groundwater, leaving all or part of their radioactivity held either chemically or physically in the soil (U.S.A.E.C. Annual Reports to Congress, Washington D.C. 1966, p. 344). The nuclear reactors, after they have become unserviceable, are most massive wastes. The reactors have an average life of 30 years and they cannot be dismantled and shifted but have to be left standing for centuries where they will silently leak radioactivity into air, water and soil.

There is no foolproof method of disposal of radioactive wastes. In many cases the radioactive wastes are sealed in steel drums or concrete blocks and sunk into sea to await their natural decay. The pollution of aquatic environment by radioactive materials poses a serious problems as radioactive isotopes can produce not only immediate effects on living organisms but, even more important, can produce mutations in the genetic materials which would have serious consequences for the later generations. Radioactivity can reach the aquatic environment from a variety of sources such as nuclear fallout, nuclear powered ships and submarines, nuclear power plants, laboratory experiments with medical uses of radioisotopes and so on. Desalination and Effluent Engineering Division at BARC has developed high efficiency particulate filters for air pollution control from hazardous contaminants, such as radioactive isotopes from nuclear installation and compounds of such toxic elements as berylium and tellurium.

DISEASE DUE TO RADIATION POLLUTION

The effects of radioactive elements are difficult to judge unless they are in such concentrations as to give acute effects. Malformation of body at birth, abnormality in organ development are some of the effects observed in the laboratory animals.

On the basis of rate of effect, radiation pollution can be mainly classified into acute and chronic effects:

Acute effects

Accidental exposure to radiation causes Acute Radiation Syndrome (ARS). Exposure of small areas leads to localised tissue damage. Damage to the blood vessels in the exposed areas, causes disturbed organ function; and at higher doses, necrosis (localised tissue death) and gangrene. Consequences may include degeneration or destruction of the irradiated tissue and the initiation of cancer.

Chronic effects

This is also known as late effect, since the effect of radiation on the body is manifested after some duration of time. Some of the pioneer workers on radioactivity, such as Madam Marie Curie, who discovered radium, succumbed to leukaemia. The late effects of radiation are manifested in the form of chronic diseases of kidneys, lungs, bone marrow, eyes, etc. Statistically significant increases in leukaemia and of cancers of thyroid, lung, and the female breast have been demonstrated in populations exposed to relatively high doses.

CONTROL OF RADIATION POLLUTION

Radiation pollution could be a fatal health hazard and it can be reduced by the following ways:

- Proper maintenance of nuclear power plants by well trained experts to avoid the occurrence of Chernobyl like disasters
- Proper disposal of radioactive wastes from the nuclear reactors
- Proper use of radiation technology in hospitals to avoid injuries to technicians and patients
- Proper disposal of radioactive bio-medical wastes

THERMAL POLLUTION

Heat or thermal pollution is common to both nuclear and conventional sources of energy production. It can be both an air and water pollutant. Waste heat not only causes widespread climatological changes but also it can damage aquatic and terrestrial life. The thermal discharges are usually favourable to bacteria and pathogens. Rapid temperature changes produce thermal shocks and result in immediate death of fishes. Temperature of the environment controls reproductive cycle, digestion rates, respiration rates and many chemical activities taking place in the body. As the temperature of water increases its viscosity decreases, the vapour pressure increases, and solubility of gases in water decreases. These changes have important effects on aquatic life.

NOISE POLLUTION

In addition to disturbances caused by air pollution, a new kind of disturbance has come into prominence. This is noise or sound pollution. Noise is generally defined as an unwanted or undesirable sound and the release of unwanted sound into the atmosphere is called noise pollution. High pitch sound produced by automobiles, machines, power plants, trains, helicopters, aeroplanes, jets, rockets, public broadcasting systems, T.V., sudden ratting of doors and windows, explosion of bombs, sound of crackers etc. are potent sources of noise pollution. Noise provides alarm system in man's physical environment. It has been regarded as urban pollution. It has become serious in urban areas due to technologies increasing population, urbanization and industrialisation. It is associated with almost all human activities. In 1990, Nobel laureate Robert Koch said, "a day will come when man will have to fight merciless noise as the enemy of health and that day is not far off". The noise should be taken not only as a factor of annoyance but also as a lethal partner of life. According to Dr. Vern O. Knudsen of university of California, noise, like smog, is a slow agent of death. If it continues unabated for the next 30 years it could become lethal to an alarming degree.

Sound is a form of energy. Man is not only affected by the sound energy but also by sound pitch or frequency and intensity. Intensity of sound is measured in decibels (d = deci meaning 10, B for Bell, name of scientist Alfred Grahm Bell). Technically a dB is the amount of sound pressure that equals to 0.0002 microbars (energy of about 10^{-15} watts). One dB is the finest sound intensity which can be perceived by human ear and 180 dB is the maximum tolerance limit.

The frequency is defined as the number of vibrations and is denoted by Hertz (hz). One Hertz (hz) is equal to one vibration per second. Man can hear sound from 16 hz to 20,000 hz. Less than 16 hz second is infra-audible and sound above 20,000 hz is ultrasonic, some animals however can perceive sounds that are not andible to man. Loudness of sound can also be expressed in sones. One sone is equal to 40 dB at a pressure of 1,000 hz.

The loudness of noise is measured on a logarithmic scale called "decibel". Mathematically the decibel (dB) can be expressed as follows:

$$I \text{ dB} = 10 \log_{10} (P/p \text{ ref}) \text{ or } 10 \times \log 10 \text{ p/p ref.}$$

where, P = measured sound pressure in dynes /cm and, p ref. = reference sound pressure which is equal to barely audible sound taken as 0.002 dynes /cm.

As the decibel is measured on a log scale, rise of every 10 decibel would cause 10 fold change in sound intensity. For example, a sound level of 100 decibels has an intensity 10 times greater than a sound level of 90 dB. According to World Health Organization (WHO), a level of 45 dB is considered a safe noise level for city. By international standard, a noise level up to 65 dB may be taken as tolerable. Noise level above 80 dB causes noise pollution. Noise level above 100 dB becomes uncomfortable and beyond 120 dB it becomes painful. Sudden or intermittent noise is more harmful than continuous noise. According to a report of National Physical Laboratory, India, the noise level is increasing at the rate of 1 dB per year.

Some important standards developed for noise levels in India are given in Tables 13.20, 13.21, 13.22 and 13.23.

Table 13.20. Intensity of sound noise sources and human perception.

Noise Source	Intensity of sound (db)	Human perception
Threshold of hearing	0	Threshold of hearing
Breathing	10	Just audible
Sound of leaves in trees	20	Very quiet
Whispering	30	Very quiet
Normal conversation	30-40	Quiet
Homes and Restaurent	45-50	Quiet
Loud conversation	65	Moderately loud
Lawn mower	60-80	Moderately loud
Vacuum cleaner	80	Moderately loud
Traffic noise	60-90	Loud
Heavy trucks	90-100	Very loud
Thunderstorm	110	Very loud
Rock music	120	Uncomfortably loud

Jet take off (100 m distance)'	120	Uncomfortably loud
Jet engine (at 25 m distance)	140	Painfully loud
Rocket engine	170-180	Painfully loud

Table 13.21. Noise limits in different areas

Category of Area	Noise Limits in dB	
	Day hours	Night hours
Industrial area	75	70
Commercial area	65	55
Residential area	55	45
Silence zone	50	40

Table 13.22. Noise limits for different automobiles in 1992

Category	Standards (dB)
1. Motor cycles, scooters, three-wheelers	80
2. Passenger cars	82
3. Passenger or commercial vehicles up to 4 MT	85
4. Passenger or commercial vehicles above 4 MT and up to 12 MT	89
5. Passenger or commercial vehicles exceeding 12 MT	91

Table 13.23. Damage risk criteria for hearing loss
(According to Occupational Safety and Health Administration (OSHA)

Noise level (in dB)	Maximum Allowable duration (hours per day)
90	8
92	6
95	4
97	3
100	2

Effects of Noise Pollution

Human response to noise varies from man to man according to age and temperament. It may vary even in the same individual from time to time because of change in health, fatigue and other conditions. The effects of noise on human beings can be related to physiological, behavioural and psychological responses.

When we hear a loud noise the blood vessels constrict, the skin pales, the pupil dilates, the eyes close, the voluntary and involuntary muscles tense, gastric secretions reduce and the blood pressure increases. The immediate reactions to noise which starts at noise level of 30 to 70 dB include constriction of peripheral blood vessels with a consequent increase in blood flow to the brain, change in breathing rate, change in muscle tension and gastrointestinal mobility. As these reactions can also be carried out by some other factors, they are called non-auditory reactions. At slightly higher noise level there is "startle effect" produced due to change in pulse rate and blood pressure, release of stored glucose from lever into the blood stream and the excess secretion of adrenalin harmone because of which fear reaction develops in body. Some of the other effects of noise pollution are given below:

1. The short-term auditory effects of noise include some temporary loss in hearing at high noise levels. Explosions or other high intensity sounds may cause immediate deafness by rupturing the ear drums or damaging the cochlea. Continuous exposure at high noise level may cause some permanent damage.

2. Prolonged exposures to noise are known to lead to gradual deterioration of internal ear and to subsequent hearing loss or deafness. Scientists are of the opinion that constant occupational exposure to sound at higher frequency is dangerous for human beings.
3. Noise also clearly produces annoyance in human beings and can interfere with proper rest and sleep. Annoyance seems to increase with the loudness of the sound.
4. Noise also produces psychological effects in human being.
5. One important effect of noise is the vasoconstriction reflex in which small blood vessels of the body constrict and reduce the flow of blood.
6. Physiological effects of noise include dilation of the pupils, paling of skin, tensing of voluntary muscles, diminishing of gastric secretions, increase in diastolic blood pressure and the sudden injection of adrenalins into blood stream which increases neuromuscular tension, nervousness, irritability and anxieties. It can adversely affect the development of unborn babies.
7. Impulsive noise and thunders are worse than continuous noise which, if loud enough, may cause sudden hearing loss. Sudden noises also startle persons and lead to accidents. Surpersonic booms produced by objects travelling faster than the speed of sound produce shock waves which deviate from atmosphere pressures.

DISEASES DUE TO NOISE POLLUTION

Recently there has been growing evidence that noise in the range of 90 decibels and above may cause temporary or irreversible changes in the autonomous nervous system. Noise may be an important factor in many stress-related diseases, such as peptic ulcer and hypertension.

A brief summary of noise induced diseases are given below:

Deafness or impaired hearing: This may occur due to continuous exposure to noise level of more than 90 dB. It may be temporary or permanent.

Auditory fatigue: Continuous humming sound may cause side effects such as whistling and buzzing in the ears.

Irritation and annoyance: Noise, sometimes, leads to emotional disturbances and makes people loose their temper.

Work efficiency: It has been observed that people work efficiently in a quiet environment and noise reduces the efficiency of work.

Other health effects: Noise is also associated with:

Rise in blood pressure (hyper tension)
Headache and giddiness
Sweating, nausea and fatigue
Difficulty in breathing
Disturbed sleep pattern
Psychological stress

CONTROL OF NOISE POLLUTION

Noise pollution can be checked by the following means:

- Noise-generating factories should be located far away from the populated areas.
- Proper maintenance of machinery checks noise pollution.
- Use of pressure horns in automobiles in the densely populated areas should be banned.
- Loudspeakers and mikes should be banned from 10 pm to 6 am.
- Firecrackers producing loud noise should be banned.
- Televisions, music systems. VCDs should be played at a low volume.

There is no separate law for noise pollution control in India. It has been suggested that cultivation of thick vegetation along the roadsides, railway tracts and industrial establishments may considerably reduce noise pollution.

IMPACT OF POLLUTION ON VEGETATION

Many chemicals, fertilizers, pesticides combustion of coal, and waste treatments release toxic substances into the environment that are taken up by the plants from air, water and soil. Atmospheric pollutants particularly SO_2, halides (HF, HCl), ozone, carbon monoxide and peroxiacetyl nitrate (PAN) produced from automobiles, industrial fumes and strong radiations are dangerous to plants. Harmful substances that reach to plants through the air are SO_2, nitrogen oxides, hydrocarbons, dusts and smokes. Plants growing in water are severely affected by toxic chemicals like cyanides, chlorine, hypochlorate, phenols, benzyl derivatives and heavy metal compounds of sewage.

Effects of different kinds of pollution can be determined by the nature of pollutants, their concentration and the period of exposure. Under exposure to high concentration, plants suffer acute injury with externally visible symptoms, such as chlorosis, discolouration, necrosis and death of entire plant. Besides morphological changes, chemical, biochemical, physiological and fine structural changes also occur in plants.

The damaging effects of air pollutants, on vegetation have been recognised by many workers. In general, air pollution decreases the yield of all crops by affecting their photosynthetic activity and growth. Pollution damage can also be recognised by the accumulation of toxic material in the plant, changes in pH, reduced or increased activity of certain enzymes, increase in compounds with SH groups and phenols, lowered ascorbic acid level in the leaves, depression of photosynthesis, stimulation of respiration, low dry matter production, changes in permeability, disturbances in water balance and reduced fertility under prolonged exposure. The disturbances in metabolism are due to chronic injury with irreversible consequences. Plants show reduced productivity and yield and quality is lowered and ultimately they die. The symptoms of pollution affected plants are varied and unspecific. A particular pollutant affects different plants in very different ways and a particular symptom can be produced by a variety of substances. The influence of external factors (pollutants) on plants depends upon the species, stage of development and the organ and tissue involved. Morphological alteration of a plant and floristic composition of a plant community are commonly used to indicate changes in the environment. According to Van Haut and Stratmann (1970), visible plant symptoms are most commonly used to indicate the responses of plants to pollutants. Jacobson and Hill (1970) have studied the effects of common pollutants on plants. It is possible that any part of a plant body, if it responds specifically or characterstically to any pollutant, can be used for its indication. M.U. Beg (1980) from Industrial Toxilogy Research Centre, Lucknow has reported the responses to air pollutants as a biological indicator, taking several parameters into consideration such as, seed germination, growth of plant, development of lateral branches, expansion and colour changes in leaf, flower and fruit formation, decolouration of flower, loss of physiological control, mineral composition, chemical constituents of cells, enzymatic activity and pollen germination.

Germination and General Growth

Seed germination has been used by many workers to monitor pollution responses. Several growth parameters such as percentage of germination, seedling survival, seedling height, cotyledonary expansion and fresh and dry weight have been taken as criteria to assess plant response to a specific pollutant. *Phaseolus vulgaris* has been grown in smoke free and smoke affected region by Sorauer (1899). The toxic effect of thiosulphate has been indicated as germination inhibition in many plants. Houstan and Dochinger (1977) have evaluated germination inhibition in relation to pollution by sulphur dioxide and ozone. The effects of lead, cadmium, NO and CO have been studied on many plants.

Some plant species are sensitive to pollution *e.g., Polygonum, Rheum, Vicia, Phaseolus* and *Capsella.* Generally, the plants' response to pollutants is characteristic rather than specific. Stunting of corn, sweet potato and rye has been reported due to high toxicity. Reduction in root length, shoot length, numbers of tillers, leaves, ears and grains in wheat have been reported under conditions of cement dust pollution. Similarly plant height, number of leaves and bolls per plant are reduced in cotton exposed to particulate pollution. Inhibition of lateral growth of forest trees is caused by lime stone dust. Pine trees do not flourish in SO_2 polluted areas. It has been noticed that leaf is the most sensitive organ to pollution. The pollution indicator value of leaf has been exploited by many workers in response to a variety of conditions. Leaf injuries are a characteristic symptom to various pollutants. The characteristic symptoms on leaf include pigmentation, chlorosis, yellowing, necrosis etc. The leaves of dicotyledons generally exhibit spotted markings between the veins while monocotyledons usually show necrotic streaks between parallel veins. Injury may also occur along the margin and tips. Reduced expansion of cotyledonary leaves in response to pollution has been observed in several cases. Decrease in dry weight of leaf, decrease in leaf thickness, cell size, loss of leaves and early senescence may be due to smoke and SO_2 pollution. Yunus and Ahmad (1980) have observed that leaves in the polluted area of cement factory showed higher stomatal and trichome densities, smaller epidermal cells and trichomes as compared to leaves obtained from unpolluted atmosphere.

Biochemical and Physiological Changes

Among the biochemical aspects, the most important parameter is pigment analysis. Chlorophyll a and b have been measured as index of response to different types of pollution. In *Cassia* and *Cynodon*, 50 per cent reduction of chlorophyll has been observed while in *Saccharum* the pigment is least affected. Chemical estimation like proteins, aminoacids, soluble sugars, sucrose, starch, reducing sugars, vitamin C, riboflavin, thiamine and carbohydrate are used to indicate foliar sensitivity to air pollution.

Physiological activities as opening of stomata and rate of photosynthesis can also be used as indicator of pollution. Photosynthesis as a parameter has been used for mixed exposure to SO_2, NO_2 and dust.

Enzymatic parameters are also used to indicate the presence of particular pollutant. Peroxidase was found to be most sensitive indicator of pollution in the absence of visible injury. Thus on the basis of enzyme activity, the susceptible species of plants can be identified. Many workers have reported that enzymatic activity is retarded due to air pollution. Other common enzymatic parameters used are ribulose diphosphate carboxylase, glutamate-pyruvate transaminase, glutamate-oxaloacetate transaminase etc.

Effects of Different Pollutants on Vegetation

Air pollutants are absorbed by the plants in appreciable amount. The absorption is affected by the nature, concentration of pollutant, exposure time, plant species, age of the plant, nutrient supply and other environmental factors. The effects of common air pollutants on plants are described below:

Sulphur dioxide. SO_2 is a major air pollutant in the areas where large quantities of fuel are burnt. Leaves of green plants are able to absorb SO_2 through their stomata. The absorbed sulphur is incorporated into various organic compounds.

SO_2 is one of the most phytotoxic pollutants. Foliar injury is a prominent symptom recognised in the leaves. Its effect on plants varies from species to species depending on genotype and physiological stages of the plants, their age, climatic condition, concentration of pollutant and duration of exposures. The most recognised symptoms are interveinal chlorosis, necrosis in plants, brownish colouration in the tip of pine needle. Lower concentration over long period leads to chronic leaf injury such as gradual chlorosis. Trees are injured followed by shrubs and herbs.

Some of the plant species, especially the sensitive ones are unable to regenerate in the area and gradually get eliminated from the soil. SO_2 also reacts with cell through stomata causing injury or death of tissues. Flower buds of *Gladiolus* are very sensitive and they exhibit necrosis at the tips and margins. SO_2 also affects photosynthesis, water relations and enzymatic system of plants. Beg and his team (1982, 1985) at ITRC, Lucknow have tried to investigate the mechanism of phytotoxicity of SO_2 and suggested certain morphological and physiological traits in plants as monitors of air pollution. Scientists of ITRC, Lucknow have evolved a simple and quick technique of screening the forest species according to their reactivity. On the basis of their experiment they concluded that the plant spp. like *Eucalyptus, Ficus infectoria* (Paker), *Dalbergia sissoo* (Shisham) and *Syzygium cuminii* (Jamun) are resistant where as *Ficus bengalensis,* (Bargad), *Ficus religiosa* (Peepal), *Bombax melabaricum* (samel) and *Terminalia arjuna* (Arjun) are susceptible to SO_2. Characteristic symptoms were noticed on *Aegle marmelos* (Bel) and *Holoptelea integrifolia* (Chilbil) when trees are exposed to 6.5 PPM SO_2 for four hours. Interveinal necrosis, marginal necrosis and leaf curling were common visible symptoms. These symptoms appeared after 48 hrs of exposure.

At low concentrations of SO_2 no significant reduction in chlorophyll and carotenoid contents were noticed in forest trees. But at higher concentration of SO_2 the reduction in total chlorophyll and carotenoid contents was noticed. Thus, it can be inferred that forest species can be proved as an effective measure for purifying the environment.

Oxides of nitrogen. Oxides of nitrogen are produced in internal engines where oxygen and nitrogen are subjected to heat and pressure and also by incineration of organic wastes and in petroleum refining. These pollutants produce marked injury on affected plants. Sometimes there is a combined effect of SO_2 and oxides of nitrogen which produce more injury than either of them alone. Besides leaf injury, reduction of growth occurs in sensitive plants. NO_2 causes leaf injury and reduction of growth in several sensitive plants.

Fluorides. Fluorides are emitted largely from aluminium and phosphate plants. It is widely spread in the earth's crust as a natural component of soil, rocks and minerals. When these substances are heated or treated with acid, toxic fluoride are released into the atmosphere. This becomes phytotoxic above the critical levels. In recent years, the losses to agriculture from fluoride have greatly increased with the vast expansion of industries. This pollutant damages plants if deposited on the leaf surfaces. Thus reduces the photosynthesis. It also blocks stomatal pore, thereby reducing respiration, photosynthesis and transpiration and finally plants become weak and ultimately die. On account of fluoride pollution, leaves show marginal necrosis as well as interveinal chlorosis. The fluoride accumulated in plants enters the food chain through herbivorous animals and passes into the soil through animal wastes.

Ozone. Ozone is a natural component of atmosphere where it filters out dangerous ultraviolet radiation. In urban localities where there is high population density, hydrocarbons and oxides of nitrogen are emitted into the atmosphere from automobile exhaust which in presence of sunlight interact to form large amounts of ozone which causes serious injury to vegetation.

Ozone injury on plants range from collapse of extensive areas of the leaf blade to discrete necrotic spots on upper surface, chlorosis, abscission of leaves, fruits and suppression of growth without visible symptoms of injury. Extent of injury and characteristics symptoms vary widely with type of plant, concentration of toxicant, duration of exposure and environmental conditions. Visible injury may occur on the most sensitive plants.

Ammonia. It has been observed that fumigating trees with ammonia gas increases the ammonium content. Leaves of soybean, sunflower, cotton and maize are able to absorb ammonia from the atmosphere. In maize, a part of absorbed ammonia is metabolised into aminoacids. It is experimentally demonstrated in rye grass and several other crops too. Some hydrophytes *e.g., Ceratophyllum demersum, Potamogeton pectinalis* and *Elodea* spp. also absorb NH_3 through their roots because ammonia is highly soluble in water. The absorbed ammonia is translocated to the leaves for assimilation.

Carbon monoxide. Carbon monoxide is a product of incomplete combustion. It is deadly poisonous at high concentration and is one of the important gaseous air pollutants. Human activities result in the production of nearly 250 million metric tons of CO annually. It is taken up by plants including algae.

Particulates. Particulates or dusts come under major pollutants which remain suspended in the air for a short time. Particulates are produced from many sources as cement factory, lime processing plant, refuse disposal and agricultural practices. Heavy metals arise from factory affluents and combustion. Rao and his associates (1980, 1981, 1985) have studied the responses of certain crops to particulates and developed certain methods to mitigate pollutant including its injury to plants. These particles become deposited on the plants especially on the leaf surfaces. Thus leaves become coated with dust which block the stomata causing the reduction of photosynthesis followed by reduction in food production. Transpiration and respiration processes are also regulated.

In the coal dust polluted areas mango and lemon trees show morphological changes leading to loss in productivity. It has been seen that the fallout of fine coal particles especially at the time of flowering significantly hampers the process of pollen germination and fertilization which are essential for fruit setting. Trees in such areas show decortication of bark, defoliation of branches and stunting of growth. Similarly, cement dust released from cement factories leads to irreparable degradation of soil characteristics and plant's structure and function. It is likely that cement dust affects plants through leaf encrustation, stomatal plugging, and changes in leaf reaction.

Control of Pollution through Vegetation (Plants as Ameliorator of Environmental pollution)

Plants, being the most important component of the ecosystem must be given top priority in the environmental planning and management. Trees have ability to reduce the amount of pollution in air. Plantation is extremely important in polluted and water catchment areas including urban and industrial areas. Many instances are known which indicate that the trees are capable of making our environment more pleasing and livable. Trees also provide an aesthetic link between man and his environment. Trees play an important role in enhancing the quality of environment by influencing the life supporting systems. They restore ecological balance of all ecosystems, maintain biological diversity, act as catchment for soil and water, prevent floods and check or reduce the noise pollution. The capacity of plants to absorb aerosol (very small particles of solid or liquid that remain suspended in the air) or gaseous pollutants from the air has not been thoroughly investigated but few studies have shown that trees can effectively filter certain aerosols. Researches based on the foliar accumulation of chloride and fluoride have also stressed the ability of plant leaves to retain atmospheric aerosols. Plants also improve the quality of air by absorbing or altering gaseous pollutants from the atmosphere. For example, leaves exposed to low level of SO_2 may transform the gas to SO_4. SO_4 is less injurious to plants. Thus plants, in general, act as mitigators of air pollution.

The surface of vegetation provides a major filteration and functions as sink to mitigate pollutants from air. At present there is great crisis of CO_2–oxygen balance in atmosphere. Only green plants have capacity to utilize CO_2 in the process of photosynthesis and release free oxygen in the atmosphere. Green plants not only take up CO_2 and release O_2 but their leaves retain some pollutants. Dense and tall trees also filterout dust. Data collected from Lucknow showed high dust trapping potential of different plants. In this regard peepal (*Ficus religiosa*), mango (*Mangifera indica*) and ashok (*Polyalthia longifolia*) trees were found to be better dust collector than amaltas (*Cassia fistula*), Kachanar (*Bauhinia variegata*) and gulmohar (*Delonix regia*).

Enough evidences are now there which show the role of plants as sinks for gaseous pollutants.

For example, an oak (*Quercus suber*) tree has been found to reduce about 20 per cent ozone concentration. It has been shown that alfalfa canopy removed 1/4 ton of nitrogen oxide or SO_2 per square mile per day. Tree plantation is economically important for protection against environmental pollution. According to an estimate a peepal (*Ficus religiosa*) tree is reported to replace 2,252 kg polluted air with 1,713 kg of O_2 during its life time which is sufficient for 60,000 people. Hence real value of medium size tree comes to about Rs. 15.7 lac in 50 years of life time.

VALUE OF A TREE

Trees play a significant role in the productivity of land and soil conservation. The value of an average sized peepal tree of 50 years old can be evaluated as follows.

Control of soil erosion and maintenance of fertility	Rs. 2,50,000
Production of Oxygen	Rs. 2,50,000
Protein conversion	Rs. 20,000
Control of atmospheric pollution	Rs. 5,00,000
Recycling of water and control of humidity	Rs. 3,00,000
Shelter for birds and other animals	Rs. 2,50,000
Total	Rs.15,70,000

The use of vegetation in filtering out the dust, soot and particulates from the atmosphere has been widely accepted. Ahmad *et al* (1988) at National Botanical Research Institute, Lucknow have initiated studies on the impact of air pollution on vegetation with following objectives :

1. To study the nature and extent of damage caused to vegetation by common air pollutants.
2. To determine the role of plants in indicating the level of pollution in a particular area.
3. To identify pollution tolerant/resistant plant species to be used as green belts around industrial areas to mitigate air pollution.
4. To determine the dust trapping/filtering potential of common trees and shrubs.

Thus pollution is a problem which is going to increase with the increase of population in the coming years but vegetation can reduce it upto some extent if it may not altogether eliminate. Pollution can be checked through vegetation by adapting the following measures :

1. Green belts should be planted around urban and industrial areas with plants that are pollution resistant to help in purifying the atmosphere by regulating temperature, humidity and air movement and also to reduce damage caused by air pollutants. Green belt around the cities will also increase the aesthetic sense.
2. Screening of Indian plants should be done in order to replace pollution tolerant species.
3. Research laboratories and organisations should carry out regular monitoring of environmental pollution through proper vegetation in metropolitan cities and industrial towns.
4. Many forest tree species have been proved as an effective measure for purifying the environment.

Based on the vegetational survey in response to pollution some tolerant species have been enlisted as suggested by Ahmad *et al,* 1988 :

Herbs

Achyranthes aspera (Latjira)
Amaranthus graecizans (A variety of Chaulai)
Argemone mexicana (Sialkanta)
Solanum xanthocarpum (Bhatkataiya)

Shrubs

Bougainvillea (Baganvilas)
Calotropis procera (Madar)
Cassia sophera (Kasunda)
Clerodendron infortunatum (Bhant)
Croton sparsiflorus
Ipomoea fistulosa (Behaya)
Lantana camara (Ghaneri)
Murraya exotica (Kamini)
Nerium odorum (Kaner)
Ricinus communis (Arand)
Tabernaemontana coronaria (Wild Chandni)

Trees

Acacia arabica (Kateria Babul)
Aegle marmelos (Bel)
Albizzia lebbeck (Siris)
Syzygium cuminii (Jamun)
Tamarindus indica (Imli)
Zizyphus mauritiana (Ber)
Alstonia macrophylla (Chatian)
A. scholaris (Chitaman)
Anthocephalus cadamba (Kadamb)
Artocarpous heterophylla (Kathal)
Azadirachta indica (Neem)
Butea frondosa (Palas)
Casuarina equisetifolia (Jhau)
Citrus medica (Lemon)
Dalbergia sissoo (Shisham)
Delonix regia (Gulmohar)
Diospyros melanoxylon (Tendu)
Eucalyptus citriodora
Ficus bengalensis (Bargad)
F. infectoria (Pakar)
F. religiosa (Peepal)
Holoptelea integrifolia (Papri)
Lagerstroemia sp. (Sawani)
Madhuca indica (Mahua)
Mimusops elengi (Maulsri)
Moringa oleifera (Sahjan)
Phyllanthus districhus (Harfarwari)
P. emblica (Amla)

Pithaclobium dulce (Jangal Jalebi)
Polyalthia longifolia (Ashok)
Psidium guajava (Guava)

CONTROL OF ENVIRONMENTAL POLLUTION (ENVIRONMENTAL MANAGEMENT)

The recent growing concern for the quality of our environment has strengthened the concept of environmental monitoring which requires regular periodical measurements of environment followed by timely assessment and interpretation of data obtained. Reliable monitoring data are necessary to know pollution problem.

Environment is deteriorating day by day due to industrial pollution, toxic chemicals, automobile emission and natural resource depletion. Pollution in its various forms is increasing tremendously. Environmental management aims at the study of all these problems and checking pollution so that the environment satisfies the basic human needs at the minimum and more, if possible, for an indefinite future. The main aim of environmental management is overall development of environment. Purpose of environmental management is to see whether air is fit for breathing, water is fit for drinking and soil and sea resources are fit to provide all needs. One also expects that people are well fed, well housed, well educated and enjoying happy and healthy life.

Environmental problems are so diverse and diffused that virtually any activity of civilization interacts with the environment. Many environmental pollution problems are local in character and they can be controlled by creating environmental consciousness in each and every citizen. People should be told about the importance of clean atmosphere as well as about the consequences of different types of environmental pollutions. Besides, action is also needed at national level and guidelines may be established internationally by the United Nations Joint Committee of Experts. Some of the effective and practical control measures for minimising environmental pollution are outlined below :

1. Combustible solid wastes should be burnt in incinerators. This method does not solve the problem in a real sense because in this, solid waste is being converted into gaseous wastes causing air pollution. Unless it is properly controlled, incineration may cause more nuisance.
2. Solid organic wastes including faecal matter and wastes from tanneries should be converted into compost manure at the places far away from the cities and human dwellings. The composting should be done in pits or in heaps adequately covered with layers of soil at least 8-10 cm thick to pervent fly breeding and rat menace which are important carriers of various diseases.
3. Non-combustible solid waste materials like ash, rubbish, tins, glass pieces if not recoverable for usual purposes should be disposed of by landfill method in low-lying areas.
4. Anaerobic septic tank treatment can be used for individual houses or small communities. Besides, aerobic biological treatment systems including trickling filters, activated sludge treatment and oxidation ponds can also be used for liquid wastes or sewage disposal.
5. Automobiles must be either made to eliminate use of gasoline and diesel oil or complete combustion is obtained in the engine so that noxious compounds are not emitted. The automobiles, trucks and other transport systems must have an antismog device. In some countries factories are using devices like scrubbers, cyclone separators or electrostatic precipitators to minimise pollution.
6. There should be cut back in the use of fertilizers, herbicides, pesticides and other agrochemicals as far as possible.
7. Excessive and undesirable burning of vegetation should be stopped.

8. Sponges and towels should be used in place of paper towels and also the use of paper cups and plates and similar materials should be stopped.
9. Little use of electric appliances and motor-run appliances will reduce thermal pollution.
10. Washing soda and scouring pad should be used instead of detergents.
11. Waste management is based on principle of '3Rs' *i.e. Reduce, Recycle* and *Reuse*. Used boxes, bags, plastics and bottles should be reused whenever possible.
12. Since about 40% of the phosphates in water pollution comes from detergent, it has been suggested that only detergents low in phosphates should be used.
13. Shampoos, lotions and similar products should not be bought in plastic bottles. It has recently been suggested that use of plastic containers and glasses may cause cancer.
14. Smoking should be stopped (there is 5,00,000 tonnes tobacco pollution annually).
15. Proper attention should be given by the government to make people realise the implications of environmental problem,
16. Legislation against pollution should be strictly implemented.
17. International action is needed to deal with the problems presented by highly toxic pollutants like lead, mercury, organochlorine pesticides released in to the atmosphere and carried far beyond the country of origin as well as carried down to the sea by rivers. Successful action to improve environmental qualities depends mainly on the acceptance by industry and local authorities of the need to reduce greatly both quantities and toxicity of certain wastes at present being discharged into the sea and the modernization and expansion of sewage disposal systems.
18. Environmental education is the best programme to deal with the environmental problems. It is most fundamental in our efforts to combat and control pollution, over-population and misuse of natural resources. Environmental education includes the following objectives which are based on the major outcome of the 1975 workshop on environmental education.
    (*i*) awareness of the problems.
    (*ii*) Providing knowledge to deal with the problems.
    (*iii*) Developing new attitudes towards environmental problems.
    (*iv*) Developing skills for solving environmental problems.
    (*v*) Providing development or evaluation ability in these areas.
    (*vi*) Increasing participation and involvement of public.
    (*vii*) A broad recognition of the facts that we are all responsible for helping to solve environmental problems.
19. Recycling of wastes and sewage should be done.

The work on pollution control and abatement techniques is being carried out by Desalination and Effluent Engineering Division of Bhabha Atomic Research Centre (BARC). There is special Directorate for Radiation Protection (DRP), an all-India organization, which keeps watch on exposure of the users or public at large to irradiation and provides proper instructions to users for using radioactive materials. Besides these, a National Environmental Engineering Institute (NEEI) is established at Nagpur to execute studies on different aspects of pollution. This institute has regional offices, one each in Tamil Nadu, Calcutta, Mumbai, Delhi, Hyderabad, Jaipur, Kanpur and Ahmedabad. One more all-India organization, Industrial Toxicological Research Centre (ITRC), has been established at Lucknow which is concerned mainly with the study of toxicological and environmental problems.

Action Committee of 68th session of Indian Science Congress held at Varanasi (1981) recommended the following points on environmental management :

1. Immediate implementation of acts against air pollution, water pollution and noise pollution. (Now these acts have been framed).
2. There should be specific standards for the use of agricultural chemicals such as insecticides, pesticides and fertilizers.
3. Action is necessary against food adulteration, drug control and against factories producing effluents.
4. Department of Environment should be empowered for initiating legislative measures by the Parliament. Department should have control of soil, air and water environment and should maintain a strict vigilance over metropolitan and industrial areas. There should be regular monitoring of air and water quality. Penalty should be imposed on offenders not adhering to set standards against air, water, noise and other sources of pollution.
5. Environmental protection should be included in the concurrent list of constitution.
6. Government and voluntary agencies should be provided special incentives for successful anti-pollution researches.
7. Department of Environment should carry out publications on health safety and set directive criteria for different pollutants.
8. Environmental education should be made compulsory in the curricula of schools, colleges and universities. This has already been recommended by. The Supreme Court of India in 2005.
9. Mass communication media should be adopted for information related to environmental problems.
10. Scientific societies NGOs, Associations and Academies should create a temper of environmental problems in the rural and urban areas.
11. Plantation of trees suited to different climates should be adopted. Deforestation must be legally checked.
12. Wildlife board and Environment cell should be created in each state. Central and State Pollution boards have already been established.
13. Recycling plants for municipal wastes and sewage should be established. In many cities it has already been implemented.
14. Anti-pollution tax should be imposed on industries for discharging the effluents.
15. Reclamation of land to its original state should be the responsibility of the users. There should be legislation against violation for this.
16. Encouragements and incentives should be given to voluntary movements as "Chipko movement" for the protection of plants.
17. In agriculture, biological control of pests should be applied to replace wide use of insecticides, pesticides and fungicides. Biological nitrogen fixation should be adopted.
18. Several Biosphere Reserves containing endangered or rare or commercially important species should be established as early as possible.
19. Gene Pool Reserves and Gene Sanctuaries should be demarcated and protected in their natural environment to preserve wild germplasm from extinction and conservation.
20. For conservation of flora and fauna the list of endangered species should be continuously renewed.
21. Aero-biological centres should be established in different parts of India.
22. Brick fields should be prohibited by law near agricultural or populated areas.
23. Cleaning of choked lakes and rivers should be done regularly.

24. Different aspects of alternative energy resources should be created as biomass, energy plantation, fuel energy etc. Social forestry should be encouraged.
25. A suitable agency should be established for conservation and monitoring of marine resources and providing protection against pollution hazards.
26. Satellite and space technology may be adopted for weather forecasting, agriculture resource inventories and monitoring of air borne particles.
27. Airport areas should be separated to minimise pollution from noise.
28. Genetic counselling units should be established at different regions of India to monitor genetic disorders.
29. Scientific, administrative and social measures must be adopted to check pollution explosion.
30. At different levels of planning, the involvement of ecologists is essential. All programmes for development should be undertaken without or with the least disturbances to the natural ecosystem.

DIFFERENT LEGISLATIONS FOR ENVIRONMENTAL PROTECTION

Environmental legislation is one of the most important means for improving the environment and for controlling pollution. This is so because law is the only effective instrument for social justice, when other aspects of environmental protection are not followed.

The first international attempt on legal provisions regarding environmental concerns was made when a conference on the human environment was held at Stockholm (Sweden) in 1972, in which 113 countries participated. The recommendations evolved the principles and action plan to control and regulate the human environment. India also participated and was one of the first countries to enact the following legislations :

1. Water (Prevention and Control of Pollution) Act, 1978, 1988.
2. Air (Prevention and Control of Pollution) Act, 1981, amended in 1987.
3. Environment (Protection) Act, 1986.

India was also the first country to impose a constitutional obligation on the States and citizens to protect and improve the environment as one of the primary duties. To exercise its powers, the Government of India has constituted the Central and State Pollution Control Boards (**CPCB/SPCB**). Different rules formulated by these boards are:

1. Water (Prevention and Control of Pollution) Rule, 1975.
2. Air (Prevention and Control of Pollution) Rule, 1982.
3. Environmental (Protection) Rule, 1986.
4. Hazardous waste management Rules, 1989.
5. Manufacturing, Storage and Import of Hazardous Chemicals Rules, 1989.

Other Acts regarding environmental protection, some of which were in practice before independence and were subsequently amended are:

1. Indian Forest Act, 1927.
2. Motor Vehicle Act, 1938 amended in 1988.
3. The Factories (Pollution and Pesticides) Act, 1948.
4. The Insecticide Act, 1968.
5. The Water (Prevention and Control of Pollution) Act, 1974, 1977. Under this Act, pollution of water and water bodies has been made a criminal offence under Section 277 of the Indian Penal code.

6. Air (Prevention and Control of Pollution) Act, 1981. This Act provides a declaration that "No industrial plant shall be operated in an air pollution controlled area, by any person, without the previous consent of the State Board.
7. The Environment (Protection) Act, 1986 passed by Parliament. This Act declares – "Persons responsible for the discharge of any hazardous substance in excess of the prescribed norms will have to immediately intimate the fact of such occurrence to the concerned authorities".
8. The Hazardous Waste (Management and Handling) Rules, 1989.
9. Biomedical Waste (Management and Handling) Rules, 1998.
10. Batteries (Management and Handling) Rules, 2001.

Thus, India has some legal provisions to guide the public and the authorities for discharging their duties sincerely and effectively towards reducing pollution and improving the environment.

QUESTIONS

1. What is pollution? What are the various pollutants causing environmental pollution? What are their effects?
2. What is air pollution? In what ways man is affected by air pollution?
3. What is water pollution? Describe various kinds of water pollutions and their ecological effects on the aquatic organisms and man.
4. What is environmental management? What are the various control measures for minimising environmental pollution?
5. Write short notes on:

   Thermal pollution, Noise pollution, Radiation pollution, Hydrocarbons, Pesticides and their effects, Aerosols, Agricultural pollution.
6. Write short notes on :

   (*a*) Air Quality Index (IQI)
   (*b*) Sewage disposal
   (*c*) Potable water
   (*d*) BOD
   (*e*) CPCB
7. What is meant by Acid rains? Discuss its causes and impacts on vegetation.
8. Describe different legislations for environmental protection.
9. Define (*i*) the green house gases, and (*ii*) photochemical smog
10. Define (*i*) Aerosol, (*ii*) BOD; (*iii*) Pollution
11. What are the impacts of pollution on vegetation ?
12. What are the morphological and biochemical changes in plants caused by various pollutants.
13. Discuss the role of vegetation in control of environmental pollution.
14. Plants act as ameliorator of environmental quality. Discuss.
15. Give a list of plants which are pollutant tolerants.

14

Global Warming (Green House Effect)

The world is facing ecological problems like environmental pollution, increase in population, deforestation, energy crisis, soil erosion and so on. An outcome of the atmospheric pollution is warming of the earth's environment. Warming of the earth atmosphere is a serious problem and in twenty-first century this may be a problem of global concern.

There is a clear indication that the temperature of the earth has increased slightly during the past 50 years. It is clear from the temperature records of the last 50 years illustrated in the following Fig. 14.1.

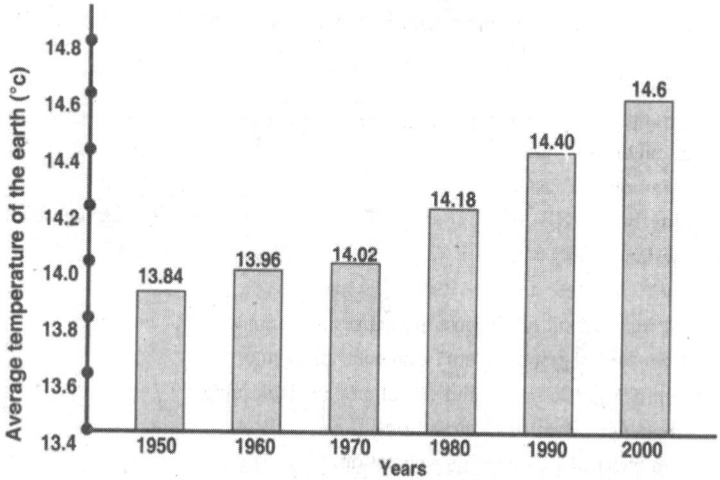

Fig. 14.1. Average temperature of earth recorded in different years.

This increase in temperature of atmosphere is very dangerous for living organisms. Though there are numerous causes for the global warming, the following two factors are considered to be very significant :

1. Industrialization
2. Deforestation

During the last two decades there has been tremendous growth of industries in the world. These industries release toxic gases, chemicals and effluents in huge quantities into the environment. The industrial wastes produce many harmful effects and ultimately disturb the ecosystems. The toxic gases and effluents released into the environment cause pollution of air, water and soil.

Deforestation is mainly due to increase in population, industrialization, urbanization and increasing agricultural activities. Deforestation has led to soil erosion, wasteland development and extinction of certain species from the flora and fauna. Deforestation is taking place at an alarming rate. For healthy environment one-third area of the earth should be covered by the forest but the total forest cover in India is much less than 33 per cent. Data provided by the Remote sensing Technology reveal that only 17 per cent area of India is covered by the forest. After 1950, about 4.2 million hectares of forest were destroyed and since 1980 every year about 1.5 million hectare forest area is either converted into agricultural land or used for some other purposes. Due to two aforesaid reasons the concentrations of CO_2 gas is increasing in atmosphere. Carbon dioxide at normal concentration (0.03%) is not considered as a pollutant but its concentration above 0.03% in the atmosphere has an adverse impact on our climate and living organisms.

Besides CO_2, some other gases present in our atmosphere are also responsible for raising the temperature of the earth and its environment.

GREEN HOUSE EFFECT

It is a universal truth that the sun is the main source of energy on the earth. The sun is continuously radiating energy in the form of electromagnetic waves into the space. It has been calculated that one part out of two billion parts of total solar energy radiated from the sun reaches the earth because of small size of the earth and very long distance from the sun. The solar energy reaches to the earth atmosphere in three forms :

(*i*) Ultra-violet radiation

(*ii*) Visible light (VIBGYOR spectrum)

(*iii*) Infra-red and radio waves

Cosmic radiations do not reach to the atmosphere of earth. When the solar radiations first fall on the top of atmosphere the harmful ultraviolet radiations are absorbed by the ozone layer in the stratosphere and are prevented from reaching the earth surface. Thus ozone layer acts as a protective layer in the stratosphere. The visible light and infra-red rays then pass through the carbon dioxide layer in the lower region of the atmosphere (troposphere) and ultimately fall on

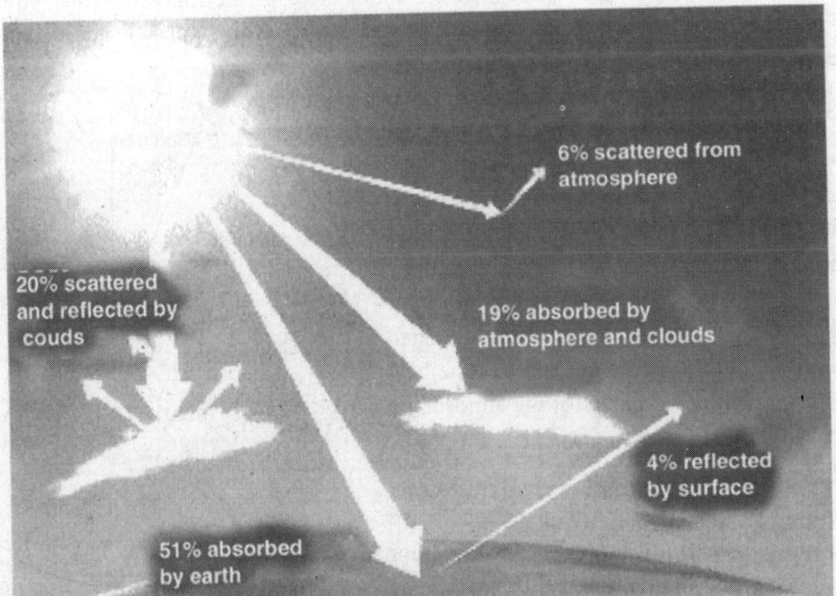

Fig. 14.2 Green house Effect

the earth surface. Since infra-red radiation has heating effect, it warms up the earth, its atmosphere and various objects. A part of infra-red rays falling on the earth surface is reflected back into the outer space. Infra-red radiation coming to earth from the sun are of short wavelength but the infra-red rays reflected from the earth and its various objects are of long wave lengths. The infra-red radiation reflected from the earth cannot escape out from the carbon dioxide layer present in the atmosphere. CO_2 and some other gases like methane, nitrogen oxides and chlorofluoro carbons (CFCs) act as the glass panels of a green house or window panes of a car which allow the solar radiations to pass through and heat the surface of earth but do not allow the heat radiated from earth is surface to pass through thereby trapping it in the process. These gases have the ability to absorb infra-red radiation reflected from the earth surface. Therefore, the blanket of these gases in the atmosphere traps all the infrared rays which produce heat on the earth surface. The heating of earth's atmosphere due to trapped infrared rays reflected from the earth surface is called *Green House Effect* which causes **global warming**.

Table 14.1 Major Green house gases from human activities

Green house Gas	Sources
Carbon dioxide	Burning of fossil fuel, especially coal, deforestation, plant burning
Methane	Rice paddies, termites, landfills, coal seams, gas leaks from oil and gas production and pipelines
Nitrous oxide	Fossil fuel burning, fertilisers, animal wastes
Chlorofluorocarbons (CFCs)	Air conditioners, refrigerators
Hydrochlorofluorocarbons (HCFCs)	Air conditioners, refrigerators
Hydrofluorocarbons (HFCs)	Air conditioners, refrigerators
Halons	Fire extinguishers

The rise in temperature of earth's atmosphere caused due to green house effect depends on the amount of carbon dioxide present in the atmosphere. At normal CO_2 concentration (0.03%) in the atmosphere the surface temperature of the earth remains constant due to energy balance of the sun rays which strike on the earth, heat it and then radiate back into the space. This is called *Energy Budget*. But when there is increase in CO_2 concentration in the atmosphere, the thick layer of CO_2 prevents the heat from being radiated out into space. This layer of CO_2 thus functions as a glass panel of green house or glass window of a motor car which allows the sunlight to filter through it but prevents heat from being radiated back into the outer space. Thus most of the heat energy absorbed by CO_2 layer of the atmosphere warms up the air in the troposphere.

The gases responsible for raising the temperature of the earth and its atmosphere are referred to as **green house gases.** The different green house gases and their percent ratios are as follows (Fig. 14.3).

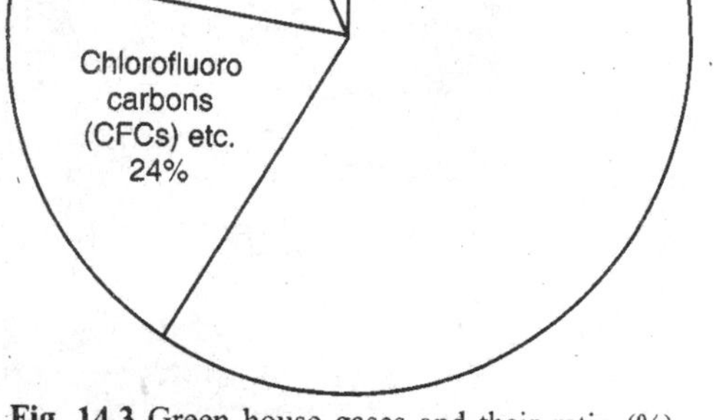

Fig. 14.3 Green house gases and their ratio (%)

1. Carbon dioxide — 55 %
2. Methane — 15 %
3. Chlorofluoro carbons (CFCs) — 24 %
4. Nitrogen oxides (N_2O and NO) — 6 %

SOURCES DISCHARGING GREEN HOUSE GASES

The various sources responsible for discharge of green house gases into the earth's atmosphere and their contribution (in percent) are presented in the Tables 14.1 and 14.2.

Table 14.2. Sources of green house gases

Gas	Source of discharge	Percentage of discharge	
Carbon dioxide	Burning of fossil fuels	81%	100%
	Destruction of forests	17%	
	Cement plants	2%	
Methane	Rice cultivation	30%	100%
	Domestic animals	20%	
	Burning of vegetation	17%	
	Landfills	12%	
	Coal mines	9%	
	Natural gases	12%	
Nitrogen oxides	Burning of fossil fuels	43%	100%
	Use of fertilizers	25%	
	Burning of vegetation	25%	
	Cultivated land	7%	
Chlorofluoro carbons (CFCs)	Industrial activities	100%	

Carbon dioxide gas plays the following important roles in the environment:

1. It produces green house effect causing heating of the earth and its atmosphere. (Table 14.3).
2. Carbon dioxide is a raw material for photosynthesis in green plants.
3. The gas dissolved in water reacts with lime to form carbonate rock.

Carbon dioxide gas is confined only to lower level of atmosphere (troposphere). The main sources through which the CO_2 is released into the atmosphere are as follows :

(*i*) Burning of fossil fuels
(*ii*) Decay of dead organisms
(*iii*) Respiration of living organisms
(*iv*) Automobile exhausts
(*v*) Thermal power plants
(*vi*) Cultivation of land
(*vii*) Eruption of volcanoes
(*viii*) Domestic cooking
(*ix*) Forest fires.

Carbon dioxide present in the earth's atmosphere not only checks infra-red radiations from being reflected back into the space but also produces heat effects. The average temperature of the earth remains almost constant because of the balance between the amount of solar energy reaching the earth surface and the amount reflected back into the outer space from the earth surface. Sometimes, this balance is disturbed.

CO_2 is emitted into the atmosphere is equivalent of more than 6,000 tonnes of carbon per year. About 194 Gt. (1 giga tonne = 1000 million tonnes) CO_2 is produced by biosphere annually. 6 Gt CO_2 is released from burning fossil fuels and rest comes from burning forests and about 16-9 Gt carbon is added to atmosphere through human activities. Small amount of CO_2 is released into atmosphere by volcanoes. About 1.7 ppm (parts per million by volume) of CO_2 is added to environment every year. A part of atmospheric CO_2 is consumed in photosynthesis and respiration and decomposition release it again into the atmosphere. In the recent past the average volume of CO_2 in the atmosphere has increased from 316.8 ppm (part per million) in 1960 to 366.7 ppm in 2000 (Table 14.3) and in the beginning of 21st century it may reach to about 400 ppm. According to an estimate, by the year 2040 the concentration of CO_2 may reach upto to 500 ppm. There is an indication that if CO_2 concentration becomes double, the average temperature of earth will rise by 2°C.

Table 14.3. CO_2 concentration and average temperature of the earth

Year	CO_2 concentration in atmosphere (ppm)	Average temperature (°C)
1950	—	13.84
1960	316.8	13.96
1970	325.5	14.02
1980	338.5	14.18
1990	354.0	14.40
2000	366.7	16.60

In addition to CO_2, some other gases like nitrogen oxides (N_2O) and (NO), methane (CH_4), carbon tetrachloride, (CCl_4), Chlorofluorocarbons (CFCs) are also responsible for increasing the temperature of the earth and its atmosphere.

It has been observed that carbon dioxide alone increases the temperature by 50 per cent, CFCs by 20 per cent and other air pollutants by 10 per cent. More recently, the role of methane in atmospheric temperature changes received much attention. Methane has a global warming potential about 25 times more than CO_2 and accounts for about 15 per cent of the global warming (Brown, 1994).

Annual global emission of methane in 1997 was estimated to be 7.1 tonnes from land fill sites. According to an estimate, the total concentration of CH_4 was 1630 parts per billion (ppb) in 1984. The main sources of methane are sheep and cattle dung, wet lands, biomass burning mining of coal and mineral oils paddy fields and bacterial biomass. Roots of paddy plants absorb methane from muddy soil and translocates that along with solutes to aerial parts through vascular tissues from where it is released in the atmosphere.

Impacts of Global warming. Global increase in temperature of earth will have serious consequences on human society, agriculture, plants and animals. These are discussed below:

1. **Sea-level rise :** Global warming is likely to result in a rise in sea-levels that could threaten many coastal areas around the world. It will also melt much glacier, especially around Greenland. Sea-level changes will complicate life in many coastal regions. A 100 cm rise

could submerge 6 per cent of the Netherlands, 17.5 per cent of Bangladesh and many islands. Countries will have to spend huge amounts of money to protect their shorelines, while poor countries may be forced to evacuate low-lying coastal regions. New marshes will also form in many places.

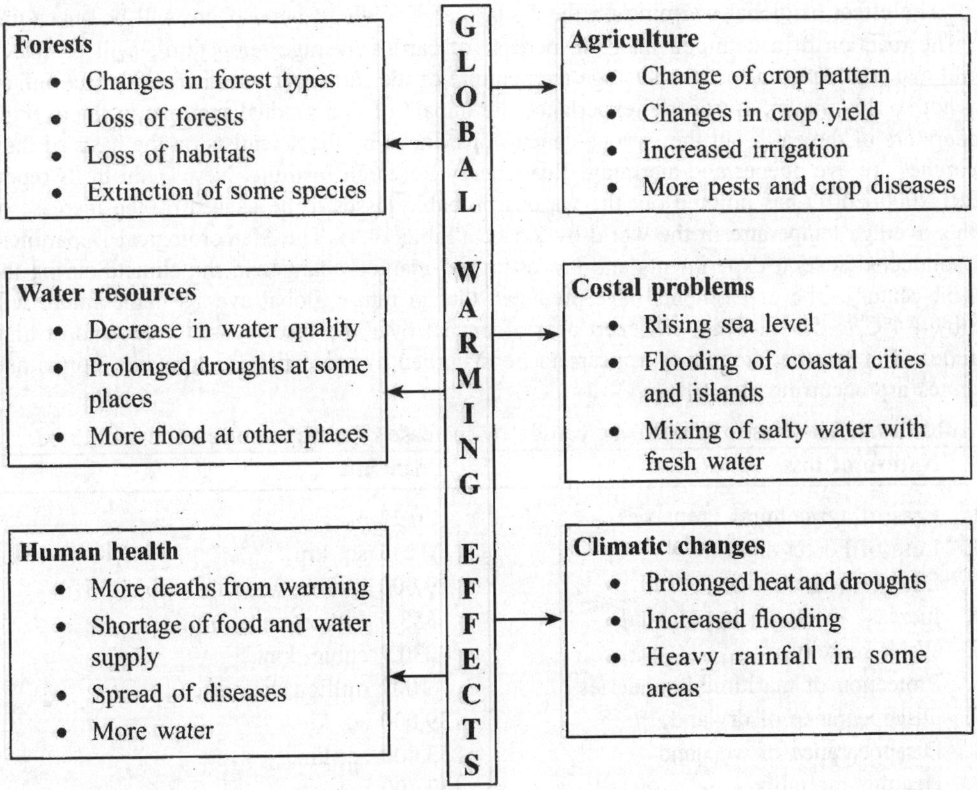

Increase in the flow of rivers and change in rainfall pattern.

Occurrence of more cyclones and hurricanes.

Fresh water will be contaminated with the salty water of sea.

Human health : As a result of global warming human populations will be displaced. In a warmer climate people will get more frequently sick. Diseases now found in the warm tropical areas will spread to new regions. Many diseases are transmitted by mosquitoes and other animals when these animals will move into colder regions. Other tropical diseases including dengue fever, yellow fever and encephalitis may spread into new regions. It is also predicted that there will be rising incidence of allergies and respiratory diseases, as warmer air is more charged with pollutants, fungal spores and pollens.

Effect on agriculture : It has been speculated that due to global warming the world food production might not be seriously affected. However, since people are now adapted to certain climatic conditions, a drastic change would adversely affect them. The main reason for this would be the shifting of agricultural zones. For example, Southern Canada may benefit from more rainfall and a longer growing season. At the same time, the semiarid tropical farmlands, in some parts of Africa, may become further impoverished More insects and plant diseases may also afflict crops and woodlands.

7. The forest vegetation will not be able to adapt with the changing temperature and may face destruction.
8. As a result of global warming, the temperature of southern Europe and central America will rise and affect the rainfall in these areas.
9. The effect of global warming on the Tropics and Southern hemisphere will be uncertain.

The reaction of a common man on increase of earth's average temperature will be that it would make no difference if the average temperature of the earth increases by 1.2°C. But, infact, it is not so. No matter, we do not experience the impact of this gradual increase in the average temperature of the earth but the various conclusions drawn by the scientists on the basis of their researches are eye opener and alarming. Tata Energy Research Institute, New Delhi in its report (TERI 2000–2001) has pointed out the various probable losses to be caused by the increase of earth's average temperatre in the world by 2.5 °C (Table 14.4). The Meteorological Department has conducted several experiments and has estimated marked changes in the climate during the present century. The experiments have predicted that in future global average temperature will rise by 1.4°C to 5.8°C. This will exert a great impact over the land, coastal areas and at high altitudes. The weather frequency appears to be disturbed as more floods, droughts, storms and cyclones are occurring.

Table 14.4. The losses likely to be caused by increase of earth's temperature by 2.5 °C.

	Nature of loss	Amount
1.	Loss of agricultural Properties	0.23 %
2.	Loss of Forest area	1235 sq. km.
3.	Loss of Fisheries	68,29,000 tonnes
4.	Increase in electricity demand	353.9 trillion units.
5.	Water scarcity	230.7 cubic. km.
6.	Protection of maritime bounderies	1007 million dollers.
7.	Disappearance of dryland	139,000 sq. km.
8.	Disappearance of wet land	253,000 sq. km.
9.	Health / mortality	137,700 persons
10.	Migration	27,34,000 persons
11.	Losses caused by storms	
	(*i*) Human loss	8000 Nos.
	(*ii*) Other losses	630 million dollers.

Ozone Depletion and Ozone Hole

Ozone gas is found on the upper surface of stratosphere at the height of about 50 km in the form of a layer called ozonosphere. The main functions of ozone layer are as follows :

1. It regulates the weather and climate of the earth.
2. It filters away the ultraviolet radiation and protects the earth from harmful effects of UV rays. Thus it acts as a protective layer.

Earth receives ultraviolet radiation in three forms :

(*i*) UV-A in the wavelength range of 315-400 NM

(*ii*) UV-B with wavelength from 289-315 NM and

(*iii*) UV-C with wavelength from 100–280 NM

Out of these, UV-B in the wavelength range from 250-315 NM is most dangerous. It affects the living organisms in many ways. Sunburn, ageing, wrinkling of skin, cataract of eye, destruction of protein, mutation of genes leading to skin cancer or melanoma, etc. are some of the common effects of UV radiation on human population.

Ozone is an unstable gas and for its formation and destruction both UV rays are essential which come from the sun. The following reactions are involved :

$$O_2 \xrightarrow[\text{UV rays}]{} O + O$$

$$O + O_2 \xrightarrow[\text{UV rays}]{} O_3$$

$$O_3 \xrightarrow[\text{UV rays}]{} O + O_2$$

This is continuous process. Thus ozone is in dynamic equilibrium. In the processes of forming and breakdown of the ozone molecule a part of the harmful ultraviolet rays is converted into infra-red rays which is not lethal to man. In addition, ozone prevents the escape of UV-rays into the earth's atmosphere. Thus a balanced chemical reaction is set in. The retarded rate of formation and increase in destruction of ozone molecules allow UV-rays to escape into the atmosphere. This leads to less ozone formation which is commonly known as **ozone depletion**. The man made nontoxic, inflammable and highly stable chemicals Chlorofluorocarbons (CFCs; CCl_3F and (Cl_2F_2) as well as CCl_4, CH_4 etc. are main chemicals which check the ozone formation (Table 14.5). The main damaging among them are CFCs. Since their discovery in 1928, CFCs have been used as cooling materials in refrigerators and air conditioners, propel aerosal sprays, to clear computer chips and to make foam rubber and polystyrene containers. The British chemist James Loverlock developed a device to detect CFCs in late 1969s and reported that CFCs were a great problem in the environment. He found that CFCs once released in to the atmosphere remain intact and they go on spreading indefinitely mixing with other molecules in air until they are dispersed over the globe.

Table14.5 : Important ozone depleting chemicals

Chemical	Atmospheric lifetime (years)	Applications	Share of contribution to depletion (%)
CFC - 12	139	Refrigeration, foams, air conditioning	45
CFC - 11	76	Foams, refrigeration	26
CFC - 113	92	Solvents	12
Carbon tetrachloride	67	Solvents	8
Methyl chloroform	8	Solvents	5
Halon 1301	101	Fire extinguishers	4

Two US chemists, Rowland and Molina reported in 1974 that USA alone was producing 380 million kg of CFCs and when released in atmosphere they remain intact for 50 to 100 years and slowly drift towards stratosphere 12-50 km above the surface of the earth. These carbon compounds of the stratosphere in presence of ultraviolet rays coming from the sun break and produce chlorine in nascent state which breaks the ozone molecules and reduces ozone content of the atmosphere. One atom of chlorine is sufficient to destroy a lac of ozone molecules and since this process is cyclic, the damage being caused is tremendous.

The reaction is given below :

(*i*) $Cl + O_3 \rightarrow ClO$ (chlorine monoxide) $+ O_2$

$ClO \rightarrow Cl + O$

Chlorine atom liberated at the end of reaction breaks another ozone molecule in cyclic chain.

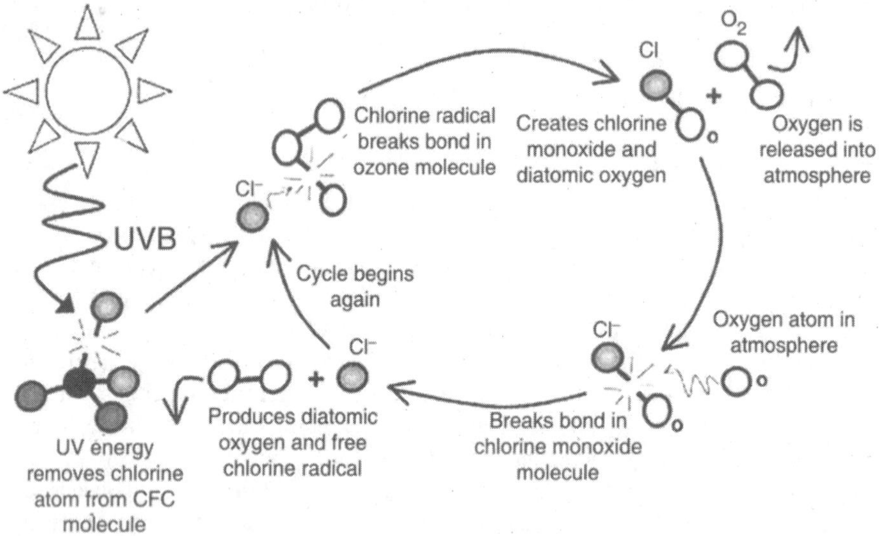

Fig. 14.4 Figure showing action of the chlorofluorocarbons destroying ozone.

Ozone hole. Ozone hole means complete disappearance of ozone layer over a part of the atmosphere (Fig. 14.5). Ozone hole is formed by intense UV activity on CFCs. It was first detected in 1975 but its worst effect was noticed in 1985 by some British scientists. They saw ozone hole over Antarctica. Later on, the presence of ozone hole was confirmed over the North Pole also. It starts appearing at the end of winter, shows maximum display in spring and disappears in summer. This is because in winter the air is heavy and industrial gases like CFCs get concentrated over an area in Antarctica where the maximum damage of ozone layer has been noticed. In Antarctica CFCs bearing winds become frozen to form ice crystals in winter. During spring when ice melts the trapped CFCs and ClO are freed at once which, in turn, release nascent chlorine to attack ozone. The overall sequence of reactions is illustrated in Fig. 14.4. Consequently, the hole becomes large during the spring.

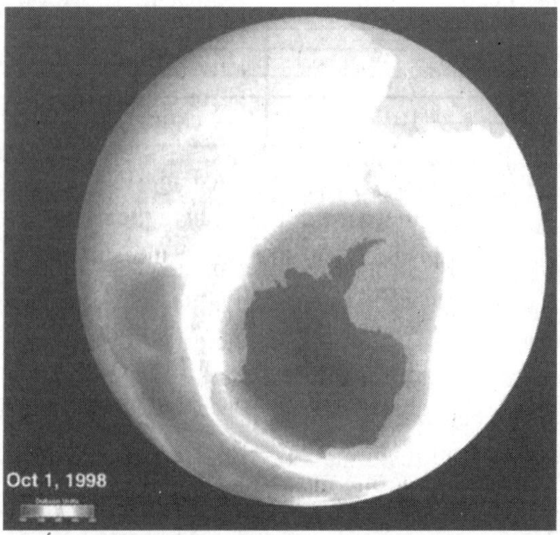

Fig. 14.5 Ozone hole above Antarctica

GLOBAL CONCERN

Global warming is the Global concern. In this context, the representatives of many countries assembled and held a conference at Montreal in 1987 called **Montreal Protocol.** It was decided in the conference to limit the use of CFCs and cut level of CFCs production to 50 per cent by 1999. But India and China opposed and refused to sign this protocol and this was supported by many developing countries on the ground that their per capita consumption of CFCs was only 1/100th as compared to developed and industrialized countries. Later on, in **London Protocol** (1990) India, China and several other countries demanded more time to reduce CFCs consumption. The developed countries agreed to stop CFCs consumption completely by 2000 and for developing nations the time limit was extended to 2010. Developed countries have also agreed to establish a multilateral fund under Montreal Protocol to provide financial support to developing countries for that purpose.

In recent years, few substitute of CFCs have been explored. Two viable substances in the form of HCFC and HFC-34 are considered to be harmless for ozone. Out of these two, HFC-34 is less harmful because of absence of chlorine in it, although it is expensive, inflammable and toxic in nature. There are some other substitutes such as, Butane and Propane which are less expensive, less harmful and freely available. The ozone depletion and its impact was discussed in detail in London agreement at global level. In 1988, the world conference on the changing atmosphere was held in Toronto (Canada). There it was decided to reduce 20% carbon dioxide emissions by the year 2005. Then in 1990 the second world conference on climate was held in Geneva. In 1992, the Earth Summit was held in Rio de Janeiro but with no concrete results because of interference of the industrialised countries. The United States and Japan have not been able to decide so far to reduce CO_2 emissions although they verbally declare their commitment to the Toronto agreement every now and then. The European countries, however, have set the goal of stabilizing CO_2 emissions at the level of 1990 by the year 2000.

At present on the basis of Montreal Protocol India has established a *ozone cell* under the Ministry of forests and Environment with the help of UNDP (United Nations Development Programme). The ozone cell has prepared a document under National Programme for phasing out of ozone depleting substances (ODS). This document has recommended two billion dollars for India to phase out "ODS" completely. This includes the cost of new technologies and expenses for creating public awareness and coordinating the activities at National level.

During the 6th meeting of Executive Committee of the Multilateral Fund held in Nairobi in October 1994, the committee decided to create a fund of 4 million dollars as a contingency grant for India to undertake the programme and extension of technical assistance necessary for the purpose. On this ground, India has developed "R–22" and "R–134" as CFC's substitutes which are presently at trial stage.

CHECK ON GLOBAL WARMING

Global warming can be checked by reducing the concentration of CO_2, methane, nitrogen oxides and CFCs in the atmosphere by adopting the following measures :

1. Plantation of trees at large scale.
2. Increasing the diffusing capacity of oceans for CO_2.

In the photosynthetic process green plants use CO_2 to prepare their food material and release O_2 as a by-product. Therefore, planting trees on waste land and replanting of new trees in place of destroyed trees are most important. If one-third part of the world is covered by the forest then CO_2 will be stabilized. Thus green plants cause decrease of CO_2 concentration through photosynthesis.

Ocean plays a significant role in reduction of CO_2. Carbon dioxide is continuously dissolving in the oceans which is stabilized in the form of carbonate rocks at the bottom of sea. According to an estimation half of the CO_2 evolved is absorbed by the seas and their phytoplankton (free-floating plants). There is a new method to reduce the CO_2 concentration in environment. According to J.H. Martin the growth of phytoplankton can be enhanced by spraying iron supplement in the ocean. Phytoplankton take CO_2 from sea water in greater amount due to which CO_2 will decrease in the oceans. Therefore, more CO_2 will dissolve in the oceans resulting in the decrease of CO_2 concentration in the atmosphere. In 1955, 454 kg ferrous sulphate was sprayed in 1600 square km area of pacific ocean. After few days there was plenty of growth of phytoplankton in sea water of that area and due to that the colour of water became green in place of blue. About 6-21 per cent CO_2 can be reduced by this method.

United Nations Conference on Environment and Development (UNCED) or Earth Summit at Rio de Janeiro was held from 3-14 June, 1992 in which more than hundred countries participated. In Earth Summit, Global Warming was one of the important topics. In the Earth Summit, USA and other developed countries including EEC signed convention on climate change, but they refused to make a specific time bound commitments to bring down the levels of emission of carbon dioxide and other greenhouse gases.

The Second Earth Summit was organised in 2002 by UNCED in Johannessburg (South Africa) in which 110 countries participated. The main theme of the summit was sustainable development. In the protocol of the summit it was reiterated that the activities responsible for global warming should be checked and the use of CFCs should be completely banned by the end of 2010.

QUESTIONS

1. What is green house effect? Discuss the causes of global warming and various remedial measures to check it.
2. Write brief notes on the following :
   (*i*) Ozone depletion and ozone hole
   (*ii*) Causes of ozone depletion
   (*iii*) Impacts of CFCs (Chloroflourocarbons) in global warming.
3. Define :
   (1) Green house gases
   (2) Global warming
4. Explain how $CFCl_3$ molecule destroys Ozone molecule.

15

Conservation of Natural Resources

All the materials and energy essential for the survival and welfare of living beings including humans are provided by nature. They are called **natural resources**. A thing becomes resource only when it is used by humans to perform a function.

Man lives in nature and depends on the resources of nature. The sustinance and welfare of mankind depend upon the exploitation of different natural resources. The utilization of soil, water, minerals, coal, electricity, oil, gas and nuclear energy is very important for the development of nation. These resources have changed the level of living standard of man.

Of the world's total population of six billion, one billion in U.S.A. and Europe alone use 84% of world's total energy. Three billion people of India, China, Brazil and few other countries use only 15%. India contains the world's second largest resource of coal and third and fourth largest resource of manganese and iron. Fossil fuels (coal, petrol, and natural gas) on which modern industrial centres are based, are limited. At present rate of consumption, the fossil fuel may be drastically depleted leading to severe energy crisis. Coal reserves of the world are higher than petroleum and natural gas and they may last longer. The leading coal producing countries are China, U.S.A., U.S.S.R., Germany, U.K., Japan, India, Poland, France and Czechoslovakia. The major oil producing countries are U.S.A., U.S.S.R., Kuwait, Saudi Arabia, Iran, Iraq, Nigeria, Libya, Arab Republic and Indonesia. New discovery of oil has been made recently in the sea beds of Mumbai.

India is facing an ecological crisis and is degrading her natural resources day by day. Now the shortage of natural resources is a matter of international concern. There is increasing deficiency of energy, metals, coal, non-fuel and non-metallic materials. With regard to fuels there is great concern over the huge outflow of foreign exchange and every year enough oil is purchased from the Middle-East countries which are major sources of petroleum. The developed nations of the world have created abundant resources but the developing nations are importing many of these from foreign countries.

We are aware of the fact that earlier the human being was essential part of the nature and human society had impact on the other components of the biosphere. However, with the advancement of social and cultural revolution the conflict between man and nature started. Due to its unending greed, man has destroyed the nature to the maximum for his little gain and made himself the master of nature. However, all efforts to have mastery over the nature have increased his further dependence on nature. On account of such unthoughtful and ruthless exploitation the human society has vastly modified the ecosystems in different parts of the world and has brought undesirable changes in the natural habitats. Consequently, some natural stocks of plants and animals have disappeared. About 1000 species are currently threatened with extinction or are dangerously rare. The human encroachment of nature has deprived the world civilization of

130 mammal species and has endangered more than 250 species. Out of the total approximately 0.3 million species of plants in the world, over 20,000 are in the category of either endangered or threatened with extinction.

It is estimated that over 1000 animal species and 20,000 flowering plants are likely to be endangered globally.

Food, shelter and clothing are the primary requirements of man. Early human society has used natural resources relatively in much less quantity to cover his wants. Among the most essential requirements is a well cooked food. We know that cooking requires energy. The simplest source of energy available for cooking since the early human history is fuel wood. According to an estimate, about 60.5 per cent of Indian people's fuel wood consumption is fire wood and other agricultural wastes. According to Government of India's "Fuel Wood Policy Committee", the annual demand for fuel is roughly 133 million tonnes which is expected to increase in the next 50 years.

There is a great controversy over the sufficiency of the mineral resources to retain domestic and economic development all over the world. According to "Resources and Man" (1969), "true shortages exist or threaten for many substances that are considered essential for current industrial society." Mercury, tin, tungsten and helium, are known examples and the prospective resources of these substances will be nearly exhausted by the end of this century or early in the next, and new sources or substitutes to satisfy even the relatively short-term needs will have to be found out.

One of the most serious aspects of the problem arises from the impact of economic efforts of mankind especially for the purpose of maintaining its existence, protection, survival and betterment of the standard of living. The basic needs which induced the human beings to spoil the natural resources for their welfare have finally led to a situation which threatens to be disastrous. Now in almost every advanced country, the overpopulation has been a vital concern along with the atmospheric pollution. Every effort is being made to save mankind from self-destruction. The situation has become so serious that it is necessary to take some precautionary steps so that the complications may not become worse further. There are various problems which have arisen due to industrial and agricultural developmental activities. Many of these problems are due to mismanagement of natural resources and their impacts are not localized but are universal in nature. It has been estimated that the quantity of CO_2 will be doubled in 23 years and the oil and natural gas resources will be no more available after 50 years and coal will be almost consumed within 150 years.

There are four basic reasons for the depletion of natural resources :

1. Rapid population increase,
2. Pollution,
3. High consumption of resources, and
4. Deterioration of land.

1. **Rapid population increase.** There has been a tremendous increase in India's population and it has now crossed 103 crores (1.03 billion). An increase in population will decrease all types of natural resources and result in environmental pollution. Ultimately, there will be short supply, as well as deterioration in quality of natural resources. This is because increase in population will increase the demand of natural resources and environment.

At present, the world population is increasing by two per cent every year. The industrialized countries have annual growth rate of 0.5 to 1 per cent and on the other hand the developing countries have the growth rate of 2 to 3 per cent. The per capita use of energy and mineral resources shows a difference between the developing and developed countries of the world. The developed countries consume less but their resources are enough. The population and per capita consumption have a considerable impact on the environment. The world cannot meet the continuously increasing demand for natural resources.

2. **Pollution.** We are deteriorating our environment due to increasing population and industrial revolution. We are polluting atmosphere, lakes, streams, rivers by sewage, industrial wastes, heat, radioactive materials, detergents, fertilizers and pesticides. Besides these, we are releasing a number of toxic materials into our surroundings. The uncontrolled and indiscriminate use of pesticides has disturbed the entire food chains by which animals including man are affected. It has been estimated that average individual has about 7 parts per million (ppm) DDT in his body which affects in long-term. Recent researches have revealed that this proportion of DDT in our body has deleterious effects on heart and liver and higher concentration may cause several other diseases including cancer. Many gases, *e.g.*, carbon monoxide, sulphur dioxide, carbon dioxide and nitrous oxide are known to cause respiratory troubles. The unplanned and uncontrolled industrial growth may adversely affect or destroy the health of the society.

3. **Consumption of materials.** Due to tremendous increase of population, most of the natural resources are being rapidly consumed. This high rate of consumption has disturbed our ecosystems. But, on the other hand, many of the natural resources are essential basic human needs. Many industries require raw materials which are essential for the advancement of the country. However, their rapid consumption will affect adversely the quality of our environment either by unwise use of natural resources or by increasing pollution.

4. **Deterioration of land.** Due to excessive consumption of minerals of the soil by cropping or soil erosion or other natural events, fertility of soil is lost and the land deteriorates gradually. Sometimes drought also results in deterioration of land and many nutrients of the top soil are destroyed and soil fertility is lost. As a result of cropping, the cycling of soil mineral nutrients is greatly reduced.

Erosion has also depleted soil fertility because most of the minerals remain in the upper part of the soil and they are easily removed by wind or washed away by water. Sometimes water erosion takes its toll of fertile soils.

Man has also deteriorated agricultural land and ultimately caused the loss of national economy. It is commonly seen that man cannot degrade one part of his environment without simultaneously affecting other parts. For proper economic development lands for cropping, forest, recreation, transportation and wildlife are needed but their availability is reducing day by day. Therefore, integrated policy of resource management should be practised, otherwise unexpected future shortage might upset the national economy.

DIFFERENT TYPES OF NATURAL RESOURCES AND THEIR CONSERVATION

Basically the natural resources are of two types :

1. **Renewable natural resources.** These resources can be replenished and do not change the ecological balance. The cut trees can be grown again, soil forms again and animals reproduce themselves.

2. **Non-renewable natural resources.** Resources which once used up, will be exhausted forever.

Overexploitation of natural resources that are limited in stock may cause scarcity and even renewable resources, such as, water, forests, if over-exploited, require long period of time for replenishment. It clear that if the man wants to survive on this planet he must conserve the natural resources rather than merely exploit them. It does not mean mere preservation of the resources without using them. We should use the resources wisely and judiciously without wasting them.

Different types of these natural resources and their conservation are discussed under the following heads :

Mineral resources

Now it is known that the world's resources of minerals, oils, coal and natural gases are limited. Mineral resources are of several types : metallic minerals and non-metallic minerals.

Non-metallic or industrial minerals include a wide variety of substances which comprise the building materials such as rock, sand, gravels, cement and clay. The non-metallic minerals are fertilizers which are essential to increase agricultural yield. Large amount of nitrogen in the form of nitrates is available in the natural deposits but now it is possible to fix nitrogen synthetically from air. Thus, the deficiency of nitrogen resources is compensated. Phosphorus is obtained from phosphate rocks. Although the phosphorus resources are abundant, yet not evenly distributed. Potassium is also quite abundant in the world.

Among metals, iron, nonferrous metals, silver and gold are important. Among these, iron is the most important element because it is the main component of steel alloys. Chromium, cobalt, magnesium, molybdenum, nickel, tungsten and vanadium are included under ferro-alloy elements.

So far as the causes of mineral loss from the soil are concerned, soil erosion and cropping are considered to be the most significant. Most of the soil nutrients remain in the upper part of soil and many minerals of upper soil are carried away by rapid winds or running water and thus the soil becomes deficient in fertile minerals. Soil is the most important resource of nature, it is essential for human existence and provides the basic requirements to man.

At present, there is a great loss of minerals all over the world which should be considered seriously. Further, demands for mineral resources must be thought of seriously. Although it is difficult to make exact estimate, yet there should be proper relation between demand and supply. Sudden shortage of mineral resources results in poor economy. It has been estimated that for a number of important minerals the resources are limited. So the new resources must be supplied immediately otherwise the minerals will be exhausted within a few years. Due to excessive consumption new minerals should be added rapidly. Now it has become essential for resource-producing undeveloped countries to make some Act to maintain control over their own natural resources. Even oil rich countries have formed an International Organisation of Petrol Exporting Countries (OPEC) which produces more than 50% oil of the world. It has been calculated that out of 19 important minerals 9 would be exhausted in 10 years and coal, iron and aluminium will be consumed by 2100. Therefore, entire quantity of world's mineral reserves should be replaced time to time.

There are three important conservation approaches which should be taken into consideration :

(*i*) to reduce wastes and to minimise demand,

(*ii*) to change the way of life, and

(*iii*) to increase reclamation and recycling of materials.

Solid wastes should be reutilized for their energy content and it is possible to recycle the materials. The total demand can be met with a decrease in the consumption of new materials and increasing the amount of reclamation.

FOREST RESOURCES

Forest is an important natural resource. It is the most important natural habitat for wildlife. It is also utilized by farmers for commercial and recreational purposes. Many herbivores find shelter and carnivores search their prey in the forest. Many wildlife store food supplies and breed in the forest. Besides this, forest plays most important role from commercial point of view. It is the source for a large number of products useful to man. It provides raw materials for many products of daily use. It feeds several industries which depend on wood products. Turpentine oil, paints, resins and printing paper industries get raw material from the forest. Man not only benefits from forest, but also gets recreation. Forest also provides sanctuary for the modern city dwellers. Large number of people visit the forest for peace, beauty and recreation. Forest based cottage industries, such as bee-keeping, bamboo mat and basket making provide means of livelihood to the tribal people. Sal is a most important source for timber industries. Forest also provides raw materials for pulp and plywood industry.

Green plants of the forest are food-producing organisms and are primary producers of the "food chain". They trap energy from the sun and use it to transform CO_2 from the air, together with water and nutrients from the soil into food substances like starch, sugars, through the process of photosynthesis. These foods are stored in the fruits, nuts, seeds, nectar and wood. Therefore, forest serves as an energy reservoir, trapping energy from sunlight and storing it in the forth of a biochemical product.

Forest plays a most important role in keeping the atmosphere balanced by consuming CO_2 and releasing O_2, the latter is essential for animal life. So removal of plants and trees would disturb the composition of natural air. An acre of forest absorbs 4 tonnes of carbon dioxide gas and releases 8 tonnes of oxygen into environment.

If a forest is cut down, energy stored in the wood is lost and also most of the nutrients of the system are lost. Such deforestation leaves a poor soil which can support agriculture for only a short time, because the harvesting of the first few crops removes the remaining nutrients and renders the soil useless. Deforestation shows other disastrous results also. Removal of trees exposes the surface of the land resulting into erosion. Soil is then washed away. In many places soil erosion occurs ten times more rapidly than usual, once the trees are removed. Soil erosion is particularly more on hill slopes where heavy rain sweeps soil downhill to choke rivers. The reduction of forests later affects rainfall and thereby restricts the availability of a most important natural resource, the rain water.

In natural forests, the tree roots bind the soil and about 90 per cent of the water falling on the forests is retained either in humus or in the plants tissue. The forest thus acts as a soaking device and plays a vital role in the hydrological cycle. The rain water thus soaked up is gradually released over the days and weeks which supply to streams and rivers even during dry seasons. Hence, it is important to retain forest cover in upland catchment areas as an alternative to flooding the whole barren and uncultivable area. The washed away top soil silts river beds—, and reservoirs reduce the water holding capacity and flood in the surrounding area is a natural calamity. It has been estimated that in India 60,000 million tonnes of top soil is carried away annually by rain water from deforestated area.

Now a days the tendency of deforestation is increasing. Man is cutting forest to get temporary benefits but there will be a tremendous loss in due course of time. Now, due to gradual destruction of forest, wildlife is disappearing and their number is becoming reduced for which government should consider various aspects of forest management. Forest management programmes should be motivated by forest employees and general public to increase yield, to avoid forest cutting and to prevent forest fires. Whenever plant or timber is cut, that area must be reforested. Similarly, any forest which has been destroyed by insects, diseases, hurricanes that area should be reforested. The primary aim of the forest service is to make the greatest number of forest resources available to the maximum number of people.

Shifting cultivation is another practice which destroys the forest. Many farmers destroy the forest for agricultural purpose and soil is exhausted. Therefore, farmers should use the same land for cultivation and they should apply better farming methods so that soil fertility remains restored and the soil can be used again.

Most severe damage of the forest is due to attack of insects, pests and fungi. Many destructive forest diseases are due to parasitic fungi, rusts, viruses and nematodes. Young seedlings are destroyed due to attack of nematodes. Many diseases such as heart rot, blister rust, oak wilt, phloem necrosis and Dutch elm diseases are common in the forest. The forest diseases can be controlled upto some extent by the following methods :

1. By eradication of alternative hosts.
2. By using suitable antibiotics,
3. By DDT spray,

4. Sterilization, and
5. By using resistant varieties.

Besides above, the following measures of forest management have been recommended :

1. Reforestation.
2. By improving the quality of timber.
3. By converting wasteful cutting into quality yield harvesting.
4. By increasing forest protection.
5. By increasing forest area.
6. By developing fast growing trees.
7. By controlling harmful forest agents.
8. By developing the better tree varieties.
9. By using disease resistant varieties.

In India about 75 million hectares area is occupied by forest alone which is about 23 per cent of the total land. In India forests have not so far played a significant role in improvement of economic condition of the country. Proper scientific management, conservation and utilization of forest wealth are likely to increase their resource value and utility in the future. At present there is an urgent need to conserve the existing forest resources and to expand the forest area. The most important task is to restore the vegetation cover which is destroyed through our false policies. New forests and wood lots should be created to meet the daily demand of fuel and fodder and to provide more habitat for the wildlife.

The National Commission on Agriculture is giving serious thought to the problem of deforestation and has recommended introduction of "Social Forestry", *i.e.,* to create multipurpose village wood lots in rural India. Social forestry may be defined as an additional aid to wildlife conservation. *"Social forestry is a concept, a programme and a mission which aims at ensuring ecological, economic and social security to the people, particularly to the rural masses especially by involving the beneficiaries right from the planning stage to the harvesting stage. It aims at mixed production system of wood, fibre, fodder, grasses, fruits and other raw materials for self-consumption and cottage industry".*

Different components of social forestry programme are :

1. Protection and afforestation of degraded forests in the vicinity of human habitations.
2. Creation of village wood lots on community lands and government waste lands.
3. Block plantation.
4. Agro-forestry (trees along with agricultural crops) on marginal and sub-marginal farm lands.
5. Tree planting around habitation area, field boundaries and pasture development.
6. Tree planting in Urban and Industrial areas for aesthetic purposes, purification of polluted air, absorption of CO_2, release of O_2 and control of noise pollution.
7. Control of soil erosion by planting trees or shrubs.
8. Strip plantation along road sides, canals and rail lines.

JOINT FOREST MANAGEMENT

Participation of local public is needed in forest management programmes. Local people, farmers, students and women help in greening an area, if they get some economic benefit from forest conservation. There must be coordination and mutual cooperation between local communities and the Forest Department. This joint venture of economic forest management may be said as

Joint Forest Management (JFM). This step started first, in 1972, in West Bengal. The concept of **social forestry** is also the same. In this programme, local population and Forest Department are involved in plantation. Joint Forest Management (JFM) programmes involve an agreement between the local people and the Forest Department. The JFM programme cares more for the local population and their rights and benefits from the forest resources are safe. Under this programme, a committee is formed from the local public, called Forest Protection Committee. This committee helps in restoring the green cover and protects the area from being over-exploited.

Mining operations, dam construction and timber cutting, etc are some of the requirements of a developed country. But in mining and over-exploitation of timber, the ecological damage to forests is unimaginable. It has been seen that the forests are often found in the regions where there is rich mineral resources. Forests also cover the steep embankments of river valleys, which are more suitable to develop Hydel and Irrigation Projects. Thus, there is a conflict between ecologists and the Mining and Irrigation Department. But it is to understand that long-term ecological benefits cannot be overlooked for short-term economic profits that lead to deforestation. In the forest areas, where development projects are approved, these dislocate and displace thousands of tribal population along with their houses and properties. The Joint Forest Management looks to resolve such sensitive issues.

If above programmes are carried out effectively the basic needs of rural people such as fuel, fodder, fibre and timber could be met easily and it will ensure ecological security like protection against wind, erosion, polluted water and air and availability of desired habitat for the wildlife. Social forestry can play a significant role to check flood and drought which affect 34 and 68 million hectares of land respectively in India.

WILDLIFE RESOURCES

Nature has gifted us a priceless biotic wealth, **the wildlife** which is a thing of beauty. It needs to be preserved rather than destroyed. In broadsense, the wildlife involves animals living in a natural, undomesticated state and uncultivated plants and microbial communities living within their natural environment.

In the modern sense, according to Dr. Mahajan (1961), wildlife means, life in any form (plant/animal) existing in natural surrounding. Wildlife as a natural resource is an essential component of ecosystem. It is of high importance to human society. It contributes to the cycling of matter, flow of energy and the soil forming processes. It has a significant role in the stabilization of ecosystems by natural control and regulation of populations (Bioindicator). It acts as gene reserve. Wild fauna has a positive role in formation of ecosystem by pollination, seed and fruit dispersal and selective breeding etc. Wildlife provides recreational and economic benefits to man. Recreational and economic benefits are closely related to each other. For instance, fishing and hunting provide entertainment and economic benefit to man.

There is a mutual correlation between plants and animal communities. The composition of animal community is more or less constant and it is characterised and maintained by vegetation type which provides habitat and food for animals. Therefore, the maintenance and conservation of wild fauna depends largely on the conservation of habitat. The developmental projects such as mining, agriculture, shifting cultivation, major water resources, major engineering activities, timber extraction have created special problems of habitat and wildlife management in the tropical forests and as a result, tropical forests are diminishing rapidly. Major canals and irrigation projects have created problems of wildlife management. Selective polycyclic logging in the forest is disturbing the wildlife in the same way as cyclones and destruction of natural forests disturb wildlife.

Red Data Book

International Union for Conservation of Nature (IUCN) has published **Red Data Book** which maintains a collection of all the available data on the species threatened with extinction. Several criteria have been listed in the book to indicate the endangered state of the species, such as restricted distribution, narrow habitat tolerance, migration across international boundaries, behavioural non-adaptiveness to technical advances, over exploitation, habitat destruction, competition for food, accidental killings and so on. Wildlife includes 850 species of mammals, 1200 species and 2,100 sub-species of birds and more than 20,000 species of insects. According to the **Red Data Book,** about 277 species of mammals and 321 spp. of birds are considered to be endangered or likely to become extinct. The World Wildlife Fund (WWF) under International Union for Conservation of Nature (IUCN) is concerned with the conservation of wildlife and several projects have been launched for conservation of natural forests and wildlife. In its diversity of life forms India is the second largest country in the world.

It is absolutely necessary to protect and conserve all forms of life on this earth as they are all interdependent and form a chain. Nature has created them in such a balanced manner that if one form of life is disturbed it affects all the other lives also. The very existence of man depends upon the survival of other forms of life both plants and animals. So the destiny of humanity depends upon the survival of other forms of life.

The causes of decline of wildlife in India are many. Wild elephants were captured and trained for use in war. Rhinos were killed for making shield. Rulers of princely states indulged in killing of wild animals as their hobby. Forests were cleared for development of agriculture, industry and other developmental projects. Wild animals were mercilessly killed which resulted in decline in their number and they are facing extinction now.

The extinction of species is a general biological principle of evolution and the change of living conditions happend to be the main reason for extinction. The extinction of species is followed by replacement by the new species which are better adapted to changed conditions. This is how the species diversity and biological equilibrium is maintained in nature and the evolution continues.

The main factors for extinction of wildlife fauna are as follows :

1. Habitat distruction or contamination of the habitat which accounts for about 67% extinction.
2. Over exploitation which accounts for about 37% extinction.
3. Introduction of exotic species which accounts for about 19% of the total extinction.
4. Competition for food may account for about 7% of extinction.
5. Accidental killing which accounts for about 2% of extinction.

These figures combined are well above 100 per cent because some time more than one factor affect a threatened species.

Today wildlife species are gradually disappearing and their number is becoming reduced. Many species of wildlife have become extinct or are on the way to extinction (Table 15.1). Uptill now 106 species of mammals and 139 species of birds have become extinct due to climatic and geographical changes or hunting by man. Today the lions are limited in number. The number is around 200 in the Gir forest of Gujarat. The rhino existed in Indus valley in the vicinity of Mohenjo-Daro some 5000 years ago. The rhinos are now found in certain parts of Nepal, West Bengal and Assam.

The number of tigers is also very limited. According to 1972 census, the tigers were 1827 in India.

Table 15.1. Few Endangered Species of Birds, Reptiles and Mammals

Birds	Reptiles	Mammals
Great Indian Bustard	Gharial	Black Buck
Great Indian Hornbill	Green Sea Turtle	Lion-tailed Macaque
Brown Headed Gull	Pythons	Wild Ass
Mountain Quail	Tortoise	Sloth Bear
Peacock		Hyaena
Peafowl		Sambar
Pelican		Cheetal
Siberian White Crane		Tiger

Conservation is an intelligent and judicious management of resources towards their optimum utilization without depleting the basic stock. The protection of wildlife from unwanted destruction is called **wildlife conservation.**

Through the Convention on International Trade in threatened and endangered species of wild Fauna and Flora, a number of species such as hispid hare, pigmy hog, lion, tiger, rhino, thamin and great Indian bustards have been saved. In 1973, a project for saving tiger was started which yielded very satisfactory results. The number of tigers increased considerably and it has become difficult to maintain them in their reserves. They have started killing rural population and their cattle outside the limits of Corbett and Dudhwa National Parks of U.P. All life forms created by nature are directly or indirectly dependent for their survival on other life forms.

WILDLIFE MANAGEMENT

NEED TO CONSERVE WILDLIFE

Some of the reasons that are necessary for the conservation of wildlife are as follows:

1. Wildlife forms an essential component of any natural ecosystem. It plays significant role in cycling of matter, flow of energy, soil forming processes and it balances the physical and climatic conditions in the environment. Wildlife is our hereditary wealth which is to be transferred to our future generations.
2. They are economically important.
3. They are informative and educative. Man has learnt much from them.
4. Aesthetically wildlife made this planet attractive.
5. It is the best source of recreation.
6. Wildlife provides relaxation through its varied flora and fauna.
7. Genetic variations and interbreeding lead to evolution. It will be affected if any form of life is destroyed.
8. The yield in agriculture and animal husbandry are affected if constancy in inbreeding is restored.

Now the following steps of wildlife management have been taken into consideration :

(*i*) **Conservation of habitat.** The first step towards conserving the remaining wildlife is to conserve the different habitats in which they live, for example, tropical forest, pine forest, wet land, swamp shrubs and grasslands need protection. The conservation of wildlife also depends on the conservation of other resources. The best way to preserve wildlife is that both the habitat and animals are preserved in wildlife sanctuaries and National Parks.

(*ii*) **Species preservation.** The animals which are rare should be protected from being hunted. Generally these animals are hunted for recreation or to get various parts from their body as teeth, skin, horn or meat. The shooting, trapping or taking of game (wild animals and birds) or fish, from private property or from a place where such practices are specially reserved or forbidden, is called poaching.

Poaching is now usually done for sport or commercial profit, but it can be a serious threat to many species of wild animals, particularly those protected in Wildlife Reserves or National Parks. In Africa, the difficulty of enforcing game codes has led to the critical depletion of the rhinoceros, which is hunted for its horn. The Bengal Tiger of India and the Gorilla of Central Africa, have similarly been threatened with extinction by hunters operating illegally. River poaching has also been a problem in some countries, causing the depletion of stocks of fishes in many areas. Killing or hunting can be banned by legislation or by the protection of the animals by the forest department. If there is no proper preservation of the wildlife the animals may disappear in future due to unrestricted human predator.

(*iii*) **Introduction of exotic species.** Introduction of exotic species is another method of wildlife development and management. They may be brought over to serve as predators in controlling some pests, add colour to the wildlife community.

(*iv*) **Predator control.** Now ecologists consider that predators play an important part in the ecosystem. Predators may also serve an useful role in keeping the primary population within limits. Lack of predators results in population explosion.

(*v*) **Breeding.** Breeding is the selective control of mating in plants and animals to produce organisms that better serve human needs for food, work, sport or aesthetics. Breeding is of two types:

(*a*) **Inbreeding.** It is self-fertilisation among closely related animals. This breeding can yield suitable results in the direction of a desirable trait (character). This breeding is restricted to a small circle of animals and thus prevents the introduction of genes from outside. This process leads to a decrease in variation.

(*b*) **Out breeding or cross breeding.** It is a mating of individuals from entirely different races or different species. Out breeding increases vigour as compared to inbreeding but it is very uncommon in animals. Upto some extent, it is done in domesticated animals to increase the yield of milk, flesh or other products.

If the number of endangered species drops to low levels for several generations, then a considerable amount of the genetic variation of the original population is lost. Efforts to save endangered species also include **captive breeding** of severely endangered species, to restore or add to a breeding population (a population is a group of individuals capable of reproducing). When animal species are taken into captivity, it is important to have breeding programmes which aim to maintain as much genetic variation as possible. Captive breeding is necessary until the species can be reintroduced into its native habitat. Reintroduction may however be unsuccessful, if the natural ecosystem is on the decline.

A species, which has declined to such low numbers as to be considered endangered, may face some genetic problems, such as loss of some characters. This results in the loss of genetic diversity and reduction of a population. Once the diversity is lost, it is probably lost forever, unless it is built up again, which may take thousands of generations.

Legislation of wildlife. Nowadays government is thinking to make an Act against hunting of wildlife. Now hunting of certain wild animals has been banned such as python, elephant, lion, leopard, black buck, crocodile and gharial etc.

Through the efforts to protect and maintain wild flora and fauna in India have been made since the times of king Ashoka, the results are not appreciable and the situation is still getting worse.

The conservation of wildlife has acquired urgency because of its relevance in human survival. Since the wildlife distruction is mainly due to habitat distruction, mismanagement and over exploitation of flora and fauna, the conservation and management of wildlife require community approach (*i.e.* people's participation), cooperation among decision maker's and environmentalists. Let us think of conserving nature and natural resources for sustainable development and follow the slogans of Lord Buddha quoted by A. De Silva (1980) :

"The forest is peculiar organism of unlimited kindness and benevolence that makes no demand for sustenance and extends generously the products of its life activity, it affords protection to all beings by offering shade even to those who destroy it".

BIO-PIRACY

Many organisations and multinational companies exploit and patent biological resources or bio-resources of other nations, without proper authorisation from the countries concerned. This is called **bio-piracy.** Thus, bio-piracy may be defined as 'the theft of various natural products (plants and animals) and then selling them by getting patent or intellectual property rights, without giving any benefits (financial or technological) back to the host country or the country of origin'. For example, bio-piracy of neem, haldi (turmeric), basmati rice, etc.

Bio-piracy of neem: Patenting of seed oil from the Indian neem tree *(Azadirachta indica)* by U.S. Department of Agriculture (USDA) and WR Grace (Columbia) in 1992 was a case of bio-piracy. For generations, Indians had used neem oil as a medicine and pest-controlling agent, so the US patent on neem had created a great controversy all over the world. In May 2000, after a long legal fight, the European Patent Office revoked the patent of USDA and WR Grace.

With the emergence of modern biotechnology, one important issue has been raised, which is the legal characterisation and treatment of trade related bio-technological processes and products, commonly known as **intellectual property.** Intellectual property is the processes and products which result from the development of genetic engineering techniques. The right to protect this property is called Intellectual Property Rights (IPR). This includes rights relating to literary, artistic, scientific works, inventions in all fields, industrial designs, trademarks, commercial names, etc. IPR also includes patents, copyright, trademarks and trade secrets.

A patent is a government granted and secured legal right, to prevent others from making, using or selling the inventions covered by the patent. A patent is a personal property, which can be licensed or sold like any other property. The Indian Patent Act (1970) allows process patents, but not product patents. The duration of the patent in India is 5 years from the date of grant of patent or 7 years from the date of filling the application, whichever is less.

ENFORCEMENT OF LEGAL PROVISIONS IN INDIA

Realising the importance of the wildlife resources and in order to preserve the Biosphere Reserves, our country has taken up steps by the enactment of various Wildlife Acts (1873, 1878, 1932 & 1933). The Indian Board of Wildlife has been set up in 1952. Some other Acts have also been enacted, which are as follows:

- Creation of Wildlife Parks and Sanctuaries, 1972
- Enactment of an All India Wildlife Protection Act, 1972
- Becoming a party to the Convention of International Trade in Endangered Species of Fauna and Flora (CITES, 1976)
- Launching a national component of UNESCO's 'Man and Biosphere Programme' (1971)
- Conservation projects for individual endangered species like Hungal (1970), Lion (1972), Tiger (1973), Crocodile (1974) and Brown-antlered Deer (1981)
- The Amendment to the Wildlife Protection Act, 2002 is more stringent and prevents the commercial use of resources by the local people

The Wildlife (Protection) Act, controls wildlife conservation and protection of endangered species. The Wildlife Act has been adopted by all States. The Centre assists with financial help to the States for:

- Strengthening management and protection of National Parks and Sanctuaries
- Protection of wildlife and control of poaching and illegal trade
- Captive breeding programmes for endangered species
- Wildlife education
- Development of selected Zoological Gardens

A person who breaks any of the conditions under these Acts, shall be guilty of an offence against this Act, which is punishable with imprisonment upto three years or with a fine of Rs. 25,000 or both.

WILDLIFE SANCTUARIES AND NATIONAL PARKS

Sanctuaries and National Parks have been established in various parts of the country to protect endangered and threatened plants and animals. The Wildlife Act, 1972 provides legal protection to all the species found anywhere inside or outside forest area, but still some areas deserve special protection and treatment to preserve wild animals along with their habitat. Such areas are provided with special legal status under the wildlife protection Act and are known as sanctuaries and national parks respectively. In a **sanctuary** fauna enjoys complete protection against exploitation or destruction but flora and other forest produce can continue to be exploited. Thus sanctuary is an area where hunting is strictly prohibited but flora and forest produce can be exploited. **A national park** enjoys protection of both flora and fauna and congenial conditions are created to maintain natural habitat. A national park conserves wildlife, scenery, and natural or historical objects of national importance. Thus a national park has a higher status of protection level than a sanctuary. National park is created or abolished by legislature of province or by Parliament.

The main objective of creating such protected areas in the form of national parks and sanctuaries is to establish gene banks of all living organisms—flora and fauna to meet future exigencies. This is why the areas for parks and sanctuaries are selected in such a way that they should represent viable samples of all significant bio-geographic regions. In India about 60 national parks and more than 250 sanctuaries covering about 3 per cent of total land area and 13 per cent of total forest area have so far been established.

Some of Important National Parks and Sanctuaries of India are enlisted in the Fig. 15.1 and Table 15.2.

Table 15.2. Main National Parks/Sanctuaries of India with its associated famous fauna

State	Name of National Park/Sanctuary	Famous fauna
Andhra Pradesh	Kelameru Bird Sanctuary	Pelican and marine birds
Assam	Kaziranga National Park	One horned rhinoceros
	Manas Wildlife Sanctuary	Wild Buffaloes
Bihar	Hazaribagh National Park	Tiger, leopard, etc.
Gujarat	Gir National Park	Lion, nilgai, etc.
Haryana	Sultanpur Bird Sanctuary	Migratory birds
Jammu & Kashmir	Dachigam Wildlife sanctuary	Kashmir stags, wild goose, hungul, etc.
Karnataka	Bandipur National Park	Elephant, Indian bison
Kerala	Periyar Wildlife Sanctuary	Elephants, barking deer
Madhya Pradesh	Kanha National Park	Tiger, leopard

Orissa	Simlipal National Park	Elephant, sambar
	Kanika National Park	Mangroves and associated fauna
Punjab	Abohar Wildlife Sanctuary	Black buck
Rajasthan	Ranthambore	Tiger
	Bharatpur Bird Sanctuary	Ducks, herons
Tamil Nadu	Vedanthangal Bird Sanctuary	Water birds
	Annamalai Sanctuary	Elephants
Uttar Pradesh	Dudhwa National Park	Tiger
Uttaranchal	Corbett National Park	Tiger
	Rajaji National Park	Elephants
West Bengal	Jaldapara Wildlife Sanctuary	Rhinoceros
	Sunderban Tiger Reserve	Tiger

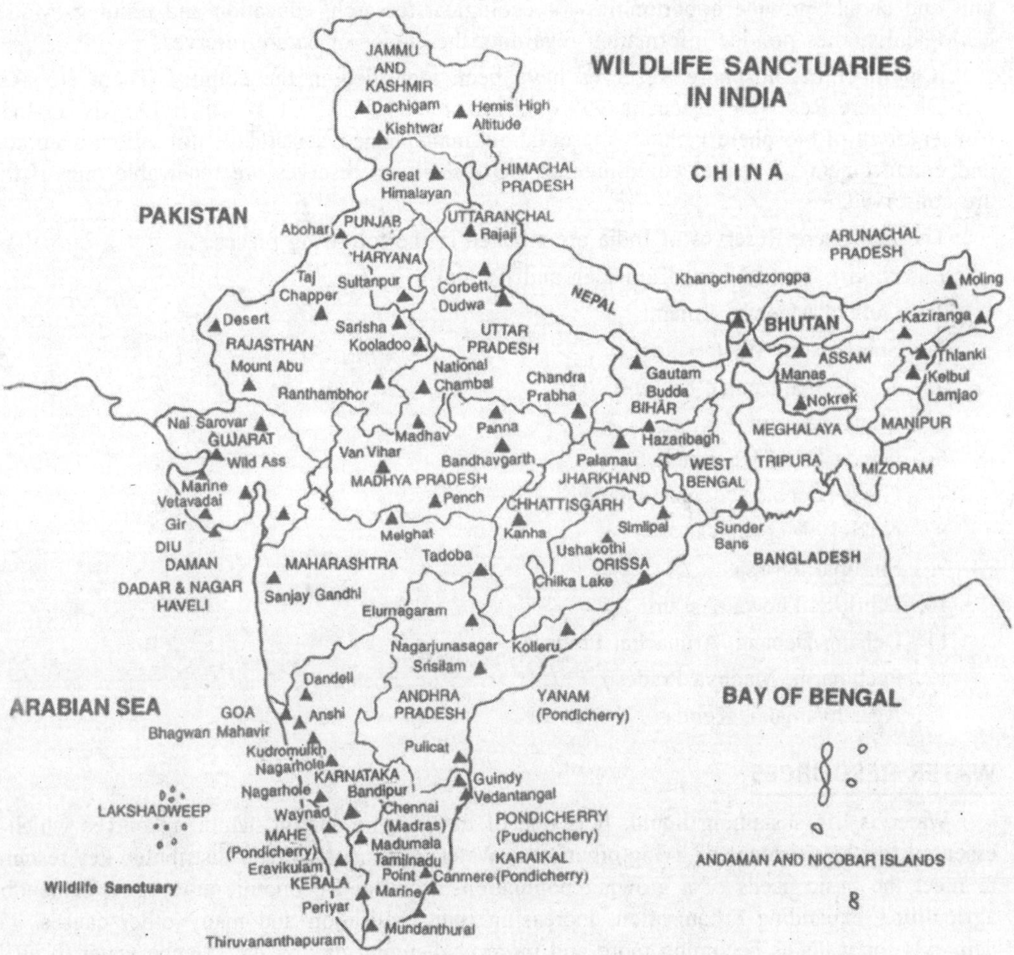

Fig. 15.1 - Map of India showing location of main National Parks and Sanctuaries

BIOSPHERE RESERVE OR BIO-RESERVE

Establishment of Wildlife Sanctuaries and National Parks form a part of the management practices for wildlife. A biosphere reserve is concerned with the preservation of overall plant and animal diversity, including domesticated plants and animals. A **Biosphere Reserve** involves the preservation of overall biological diversity. The Biosphere Reserve Programme is an integrated International cooperation for conservations of nature.

For the protection of rare or endangered animals and plants, National Parks and Wildlife Reserves have been established in many parts of the world. But now a need is felt to conserve the total assemblage of animals and plants in a balanced and viable biological community. The UNESCO under the 'Man and Biosphere Programme initiated this concept of conservation of natural areas or Biosphere Reserves.

The aim of establishing an international network of such protected areas serve as representatives of global ecosystems. The main objective of bio-reserves is to preserve for future use the diversity of biotic communities of plants and animals with natural ecosystems and to safeguard the genetic diversity of species. Each Biosphere Reserve should be large enough to be an effective conservation unit and should provide opportunities for ecological research, education and training. Modern ecological studies provide information regarding the choice of nature reserves.

A number of biosphere Reserves have been identified in the country. There are about 425 Biosphere Reserves present in 95 countries of the world, out of which 13 exist in India. Conservation of biosphere includes preservation, maintenance, sustainable utilisation, restoration and enhancement of natural environments; because living reserves are renewable only if they are conserved.

The Biosphere Reserves of India are situated at the following places:

1. Nilgiri, Tamil Nadu, Karnataka and Kerala
2. Gulf of Mannar, Gujarat
3. Sunderbans, West Bengal
4. Khangchendzonga, Sikkim
5. Manas, Assam
6. Nanda Devi, Uttaranchal
7. Great Nicobar, Nicobar Island
8. Nokrek, Meghalaya
9. Simlipal, Orissa
10. Dibru-Saikhowa, Assam
11. Dehang-Debang, Arunachal Pradesh
12. Pachmarhi, Madhya Pradesh
13. Agasthyamalai, Kerala

WATER RESOURCES

Water is life sustaining liquid. It is one of the most important natural resources which is essential for the existence of living organisms. Water is the most widely distributed key resource to meet the basic needs of a growing population, social and economic ambitions, demanding agriculture, expanding urbanisation, increasing industrialisation and many other causes. The demands for water is becoming more and more challenging day by day. Hence water in all its forms (solid, liquid and gas) should be harnessed properly.

Drinking water is one of the basic needs for all of us. Unfortunately, about 5.76 lakh villages have been facing scarcity of water. In Rajasthan, people in certain pockets have to wait for hours to collect one bucket of water from water tankers brought by the trains, trucks and tractors. In many hilly areas, the situation is the same. In many tribal and backward areas unhygienic and unsafe water collected in ponds, tanks etc. during rainy season is the only source of drinking water.

Our country is facing frequent floods and drought often at the same time in different parts because of variable nature and uneven distribution of rains. Besides these, due to large scale deforestation and soil erosion the rivers are silted up causing floods.

For sustained and increased agricultural production, industrial development and economic emancipation of the country the most vital resources are soil and water, which need to be conserved, developed and managed efficiently. In India, floods, water-logging, soil erosion, drought, salty groundwater, etc are some of the major problems of water management for agriculture and other needs.

The major problems of water management and the possible strategies of overcoming them are explained here:

Floods

Floods refer to the inundation of large parts of land by water, which otherwise remain dry, for some duration of time. Flood causes heavy loss to agriculture, livestock and property. Deforestation, overgrazing, mining, industrialisation, global warming, etc. have contributed largely in the incidence of floods.

The best solution to overcome such damage is large scale irrigation projects, which will also protect from other environmental hazards. These hazards may be in the form of increase in water logging, soil sedimentation in reservoirs, damage to forest areas, large scale growth of aquatic weed of nuisance value, displacing wildlife and degradation of valuable landscape etc.

Plantations can reduce the impact of water flow on soil erosion. For water management, Landuse State Boards were set-up in 1980 in order to protect the soil and water to enhance their productivity through proper land and water use practices.

Water Logging

A soil is said to be water logged when it is completely saturated with water, which is caused by water stagnation on flat land and low lying areas. It occurs due to excess rainfall, floods, seepage, high water table, obstruction to natural drainage, over irrigation, etc. In most of the low lying areas, wet conditions persist longer that results in delayed sowing or less crop production. Another impact is that when water dries, salts accumulate on the soil surface resulting in salinity.

Since water logging is the second biggest threat to the soil, next to erosion, it is therefore, necessary to study water table fluctuations, groundwater recharge, assessment of seepage from canals, tanks, etc.

Salty Groundwater

Most parts of the arid and semi-arid regions contain high percentage of sodium salts. Such water is dangerous for agriculture. Continuous use of such water results in the accumulation of sodium salts to produce *Usar* or alkali soil. Some of the possible solutions to remove salinity are as follows:

- Use of gypsum ($CaSO_4$)
- Use of molasses, ash and canesugar extracts
- Cultivation of salt resistant varieties
- Recharging with good quality of groundwater

Drought

Drought is a condition of abnormally dry weather within a geographic region. Some solutions to overcome drought are as follows:

- Development of additional surface water resources
- Direct pumping from streams, rivers and open water bodies
- Proper regulation of water use
- Increase utilisation of ground water resources
- Efficient distribution of canal water
- Irrigation according to requirement of crops

Watershed Management

Rain is the main source of water on the earth. When the rainwater reaches the earth surface, a part of it is evaporated and a part of it is taken by plants, a part of ground water is retained, a part is infiltrated and the remaining water flows as surface runoff through drainage systems into the sea. Each of the river system has separate drainage units which depend upon the terrain, geology, climate and landuse pattern. In order to tackle the problems of water resources and various forms of degradation and to develop these resources for sustained and increased productivity in the regions it is essential to focus our attention to natural drainage units called the watershed (Fig. 15.2). Watershed is defined as 'land area that delivers water, sediment and dissolved substances through small streams to a major stream (river)'.

Fig. 15.2 Watershed

The management of a single unit of land with its water drainage system is called **watershed management.** This technique has several components which includes soil, water and vegetation cover. The natural drainage pattern of a watershed unit, if managed properly, can bring about a year-round supply of water. There are two main steps of its conservation:

1. Construction of many long trenches and mounds along the contours of the hill to hold rainwater and allow it to percolate into the ground. This ensures that underground reservoirs are fully recharged. This is followed by growing of grasses and shrubs which hold the soil firmly.
2. The second step is to form nala plugs in the streams, so that water is held in a stream and does not rush down the hillside.

The proper management of watershed is to maintain the water yield confined to the drainage so as to derive maximum benefits and sustained productivity by proper land-use and suitable cropping patterns. The various conservation and crop management practices should be adopted in such a way that loss of soil, as well as moisture, is minimised. The main objectives of soil and water conservation are as follows:

- To conserve these resources in the upper catchments by providing suitable vegetation cover to protect soil erosion and to prevent the silting of river channels and the reservoirs, which prevent the flood.
- In case of cultivated areas, the conservation measures are different and depend on agricultural practices, economics and other aspects.

Other conservation measures undertaken for watershed management are as follows:

Engineering practices: These include land levelling, shaping, contour bunding, terracing, grading of land, bank protection, check-dams, etc.

Agronomic practices: These include protection by vegetation, such as grass cover, pasture development, contour farming, strip-cropping, crop rotation, etc.

Afforestation: Under this, plantation is done, while overgrazing and forest fires are controlled.

Integrated Watershed Management

In recent years watershed management has received wider acceptance. It incorporates several diversified programmes, such as:

- Amendment of alkali and acid soils
- Control of shifting agriculture
- Creation of State Landuse Survey Organisation

Indian Space Research Organisation (ISRO), National Remote Sensing Agency (NRSA), Food and Agriculture Organisation (FAO) and United Nations Development Programmes (UNDP), have taken more initiatives for soil and landuse survey and for planning of soil conservation programmes. Awareness was created to develop, conserve and manage the land in the wider perspective in the various sectors. It is, therefore, considered necessary to have some high level body where all matters related to soil and water resources could be discussed under one platform.

The Ministry of Agriculture, Government of India, has set up State Landuse Boards under the chairmanship of Chief Minister of States, to take up all matters about the natural resources, their utilisation and degradation of land. The Government of India has launched an integrated watershed management in the catchments of flood prone rivers with the collaboration of NRSA and ISRO.

Rainwater Harvesting

Old people said **'capture water when it rains'.** It is also true in the present context of water crisis. Today, the world is facing serious water shortage. Every drop of water is valuable because 'where there is water there is life'. There has been increased aridity in India over the last few decades. Water scarcity is likely to increase in India due to increasing population. Under extreme conditions in future, human society will be bound to use different means of adaptation due to climatic changes.

Objectives of Rainwater Harvesting

1. To reduce loss from surface runoff
2. To avoid flooding
3. To meet the increasing demands of water
4. To raise the water-table by recharging groundwater
5. To reduce groundwater contamination

Need for Rainwater Harvesting

Why does rainwater need to be harvested? It matters more today than any other time in the past. There are following reasons:

1. About 50% fresh water goes waste due to runoff.
2. More than 1 billion people lack clean drinking water globally.
3. Population increase is much faster than the increase in the amount of available fresh water.
4. Per capita availability of fresh water will further decrease in the coming years.
5. During summer and droughts, it will supplement the domestic water requirement.
6. Climatic changes also lead to increase in precipitation, evaporation, transpiration, occurrence of storms and changes in biogeochemical processes affecting water quality.
7. It is essential to reduce groundwater pollution and improve the quality of water.
8. It is a better option for providing clean and safe water particularly for drinking and other domestic uses (Table 15.3).

Table 15.3 Per capita availability of water

Year	Kilo litre per capita per year
1950	5177
2000	1820
2025	1400 (likely population 130 crores)
2050	1140 (likely population 160 crores)

Rainwater Harvesting Technology

Rainwater harvesting is the method of storing rainwater and thereby increasing the recharge of groundwater. As India since the very beginning was primarily an agricultural country, the need to harness water was felt. This is also due to fact that rainfall in our country occurs only for two to three months; therefore, water needs to be conserved for its use throughout the year. Even the ancient civilisations like Harappa, Mohanjodaro, etc. provide excellent examples of water harvesting through a network of tanks and reservoirs. Some of the old forts like Jaigarh Fort near Jaipur and Fatehpur Sikri near Agra, also provide good examples of water storage through rainwater harvesting.

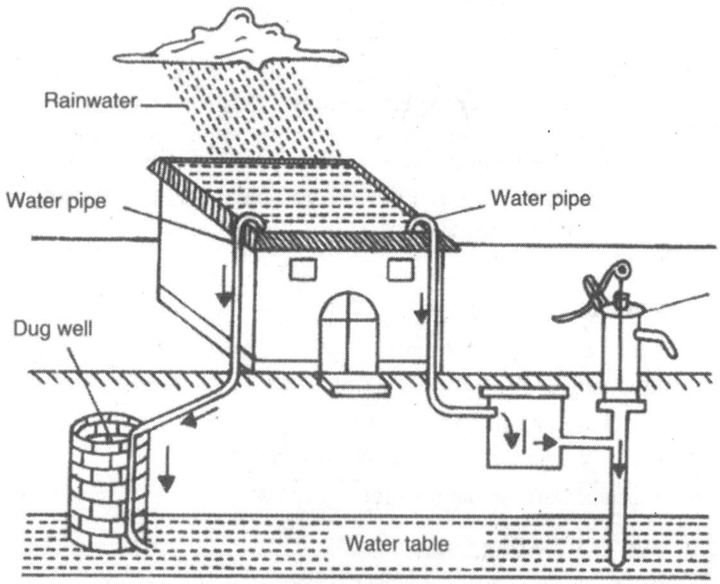

Fig. 15.3 Rainwater Harvesting

Several techniques are in practice to recharge groundwater. One method is to manage rainwater in such a way that it is used at the source. If as much water as possible is collected and stored, it can be used after the rainy season is over. This method has been traditionally practiced in dry areas. Simple local techniques such as ponds and earthen embankments can help in the harvesting and storage of rainwater. Rural and urban water use, restoration of streams for recreation, fresh water fisheries and natural ecosystems, etc. need rainwater harvesting. Local practices for rainwater harvesting can provide sufficient amount of water.

One hectare of land in an arid region with 100 mm rainfall annually could yield one million litres of water per year through rainwater harvesting.

Deep wells may provide a source of clean water, but it is possible only in the rural areas. Traditional systems could become more efficient if scientific attempts are combined to enhance their productivity.

Other methods are refilling of dug-wells, recharging of hand pumps, construction of percolation pits, trenches in the agricultural fields, bunds and check-dams etc (Fig. 15.4). The above practices have been used since long in India. Now there are advanced techniques of water harvesting systems, such as canals, tanks, embankments and wells. In hilly areas, rainwater harvesting has been practiced in rooftops and springs with the help of bamboo pipes. In arid and semi-arid regions, wells and step-wells were constructed to tap groundwater. Construction of tanks has been a very popular method in recent years to conserve rainwater.

Rainwater harvesting treatment is very important in the areas where pollution is alarming. It is now possible to use nano-filtration for the removal of hardness, natural organic materials, pesticides, bacteria, viruses, salinity, nitrates, arsenic and other pollutants.

Weather and water policy should be integrated and may be streamlined to promote rainwater harvesting in the water stretched regions of the world. In the urban areas, water resources are fast depleting due to population increase and unrestricted use of water.

Fig. 15.4 Construction of check dams for rainwater harvesting

Modern Rainwater Harvesting

Availability of water has become a major problem in the urban areas due to high density of population. In many cities, water is available only for an hour or so and that too in a trickle. The urban areas are mostly dependent on groundwater due to which its level is falling day by day. The problem is further compounded by the fact that most of the urban land is covered by concrete structures making hindrance to groundwater recharge.

To overcome this problem, rooftop rainwater harvesting has proved to be an effective method. In this method, the rainwater falling on rooftops, which otherwise flows down the drain, is diverted to an underground tank for future use. Rainwater can also be diverted to dug-wells or pits for recharging groundwater (Fig. 15.3). In Rajasthan, tremendous work has been done by Sri Rajendra Singh to collect rainwater by constructing check-dams. He was awarded the Magasaysay Award for his commendable work.

Top 10 Water-Rich and Water-Poor Countries

Water-Rich Countries: Iceland, Surinam, Guyana, Papua New Guinea, Gabon, Solomon Islands, Canada, Norway, Panama and Brazil.

Water-Poor Countries: Kuwait, Egypt, United Arab Emirates, Malta, Jordan, Saudi Arabia, Singapore, Moldavia, Israel and Oman.

MARINE RESOURCES

The oceans cover nearly three-fourth of earth's surface. Ocean has been the source of many needs of man from the time immemorial. The rapid growth of human population and the advancement of industrialization have exerted great pressure on the resources and the environment of oceans. According to recent data, about 35 km^3 of sewage and 3.5 km^3 of industrial wastes are released annually into the coastal waters of India. In addition, the offshore oil installations have added another stress on the ocean environment. The recent exploration of the oil well in the "Bombay high" invoked a great concern about growing dangers of pollution of the ocean water. The increasing exploitation of certain economically important red and brown algae without any proper conservation and indiscriminate mining of coral stones for industrial purpose, eliminating the substratum for the sea algae to grow are the serious problems which need our immediate attention.

Conservation and Management of Ocean Resources

Land resources are depleting at a tremendous rate and in view of this, humans have started the exploitation of the natural resources from the oceans. Oceans have both biotic and abiotic resources, which may be in the form of renewable and non-renewable resources. Food resources are renewable which are obtained from the plants and animals of the seas whereas most of the mineral resources are non-renewable.

Fig. 15.5. Marine resources (Seaweeds and corals)

Biotic Resources

Biotic resources of the seas include fishes, crustaceans, molluscs, corals, reptiles and mammals. Among plants, seaweeds including marine algae, mangrove forests and some other plants are found that are economically important (Fig. 15.5).

Seaweeds, which were previously considered a nuisance, could offer an alternative source for food because of their high nutritive value. Seaweeds are also used as pharmaceuticals, textile, fodder, fertilizer and biomass. It has been estimated that the total annual yield of seaweeds is about 1 lakh tonne. It has been estimated that the total annual yield of sea weeds in the country is 50,000 tons and it may reach to one lakh tons if all the precautions are taken in utilizing the resources. Planktons (phyto-planktons and zooplanktons) provide rich nutrition. Scientists are now interested in evolving methods of directly converting the plankton into food so that the food supply may be considerably enhanced and utilized for overpopulated world.

Mangrove forests are more common in the sea coasts. These act as guard. The reclamation of saline soil by enriching the coastal soils with organic matter thereby facilitating other plants to grow and the control of soil erosion in coasts are the main contribution of mangrove forests. It is a matter of great concern that these forests are being damaged at an alarming rate which expose sea coasts to the tides and cyclones. An important step in this direction is to replace the slow growing mangroves by fast growing trees along the coasts and thus provide an alternate supply of fuel to the people.

Abiotic Resources

Among abiotic resources, energy in the form of waves, tides, thermal ocean currents; petroleum, oil, gas and mineral substances in various forms are available. Thus, the oceans are a great source of minerals to meet the growing demands.

The harnessing of the ocean resources is of great importance today, as humankind will get alternative sources of food and energy when our land resources will be in short supply. Therefore, it is necessary to have a plan of action based on sustained utilisation but not exploitation. Without proper management, the resources can be threatened and countries can fail to achieve the economic and social benefits. Therefore, considering the importance of marine resources, great care must be taken to preserve them and their environment. The following methods for conservation and management of marine resources should be followed:

1. There is an urgent need to protect the mangrove forests for maintaining the ecological balance in coastal areas.
2. Seaweeds, corals and turtles are very useful components of the marine ecosystem as they maintain a balance in the environment. The interrelationship of such flora and fauna is of great importance in the management of biotic as well as abiotic components such as, certain mineral resources.
3. It is necessary to initiate researches on the various oceanic minerals and subsequently plan for monitoring the mining and extraction on keeping in view the international laws.
4. It is necessary to evolve inter-disciplinary action to encourage and support the collaboration among scientists from the different areas.
5. The weather and climate information should be relayed through media like television, radio, news bulletins, etc.
6. There should be integrated efforts for ensuring effective treatment of the effluents before disposing them into the sea.
7. In the coastal areas and estuaries, mangroves can harbour large quantities of prawn, etc. Hatcheries can be established for prawns in the marshes of mangroves and estuaries for sustained yield.
8. The participation of the student and the coastal population is necessary to make them aware about oceans and ocean resources.

It is a well established fact that the ocean surface of 1 cm thickness known as **surface micro-layer** plays a crucial role in air-sea interaction. Detailed study of this layer is important in meteorology, agriculture and pollution. This micro-layer transfers not only energy and water vapour but also rich nutrients like N.P.K. into the atmosphere. Oil spills and other pollutants over the ocean concentrate in this micro-layer and finally enter the food chain through marine organisms.

Coral Reef

Coral Reef is a ridge or elevated part of a relatively shallow area of the seafloor. It is formed by a rocklike accumulation of calcareous (calcium-containing) skeletons of coral animals, calcareous red algae and molluscs, which form an important marine habitat. Built up layer-by-layer, by the corals growing on top of the skeletons of past generations, coral reefs grow upward at the rate of 1 to 20 cm per year. Coral reefs are one of the most productive and diverse ecosystems on the earth (Fig. 15.6)

Coral reefs, which are the largest, oldest, most diverse and most beautiful communities of plants and animals, have been severely damaged. If human activities are not controlled, it is estimated that 60% of the world coral reefs could collapse in the next twenty years.

Fig. 15.6 Coral reefs with variety of living resources

The corals are the main source of calcium carbonate and are also used in the preparation of lime, cement, calcium carbide, building blocks and construction of roads, etc.

Coral reefs provide habitat to a vast variety of marine plants and animals. About 1 million species are associated with coral reefs. These organisms are the source of many useful chemicals and medicines. Coral reefs have long been the main source of protein for coastal people. With the modern preservation methods, fish and other food organisms from the coral reefs are now eaten by many people.

The excessive exploitation of coral reefs is alarming. It has caused severe destruction of both in the habitat and reef-building organisms. In recent years, attempts are being made to protect some of the coral reefs by declaring the areas as National Parks. Gujarat and Tamil Nadu have established National Parks for protecting the marine ecosystem in the Gulf of Kutch and Mannar respectively.

The task of harnessing the potential of ocean resources for development purposes is of great importance. We should pay our attention to the development of necessary resources of ocean when our land resources are in short supply. Now much awareness is needed by the National government and International agencies including United Nations about conservation of ocean's wealth. While studying the ocean environment, it is necessary to know about the physics, chemistry, biology and geology of the ocean. It is essential that in ocean-oriented programme, lawyers, economists, businessmen, scientists and social workers should be involved in the policy making decisions.

FOOD AND AGRICULTURE RESOURCES

Man needs energy for his various activities as movement, growth, respiration, reproduction and many others. For physical and metabolic activities energy is essential. In our body many biological activities are going on which are collectively called metabolism. These activities need energy which is obtained from food. A food is a substance which is mixture of nutritive and non-nutritive substances. The main components of food are carbohydrates, fats, proteins, minerals, vitamins and water. These are supplied through cereals, pulses, oils, eggs, sweets, milk, meat, cheese and fruits.

Food is classified mainly on the two bases :

1. On the basis of work.
2. On the basis of nutrients.

On the basis of work food has been classified into three groups :

(*i*) **Energy-giving food.** This includes fat and carbohydrates. This food provides high energy which can be obtained from cereals, sugars, animal fats, vegetable oils etc.

(*ii*) **Body-building food.** This includes proteins and minerals. These are found in pulses, milk products such as cheese, fish, meat, eggs etc. These provide energy which is essential for body-building, growth and repairs of tissues.

(*iii*) **Protective food.** This food protects the body from different diseases and regulates different functions of body. This food is rich in minerals, salts and vitamins. This type of food is available from green vegetables, fruits, milk and eggs.

On the basis of nutrients, food may be in the form of carbohydrates, proteins, fat, vitamins and minerals.

Our food contains different nutrients and each nutrient performs its own function. We need proteins, carbohydrates, fat, minerals, vitamins and water in proper proportion. Deficiency or excess of any of these causes disorders in the body. Thus a balanced diet is that which provides proper amount and proper proportion of nutrients. Balanced diet of an average adult man is given below in (Table 15.4).

Table 15.4. Balanced diet for adult man

Food	Weight	Food	Weight
Cereals	400 gm	Milk and milk products	284 gm
Pulses	85 gm	Flesh (meat)	125 gm
Other vegetables	85 gm	Sugar	57 gm
Root vegetables	85 gm	Fat (Ghee, oil)	57 gm
Fruits	85 gm		

A healthy youngman who does physical work needs 3,000 calories daily. Such diet should contain 90 gm protein, 90 gm fat, 450 gm carbohydrate and sufficient amount of vitamins, minerals and water. Nutrient value of different foods is given below (Table 15.5).

Table 15.5. Nutrient value in different sources of food

	Food	Protein	Fat	Carbohydrate	Salt	Water
1.	Milk	3.5	3.7	4.9	0.7	87.0
2.	Meat	15.1	14.7	—	0.8	69.4
3.	Fish	16.0	5.0	—	1.0	78.0
4.	Wheat	11.4	1.0	75.0	0.5	12.0
5.	Rice	1.7	0.4	76.0	0.4	15.5
6.	Maize	8.4	4.7	72.0	1.3	13.6
7.	Potato	2.0	0.1	21.0	1.0	75.9
8.	Sugar	—	—	100	—	—
9.	Egg	13.2	10.3	—	0.9	75.6

Today food demand is increasing rapidly due to overpopulation and increased consumption of food. Effort is being made to increase food production all over the world. In many parts of the world a number of ecologically unsound practices are being applied that have resulted in short-term production gains at the expense of long-term food output. Overgrazing, burning of vegetation, deforestation, and expansion of agriculture are some known examples.

Most of us get food mainly through agriculture. Cultivation of crop plants is done by two methods :

1. **Shifting cultivation.** This is very old agricultural practice. In this case natural ecosystem is manipulated and cultivation is done in the limited area for short period. In this system cultivated area is temporarily covered by wild vegetation. This type of cultivation is generally practised in the mountain area where the rainfall is heavy. Most of the tribal populations depend primarily upon shifting cultivation.

There are many defects of shifting cultivation. Natural ecosystem is disturbed, yield is poor due to uncared cultivation and crops are infected by many diseases.

The following control measures have been taken regarding shifting cultivation :

(*i*) Proper management of settlement of tribal population should be made because migration of tribal people needs shifting cultivation and facilities of farming should be provided to them.

(*ii*) Now agriculture should be clubbed together with silviculture so that land use involves an integrated programme of agriculture, horticulture and forestry.

(*iii*) Annual crops should be replaced with perennial crops.

(*iv*) Rotation of crops with perennial crops and then with forest trees has given better results in the sloppy area of hills.

2. **Sedentary cultivation.** In this case natural ecosystem is permanently changed. In this system previous plants are removed and then domesticated plants are grown. In dry places soil is gradually depleted with minerals and nutrients which may be restored by adding fertilizers or green manures.

Deforestation, overgrazing, fire and deserts are the results of mismanagement of agriculture. Soil erosion and lowering of water table are also the results of climatic and geological factors.

Extension and intensification of agriculture are the important steps to increase the yield which have solved the problem of shortage of food and malnutrition. For extension we need better irrigation, improved varieties of crops, agricultural instruments and better roads. Flood control, proper drainage, erosion control, mechanisation, fertilizers, pesticides and improved varieties of crops are the important steps to increase yield.

Besides these methods, we should select some alternative sources of food. Deficiency of proteins should be compensated from animal source rather than plants. Fishes and eggs are the important sources of protein. Now-a-days everywhere fisheries departments have been established to meet the high demand of fishes. It also gives enough foreign exchange. Poultry breeding programme is very significant now-a-days because the consumption of eggs is increasing tremendously day by day. Under poultry, chickens and ducks are included. Some algae, mushrooms, grasses and some aquatics are also rich in protein contents which are consumed by man.

There is a tremendous loss in yield due to plant diseases which are caused by bacteria, viruses, mycoplasma and pests. It has been estimated that there is 15–20 per cent loss in yield due to plant diseases alone. Sometimes sudden outbreak of a disease can cause severe loss in the production. Plant diseases are the main problems and barriers to crop production. The damage caused by pests is tremendous and sometimes it is irreparable.

ENERGY RESOURCES

The energy crisis is a global problem today. The survival of the man will be difficult if the energy problem is not solved on the priority basis. Many scientific organisations as Department of Science and Technology, Department of forestry and environment, Department of Non-conventional energy, Department of Oceanography, Fuel Research Institute and several other regional, National and International Institutes are engaged in researches on this current problem.

There are several energy sources in the world such as coal, oil (petroleum and its products), natural gas, electricity, nuclear energy, solar energy, thermal and wind energy. Nuclear energy is supposed to be the latest source of energy. Now efforts are being made to harness energy from solar radiations, nuclear fission, wind, thermal power, tides of the ocean and biogas etc. Fossil fuel reserves are limited and it is estimated that the stock of the mineral oil will lost in 21st century if it continues to be used at present rate. Thus, there is a need to exploit various sources of energy.

The energy consumption is maximum in developed countries as compared to developing nations. The per capita daily consumption is highest in America and then come Canada, U.K., Germany, Soviet Russia, Japan, Mexico, Brazil and India in decreasing order. United States of America alone consumes about 35 per cent of the world total energy. Soviet Russia consumes about 16 per cent and India consumes only two per cent of the total available energy of the world. The following Table 15.6 presents a comparative idea of energy consumption by some countries :

Table 15.6. Energy consumption in some countries
(Data based on Scientific American, 1971 and UN Demographic yearbook, 1974)

Name of country	World energy consumption in per cent	Per capita daily consumption
USA	35	230
Canada	20	165
UK	26	145
Soviet Russia	16	85
Japan	3	40
India	2	6

In the world coal reserves are much bigger and will lost longer than the petroleum reserves. The main coal producing countries are China, Russia, USA, UK, India, Germany, Japan, Polland, Czechoslovakia and France. The oil producing countries are Saudi Arabia,

Iran, Iraq, Nigeria, Libya, Arab Republic, Indonesia, USA and Russia. In India new resources of oil have been explored recently in the sea bed of Mumbai high and Gujarat.

According to latest information India has coal reserve of about 1,31,000 million tons which is sufficient for coming 1000 years if the current rate of production and consumption continues. The annual crude oil production in India is 8 million tons as against its requirement of 20 million tons which is much less than its total requirement. The annual oil consumption which was 33 million tons by 1980-1981 is increasing at fast rate and has now become more than double. The oil and Natural Gas Commission in India is exploring new oil and gas fields.

Agriculture, industry and automobiles are the main sectors where the consumption of oil is maximum. The tremendous increase in population would need more food and that need can be fulfilled by increased oil production. The cost of petroleum is continuously increasing in the world due to which India is greatly affected. Therefore, some suitable alternative sources of energy will have to be developed to replace petroleum. Now efforts are being made to find out new sources of fertilizers which are not petroleum based. Among them, biofertilizers, genetic engineering techniques, sewage and some nitrogen fixing blue-green algae have been found successful. Nitrogen fixing blue-green algae are used as biofertilizers in the paddy fields without any harmful effects on our biosphere. Serious efforts are being made now to produce natural gas for fuel purpose. The object of the process is to produce synthetic methane from coal. Burning of coal releases CO_2 gas into the atmosphere and causes environmental pollution.

DIFFERENT SOURCES OF ENERGY

ENERGY FROM PLANTS

Green plants by virtue of possessing special pigments are able to absorb and use a portion of solar energy for the synthesis of food. The process is called photosynthesis. By this, organic substances are synthesized from carbon dioxide and water using the light energy absorbed by the special pigment chlorophyll. In this process solar energy is converted into chemical energy or food energy and molecular oxygen is produced as a by-product. All other organisms including man directly or indirectly depend upon the energy accumulated in green plants in the form of organic food.

The total solar energy reaching the earth is 3×10^{24} J per year. Out of this, green plants utilize about 0.1 per cent resulting to net annual production of 2×10^{11} tonnes of organic matter utilizing 3×10^{12} J energy. The total solar energy received in India is 6×10^{13} MWH, with 300 days of average Sunshine per year. Thus, there is a vast scope for harvesting solar energy through improved photosynthetic efficiency. The increase in photosynthetic efficiency will lead to an increase in food, biomass and energy. The fossil fuels (coal, oil and natural gases) that we use today resulted when ancient wide spread dense forests were fossilized and converted into major coal beds during the carboniferous period about 300-350 million years ago. Thus photosynthesis is a vital process in the living world of this planet. A detailed study of the biophysical, and biochemical aspects of this process is urgently needed for improving the photosynthetic efficiency and productivity to meet the future food and energy requirements. Photosynthetic efficiency is much greater in algae and bacteria. It is 16% in algae and bacteria as compared to 1-2% in land plants.

Energy Plantations

Fuel wood has been the primary energy source for mankind from the beginning of civilization and still continues to be the main source of energy in the developing countries. Plant based energy is obtained by energy plantations which can produce biomass from selected trees in the shortest possible time at a low cost. These plants yield solid, liquid or gaseous fuel through burning, gasification, digestion etc. Such plantations can be raised in both hills and plains particularly on marginal land. It has been calculated that about 4 tonnes of dry matter can be obtained in 0.25 hectare land which is sufficient for an average family. According to Khoshoo, 1983 around 1.86 million square km of marginal or stripland is available in India for such plantation which would yield around 1,400 million trees. Water hyacinth (Jalkumbhi) when grown in water enriched with sewage, may produce 8–10 tons of plant material/acre which may yield 3500–7000 cubic metre of methane.

Plants suitable for Energy plantation

Tree species

Ailanthus excelsa
Anogeisus latifolia
Millingtonia hortensis
Trema orientalis
Givotia rottleriformis
Duabanga sonnertides
Erythrina indica
Cedrela toona
Casuarina equisetifolia
Dipterocarpus indicus
Eucalyptus globulifera
Ficus bengalensis, F. religiosa, F. elastica.
Ficus glomerata
Aleurites moluccana
Butea monosperma
Moringa tinctoria
Stereospermum angustifolium
Aguilaria agallacha
Odine odier
Abies pindrow
Grevelia robusta
Dendrocalamus strictus

Shrubs

Jatropha glandulifera
Clerodendron enerme
Altantia monophylla
Memecylon edule
Kydia calyxeina
Ipomoea carnea
Tecoma gracillis
Dodonaea viscosa
Macaranga peltola
Grewia latifolia

Herbs

Phaseolus trilobatus
Psophocarpus tetragonoloba
Sesbania aegyptiaca
Cymopsis tetragonoloba
Crotalaria juncea
Glyricidia maculata
Ipomoea batatas
Butea vulgaris
Bambusa arundinacea
Cucumis melo
Alternanthera sessilis
Cymbopogon martini
Amaranthus paniculatus

Aquatic plants

Eichhornia crassipes
Juncus indicus
Typha latifolia
Polygonum glabrum
Lemna, Wolffea, Pistia
Salvinia, Hydrilla.

Now-a-days extensive cultivation of alga *Spirulina platensis* is being undertaken in India. This alga contains proteins, carbohydrates and lipids in varying amounts and is a rich source of hydrocarbons. It has been calculated that a 200 square miles of algal farm could provide enough petroleum substitute (30,000 barrels/day) (Khoshoo, 1983). *Spirulina platensis* contains 60% crude protein rich in Tryptophan, Lysine and Vitamin B12. Recently, the NBRI, Lucknow has started its culture on mass scale in sewage. Dabur Research Foundation and some other pharmaceutical companies in India prepared *SPIRULINA* health capsule which is said to be richest source of iron, beta-carotin, vit. B12, glutamic acid and a variety of other essential aminoacids. It cures heart ailments, stress and builds fitness. The use of blue-green alga as a self-generating source of nitrogen in rice field is well known.

Now-a-days much emphasis is being laid on the cultivation of sea weeds. Many varieties of sea weeds are used for various purposes such as food, pharmaceuticals, textiles, fodder, fertilizers and biogas. It is estimated that the total annual yield of sea-weed is about 50,000 tonnes and if all resources are explored it can produce tremendous energy.

Some of the important plants for energy plantation which are extensively grown in India are *Leucaena leucocephala, Casuarina,* Poplar tree and *Eucalyptus.* Recently an exotic plant *Palonia* has been introduced from China for agroforestry purpose. This tree species has good adaptability in Indian soils and is fast growing one. *Leucaena leucocephala* possesses all the qualities as wide adaptability, high genetic variability, hardyness, resistance to diseases and pests, adaptability to different edaphic conditions, requiring only marginal inputs of fertilizers, high regeneration potential and easy propagation, wood with high caloric value and ability of wood to burn without sparks and smoke, nitrogen fixing ability by virtue of root nodules, fast growth and high yield of biomass per unit area. It is used as a source of fuel, fodder and fibre.

Rice bran oil is a rich source of edible oil. In Japan bran oil is used in cooking. Paddy contains about 25 kg of oil/tonnes. In India the annual production of rice bran oil is about 80,000 tonnes.

The plantation of energy plants would result in an inexhaustible source of renewable energy. The agro-industrial products, their by-products and their residues can play a very important role in the production of energy. Biomass from corn, sorghum, sugar cane, cassava also provide base material for production of alcohol and gases. The biomass available from oil seeds, fruits and vegetable wastes, cowdung, waste from the sea weeds, rice husk and a variety of other waste organic matter can serve as a suitable medium for the production of gases.

Biomass obtained from biotic sources is an important source of energy. Biomass collected in the form of wood, roots, foliage, algae etc. from agricultural land or forest can be converted into energy source either in gaseous, liquid or solid forms through diverse ways. Biomass from natural forests and energy plantations will serve as dependable and renewable energy sources in future. For increasing forest biomass fast growing tree species are planted, harvested and regenerated on short-term rotation of 2-4 years.

The fast growing trees as for example, *Eucalyptus, Zizyphus, Prosopis, Popler, Albizzia, Acacia* etc. are not only important from silviculture point of view but are known to reclaim the waste land and check soil erosion as well. The reclamation of waste land or usar land can be coupled with biomass production. *Acacia, Terminalia, Prosopis, Ficus* etc. can be raised on saline and alkaline soils.

Among the various non-conventional sources of energy, forest biomass plays a significant role in solving the fuel wood crisis. Recent data reveal that one seventh of world's fuel need is met by forest biomass. This figure is equivalent to 20 million barrels of oil per day.

Many countries, for example Bangladesh, China, France, Germany, India, Indonesia, Japan, Pakistan, Sweden, UK, USA are actively engaged in research in the field of bioenergy. Several International agencies as Food and Agricultural Organization (FAO), International Development Research Centre (IDRC), United Nations Educational, Scientific and Cultural Organizations (UNESCO), United Nations Environmental Programmes (UNEP), United Nations Industrial Development Organization (UNIDO) and World Health Organization (WHO) are also engaged in Bioenergy research and development programmes.

Importance of Fuel Wood Plantations

Plants absorb and convert solar energy into chemical or food energy through photosynthesis which is stored in their body. When the parts of the plants are burnt that stored energy is released in the form of heat. When plants are grown for producing fuels for cooking and various other purposes, the process is called fuel wood plantation. For fuel wood plantation such tree species are selected as are fast growing. The advantages of fuel wood plantations are as follows :

1. The fuel wood would save enormous quantity of cowdung which is applied as natural manure or farmyard manure (FYM) or compost.
2. Fuel wood plantation would provide an inexhaustible and renewable source of energy.
3. Use of fuel wood produces non-hazardous mineral-rich ash which can be used as fertilizer.
4. It causes little or no pollution.
5. Plants create pleasant and beautiful surrounding and provide safe habitat for wildlife.
6. There is little or no energy input in fuel wood plantation.
7. Fuel plants also provide us fruits, seeds, edible leaves, forage, tannins, gums, dyes and several other useful substances.
8. Soil erosion is checked through plantation.
9. There is no problem of waste disposal.

PETROPLANTS (HYDROCARBON PLANTS)

Several plants are known to yield liquid hydrocarbons, the substitute for liquid fuels. Such plants are called Petroplants. The hydrocarbons present in such plants can be converted into petroleum hydrocarbons of high molecular weight (10,000). More than 385 plant species belonging to Euphorbiaceae, Asclepiadaceae, Apocyanaceae, Convolvulaceae, Sapotaceae and some other families have been screened for hydrocarbon contents. Now, the efforts are being made to increase the biomass of such plants and convert their hydrocarbons into petroleum products. Indian Institute of Petroleum, Dehradun and National Botanical Research Institute Lucknow have undertaken research programme in this area.

A large number of species belonging to Euphorbiaceae secrete latex containing 30% hydrocarbon. *Hevea brasiliensis* (rubber tree) is important source of hyrdocarbon. These plants may be utilized either to get diesel fuel or high quality liquid fuel. *Euphorbia abyssinica, E. resinifera, E. lathyris* produce good amount of biomass. *E. hirta, E. splendens, E. pulcherima* also have enough fuel potential. There are few trees such as, *Eucalyptus, Agathis dammara, Canarium ovatum* which produce cheaper fuel. *Jatropha curas* tree yields 2 kg of seed oil per plant/year. The oil obtained from this plant is expected to replace the conventional diesel fuel (Khoshoo, 1983). Some other common plants producing hydrocarbons are listed below.

Some Plants producing Hydrocarbons or Petroplants

Thevetia nerifolia
Calotropis procera
Vitis quadranguris
Sansevieria
Aloe vera
Euphorbia nerifolia
Jatropha curas
Wrightia tomentosa
Plumeria alba
Allamonda cathartica
Calotropis gigantia
Cryptostegia grandiflora
Artocarpus integrifolia
Bassia latifolia
Argemone mexicana
Givorta rotalariformis
Ceropegia tuberosa
Pedalium marex
Agave americana
Euphorbia hirta
E. royleana
Tabernaemontana cornaria
Plumeria acutifolia
Nerium odorum
Asclepias curassavica
Stapelia grandiflora
Mimusops elengi

Besides higher plants, some algae also yield hydrocarbons. *Botryococcus braunii,* an unicellular green alga contains about 70% hydrocarbons. The algal hydrocarbons is very similar to crude oil. This alga can thus be exploited for production of hydrocarbons. For large scale production of hydrocarbons it is therefore necessary to increase the algal biomass. Hydrocarbons are recovered from the algal cells by centrifugation. *Chlorella pyrenoidosa* is another fresh water alga which can yield hydrocarbons. When this alga is subjected to hydrogenation for about an hour at high temperature 50 per cent of algal biomass is converted into oil with a little amount of ammonium carbonate.

OTHER RESOURCES OF ENERGY

Energy consumption and its qualitative trends characterise the life style of a country. The highly industrialised countries depend upon different sources of energy such as coal, oil, natural gas, hydroelectric power and fission fuels for their developmental activities. In developing countries, on the other hand, there is more dependance on non-conventional sources of energy like fire wood, animal wastes and agricultural wastes rather than the fossil and fission fuels. Energy is consumed in different sectors as industry, transport and agriculture. The following table 15.7 gives the details of total energy consumption in different sectors in India:

Table 15.7. Energy consumption in different sectors in India
(According to Fuel Policy Committee Report, 1974)

Sector	Energy consumption in India (Kilo calories × 10^{12})	Percentage of energy used
Industry	345	22.4
Transportation	160	10.4
Govt. Commercial	35	2.3
Agriculture	30	2.0
Urban domestic	83	5.3
Rural domestic	890	57.4

Various sources of energy, *i.e.*, coal, hydroelectric, thermal power, nuclear energy, wind energy, biogases etc., are discussed here as under :

Coal. India has vast coal reserves which may be sufficient for coming 1000 years. Coal is used as a fuel in railways, thermal power plants, domestic cooking and many industries. Nevertheless, the use of coal often causes environmental pollution. The emission of smoke, ash and other particulate matter (fly ash) from coal burning poses a serious threat to human health.

Nuclear energy. The generation of electricity from radioactive uranium and plutonium fuels by nuclear power plants has solved energy problem to some extent but at the same time it is creating several problems. Many harmful gases are released by atomic power plants and radioactive wastes are posing a serious threat to the environment. In spite of development of nuclear waste management technology the atomic power plants are not as safe as claimed by atomic scientists. Even slight leakage of radioactivity in the coolant water will tend to accumulate and amplify in plants, animals and human beings and cause mutations, body deformity, many physiological disorders and even death. It will be fatal to human beings if misused. Proper use of x-rays, γ-rays may be beneficial to human and agriculture, but their misuse proved to be deleterious. Recent Pokhran nuclear explosion in 1998 is well known. A small quantity of radioactive material can produce an enormous amount of energy. For example, one tons of Uranium^{235} (U^{235}) produces energy equal to that produced by three million tons of coal or 12 million barrels of oil. Presently there are 300 atomic power plants in the world. The principal atomic power producing countries are USA, Soviet Russia, UK, France, Japan, Germany, China and India. India ranks 7th among the atomic power producing countries of the world.

WIND ENERGY

Air in motion is called wind. Wind is created when air that has been warmed over sun-heated land rises, leaving a vacuum in the space it once occupied. Cooler surrounding air then rushes in to fill the vacuum. This movement of rushing air is then known as wind.

Wind Energy is the kinetic energy associated with the movement of large masses of air in the shape of winds blowing across the earth's surface. Ancient civilisations first used wind energy when they sailed boats in the rivers and oceans. The ancient Persians were the first to use wind energy on land by developing the first windmill, sometime in the 7th century. By the 12th and 13th centuries, windmills were used on a large scale in many European countries. Windmills were mostly used to mill grain, pump water, saw timber and provide other forms of mechanical energy. Since the late 20th century, wind energy is being used to generate electric power.

India has an estimated wind power potential of 20,000 MW. About 100 sites having a potential of more than 4500 MW have been taken up for wind power generation. These are located mostly in Tamil Nadu, Karnataka, Andhra Pradesh, Gujarat, Kerala, Madhya Pradesh and Lakshadweep. The country's largest wind farm is located near Kanyakumari.

Harnessing of Wind Energy

Modern wind energy systems consist of the following three basic components:

- Tower on which the wind turbine is mounted.
- Rotor that is turned by the wind.
- *Nacelle,* which houses the equipment, including the generator that converts the mechanical energy in the spinning rotor into electricity.

Fig. 15.7. Wind Farm

By connecting a spinning rotor (an assembly of blades attached to a hub) to an electric generator, modern wind turbines convert wind energy turning the rotor into electrical energy. Small wind turbines are capable of generating between 50 watts and 60 kilowatts of power, and use rotors ranging in diameter from less than 1 m to 15 m. Large wind turbines are huge structures with rotors having diameters between 60 m and 100 m and are capable of generating 2 MW to 3 MW of power. The height of poles raises between 10 m to more than 100 m. Wind turbines can be installed in single units, in clusters of two to ten turbines and in large groups, called **wind power plants or wind farms.**

Biogas. Biogas production technology offers a low cost alternative source of energy. Lacs of gobar gas plants installed in rural areas are the important and cheapest source of energy without any kind of pollution. The enormous quantity of animal dung (gobar) available in India may produce about 22,425 million m^3 gobar gas. Animal dung is collected in a concrete tank or steel chamber. The chamber is covered by a metallic lid to make it air tight. In the absence of air anaerobic bacteria, decompose the dung and produce biogas or gobar gas (Fig. 15.8). Gobar gas is used mainly in cooking and lighting. Besides gas, waste (slurry) of gobar gas plants can yield about 206 million tonnes of organic fertilizer every year. Biogas is a mixture of several gases such as methane, CO_2, N_2, O_2 as given in Table 15.8.

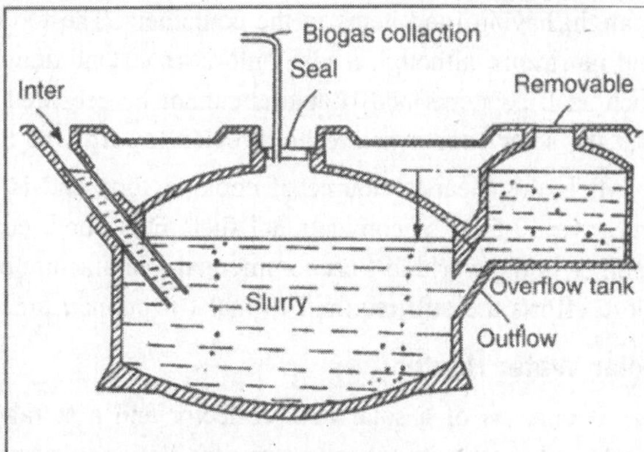

Fig. 15.8. Biogas plant

Table 15.8. Composition of biogas

Name of gas	Percentage
CH_4	50 – 68
CO_2	25 – 36
N_2	2 – 7
H_2	1 – 5
O_2	0 – 0.1
H_2S	rare

SOLAR ENERGY

Sun is the ultimate source of energy on this earth. The amount of solar energy reaching the earth is tremendous. According to an estimate, the amount of energy received from the Sun each day is 600 times greater than the amount of total energy produced each day by all other combined energy sources. For example, the Sun provides us energy equivalent to the burning of 120 trillion tonnes of coals every year which is twenty times more than the world's total reserves.

The energy released from the Sun is due to the process of **thermo-nuclear fusion** continuously taking place in the sun. As the Sun will continue to shine for billions of years, it will continue to provide us with this abundant form of energy. If we can harness even some percentage of solar energy, it will go a long way in solving most of our energy requirements.

Solar Cooker

Solar cooker is a kind of insulated box with a black polished base. There is a big mirror fitted at some angle to reflect the sunrays at a particular point to give more heat for cooking (Fig. 15.9). The solar cooker is kept in bright

Fig. 15.9. Solar cooker

sunlight having food items in the containers. The food prepared by the solar cooler is very tasty and nutritious, although it takes more time. One drawback is that foodstuffs requiring high heat such as fried items and chapattis cannot be prepared in the solar cooker. Another drawback is that the solar cooker has to be directed towards the Sun after regular intervals of time.

It is the cheapest source of cooking food and is a boon for the rural areas. These cookers can save 30-50% of conventional fuels like wood, coal, LPG, kerosene, etc. India was the first country in the world to start commercial manufacture of solar cookers in the year 1982. However, more efforts are still required to make it popular among the masses.

Solar Water Heater

It consists of a solar heat collector and a storage tank. The heat collector comprises of a blackened sheet with tubes to carry the flowing water, front glass cover and insulated outer base. The blackened sheet of the heat collector gets heated when sunlight falls over it through the glass cover. It can raise the temperature of water upto 90°. The collector is connected to a storage tank from which water is drawn for use. Such water heaters are used in domestic buildings, hotels and industries Fig. 15.10.

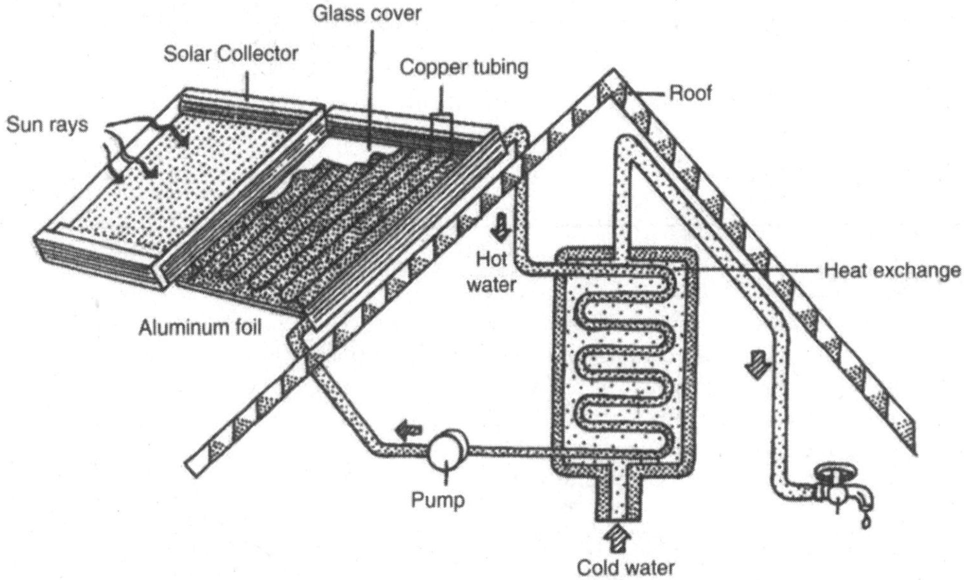

Fig. 15.10 Solar water heater

Solar Desalinisation Unit

It is similar to solar water heater but here the water is kept stagnant. Water is kept in a blackened metal box with a sloping transparent roof, which is exposed to solar radiation. As the water gets heated, some of it gets evaporated and the vapours condense into pure water drops. This can be used to purify salty water for drinking purposes and also to provide distilled water for industrial uses.

Solar Air Heater

It is similar to solar water heater but instead of water, air is heated. It is used for drying food grains, vegetables, etc. and also for industrial purposes.

Solar furnace

A solar furnace consists of thousands of small mirrors, arranged in such a manner so as to concentrate the solar energy. The largest of these is located at Odeillo in France, which uses 9600 reflectors with a total area of approximately 1860 sq m (about 20,000 sq ft) to produce temperatures as high as 4000°C. Such furnaces are ideal for research that require high temperatures and contaminant-free environments for example, materials research.

Solar power plant

The principle employed is similar to that of a solar furnace. Here, the high temperature generated, is used for boiling water to produce steam. The steam is then used to drive a turbine for the generation of electricity. Such a power plant has been established at Gurgaon, with a capacity of 50 KW (Fig. 15.11).

Fig. 15.11

Conversion of Solar energy into electricity

In solar power plants, the radiant energy of the Sun is first converted into heat and thereafter into mechanical energy and electricity. However, in such cases, the efficiency becomes low due to energy losses at each stage. There are certain materials that can convert the solar energy directly into electricity. It is done through **solar or photovoltaic cells.**

Solar or Photovoltaic cell: Solar cells are mainly made from thin slices of silicon and gallium or other semi-conductor materials, which convert solar radiation directly into electricity. In the photovoltaic principle, light excites electrons to move from one layer to another through semi-conductive silicon materials. This produces an electric current. A number of solar cells are connected together to form **solar modules** that can generate sufficient amount of power. Current use of solar cells is limited to remote areas that are not connected by direct power lines and devices such as satellites and spacecrafts.

Many villages in India have been provided street lighting through such solar cells. India has great potential as far as solar energy is concerned, because most of the areas of the country remain hot and sunny for a long period of time.

HYDEL ENERGY

Mankind has used the power of flowing water since long for a number of purposes, such as to grind grains, saw wood, etc. However, today hydel energy refers to the generation of electricity from the energy of flowing water. The flowing water supplies energy to turn the turbines and generate electricity. Since water is used as an energy source, it is also known as **hydroelectricity.** Hydroelectric power plants are commonly operated by constructing artificial reservoirs. The reservoirs are built by constructing **dams** across the river, thus impounding the water. The impounded water is the potential energy source. However, in some mountainous regions, where the streams have steep gradients, hydroelectricity is generated without the construction of reservoirs (Fig. 15.12).

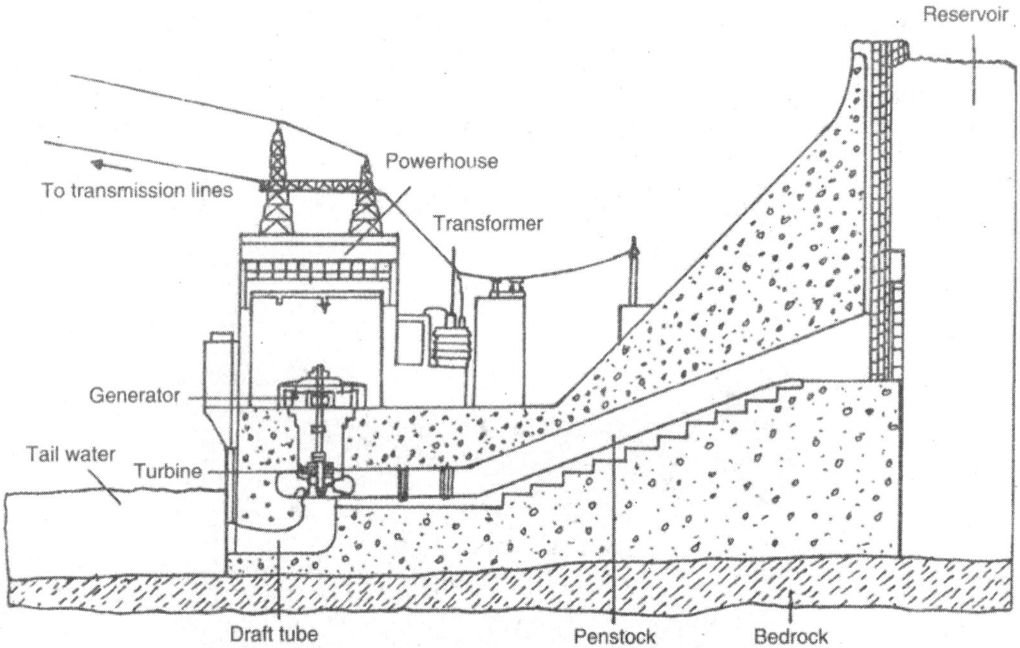

Fig. 15.12 Generation of hydroelectric power

In India, about 25% of the total energy requirements are met through hydroelectricity. India has tremendous potential for generating hydel power, as the country possesses a vast network of rivers throughout its length and breadth. Moreover, the mountainous regions of the country can provide us with enough hydel power. The total hydroelectric potential of India is estimated to be around 150,000 MW, of which less than one-sixth has been exploited so far. The important hydroelectric power plants are situated in Punjab, Andhra Pradesh, Karnataka, Kerala and Orissa. Major hydroelectric projects in the form of Narmada Dam and Tehri Dam are being developed in the states of Madhya Pradesh and Uttaranchal respectively.

Harnessing of Hydel Energy

Hydroelectric dams harness the energy of water released from the reservoir to turn hydraulic turbines. The turbines convert the energy of the falling water into mechanical energy, which is used to generate electricity. Dams designed to generate electricity deliver water to a building, called a **powerhouse,** which contains highly specialized power-generating equipments. Large pipes carry water from the reservoir down into the powerhouse. Water exits through small openings, which concentrate the flow and direct it onto the blades of a large hydraulic turbine. The force exerted by the falling water rotates the blades and drives the shaft of an electric generator. The shaft spins giant magnets in the generator, which create an electric current. Power lines then transmit the current to the consumers.

OCEAN ENERGY

Another alternative source of energy is from the **rise and fall of tides in the ocean** and hence it is also called **tidal energy.** Tides are caused due to the gravitational pull of the moon and due to the Earth's rotation. The tidal movement of water has a lot of energy, which is now being looked into for producing electricity. Such plants have been installed at the Bay of Fundy, Canada and La Rance in France. In India, three potential sites have been identified, viz. the Gulf of Kutch, the Gulf of Cambay, both in Gujarat; and Sundarbans in West Bengal.

HARNESSING OF OCEAN ENERGY

The principle employed for harnessing tidal energy is the same as that in a hydroelectric plant. A minimum range of 8 m in the height of tides is necessary for harnessing tidal power. To harness tidal power, dams are built across the entrance to a bay or estuary, thus creating a reservoir. As the tide rises, water is initially prevented from entering the landward side. During high tide, the water is allowed to flow through the dam into the reservoirs. This water is allowed to run out through a turbine that generates electricity. During low tide, the water, stored in the reservoir, is allowed to flow back into the sea, which is now at a lower level; thus again moving the turbines.

Ocean Thermal Energy Conversion (OTEC)

The surface temperature of ocean water in the tropical region is around 28°C. However, at a depth of around 600 m, the temperature of water falls to 1°C. This difference in temperature between the surface and deeper parts of the ocean can be used to generate electricity. Equipments have been designed that bring the surface warm water and the cold water through pipes. The warm water is used to evaporate liquid ammonia into vapour. The vapours come out at high pressure, which moves the turbines to produce electricity. The ammonia vapour is again condensed back into its liquid form by passing it through cold water, thus completing the cycle. Although, theoretically, there exists a great potential in OTEC, it remains to be seen whether it is commercially feasible or not.

GEOTHERMAL ENERGY

Geothermal implies the 'heat of the Earth'. Geothermal energy is manifested on the Earth surface in the form of volcanoes, geysers and hot water springs. The temperature in the interior of the Earth is very high and all the rocks are in the molten state, called magma. If the magma reaches the surface, it forms volcanoes, but most of the molten rock stays underground, creating huge subsurface regions of hot rocks.

In certain areas, water seeping down through cracks and fissures in the crust, comes in contact with this hot rock and is heated to high temperatures. Some of this heated water circulates back to the surface and appears as hot springs and geysers. However, the rising hot water may also remain underground, forming geothermal

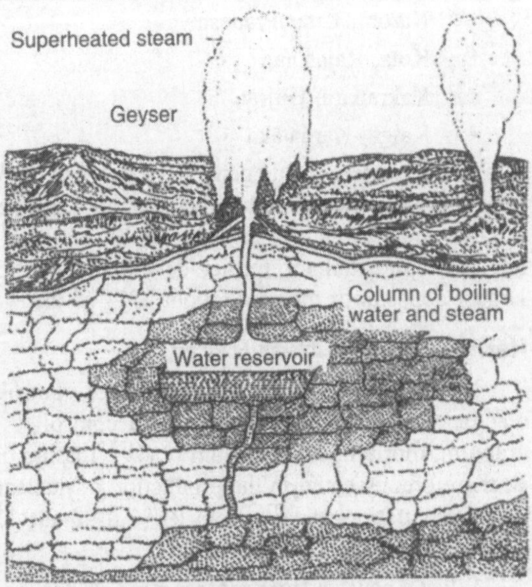

Fig. 15.13 Geothermal energy

reservoirs. Geothermal reservoirs, which may reach temperatures of more than 350° C, can provide a powerful source of energy.

The natural steam and hot water obtained from geothermal reservoirs are exploited for various purposes including power generation. Italy, New Zealand, Russia, Japan, Iceland and USA are the main countries exploiting this resource. In India, geothermal energy in the form of hot springs occur widely especially in the Himalayan region, such as Manikaran, Badrinath, Kullu, Puga Valley (Ladakh), etc. Thus, geothermal energy forms a good resource for future use. Presently, a geothermal power plant is being established at Tatapani village, Ambikapur district, Chhattisgarh.

Harnessing of Geothermal Energy

Geothermal reservoirs, which are shallow (about 5 km deep), can be reached by drilling a well. The hot water or steam from these wells can be used to turn turbine generators to produce electricity. A power plant that uses this natural source of hot water or steam is called a geothermal power plant. In addition to generating electricity, geothermal water is used directly in spas to heat greenhouses, for industrial processes and for space heating in homes and other buildings. People in over 35 countries have developed geothermal water for such purposes (Fig. 15.13).

NUCLEAR ENERGY

Nuclear energy is the energy stored in the atomic nucleus. Two nuclear processes can be used to release this energy that binds the nucleus: Nuclear fission and nuclear fusion. **Nuclear fission** is the splitting of the atom's nucleus into smaller fragments and **nuclear fusion** is the combining of atomic nuclei. In both the processes, a tremendous amount of energy are released. To produce energy, controlled nuclear fission reactions are undertaken in the **nuclear reactors.** Producing energy through nuclear fusion has not yet been commercially undertaken; as such reactions require a temperatures of approximately 100 million degrees Celsius. The radiant energy of the Sun is because of the continuous fusion reactions taking place in the Sun.

India is also generating nuclear energy through six nuclear power plants, which are:

- Tarapur, Maharastra
- Kalpakkam, Tamil Nadu
- Narora, Uttar Pradesh
- Kota, Rajasthan
- Kakrapara, Gujrat
- Kaiga, Karnataka

These nuclear plants generate about 2500 MW electricity per year, which is about 2% of the total energy produced. In comparison, nuclear energy contributes 20% of the energy requirements in USA and 75% in France. In India, nuclear energy is obtained from uranium and thorium. India has great potential for further development of the nuclear energy sector as it has the world's largest deposits of thorium (about 50% of the world).

Harnessing of Nuclear Energy

Nuclear energy is harnessed through controlled nuclear fission in **nuclear reactors.** In the reactor, neutrons are used to cause a controlled fission of certain radioactive elements such as uranium, thorium, etc. Uranium^{235} (U^{235}) is the only naturally occurring fissionable material and is, therefore, essential in the production of nuclear energy. The processing of uranium to increase the concentration of U^{235} from 0.7% to about 3% is called **enrichment.** This enriched U^{235} is used as fuel in the fission reaction.

When a neutron from another atom strikes the nucleus of U^{235} atom, the nucleus of U^{235} splits. This releases a huge amount of energy, fission fragments (barium and krypton) and more neutrons.

The neutrons that are released, strike the nuclei of other atoms of U^{235} and also cause them to undergo fission. This, in turn, releases more energy and more neutrons in a **chain reaction.** The reactors contain the **fuel rod** containing uranium and **moderators** that slow down the speed of neutrons. Fast moving neutrons are less effective at splitting atoms than slow moving neutrons. Water is commonly used as a moderator (Fig. 15.14).

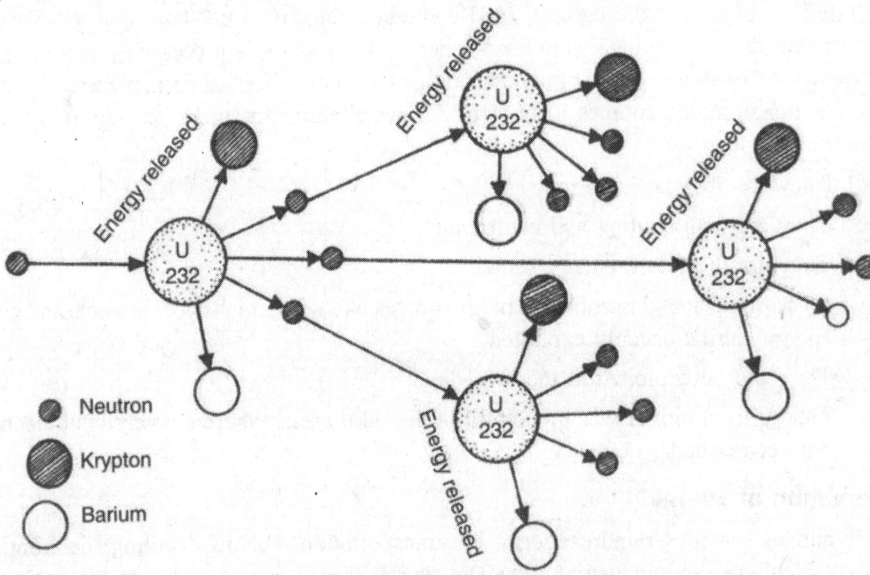

Fig. 15.14 Nuclear fission chain reaction

In the production of electricity, the heat produced in a nuclear reactor is used to convert water into steam as in the case of thermal power plants. The steam moves the turbines and generates electricity. After passing through the turbine, the steam is condensed and is returned to the reactor to be heated again (Fig. 15.15).

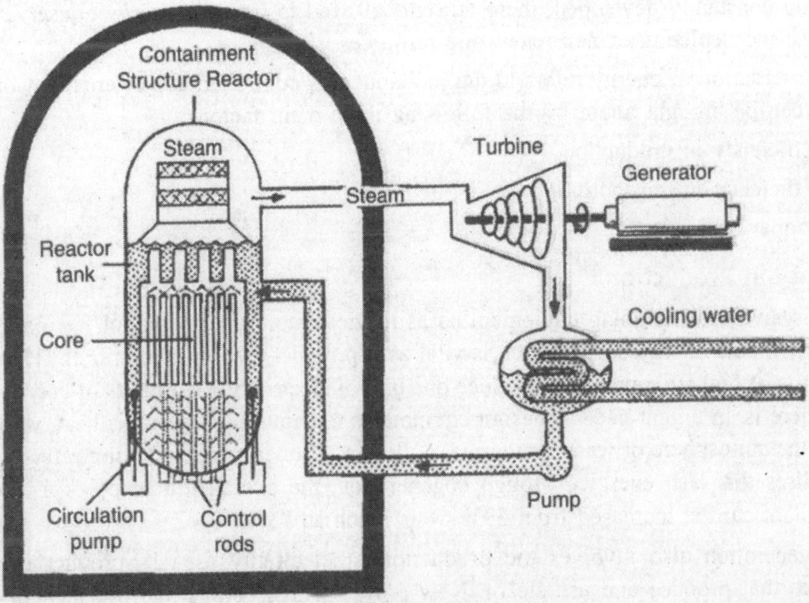

Fig. 15.15 Functioning of nuclear reactor

Need to Promote Non-Conventional Energy Sources

The power hungry world has been clamouring for more and more energy for industrial and domestic purposes. The energy requirements of the world are bound to increase further, as population continues to explode and the developing countries make further progress in industry and lifestyle. This has led to the depletion of the conventional sources of energy at a much faster rate than before. If the present trend continues, the day is not far when the world will be stripped of all the conventional sources of energy. Thus, there is a pressing need to explore tap new sources of energy so that future generations do not inherit a dark planet. In this regard, non-conventional energy sources need to be further explored, as there are certain advantages in their promotion such as:

- They are abundant and renewable.
- They are nonpolluting and ecofriendly.
- They are the future energy source.
- Some of the non-conventional energy sources like wind, mini-hydel projects and geothermal energy can be cheaply exploited.
- They can be exploited at the local level.
- Energy from biogas will be helpful in the rural areas, where conventional electricity has not yet reached.

Conservation of Energy

All human societies require energy for transportation, heating, cooling, cooking, lighting and for agricultural and industrial use. The world energy supply depends on many different resources. Energy has become a prerequisite for modern day living and it influences all aspects of development of a nation. The developed countries are the major consumers of energy, but the developing countries also require more and more energy for their economic development.

Fossil fuels account for more than 90 per cent of global energy needs, but the problem is that they are non-renewable, i.e. they can be depleted. Though non-renewable sources of energy are being constantly developed, there is a global need to increase energy conservation, so as to minimise the depletion of non-renewable resources.

Conservation of energy refers to the judicious and controlled use of energy. Conservation of energy can be brought about by the following three main factors:

- Efficiency in production
- Efficiency in transportation
- Proper utilisation

Efficiency in Production

This involves designing equipment so as to yield more energy output from a given amount of energy input. Energy conservation is vital as it provides saving in a ratio of 1:3. For example, three units of fuel are required to produce one unit of electricity. Two-thirds of energy is wasted as heat. There is an urgent need to design equipments to capture and use this heat, which otherwise pollute the atmosphere or water as thermal pollution. The term **cogeneration** refers to the process that utilises this heat energy. Through cogeneration, the overall efficiency of a typical thermal power plant can be increased from 33% to as much as 75%.

Cogeneration also involves the production of electricity as a by-product from industrial processes that produce and use steam. New power plants, boilers and other energy conversion equipments should be built to ensure maximum efficiency.

Efficiency in Transportation

The transport sector accounts for the major consumption of petroleum, especially in the developed countries. Increase in the prices of petroleum has encouraged the conservation of energy in the transport sector, which can be brought about by the following measures:

- New engine technologies to get more mileage per litre of petrol
- Using more fuel efficient smaller personal automobiles
- Encouraging use of engines that run on natural gas
- More reliance on electric batteries or highly efficient fuel cells
- Using other modes of transportation like bicycles and also car-pooling and public transportation
- Checking unnecessary travel by fuel consuming automobiles

Proper Utilisation

As the resources for energy are limited and generation of energy is a costly affair, it is the responsibility of every individual to make personal efforts for proper energy utilisation. Wastage of energy should be checked at every step, as wasting energy means depleting the natural resources. Energy conservation should become a habit and part of our everyday life. It can be brought about in the following ways:

- In the household, turning off the fans, switching off unnecessary lights and other appliances when not in use, can conserve energy.
- Buildings should be designed in such a way that does not consume more energy for heating or cooling; and may use the natural air and light.
- Old technology should be replaced with more energy efficient technologies.
- Huge losses of electricity occur during transmission from power plants to the consumers. This should be minimised by using better conductors and transformers.

Future Sources of Energy

Apart from the conventional and non-conventional sources of energy, research is being done to look for even more energy sources. Some of the new energy sources that can have future implications are given below:

Hydrogen

Hydrogen gas is considered as the fuel of the future. It can be used in a similar way as fossil fuels are used. However, it is still largely in the experimental stage.

Advantages

- It is a high-energy fuel.
- It is a potentially unlimited fuel supply source if produced from water; currently made from natural gas.
- If produced from water using solar energy and then used in fuel cells, it would be a zero emission technology.
- Produces no carbon dioxide.

Disadvantages

- Highly explosive, though less dangerous than gasoline.
- Costly and difficult to produce and store.

Alcohol

Certain types of biomass can be fermented into alcohols (ethanol / methanol), which can be used as fuels. These can be produced from carbohydrate rich substances like sugarcane, potatoes, sugar beet, etc and also from woods and grains. Gasohol is a common alcohol based fuel, which is a mixture of ethanol and gasoline.

Advantages

- It is a clean, non-polluting fuel.
- It can be made from natural gas and other renewable resources.
- It can be used in conventional engines with some minor changes.

Disadvantages

- Highly corrosive and can damage the engine.
- Production is not adequate to replace fossil fuels.
- It has low energy content than gasoline.

Fuel cells

As automobiles consume most of the petroleum and are also a major contributor of air pollutants, research is being done to generate power using environmental friendly technology within the vehicle itself. Fuel cells are power-generating systems that produce electricity by combining fuel and oxygen in an electrochemical reaction. Fuel cells can run on methanol, ethanol, natural gas and hydrogen.

Advantages

- Highly efficient, as chemical energy is directly converted to electrical energy.
- Harmful emissions are absent or negligible.
- Well suited for powering automobiles.

Disadvantages

- Experiments are still in their infancy.
- Use of hydrogen in fuel cells is difficult to attain under the present technology.

OPTIMISING ENERGY UTILISATION

The optimal use of available energy resources and its proper utilisation will contribute to a large extent in the conservation of energy. This can be done both by man and machines by making concerted individual efforts to stop the wastage of energy, and similarly by enhancing the efficiency of the machines and energy consuming devices. Some of the important steps that the Indian Government has taken in this regard are briefly given below:

Smokeless Chulhas

In the rural areas, biomass is the most frequently used source of energy for cooking purposes. Biomass, consisting mainly of wood, barks, sawdust, cow-dung cakes, straws, etc. are burnt in open furnaces called chulhas. These chulhas are inefficient and cause much loss of energy. Moreover, they produce a lot of smoke causing indoor pollution. To overcome this problem, new chulhas with long chimneys have been developed, which are more efficient and smokeless.

Liquefied Petroleum Gas (LPG)

Liquefied Petroleum Gas (LPG) is a mixture of liquefied gases such as propane and butane. It is obtained from natural gas or petroleum. The natural gas is converted into liquid form, under high pressure. LPG is then vaporised for use as a heating or engine fuel. LPG is now commonly used as a cooking fuel in almost all the urban households. There is an urgent need to promote its use in the rural areas, as it is more efficient, and also to save the rural population from the ill effects of wood burning.

Compressed Natural Gas (CNG)

It is presently being used as automobile fuel in most of the developed countries, and is gradually replacing conventional petrol and diesel in vehicles. It is less polluting for the environment. The Delhi Government, in 2004, made it mandatory for city buses to run in CNG.

Industry

In India, the industrial sector consumes most of the energy. So the machines and tools need to be properly maintained to make them more efficient.

Agriculture

Wastage of energy in the agricultural sector needs to be minimised by using better implements, pumps and motors.

QUESTIONS

1. What do you know about conservation? What are the main reasons of depletion of natural resources?
2. What are the different kinds of natural resources? How can their misuse be checked ?
3. Write brief notes on:
   (*i*) Mineral resources,
   (*ii*) Wildlife,
   (*iii*) Social forestry,
   (*iv*) Food resources,
   (*v*) Biogas
   (*vi*) Sanctuaries
   (*vii*) National Parks
4. What are the various steps for conservation of natural resources at national and international levels?
5. Give an account of various energy resources.
6. Define wildlife. Discuss the various steps taken for the conservation of wildlife.
7. Discuss non-conventional source of energy with special reference to India.
8. What do you mean by renewable and non-renewable sources of energy ? Mention various non-renewable sources of energy.
9. Discuss in brief marine resources of India and their conservation.
10. Write brief notes on :
    (*i*) Nuclear energy
    (*ii*) Geothermal energy
    (*iii*) Use of solar energy
11. Write short notes on:
    (*i*) Rainwater harvesting technology
    (*ii*) Red Data Book

16

Forests and Forest Management

The word forest is derived from the Latin word 'foris' meaning "out of door". It is large uncultivated tract of land covered by trees and underwoods. In other words, forest is a natural ecosystem of tree-dominated vegetational association.The occurrence of forest indicates optimal growth conditions in the area. The ecological conditions which favour the forest growth are availability of moisture throughout the year, favourable temperature conditions during growth period and decreased biotic interference. Forests are very rich in flora and fauna. Thus they store maximum information by way of diversity of species and life forms. The maximum energy transactions take place in the forests where nature produces maximum biomass in minimum space. The present status of forest areas in different continents of the world is illustrated by the following Table 16.1.

PRESENT STATUS OF FOREST WEALTH OF THE WORLD

At present the earth is occupied by 3.87 billion ha forest having biomass 421.27 billion tons. Of the total world biomass Africa shares 16.79%, Asia 17.86%, Europe 14.5%, South America 22.89%. Brazil has the largest biomass which is about 27% of the total world biomass Table 16.1 and Fig. 16.1.

Table 16.1. Distribution of forest wealth of the world

Region/Country	Percentage of the world area	Percentage of biomass of the world
Africa	16.9	16.8
Asia	14.16	10.7
Europe	26.86	14.5
North & Central America	13.16	12.3
South America	22.89	42.7
Brazil	14.06	26.99
India	2.66	1.11
Oceanic Islands	5.11	2.9

State of world forest, 2001

In developing countries like India forests play important role in providing economic wealth, maintaining ecological balance and improving the productivity of agriculture. With the rapid increase in human population more and more areas of forests are converted into agricultural land. The consequences of this are the increased soil erosion, frequent floods, drought and land slides. The role of forests in sustaining agriculture must, therefore, be realised and the conversion of forest areas into agricultural land must be checked.

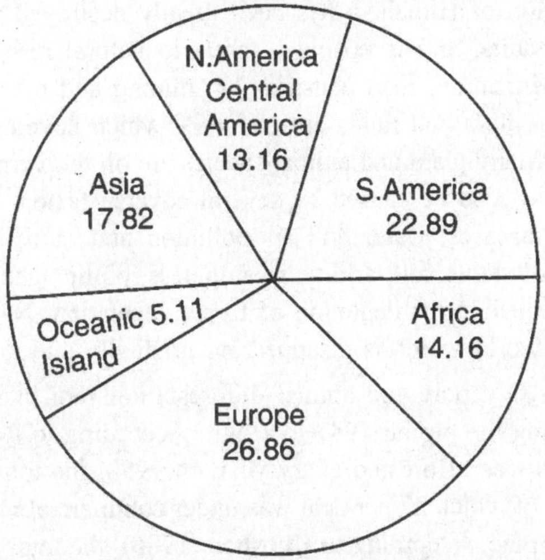

Fig. 16.1. Forest wealth (biomass percentage) of the different parts of the world.

ECOLOGICAL SIGNIFICANCE OF FOREST

Forest is an important natural resource which plays several important roles in nature. It is a store house of biological diversity. It is most important natural habitat for wildlife. Forest is composed of a large flora and fauna. It maintains ecological balance and biological diversity, acts as catchment for soil and water, and provides food, fuel, fodder, fibre, shelter and timber to man. It supports industrial and commercial activities and provides job opportunity to a large number of people. It maintains life support systems essential for food production, health and all round human development. Forest is best site of recreation, meditation, peace and it provides aesthetic sense. It conserves water and soil moisture. Development and conservation of forest will improve the living standard of local people and tribals. Forests control air and noise pollution to a great extent. Forest prevents land slides in the hills and flood havoc in the plains. It is harbour of genetic resources.

Green plants of the forest are primary producers of the food chains. In forest ecosystem they trap solar energy and transform CO_2 absorbed from the air and water and nutrients absorbed from the soil into food substances as starch, sugar and protein through the process of photosynthesis. Food materials are stored in the different parts of the plants such as fruits, nuts, legumes, seeds and wood in various forms. Thus forest acts as energy house by taking energy from the sunlight and converting that in to potential or biochemical form. Forest plays a significant role in keeping the balance of atmospheric gases by consuming CO_2 and releasing O_2. Oxygen is essential for living organisms. Thus the removal of forest will disturb the composition of natural air. In forests, the roots of the trees bind the soil and increase the porosity and water absorbing capacity of soil. About 90 per cent of the water falling on the forests is retained in the humus or in plant tissue. Thus forest plays a major role in the water cycle or hydrologic cycle. It maintains soil fertility, regulates earth's temperature, checks soil erosion, land slides, shifting of sand and flood. Thus it helps in the improvement of environment.

DEFORESTATION

Destruction of forest is called deforestation. Deforestation has serious effect on human life and environment. According to a survey of forest department, India has about 75 million hectares of forest area. Recently collected satellite imagery data have revealed that only about 17 per cent area is covered by forest. India is loosing 1.3 million hectare of forest every year. In the hilly region deforestation is so acute that economy and ecology of the area are severely affected.

The original vegetation of Himalaya has been greatly destroyed which has resulted in gradual loss of the natural habitat and is posing a threat to natural resources. Due to overpopulation, industrialization, urbanization, road construction, mining and other developmental activities the natural habitats of the flora and fauna are disturbed which have caused tremendous pressure on the living resources. Many plant and animal species are on the verge of extinction or endangered. Forest destruction may also be caused by several adverse factors as land slides, drought, flood, storm, earthquakes, diseases, water and air pollution and human interferences. Other adverse factors such as lack of stable soil, aridity, swampiness, biotic agencies, commercial exploitation, etc may also be responsible for depletion of forest vegetation. Natural diversity of India is one of the richest in the world which is disappearing gradually due to aforesaid factors.

According to FAO report, the annual deforestation rate in India was 0.6 per cent (0.34 million hectares during the period 1981 to 1990). According to Ravindra Nath and Hall (1994) 1.44 million hectares was afforested every year. In 1990, the total afforested area in India was 70.6 million hectares of which 27 per cent was under commercial plantations consisting mainly of *Eucalyptus,* teak and pine. According to Khoshoo (1986), the total area under forest in the world was about 7000 million ha. in 1900. By 1975 it was reduced to 2890 million ha. The destruction of forest cover in the ecologically sensitive Himalaya region has already started showing adverse impact in the form of increasing shortage of water, recurrent land slides, increasing flood, high sedimentation in the rivers, shortage of fuel and fodder and decrease in grazing land. Due to deforestation the life supporting systems are disturbed. Underground water table is progressively going deeper and deeper. Large area of the land becomes affected by drought and wells, tube-wells, lakes, ponds etc. dry up sooner than expected during summer months.

In Kumaon and Garhwal Himalaya the oak forests are maintaining general environment and villagers depend to a great extent for fodder, fuel and some other necessities on these forests. But now oak forests are being destroyed to meet the ever increasing demands of the people. This has resulted marked changes in the environmental conditions. Consequently, the herbal vegetation and microbial community associated with oak are destroyed. This may lead to the loss of medicinal herbs and shrubs associated with oak. The availability of fodder will be reduced and the age-old animal link in the hill ecosystem would be broken.

At the Earth Summit in Rio de Janeiro (1992), issue related to deforestation was a major concern. The Agenda 21 of the UNCED (United Nation Conference on Environment and Development) stated: "Deforestation is a result of many causes; some natural, but others mainly due to human development, such as inappropriate land tenure systems and incentives, expansion of agricultural areas, increasing forest product demand and lack of information about and understanding of the value of forests." A review of available literature has revealed the following facts about Indian forests :

1. The sustainably extractable quantity of fuel wood from Indian forests is far below the requirement of the population.
2. Livestock population in India is greater than can be sustainably supported by the available land and forest resources.
3. The demand for industrial wood and other wood, part of which is currently being met by imports, will continue to rise with industrial and economic growth.
4. Unregulated and increased harvesting of non-wood forest produce will result in loss of biodiversity.
5. Expansion of protected area network will result in increasing pressures on the existing productive forests.

Causes of Deforestation

The different causes of deforestation are as follows :

1. **Overgrazing.** Overgrazing in forests destroys newly regenerated growth. It also makes soil more compact and impervious. Soil becomes less fertile due to destruction of organic matter. Seeds of certain species do not germinate in excessively grazed soils which results in reduction of species. Overgrazing also leads to desertification. Domesticated animals are thus deprived of their natural grazing and forage support. Overgrazing also accelerates the soil erosion. Soil erosion results in the removal of minerals and nutrients from the top soil and adversely affects the soil structure which ultimately lowers the productivity. Satellite imagery data indicate that the area under pasture land is severely degraded. The uncontrolled and indiscriminate grazing in the forests leads to degradation of forest soil and affects natural regeneration of forests.

2. **Shifting cultivation.** This is most common in North-Eastern India due to heavy water erosion, Shifting cultivation is locally called *Jhum*. Many farmers destroy the forest for agricultural and commercial purposes and when fertility of soil is exhausted due to repeated cropping, a new forest area is destroyed. Therefore, farmers should be advised to use the same land for cultivation and apply improved farming methods. It is estimated that every year about one million hectare land is degraded by shifting cultivation.

3. **Fuel wood.** Maximum destruction of forest vegetation is done for fuel wood. Of the total fuel wood nearly 85 per cent is used in rural areas and 15 per cent in urban areas. The annual fire wood consumption was 134 million tonnes in rural areas and 23 million tonnes in cities in 1987. According to an estimate of Forest Survey of India (1987), the annual demand for fire wood in the country was 235 million cubic metres. Thus fuel wood is a major factor for deforestation.

4. **Forest fires.** Frequent fires are the major cause of destruction of forests in India. Some fires are incidental while the majority of them are deliberate. According to a study conducted by the Forest Survey of India (1996), on an average, 53.1% forest vegetation is affected by fire. Data further indicate that fire destroys about 0.5 million hectares of forests annually.

5. **Timber.** Timber and plywood industries are mainly responsible for the destruction of forest trees. Thus the increased demand for timber led to a rapid depletion of forest. According to Forest Survey of India 1987, against an annual demand for more than 27 million cubic metres the permissible cut of timber from forests was only 12 million cubic metres.

6. **Industry establishment.** Sometimes factories are established after destruction of forest. Thus for a small gain there is an irrepairable loss. In this process precious plants, wild animals and rare birds are destroyed and the quality of environment is adversely affected. A factory should be established on a waste land away from the urban population. Supply of raw materials to the forest based industries is another major cause of forest destruction in the hills. The forest based industries such as Resin and Turpentine industry are also responsible for destruction of trees in the hills.

7. **Encroachment of forest.** Another cause of deforestation in India is encroachment by tribals on forest land for agriculture and other purposes. According to Forest Survey of India, about 7 million ha of forest land has been encroached for agriculture. Although such land makes a good contribution for agriculture production yet it produces environmental hazards. So it is not desirable to convert forest land into agriculture land.

8. **Forest diseases.** Many diseases caused by parasitic fungi, rusts, viruses and nematodes cause death and decay of forest plants. Young seedlings are destroyed due to attack of nematodes. Many diseases such as heart rot, blister rust, oak wilt, phloem necrosis and Dutch elm disease, etc. damage the forest trees in large numbers.

9. **Land slide.** Deforestation due to land slide in the hills is of great concern. It has been observed that, the land slide occurred mainly in the areas where developmental activities were in progress for past few decades. The construction of roads, and railways particularly in hilly terrains, setting up of big irrigation projects have caused enough destruction to forest and accelerated the natural process of denudation.

10. **Ravine formation.** The forests and cultivated lands along the sides of ravines of big rivers (Yamuna and Chambal) are facing a serious danger of soil erosion. Once the ravines are formed, they continue to destroy the vegetational cover.

11. **Increase in population.** The population of India which was 36 crores in 1951 is expected to touch 1.05 billion (105 crores) by 2010 A.D. The number of persons dependent on one hectare of agricultural land increased from 3 in 1960-61 to 5 in 1990-91. Such an increase has resulted in acute shortage of land. Besides agriculture, forest land has also been diverted into other sectors. Table 16.2 gives an idea about diversion of forest land into different sectors.

Table 16.2. Diversion of forest lands to other uses

(According to a report of Forest Survey of India, 1988 and Indian Council of Forestry Research and Education, 1995)

Land use	Total Area (in million ha) diverted into different sectors from 1951 to 1995
Agriculture	2.764
Ravines Valley Projects	0.518
Industry and townships	0.141
Miscellaneous	1.036
Total	4.696

FOREST MANAGEMENT & CONSERVATION

Forest management has been defined as the practical application of the scientific, technical and economic principles of forestry. Forest management or forest maintenance is essential to achieve the following two objectives:

1. For climatic or protective purposes which are sometimes referred to as indirect objectives.
2. For productive purposes or economic objectives which are sometimes referred to as direct objectives.

The forest protection is essential due to the following physical reasons :

1. Protection of mountain slopes and catchment areas,
2. Protection against wind,
3. Protection against erosion and ravine formation,
4. Protection of aesthetic value or recreation,
5. Protection of pasture, and
6. Protection of wildlife.
7. Protection of soil

For all these, the forest management must be such as to conserve moisture, preserve climatic factors, and prevent soil erosion. The forests have significant role in the improvement of economic condition of the country. Therefore, scientific management, conservation and utilization of forests wealth need special attention of the scientists. At the first step, the existing forests should be protected and the forest area should be expanded. The National Commission for agriculture has taken up the problem of deforestation seriously and recommended plantation through social forestry and other programmes. Thus plantation is the best method of conservation of forest. The forest plays vital role in the amelioration of environment and improves the quality of life. Therefore, considerable attention must be paid for its proper management. There is an urgent need to restore the ecological loss through afforestation in all areas including arid and semi-arid zones. This is essential for conservation of biological diversity, increasing catchment for water conservation, reducing the sediment load of rivers thus preventing flood. National Forest Policy of 1952 recommended that at least 33 per cent of the land area should be covered by forest. The forest area of different states of India are presented in Table 16.3.

Table 16.3. Forest area in different states of India and their position in the country.

State/union Territory	Percentage of forest of total Land	Position of State in the Country
Andhra Pradesh	23.20	20
Arunanchal Pradesh	61.54	6
Assam	30.708	12
Bihar	6.45	29
Chhatisgarh	43.85	9
Delhi	2.83	32
Goa	38.41	13
Gujarat	8.89	25
Haryana	3.78	31
Himachal Pradesh	63.60	5
Jammu and Kashmir	9.08	27
Jharkhand	29.04	17
Karnataka	20.19	22
Kerala	11.221	18
Madhya Pradesh	30.89	16
Maharashtra	20.75	21
Manipur	67.87	3
Meghalaya	42.34	10
Mizoram	75.59	2
Nagaland	52.04	8
Orissa	36.73	15
Punjab	5.76	30
Rajasthan	9.26	26
Sikkim	37.34	14
Tamil Nadu	17.40	23
Tripura	60.01	7
Uttaranchal	64.81	4
Uttar Pradesh	7.06	28
West Bengal	13.38	24
Union Territories		
Andaman and Nicobar Islands	86.93	1
Chandigarh	27.19	19
Daman and Diu	0.62	33
Lakshadweep	0	—
Pondicherry	0	—

Source : Forest Survey of India–2000

The forest cover of India specially of Uttar Pradesh is much less than recommended by the National Forest Policy (1952). In view of this fact, vigorous afforestation and social forestry programmes have been launched by Govt. of India to bring one third of the total land area under forest cover. One of the important aspects of these programmes is the selection of suitable land for afforestation. The best location for afforestation programme in the first instance would be degraded and waste lands which are not suitable for agriculture. This would increase the productivity of the wastelands and lead to qualitative improvement of such degraded lands. Under afforestation programme indigenous and exotic fast growing tree species are being planted on selected land. Social forestry programme will also fulfil the demands of local people. Every year plenty of seedlings are distributed for plantation. Besides this, many plans and programmes for conserving forests are being implemented by the government and voluntary organizations. Government of India has recently brought legislation to preserve forests, wildlife and biosphere reserves. Current National Forest Policy has recommended the following categories of forest to reduce environmental degradation caused by deforestation which are presented in Table 16.4.

Table 16.4. Categories of forest to reduce environmental degradation (Forest Survey of India, 1996)

Categories of forest	Area Million hectares	Percentage of total forest area
Reserve	41.65	54.4
Protected	22.33	29.2
Unclassed	12.54	16.4
Total	76.52	100.0

The various aspects of forest management are as follows :

1. Conservation or Protection forestry or Reserve forestry
2. Production or commercial forestry
3. Social forestry including afforestation of waste land

Conservation or Protection or Reserve Forestry

Reserve forests are the most important subject for conservation and scientific management. Conservation forests are also called reserve forests. The major aims of conservation forestry are protection or conservation of existing forests and restoration of the degraded forests in ecologically sensitive areas as Himalayan region, western and eastern Ghats, catchment areas, National parks, sanctuaries and biosphere reserves etc. These areas are restricted from commercial exploitation.

COMMERCIAL FORESTRY

When the afforestation is undertaken and managed intensively for much higher productivity on limited area of forest land it is called commercial forestry. The main aim of commercial forestry is to provide and fulfil the requirements of the people and industries.

SOCIAL FORESTRY

Social forestry means afforestation or developing of forests with the participation and cooperation of public to meet the requirements for domestic consumption and cottage industries. The main objects of social forestry are to provide adequate fuel wood for cooking, green fodder for domestic animals, edible fruits for improved nutrition and timber for house construction.

Forest management programme includes the following activities:

1. Reforestation.
2. Forest protection from encroachment and shifting cultivation.
3. Control of forest fires.
4. Restriction on over-grazing by domestic animals.
5. Extension of forest areas.

6. Control of excessive cutting.
7. Cultivation of useful trees on wastelands in rural and urban areas.
8. Plantation of fast growing trees.
9. Control of harmful agents.
10. Improvement of timber quality through cultivation of improved varieties.
11. Converting wasteful cutting into quality yield harvesting.
12. Legislation to check deforestation.
13. Controlling of forest diseases.

The forest diseases can be controlled upto some extent by:

(*i*) Eradication of diseased host plants.
(*ii*) Spraying of suitable antibiotics, fungicides, nematocides.
(*iii*) Spraying of effective insecticides.
(*iv*) Sterilization.
(*v*) Planting of resistant varieties.

14. Undertaking afforestation with particular emphasis on fuel and fodder plants.
15. More attention for forestry research, personnel management and data base in the field of forestry.

The Present Programmes for the Development of Forestry and Wildlife

The Ministry of Environment and Forest is implementing the following Projects :

1. Biosphere Reserves.
2. Development of National Parks and Sanctuaries.
3. Project tiger.
4. Project elephant.
5. Modern forest fire control methods.
6. Ecodevelopment around National Parks.
7. National Afforestation and Écodevelopment Board (NAEB).
8. Integrated action oriented research demonstration and projects for Himalayan regions.
9. Association and involvement of Scheduled Tribes and Rural poors in the programmes of region of degraded forests.

State Governments have also many schemes to implement which are:

1. Afforestation and regeneration of degraded forests.
2. Greening of rural and urban areas.
3. Development and conservation of minor forest produce.
4. Development of National Parks and Sanctuaries.
5. Regeneration of Sal and Oak forests.

Afforestation in Denuded Hill Slopes

Denuded hill slopes are found throughout the country. In most of the humid regions vegetational cover establishes naturally, but in drier parts of the country, the afforestation assumes special importance not only for productive reasons but also for conservation of soil and water. The incidence of grazing is one of the important causes of denudation and the protection of plants raised under afforestation is a tidius problem.

Soil preparation for afforestation work is done in the form of contour trenches, patches and pits. The selection of species for plantation is made taking into consideration the various factors prevailing in the region. The surviving indigenous or local species give a clear indication of the species which may be most suited for the area under afforestation. Thus fast growing species can survive under the adverse condition of the locality. Examples of few plant species suitable for denuded hill slopes are given below:

Chir or pine trees ______________ in moist subtropical area
Acacia modesta ______________ Dry subtropical area
A. Catechu, Olia ,,
Dodonea, Prosopis juliflora ,,
Robinia ,,
Pinus wallichiana and Deodar ________ Temperate zone

Afforestation in Ravine Lands

The ravines of Yamuna and Chambal have not only rendered vast areas of land barren but are causing serious danger to the cultivated lands.

Continuous development of ravines results in the formation of hillocks and ridges in between two adjoining ravines. These hillocks and ridges have practically no contact with the permanent water table and their own moisture retaining capacity is low, their soil becomes extremely dry during summer. Uncontrolled cutting of trees, grazing and fire which are said to be generally responsible for the formation of the ravines continue to pose a serious threat to afforestation. The following species have been successfully raised in ravine areas:

In sandy soils : *Acacia catechu, Dalbergia sissoo, Albizzia* spp, *Azardirachta indica, Pongamia pinnata, Prosopis juliflora,* etc.

In clayey soils : *Acacia leucophloea, Acacia arabica, Albizzia, Syzygium, Terminalia arjuna, Holoptalia, Haplophragma, Pongamia, Prosopis spicigera,* etc.

Afforestation in Shifting Sands

Shifting sands are found along sea coasts as well as in the interior. The inland sands are found along rivers and deserts.

Coastal sand. All along sea coast large quantities of sand get accumulated as a result of tides. Such sand deposits are found in the estuaries of big rivers. Coastal sand keeps on shifting under the pressure of strong wind. It is not capable of retaining moisture, water table is found a few meters below the surface. The winds are strong and carry salt particles with them. For most of the sea coasts, *Casuarina equisetifolia* has been found to be the best species. It is quick growing and hardy. Other species which have been successfully raised are *Eucalyptus, Pongamia pinnata, Casuarina,* etc.

Inland river sand. The inland sand generally occurs along the rivers and deserts. The soil is sandy, unstable and deficient in nutrients, water table is not deep and so irrigation is possible. Soil preparation is made by digging trenches and pits. The plant species which have been successfully grown in inland sands are *Dalbergia sissoo, Acacia catechu, Acacia arabica, Albizzia procera* and *Pongamia pinnata.*

In desert areas climate is characterised by extreme temperature, dry condition and scanty rainfall. Soil varies from place to place. In order to reduce the velocity of wind and prevent sand movement, mulching is done in April and May. For this purpose. *Crotolana bruhia, Aerua tomentosa, Leptadenia pyrotechnica* plants are used. Before the onset of rains seeds of the following species are sown along the windward side of the mulch lines:

Tree species – *Prosopis spicigera, Prosopis juliflora, Acacia arabica, Acacia senegal,* etc.

Shrubs – *Calligonum polygonoides, Ziziphus* spp, *Cassia auriculata* etc.

Grasses – *Panicum antidotate, Lasiurnum* spp.

Important tree species suitable for planting in different soil groups are summarised in Table 16.5.

Table 16.5. Important tree species suitable for planting in Soils of different Climatic regions of India

1 *Species*	2 *Sites*	3 *Soil Groups*	4 *Uses*	5 *Remarks*
A. Temperate Region **(*i*) Himalayan**				
1. *Alnus nepalensis* (units)	River banks ravines & newly formed soil	Brown hill soils (on sand stones & shales), Sandy, soil	Timber & fuel	Large tree, grown by direct sowing or transplanting.
2. *Alnus nitida*	-do-	-do-	-do-	-do-
3. *Betula* (Bhuj)	Broken marginal land	Brown hill soil	Plywood, furniture etc.	Medium sized tree grown by sowing or transplanting
4. *Cedrous deodara*	Understocked, degraded, forest blanks site (Reforestation for better quality timber)	Sub-mountain podzolic soil & Brown hill soil	Railway slippers, general carpentary	Most important durable timber species.
5. *Junglans regia*	Village marginal broken land in moist locality	Brown hill soil	Timber furniture gumstock, fruit	Large tree grown by entire transplanting and also by direct sowing.
6. *Populus nigra*	Dry valleys	-do-	Light timber matchwood, fuel ornamental	Grown by branch cuttings.
7. *Prunus cerasoides* (Padam)	Marginal land around villages	-do-	Fuel, fodder, timber wood used for religious ceremonies, good for bee-keeping	Medium tree, grown by entire transplanting and also by branch cutting.
8. *Pinus wallichiana*	Understocked, degraded and forest blanks (for reforestation and enrichment plantation)	Brown hill soil and Podzolic soil	Timber and pulp wood and resign	Very good timber spp. grown by direct sowing and transplanting
9. *Quercus lamellosa*	Marginal land	Brown hill deep soils	Agricultural implements, medicinal, tussar silk rearing	Large tree of eastern parts grown by Direct sowing.
10. *Salix alba*	Broken marginal land	Brown hill soils and Podzolic soils	Cricket bats matchwood, tool handles, fuel & fodder	Large tree grown by branch cutting.

(Contd...)

1	2	3	4	5
(ii) Southern Hill tops of Tamil Nadu and Kerala.				
11. (i) *Eucalyptus globulus*	Village common land, forest blanks, barren hill slopes	Mixed red and black soil, red sandy soil	Timber, fuel, paper pulp, essential oil	Large tree grown by planting.
12. The wattles *Acacia mearnsi, Acacia melanoxylon, Acacia dealbata, Acacia decurrens*	Marginal lands & cut up land around village, barren hill slopes	Mixed red & black soil, red sandy soil	Timber fuel, paper pulp, essential oil	Large tree grown by entire transplanting.
13. *Pinus petula*	-do-	-do-	Timber, resin, pulpwood	Large exotic pine, quick growing grown by entire transplanting.
B. Subtropical Region. (i) Himalayan Foothills and Central India hills				
1. (ii) *Albizzia lebbeck*	Open lands & along pathways	Deep black to medium black	Timber fuel & fodder	Grown by direct sowing, transplanting & cuttings.
2. *Bauhinia variegata* (Kachnar)	Marginal land & field edges	Brown hill soil	Gum, fodder, flower buds eaten, bark yields dye and medicine	Grown by entire transplanting and direct sowing.
3. *Celtis australis*	-do-	-do-	Timber, fuel, fodder, sports goods, utensils	Grown by entire transplanting and branch cuttings.
4. *Dalbergia sissoo*	Marginal land, village common land, road sides and other open denuded areas	Alluvial soils	Timber, furniture plywood, fuel & fodder	Grown by entire transplanting, root and shoot cuttings and direct sowing.
5. *Emblica officinalis* (aonla)	Grounds near homesteads and on field edges	Brown hill soil & medium black soil	Fruit, Tanin, timber, fuel, fodder	Grown by transplanting or sowing.
6. *Eucalyptus grendis*	Marginal land		Timber, paper pulp, fuel, medicine	Grown by transplanting or sowing.
7. *Ficus religiosa* (Pipal)	Avenues	Brown hill soil	Religious tree, timber for packing, cases, fuel, fodder	Grown by branch cutting.

(Contd...)

1	2	3	4	5
8. *Grevillea robusta*	Avenues, tea gardens	Brown hill soil, medium black soil	Ornamental, timber, shade tree for tea gardens	Large tree grown by direct sowing.
9. *Morus alba*	Marginal land	Tarai soil, Brown hill soil, alluvial soil	Fruits, edible leaves for silk worm feeding, timber, sports goods & fodder	Medium tree grown by entire transplanting, direct sowing or branch cutting.
10. *Pinus roxburghii*	Denuded hill slopes, understocked and degraded forest areas, forest blanks	Brown hill soil	Best timber goods, paper pulp	Large tree grown by transplanting and sowing.
(*ii*) **East Indian Hills & Parts of Western Ghats**				
1. *Acrocarpus fraxinifolius*	Marginal lands and village commons, degraded understocked forest areas	Brown hill soil	Fuel, box-wood, boards, planks	Large tree grown by entire transplanting or by direct sowing.
2. *Ailanthus grandis* (gogal)	Village commons, and road side understocked and degraded forest area	Tarai sandy soil, red loamy soil	Plywood, ornamental	Large tree grown by entire transplanting or direct sowing.
3. *Grewia elastica* (dhaman)	Field edges	Brown hill soil	Fuel, fodder, timber	Medium tree grown by entire transplanting.
4. *Michelia champaca* (Champa)	Home steads, marginal land, village commons & degraded forest area	Brown hill soil	Decorative timber, fuel, ornamental	-do-
C. **Tropical Region** (*i*) **High rainfall areas of Western Ghats, W.B. & Assam**				
1. *Anacardium occidentale* (Cashew nut)	Mainly in coastal area	Coastal sandy soil, red sandy soil & lateritic soil	Fruit, essential oil	
2. *Casuarina equisetifolia* (Saru)	Mainly in coastal areas, ideal for fuel wood lots	-do-	Timber, fuel, ornamental	Fast growing spp., grown by entire transplanting.

(Contd...)

1	2	3	4	5
3. *Dalbergia latifolia* (The rose wood)	Marginal land, field edges, village commons, coffee gardens understocked & degraded forest	Red loamy soil & sub-mountain podzolic soils	Timber, fuel	Large tree, grown by entire transplanting.
4. *Gmelina arborea*	Village commons, forest blanks	Red loamy soil	Fuel wood.	Medium tree, grown by entire transplanting.
5. *Syzygium cumini* (Jamun)	Open lands, village boundaries, & along road side	-do-	Fruits, Fuel wood medicinal	-do-
(*ii*) **Medium rainfall area of Indo-Gangetic Plains & the Peninsula (Moist tropical)**				
1. *Acacia auriculiformis*	Marginal land and sand dunes	Alluvial soils, sandy soils	Timber, fuel, ornamental	Medium tree, grown by entire transplanting.
2. *Acacia nilotica* (Babul, Kikar)	Field boundaries, open fields, marginal lands & village commons	Alluvial soils & medium black soil	Timber, fuel, fodder, tanin and gum	-do-
3. *Ailanthus excelsa*	Marginal land village commons	-do-	Timber, packing cases, fishing Boats & boards, bark, medicinal, fodder	Large tree grown by entire transplanting & seed sowing.
4. *Artocarpus integrifolia* (Kathal)	Gardens & house compounds	Alluvial soils, medium black soil, red loamy soil	Timber, fruits, fodder	Grown by entire transplanting or direct sowing.
5. *Azadirachta indica*	Open lands, house compounds & along roads & paths	Alluvial soil, medium black soil	Timber, bark & seed medicinal, pesticidal, fertilizer, fodder	Large tree grown by entire transplanting & direct sowing.
6. *Cassia siamea*	House compound, marginal land, **santalum album forest** (grown as host plant)	Red sandy soil, alluvial soil, medium black soil	Timber, fuel, ornamental	Medium tree grown by entire transplanting & sowing.

(Contd...)

1	2	3	4	5
7. *Eucalyptus hybrida*	Marginal land, wood lots in private lands, shelter belt, along road sides	Medium black soil, alluvial soil	Timber, pulp wood, fuel	Large tree grown by entire transplanting.
8. *Leucaena leucophloea* (Subabul)	Along hedges, homesteads, marginal lands and village commons	Medium black soil	Timber & Fodder	-do-
9. *Madhuca latifolia* (Mahua)	Open lands around village boundary & village commons	Medium black soil, alluvial soil	Timber, fuel, fruits, flowers, edible oil	Grown by direct sowing or transplanting.
10. *Tectona grandis* (Sagaun)	Field edges, marginal lands & village commons, reforestation areas	Red loamy soil, medium black soil	Timber, railway carriages & wagons, cabinet, furniture etc.	Large tree grown by transplanting of root and shoot cutting
(*iii*) **Dry Areas of Rajasthan & Peninsular India (Dry tropical)**				
1. *Acacia auriculiformis* 2. *Acacia nilotica* 3. *Ailanthus excelsa* 4. *Albizzia lebbeck* 5. *Azadirachta indica* 6. *Cassia siamia* 7. *Eucalyptus hybrid* 8. *Leucaena leucophloea* 9. *Madhuca lalifolica* 10. *Prosopis juliflora*	Trees suitable for afforestation in arid regions, grown by entire transplanting, other details mentioned earlier.			

INDIAN FOREST POLICY 1947-1998

Scientific management of forests in India was initiated in 1864. India's first forest policy was framed in 1894 by British administration which gave priority to commercial exploitation of forest and focussed more attention on crop cultivation than on forestry, permitting forest areas to be converted into agriculture lands. The National Forest Policy formulated in 1952 laid emphasis on increasing the forest area to one-third of the total land area. It has not been possible, however, to implement this policy. On the contrary, forest land had to be used for river valley projects, minor irrigation projects, industries, residential settlements, rehabilitation of landless people and construction of roads etc. and in the late fiftys and early sixtys forest lands were allotted to people for farming under the *"Grow more Food Programme"* which was later postponed.

The National Forest Policy provided a functional classification of forests and brought under its purview the conservation of wildlife. Forest legislation was brought on the concurrent list in 1977. The enactment of the Forest Act, 1980 put a check on the uncontrolled diversion of forest land for other purposes.

The Forest Act, 1980 was revised in 1988. The main objectives of policy are :

1. Maintenance of environmental stability through preservation and restoration of ecological balance.
2. Conservation of biological diversity and genetic resources.

3. Control of soil erosion and desertification.
4. Increase in the forest cover through afforestation and social forestry.
5. Maximum utilization of forest produce and maximum substitution of wood.
6. People's participation and active involvement of women in achieving the above objectives and in minimising pressure on forest.
7. A maximum of 33% of the total land area of the country and 60% of the hill area to be brought under forest or tree cover.
8. Total protection of tropical rain forests.
9. Control on introduction of exotic species.
10. Involvement of tribal population in protection, regeneration and development of forests.
11. Discontinuation of the practice of supplying forest produce to industry at concessional rates.
12. The use of forests for grazing and extraction to be determined with regard to carrying capacity.

Laws for Conservation of Forests

Article 51-A(g) has made it the fundamental duty of every citizen "to protect and improve the natural environment including forests, lakes, rivers and wildlife". Government of India enacted the forest conservation Acts in 1980 which impose restriction on destruction of reserved forests, and use of forest land for non-forest purposes by the states. These restrictions are contained in section 2 of the Act.

The section 2 of the Act states that—Notwithstanding any thing contained in any other law for time being in force in a state, no state government or other authority shall make, except with the prior approval of the Central Government, any order directing :

(*i*) that any reserved forest or any portion thereof, shall cease to be reserved.

(*ii*) that any forest land or any portion thereof may be used for any non-forest purpose ("Non-forest purpose" means the breaking up or clearing of forest land or portion).

(*iii*) that any forest land or any portion thereof may be assigned by way of lease or otherwise to any private person or to any authority, corporation, agency or any other organization not owned, managed or controlled by Government.

(*iv*) that any forest land or any portion there of may be cleared of trees which have grown naturally in that land or portion, for the purpose of using it for reafforestation.

The Act does not permit any mining in the forest areas. Therefore, the mining activity in the forest areas must be stopped.

SOCIAL FORESTRY

The National Commission on Agriculture has given serious thought to the problem of deforestation and recommended introduction of *"Social Forestry"* to create multipurpose wood lots in rural India. Social forestry may be defined as additional aid to wildlife conservation. Social forestry is a new revolutionary concept, a multipurpose programme and a mission which aims at ensuring ecological, economic and social security to the people particularly to the rural masses especially by involving the beneficiary's right from the planting stage to the harvesting stage. It aims at a mixed production of fire wood, timbers, fibres, fruits, fodder and other raw materials for self-consumption and cottage industry. These objectives are achieved through public participation and cooperation.

In view of increasing population and deforestation, social forestry will not only serve to meet the ever increasing demand of fire wood, fodder, timber and a variety of tree-based produce for rural cottage industries but also to maintain the ecological balance.

To meet the basic needs of villagers Social Forestry Programme was initiated in 1979 with the help of world bank. Now the social forestry is adopted in almost all the states. Under social forestry programme different species of fast growing trees are planted which will provide fuel, fodder, raw materials for cottage industry.

Under social forestry programme plantation is done in the following types of land :

1. Government land which includes roadsides, side land along railway lines and degraded forest land.

2. Semi-government and institutional lands which include wasteland under Panchayat, school and college land and land around other buildings in cooperative sector.

3. Private land.

Forest department manages for cultivation in the wasteland of village according to requirement of villagers. Care and protection of plants is done by village committee. There is agreement between village committee and forest department through which 20 per cent profit is given to village Panchayat and the rest goes to forest department. Seedlings or nursery plants are supplied free of cost to villagers and the cost of plantation is met by the village Panchayat.

The social forestry which is done on private land is called ***agriculture-forestry* or *agroforestry.*** Agroforestry is one of the important components of social forestry. In this forestry, nursery of economically important plants is prepared at Block and Panchayat levels and supplied at nominal costs to villages for agricultural and timber purpose. Such plantation is done around agricultural fields, wasteland of houses and schools.

The preparation of planned scheme to ensure rapid economic growth of large number of poor people without any capital in rural areas needs a different approach. The trees like mango (*Mangifera indica*), Guava, Mahua (*Madhuca longifolia, M. indica*), Neem (*Azadirachta indica*), Kanji (*Pongamia pinnata*), Imli (*Tamarindus indica*), Saijana (*Moringa oleifera*), Lisora (*Cordia*) spp, Ber (*Ziziphus*), Bel (*Aegle marmelos*), Babul (*Acacia nilotica*), August (*Sesbania grandiflora*), Bamboos etc. will provide necessary wood, timber, fruits, fodder and raw materials for a variety of cottage industries. Some other important plants suitable for social forestry are listed in Table 16.6.

Table 16.6. Some Important Plant species for Social forestry

Botanical names	Common name	Habitat Suitability	Remarks
Acacia catechu	Khair or Katha	River beds upto 1000 m altitude	Suitable for denuded areas
Aegle marmelos	Bel	up to 1000 m altitude	Medicinal, hardy, edible
Bauhinia purpurea	Lal kachnar	up to 1000 m	Fast growing
B. variegata	Kachnar	up to 1000 m	Fast growing, Flower buds edible, nutritious
Boehmeria regulosa	Genthi	up to 1300 m	Fast growing
Bombax ceiba	Semal	up to 1000 m	Fast growing
Betula alnoides	Kathbhuj	up to 2000 m	Plywood, suitable for ravine areas
Alnus nepalensis	Utis	up to 2000 m	Suitable for soil conservation
Dalbergia sisso	Shisham	up to 1000 m	Soil conservation
Emblica officinalis	Amla	up to 1000 m	Fruits rich in vitamin C
Eucalyptus hybrida	Mysore gum	up to 1000 m	Pulp
Ficus palmata	Pheru	up to 100 m	

(Contd...)

Castanea sativa	Chestnut	up to 2000 m	Soup delicacy
Grewia oppositifolia	Bhimal	up to 1700 m	Fodder and fibres
Morus alba	Shahtut	up to 1000 m	Sericulture, fast growing
Moringa spp	Saijana	up to 900 m	Fast growing, used in vegetables
Populus alba	Safeda	1000–1700 m	Match wood, Packing
Madhuca indica	Mahua	Plain	Medicinal and timber
Tamarindus indica	Imli	Plain	Fruit edible
Cordia spp.	Lisora	Plain	Used in sauce
Sesbania grandiflora	August, Jait	Plain	Flowers used in vegetables

FUTURE PLANNING OF SOCIAL FORESTRY

Today a new consciousness is developing in our country about the importance of forests and social forestry. Social forestry is a new catch word which aims at meeting the fuel, fodder, fruit and timber needs of the people. The programme aims at covering a total of Rs 2.15 million hectares (ha) of land of which 1.52 million ha land will be covered under social forestry programme and the rest under production forestry. The afforestation in general and social forestry programme in particular have increased substantially in recent years.

Several state governments, particularly Tamil Nadu, Karnataka, Gujarat, Haryana and West Bengal have already started new social forestry projects with the financial support of International agencies like World Bank, Swedish International Development Authority (SIDA), Canadian International Development Authority (CIDA) and the US Agency for International Development (USAID).

The Government of India has already sponsored programmes like a *"tree for every child", ecodevelopment camps* involving college students in tree planting and free distribution of seedlings to farmers. State sponsored Agroforestry schemes in Gujarat and Karnataka have resulted in plantation of trees of commercial values. The World Bank has funded extension of the existing social forestry programme to bring under tree cover some 110,000 ha private land holdings currently being used for growing crops. Despite these developments, many environmentalists are critical of government's social forestry schemes. They often question the choice of tree species, monoculture plantations and the actual beneficiaries of these schemes. Critics of these schemes like Sundarlal Bahuguna from the Garhwal district of Uttaranchal have charged that paper Mill owners and synthetic fibres manufacturers are using social forestry programme to meet their selfish end rather than helping the poor man facing energy famine.

In the hills of Uttaranchal recent afforestation pattern has changed and in place of naturally growing broad leafed deciduous tree species such as oak, Ash, Pangar etc., timber trees like Pines, Deodar and other conifers are being planted now which have industrial and commercial values. While the old stands of broad leafed trees produce a rich humus and hold the rain water well, the new trees don't play an adequate ecological role. The survival rate of new tree species is also poor because local people don't care for commercial value of those trees as they do not gain any thing from them.

There is enough scope for social forestry in the hills to cover lands with timber and other useful trees in order to check rapid run off rain water and to avoid sudden floods in the plains. Soil conservation measures should be adopted to check washing off of the top fertile soils and to increase moisture retaining capacity of soil which promote the growth of trees and crops. These two programmes, social forestry and soil conservation if launched together will prove to be beneficial for the poor masses and ensure the better living standards.

AGROFORESTRY

The word Agroforestry is a combination of two words : *Agro* = Agriculture and *forestry.* It is a part of social forestry programme under which tree plantation is done in private lands around agricultural fields, wasteland of houses in rural areas.

The main objectives of agroforestry are as follows :

(1) To reduce the dependence of farmers on forest for the purpose of firewood, fodder, timber and other products.

(2) To provide extra means for earning money or economic benefits to the farmers.

To make agroforestry programme a success, the state forest departments, Blocks, Panchayats and some private institutions are providing material help such as, the supply of seeds and nurseries of improved important tree species at nominal costs to farmers.

Advantages of Agroforestry

Agroforestry is advantageous to farmers in several ways as follows :

1. It makes available much needed fuel wood, fodder and other useful plant products to the farmers for their personal use.
2. It increases the income of the farmers.
3. Tree plantation near agricultural fields provides safety to crops from the adverse effects of strong wind, cold waves and heat waves.
4. It helps in the maintenance of fertility and productivity of soil.
5. The money earnt by the farmers through agroforestry enables them to repay the debts or loans in short period of time.
6. It provides insurance against natural calamities.

Selection of Trees for Agroforestry

The farmers select the trees for agroforestry keeping in view their personal needs. The following points are taken into account while selecting tree species for plantation :

1. **Fast growing species.** It is essential that tree species should be fast growing ones so that the plants attain tree size within a short period. The farmers hope to get good monetary return by selling the trees.

2. **Multipurpose use of the tree species.** Usually such tree species should be selected for plantation as are having several economic values so that they may fulfil the needs of farmers.

3. **Selection of trees as may not affect the light requirement of crops.** For agroforestry such tree species are selected as are sparcely branched and the branches are not dense so that sun light may easily reach the ground and become available to the crop plants growing under or near them.

4. **Selection of deeply rooted trees.** The agricultural crops absorb water and solutes from the top layer of soil. Therefore it is desirable that the selected tree species should be such as have roots penetrating quite deep into the soil to absorb nutrients and water from there.

5. **Selection of tree species according to soil conditions and climate.** For agroforestry programme such tree species should be selected by the farmers as are well adapted to grow vigorously in particular type of soil and climate of their locality.

6. Easy availability of plant species that checks soil erosion effectively.

Some Important tree species recommended for Agroforestry Programme.

Some of the tree species generally preferred by farmers for agroforestry are Babool (*Acacia catechu*), *Eucalyptus, Shisham* (*Dalbergia sissoo*), Poplar (*Populus alba*), Semal (*Bombax *malabaricum*), Kadamb (*Anthocephalus kadamba),* Mango (*Mangifera indica*), Ber (*Zizyphus* sp.), Mahua (*Madhuca indica*), Peepal (*Ficus religiosa*), Gular (*Ficus glomerata),* Neem, Bel (*Aegle marmelos*), Saijana (*Moringa* sp.), Sahtoot *(Morus alba),* Teak (*Tectona grandis*), Bamboo etc.

Now the state forest departments, agriculture department, Development Blocks, Village Panchayats all over India are making special efferts to promote agroforestry and are providing seeds and nursery of genetically advanced varieties of economically useful tree species to farmers at nominal costs. For extension and expansion of this programme meetings are being organised to make the farmers aware about agroforestry and to highlight the techniques and utility of the programme.

CHIPKO MOVEMENT

The Chipko movement was started by Mr. Sundarlal Bahuguna in Tehri-Garhwal district of Uttaranchal against ruthless felling of trees and destruction of forests by contractors. The movement gathered momentum in 1978 when the women faced police firings and other tortures. Though the objectives of the movement were broad based, the main objective was to protect the trees on the Himalayan slopes from the axes of contractors of forest. The movement was organized to oppose the ruthless destruction of nature to achieve short term gains. Mr Bahuguna emphasised the importance of trees in environment which check the erosion of soil, cause rains and provide pure air. The women of Advani village of Tehri-Garhwal tied the sacred thread around trunks of trees and they hugged the trees, hence it was called *Chipko Movement.* When any body tried to cut trees villagers faced police firing and later courted arrest in February 1978. This Chipko Movement under the leadership of Sri Sundarlal Bahuguna spread in other villages of Tehri-Garhwal. Mr Bahuguna presented a plan for the protection of soil and water through ban on tree felling in the Himalayas at the meeting of United Nations Environmental Protection (UNEP) held in London in June 1982. He emphasised that every standing green tree in the forest protects us from avalanches and land slides, purifies our atmosphere, saves our soil, water and other components of environment.

Chipko Movement is now a movement for planting of food, fuel, fibre, fodder and fertilizer yielding trees to make the people self-sufficient in all their basic needs. It would generate a decentralized and long term policy which will conserve the environment and bring everlasting peace, prosperity and happiness to mankind. Mr Bahuguna took this mission along with his dedicated workers and marched 3,000 km from Srinagar (Garhwal) to Siliguri. Mr. Bahuguna has focussed public attention for protection and conservation of forests which were being destroyed due to construction of Tehri Dam.

Tehri Dam on the Bhagirathi Ganga in Uttaranchal at the foot hills of Himalayas is big project of billions of rupees. The dam has displaced 85,000 people and has submerged the Tehri town and 100 villages. The site is prone to intense seismic activity. The 3,200 million tonnes of water could cause a major earth tremor. In the event of a disaster, Deoprayag, Haridwar and Rishikesh would be devasted and thousands of acres of agricultural land will be submerged. The efforts were made to pressurise Government of India to stop further construction of this dam because this dam could destroy the forests, wildlife, tribal habitation and disturb the ecosystem of that area. It is unfortunate that the construction of Tehri dam completed and has started functioning. Much ecological damage is apparent.

People from France, Sweden, Germany, Switzerland and several other countries have approached Mr. Bahuguna to get experience of this movement. In an International meeting held on June 5, in Stockholm to celebrate "World Environment Day" the following statement was given about *Chipko Movement.*

A powerful environmental movement has grown up on the slopes of mountains of Himalayas. Villagers have created an effective non-violent way to stop the devastation by forest industries. When the axemen come, the people form circle around the trees and they embrace the trees. This has given the movement its name *Chipko Andolan*, the tree hugging movement."

The following suggestions are made by the organisers of Chipko movement :

1. All commercial green tree fellings should be stopped forthwith.
2. No new contracts should be entered by forest departments with the industrialists to supply raw materials and old contracts should be revised, especially those made for long-term supply of raw materials at cheap rates.
3. Pine trees damaged due to extraction of resin should be given rest for a period of 10 years.
4. A massive programme for setting up biogas plants, especially in the lower region be taken up. Night soil and other refuge of the cities be utilized by bio-gas industries.
5. Every water source should be trapped to generate hydroelectric power. People should be encouraged to set up their community power houses.
6. Plantation of the trees for food, fodder, fuel, fertilizer etc, should be encouraged.
7. All branches of forest department should be integrated into one. At a later stage integration of Agriculture, Horticulture and Forest department should be considered into one "Land Use Dept".
8. There should be strong people's participation in protection of environment. Government alone cannot achieve success. Foot-marches should be organized in all districts to create general awareness in public regarding protection of environment.

In Chipko Movement there is greater people participation for soil, conservation and plant protection. The rural people have preferred their own priorities. Conservation work first began in the Chipko villages not for protecting trees but walls were built around agricultural fields to protect from wild animals. The grasses grew rapidly in the protected areas in the fields. When this benefit became clear to women, they began to organise themselves for protecting and afforesting other patches of common lands. While trees take many years to bear fruit, grasses grow faster in a protected area and can provide fodder in few months. The Chipko women have devised a simple way for sharing this produce. The head of the *Village Mahila Mangal Dal* announces once a month a particular day on which one member from each family can take away as much grass as she can. Thus the role of women in ecological regeneration is very significant.

Today many voluntary organizations in the country are involved in environmental issues, although their objects are different. Some have aim in preventing deforestation, while others are interested in afforestation. Some are interested to prevent the construction of dams. Some prevent water and air pollution. Among all these organizations Chipko movement in the Uttaranchal Himalayas, is the oldest and most famous of all the organisations which have played a major role in deforestation. There is another parallel movement in the South *'the Apiko Movement'* in the western Ghats of Karnataka started by Medha Patekar. Dams like silent valley and Bethi have already been stopped because of strong people's protests through this movement but again these are heading towards completion. *Kerala Sasta Sahitya Parishad* is another important organization which made efforts against water pollution of the Chaliyar river in Kerala by a Rayon mill. There are many others who are doing excellent work in mobilising people to prevent further ecological destruction and bring about ecological regeneration and protection.

QUESTIONS

1. "Afforestation plays an important role in the improvement of Environment." Discuss
2. Write brief notes on :
   (*i*) Afforestation
   (*ii*) Deforestation
3. What do you mean by forest management? Give a brief account of Afforestation programme in different habitats.
4. What are the aims and objective of social forestry? Discuss the role of forest department in social forestry programme.
5. Discuss the significance of forests. Mention different causes of deforestation.
6. Write notes on:
   (*i*) Agroforestry (*ii*) Reserved forestry.
7. What is Chipko Movement? What are main objectives of this movement?

17

Waste and Its Managements

Human society produces some unwanted and discarded materials which are called **wastes**. Wastes are produced from different activities such as household activities, agricultural activities, industrial activities, hospitals, educational institutions, mining operations, and so on. These sources generate different types of wastes, many of which are hazardous in nature. They cause spread of many diseases. In general, the wastes maybe categorised as follows:

Solid wastes: The solid wastes are the useless and unwanted substances discarded by human society. These include urban wastes, industrial wastes, agricultural wastes, biomedical wastes and radioactive wastes. The term **refuse** is also used for solid waste.

Liquid wastes: Wastes generated from washing, flushing or manufacturing processes of industries are called liquid wastes. Such a waste is called sewage. The most common practice is to discharge it on the ground, nallahs, rivers and other water bodies, often without any treatment.

Gaseous wastes: These wastes are released in the form of gases from automobiles, factories, burning of fossil fuels etc. and get mixed in the atmosphere. These gases include carbon monoxide, Co_2, sulphur dioxide, nitrogen dioxide, ozone, methane, etc.

SOURCES OF WASTES

Wastes produced from different sources, are classified as follows:

1. Urban or Municipal wastes
2. Industrial wastes
3. Commercial wastes
4. Agricultural wastes

Municipal Solid Waste

The wastes, collected from the residential houses, markets, streets and other places mostly in the urban areas and disposed of by municipal bodies are called **municipal solid wastes** (MSW). In general, the urban solid wastes are called **refuse.** The Municipal solid wastes are a mixture of paper, plastic, clothes, metals, glass, organic matter etc. generated from households, commercial establishments and markets. The proportions of different constituents vary from season to season and place to place depending on the life style, food habits, standard of living and the extent of commercial and industrial activities in the area.

Municipal solid wastes are collected locally and the amount collected depends upon the size and consumption of the population. The municipal wastes, their contents and sources are summarised in the Table 17.1 & 17.2.

Table 17.1 Classification of Refuse Material

Name	Contents	Source
Garbage	Wastes from kitchen, cooking and serving of food, slaughter houses, market refuse, waste from handling, storage	From households, institutions etc.
Rubbish: A. Combustible B. Non-combustible	Leaves, grasses, plants clothes, paper, leather, rubber etc. Bottles, crockery, plastic materials, glass, metals etc.	Households, restaurants, markets etc,
Ashes	Residues from fire, cooking, cinders etc.	Fuel burning and cooking by households and small industries
Street refuse	Leaves, dirt, litter, paper, plastic etc.	From streets
Dead animals	Small animals; cats, dogs, poultry etc. Large animals; horses, cow etc.	
Construction and demolition wastes	Wood, roofing and sheating scrap, rubble, concrete, plaster etc.	Buildings
Sludge	Settled solid components of sewage wastes	Sewage treatments plants, septic tank

Table 17.2 Urban Wastes Generated Daily in Major Indian Cities

City	Municipal Solid Waste (tonnes/day)	Per Capita Waste (kg/day)
Ahmedabad	1,683	0.585
Bangalore	2,000	0.484
Bhopal	546	0.514
Chennai	3,124	0.657
Delhi	4,000	0.475
Hyderabad	1,566	0.382
Jaipur	350	0.321
Kanpur	1,200	0.640
Kolkata	3,692	0.383
Lucknow	1,010	0.623
Ludhiana	400	0.384
Mumbai	5,355	0.436
Patna	330	0.360
Pune	700	0.312
Vadodara	400	0.389

Source: Central Pollution Control Board, New Delhi.

Industrial Wastes

Industrial wastes are released from chemical plants, paint industries, cement factories, power plants, metallurgical plants, mining operations, textile industries, food processing industries, petroleum industries and thermal power plants. These industries produce different types of waste products (Table 17.3). Industrial solid wastes can be classified into two groups:

Non-hazardous wastes: These wastes are produced from food processing plants, cotton mills, paper mills, sugar mills and textile industries.

Hazardous wastes: Hazardous wastes are generated by nearly every industry. Metals, chemicals, drugs, leather, pulp, electroplating, dye, rubber are some of important examples.

Liquid Industrial waste that runs into a stream from a factory can kill the aquatic fauna and also cause health problems for humans.

Table 17.3 Industrial products and hazardous waste

Products	Hazardous Wastes
Medicines	Organic solvents and residues, heavy metals (mercury and zinc)
Metals	Heavy metals, fluorides, cyanides, acids and alkaline cleaners, solvents, pigments etc.
Paints	Heavy metals, pigments, solvents, organic residues
Leather	Heavy metals, organic solvents
Oil, Petroleum products	Oils, phenols, organic compounds, heavy metals etc.
Pesticides	Organic chlorine compounds, organic phosphate compounds.
Plastics	Organic chlorine compounds
Textiles	Heavy metals, dyes, organic chlorine compounds, solvants

Agricultural Wastes

Agricultural areas produce plants and animals wastes. Excess use of fertilizer, pesticides and other chemicals used in agriculture and the wastes formed from these cause severe land and water pollution. They also contaminate the soil. Among pesticides chlorinated hydrocarbons, DDT, BHC, endrin, dieldrin, lindane, parathion, malathion and endosulphon are important which are absorbed by the soil and contaminate crops grown in the soil. Other agriculture wastes are produced from sugar factories, tobacco processing units, slaughter houses, livestock, poultry etc.

Commercial Wastes

With the advancement of modern cities, industries and automobiles, huge amount of wastes are generated daily. These include markets, roads, buildings, hotels, commercial complexes, hostels, auto workshops, printing press etc. Hospitals, nursing homes and medical institutes also release tremendous amount of wastes which are hazardous and are much toxic in nature. Many chemicals and disposable items are also produced from these units. These wastes are dumped in inhabited areas which pose much danger to human health and life and cause several types of infectious diseases.

Apart from wastes, generated from the above sources, there are certain wastes produced from mining activities and radioactive substances that cause much damage to the society and environment.

Mining: The wastes generated by mining activities disturb the physical, chemical and biological features of the land and atmosphere. The wastes include the overburden material, mine tailings (the waste left after ore has been extracted from rock), harmful gases released by blasting etc.

Radioactive substances: Although every precaution is taken in the functioning and maintenance of nuclear reactors, yet it has been observed that measurable amount of radioactive waste material escapes into the environment. Other sources of radioactive wastes are from mining of radioactive substances and atomic explosion etc.

Bio-medical wastes: Wastes, which are produced from the hospitals, medical centres and nursing homes are called **bio-medical wastes.** These wastes are highly infectious which include used bandages, infected needles, animal remains, cultures, amputated body organs, dead human foetuses, wastes of surgery and other materials from biological research centres. Pharmacies discard outdated and unused drugs; testing laboratories dispose of chemical wastes which are hazardous in the environment.

SOLID WASTES IN SPACE

Recent report of Times of India (30th January, 2006) reveals that more than 9,000 pieces of solid wastes are orbiting in Space. This solid waste hazard can be expected to get worse and critical in the next coming years.

Waste in Space is about 5,500 tones orbiting the Earth as reported by NASA Scientist. The waste particle may collide and break into more pieces. This junk can pose a great risk to commercial and research flights and other space activities. Much of the debris results from explosions of satellites. At present there is no solution to remove the waste although a part of this solid waste has been brought on the earth by spacecrafts.

Classification of Wastes

In general, the wastes are classified on the basis of their biological, chemical and physical properties and also on the basis of nature.

Biodegradable Wastes : These wastes are natural organic compounds which are degraded or decomposed by biological or microbial action. Biodegradable wastes are generated in food processing units, cotton mills, paper mills, sugar mills, textile factories and sewage. Waste of slaughterhouses is biodegradable and some part of it is used, for example, skin is used to make shoes. Most of the wastes from these industries are reused. When these wastes are in excess, they act as pollutants and are not easily decomposed and they take much time for their decomposition.

Non-Biodegradable Wastes : These are not decomposed by micrbes but are oxidized and dissociated automatically. Coal stone, metal scraps, sludge are generated from colliery operations. Refineries produce inert dry solids and varieties of sludge containing oil. Fly, ash is the major solid waste from thermal power plants. Generally, these wastes are not reused and accumulate in the ecosystem and some of it move through biogeochemical cycles. Non-biodegradable wastes also include DDT, pesticides, lead, plastics, mercuric salts etc.

Hazardous wastes : Many chemical, biological, explosive or radioactive wastes, which are highly reactive and toxic, pose a severe danger to human, plants or animal life and are called **hazardous wastes**. They are highly toxic in nature.

Hazardous wastes, when improperly handled, can cause substantial harm to human health and to the environment. Hazardous wastes may be in the form of solids, liquids, sludges or gases.

They are generated primarily by chemical production, manufacturing and other industrial activities. The important hazardous wastes are lead, mercury, cadmium, chromium, many drugs, leather, pesticides, dye, rubber and effluents from different industries. They may cause danger during inadequate storage, transportation, treatment or disposal operations.The hazardous waste materials may be toxic, reactive, ignitable, explosive, corrosive, infectious or radioactive.

1. **Ignitability:** These are wastes that easily catch fire with a flash point less than 60°C. Such fires not only present immediate dangers but can spread harmful particles over wide areas.

2. **Corrosiveness:** These comprise mostly acidic or alkaline wastes that corrode other materials. These require special containers for disposal and should be separated from other wastes as they release toxic contaminants.

3. Reactivity: These are explosive or highly reactive wastes. These undergo violent chemical reactions and are exploded to generate heat and toxic gases.

4. Toxicity: These wastes release toxins or poisonous substances and pose hazards to human health and the environment.

DIRTY DOZEN

The following chemicals currently used in agriculture and industry, are considered to be most toxic and hazardous to human and other animals. They are popularly known as ***"Dirty Dozen'.***

Pesticides : *Aldrin, Chlordane, DDT, Dieldrin, Endrin, Heptachlor, Mirex. and Toxaphene.*

Industrial Chemicals : *PCBs (Polychlorinated biphenyls) used in electrical transformers, paint additives and plastics.*

HCB (Hexachlorobenzene) used in fireworks, synthetic rubber and also as a pesticide.

By-products

Dioxins - formed during chemical production, waste incineration, car emissions etc.

Furans - Formed along with dioxins.

IMPACTS OF WASTE ACCUMULATION

Industrialization on a massive scale, increasing urbanization, advance technology in agriculture and changing life pattern have resulted in the production of huge amount of wastes. The improper waste disposal creates many ecological and social problems, for instance, accumulation of wastes in the densely populated areas, disposal of urban sewage and industrial wastes discharged into rivers etc. affect soil, air and water ecosystem. Chemical, biological and explosive wastes pose immediate or long run danger to the life of man, plants and animals.

The dumping of solid wastes is hazardous to human health. It has been estimated that about twenty-five human diseases are associated with solid waste. There is an increase in the number of rats and flies due to dumping of wastes in open places and they are the carriers of other organisms responsible for several dreaded diseases. The flies which carry pathogenic organisms are spreading diseases like dysentery, diarrhoea etc. It is estimated that about 70,000 flies are produced in one cubic foot of garbage.

Dumping of solid wastes has a number of adverse effects on all the components of an ecosystem and they also affect the aesthetic sense as well. Some of the impacts of accumulation of wastes are described below:

Spoilage of Landscape

It is a common practice to dump plastic bags, containers, vegetables, fruit peels, cans etc. in the open area without thinking about its consequences (Fig. 17.1). We need to be fully aware that improper disposal of waste spoils the beauty of the landscape.

Pollution

Dumping of wastes in a haphazard and unscientific manner has serious environmental impact. Most of the wastes contain organic compounds a number of inorganic minerals and other harmful matter which contaminate the environment and lead to:

1. Degradation of land,
2. Pollution of drinking water,
3. Destruction of aquatic life,
4. Degradation of ground and surface water used for irrigation and industries, and
5. Improper disposal of wastes cause soil, air and water pollution.

Fig. 17.1 An overflowing landfill.

Health Hazards

Human health is directly concerned with the overall quality of the environment. However, in the last few decades, due to human's desire for rapid advancement in industrialisation, agriculture and other activities, much damage has been done to the environment. Large-scale deforestation, drastic climatic changes and pollutants on land, air and water are some of the unpleasant consequences that ultimately affect the human health. It is a well-known fact that an adult healthy man is exposed everyday to polluted air through breathing and to food and water through oral intake. Our skin is also exposed to the environmental chemicals which lead to many health problems immediately or after sometime.

Health hazards due to air pollution

Hazardous air pollutants present in the atmosphere affect human health both directly and indirectly. It may be a short-term or long-term effect. The following are the adverse effects on human health.

1. Toxic gas carbon monoxide reduces the blood oxygen and formation of haemoglobin, causing injury to heart and central nervous system.
2. Sulphur dioxide and sulphuric acid both cause irritation in the respiratory tracts of humans and high concentrations of sulphur dioxide leads to severe heart and lung diseases like bronchitis, asthma, etc.
3. Nitrogen oxide at higher concentration affects respiratory organs, liver and kidneys.
4. Ozone can seriously affect the pulmonary functions.
5. Lead can cause injury in blood-formation organs and nervous system, especially impairing of brain functions of newborn babies.
6. Pesticides and radiations are other toxic air pollutants which are very dangerous for human health.
7. Metal, dusts, asbestos and hydrocarbons shorten the life span and cause deterioration of nervous system and there is additional risk of cancer.
8. In mining operation, silica and dust cause pneumoconiosis (common disease in mine workers).
9. Petrolium components can affect the blood forming organs, brain, teeth bones etc.
10. Mercury and cadmium are known to damage the kidneys and brain.

Health Hazard due to Water Pollution

Water is said to be polluted when its quality or composition is changed either naturally or as a result of human activities. Nearly 80% of the human diseases in developing countries are due to polluted water alone. The well-known impacts of water pollutants are as follows :

1. A large numbers of industrial pollutants that come to human body through drinking water and contaminated food threaten the life and health. The famous MINAMATA and ITAI-ITAI diseases took a big toll of human life in Japan due to mercury and cadmium from the industrial effluents in the aquatic ecosystem.
2. Some agrochemicals like chlorinated pesticides disposed in water accumulate in the aquatic food chains and enter the human body causing heavy infection. In coastal Karnataka, several people died by consuming crabs contaminated with pesticides.
3. Changes in water quality due to deficiency of iodine lead to goitre which has been found to be endemic in many parts of India.
4. Many water borne diseases prevalent in the Indian population, like cholera, typhoid, gastroenteritis and hepatitis are due to polluted water.
5. Excess fluorine in drinking water has caused bone and teeth diseases (fluorosis), the most severe disease is the KNOCK-KNEE syndrome in Andhra Pradesh.

Health Hazards due to Soil or Land Pollution

The accumulation of toxic chemical compounds, salts, disease-causing organisms and radioactive materials in the soil cause various health problems. The impact of waste accumulation in soil/land has shown the following major health effects:

1. The impact of land pollution on human health is indirect. The pollutants added in the soil enter the human body through water or air through the food chain.
2. Several agrochemicals like DDT, fluorine, arsenic, lead compounds and organo-phosphorus compounds are super toxic and cause symptoms like nausea, vomiting, diarrhoea, sweating, salivation and muscular tremors.
3. Some rodenticides as strychnine, sodium fluoroacetate etc. are blood coagulants.
4. Ethylene dichloride, ethylene dibromide and methyl dibromide accumulate in liver, kidney, heart, spleen and cause degenerative lesions.

Impact of Waste Accumulation on Terrestrial Life

Hazardous wastes may pollute soil, air, surface water and underground water. The oil pollutants may affect man, plants and animals. These toxic substances are transferred to different organisms through the food chain and cause a number of complications in living organisms. Some of these are as follows:

1. Many toxic chemicals, pesticides, other agricultural wastes released into the environment that are taken up by the plants from air, water and soil. Plants growing under such conditions are severely affected by these toxic chemicals.
2. Exposure to high concentration of pollutants may cause acute injuries like chlorosis, discolouration and even the death of plants.
3. Crops show reduced productivity and yield. The quality of plant nutrients is also decreased.
4. Sulphur dioxide is a most toxic pollutant which damage the crops.
5. In recent years, the losses to agriculture and animal life due to fluoride content have greatly increased.
6. Besides morphological changes, biochemical and physiological changes have also been observed in many mammals including man.

7. Too much accumulation of wastes disturb the behaviour of wild and domestic animals and also cause health problems.
8. Some highly toxic chemicals lead to genetic disorders in animals.
9. Several domestic animals like cow, buffalo, goat etc. often eat polythene and plastics bags along with food material which ultimately reach to their alimentary canal causing many disorders and even their death.

Impact of Waste Accumulation on Fresh Water

Large amount of wastes of human society are disposed of in the rivers, lakes, ponds and other aquatic bodies making the water polluted which is not fit for drinking and other domestic purposes. The impacts of waste dumping on aquatic life are as follows :

1. The toxic wastes reaching the water bodies badly disturb the aquatic life.
2. The sewage of cities is often drained into the rivers, which is dangerous to flora, fauna and human life.
3. Due to heavy accumulation of wastes into the canals, lakes and rivers, oxygen concentration is reduced considerably thus affecting the life of fishes and other aquatic populations. In extreme deficiency of oxygen most of the fishes die.
4. Sewage from municipalities, sanetoria and tanneries discharged into the rivers, canals and lakes etc. carry many species of bacteria and other microbes which cause diseases in human and animals.
5. Some pollutants for example heavy metlas, cyanides and several other organic and inorganic compounds are harmful to aquatic organisms. Many of them especially non-biodegradable ones accumulate in the body of organisms and cause long-term effects.
6. Biodiversity decreases in highly polluted aquatic habitats.
7. The DDT and other pesticides present in very low concentrations in water may accumulate to higher concentration within algae, insects and fishes. The birds or people that feed on these fishes are then exposed to very high levels of hazardous substances. In birds, these substances can affect the egg production and bone formation.

Impact of Waste Accumulation on Marine Life

One of the least known but most significant uses of the sea is as an enormous dumpsite. In the past, the oceans were able to assimilate the wastes of the civilization without noticeable adverse effects. However, industrialization and other associated developments along with sharp increase in global population have given rise to huge amounts of wastes that are now taxing the capacity of the oceans to absorb them. Human wastes ranging from the raw sewage of urban centres to junked appliances and automobiles have heavily polluted the sea shores. The impacts of waste dumping on marine life are as follows:

1. The growth of marine algae is affected.
2. Massive oil spills not only spoil innumerable beaches and estuaries but also cause widespread damage to marine life.
3. Herbicides and pesticides (especially the organo-chlorides) reach the oceans via the wind and rivers and contaminate marine water.
4. It is a matter of great concern that mangrove forests are being damaged at an alarming rate due to disposal of wastes along sea shores.
5. Thermal and radioactive pollution have disturbed the life of fishes in estuaries and coastal ecosystems. Their breeding is also affected adversely.

WASTES MANAGEMENT

Waste management is a burning problem in the urban, industrial and rural areas. Hazardous wastes contaminate the environment in various ways. Even before such wastes are safely disposed of they pollute the air, water and soil and pose a constant danger to human life. Improper handling of hazardous wastes contaminates the atmosphere and the surrounding environment. The discharge of hazardous substances into water bodies often kills aquatic life. Uncontrolled disposal of wastes on land can easily pollute the soil and groundwater. Therefore, the careful management of hazardous wastes is one of the most important concerns of the modern society. Wastes management involves the following steps:

1. Reduction of wastes
2. Recycling of wastes
3. Treatment of wastes
4. Disposal of wastes

Reduction of Wastes

The first concept of waste management involves the reduction in the quantity of wastes. This can be achieved by the following steps:

Source reduction: In this the industrial processes are modified in such a way that there is optimum use of raw materials and the amount of hazardous wastes generated is reduced to a bare minimum.

Concentration: The wastes that are generated during various industrial processes are concentrated using modern methods, such as precipitation or decantation techniques. Thus, the volume of wastes is reduced, which can be manage easily.

Segregation: The wastes are segregated according to their chemical composition, hazard potential and physical nature. This helps in their eventual disposal.

Recycling of Wastes

Many types of wastes contain some useful substances. Recycling is the method by which the wastes can be further used through reprocessing. Recycling commonly refers to two things:

Reuse: This refers to the case where certain wastes can be again used with very little reprocessing. For example, discarded bottles and containers, certain chemicals, lubricants, etc. can be put to further use with minimum processing.

Recycling: It differs from reuse in the sense that the wastes must be thoroughly treated and re-processed before it can be used again. Some recycling processes use materials for the same purposes; for example, metal scraps, cans, bottles, etc are usually melted and recast into the original material. Other recycling processes turn old materials into entirely new products; for example, wood remains are used in the paper industry. Recycling encourages awareness and responsibility for waste management.

Treatment of Wastes

Hazardous wastes can be treated by a number of processes to change the physical and chemical composition of the wastes so as to reduce and neutralise their toxicity. The treatments vary depending on the physical state and chemical nature of the hazardous wastes. The treatment of wastes can be divided into three broad types:

Physical treatment: Physical treatment of solid wastes includes mechanical separation and segregation so as to facilitate the disposal. In case of liquid wastes, physical treatment mainly involves phase separation. This involves the process of sedimentation, whereby sand, silt and other particulates settle at the bottom of the tank. Extremely fine colloidal particles are combined by the process of coagulation and then settled.

Chemical treatment: The hazardous wastes are treated with different chemicals, so as to neutralise them and remove their toxicity. This treatment involves the modification of the chemical properties of the wastes through oxidation, reduction, precipitation, neutralisation, etc.

Biological treatment: This involves the action of microorganisms on the wastes products. As you know, the microorganisms like bacteria and fungi are known as nature's scavengers which break down the organic substances present in the wastes.

Biological treatment involves the processes of:

1. Conditioning : The wastes are exposed to the atmosphere for a specific period.
2. Digestion : The action of microorganisms breaks the complex organic compounds into simpler forms.
3. Composting : The organic matter is converted into useful manures.

Disposal of Wastes

The final step in hazardous waste management involves the safe disposal of wastes. However, there are certain steps that are necessary before disposal which are as follows:

1. Collection of wastes
2. Provisional storage of wastes
3. Transport of wastes

After the above pre-requisites are fulfilled, the wastes are finally disposed of depending on their nature and chemical properties.

Processes involved in Solid Waste Management

The different steps involved in the management of solid waste from its generation to final disposal are as follows (Fig. 17.2).

Waste generation: Material that is discarded because it has served its purpose or is no longer useful; becomes the waste.

Storage and Processing: The activities which are associated with the handling, storage and processing of solid waste at the point of generation.

Collection: The activities concerned with the collection of solid wastes at specific locations.

Transfer and Transport: Activities involve the transfer of wastes from the collection points to the vehicles and then transport of wastes to the disposal sites.

Processing and Recovery: Methods and facilities that are used to recover the wastes for recycling and other treatments.

Disposal: The final step is the disposal of solid wastes to a landfill site.

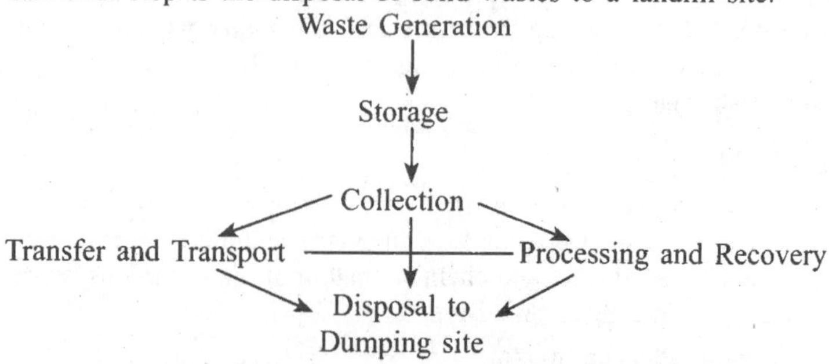

Fig. 17.2 Solid waste management system

Solid wastes are mostly generated in the urban areas and hence, solid waste disposal is primarily an urban problem. Huge amounts of wastes are produced by different activities, which need to be properly handled. The common methods of solid waste disposal include:

- Open dumps
- Sanitary landfills
- Ocean dumping
- Drainage
- Incineration
- Composting

Open dumps : Open dumps are the oldest, simplest and most common method of disposing solid wastes. Open, unregulated dumping is the predominant method of waste disposal in most of the developing countries. In developed countries, however, open dumping is banned. Open dumps are located wherever land is available. The wastes are dumped on the vacant lands which take the shape of huge piles. Sometimes, the wastes are ignited and burnt.

Disadvantages: Though open dumping is a cheapest method of solid wastes disposal, it has the following drawbacks:

1. Open dumps are aesthetically revolting.
2. They breed pests and microorganisms.
3. Pollute the air, surface water and groundwater (by leaching).

Sanitary landfills : The developed countries now dump their solid wastes in sanitary landfills. A sanitary landfill is a disposal site that does not create a hazard for public health and where solid waste disposal is regulated and controlled (Fig. 17.3). The waste dumped into the sanitary landfills are compacted to reduce the volume and then covered with a layer of soil at the end of each operation. This soil cover discourages breeding of pests, decreases offending odours and minimises leaching.

Precautions: Before operating a sanitary landfill the following precautions are required:

1. Proper site selection for a landfill is important.
2. Landfill should be located far from rivers, lakes, flood plains and groundwater recharge zones.
3. The topographic relief, depth of groundwater table and the soil and rock type should be taken into consideration.
4. The bottom of the landfill should have an impermeable lining of clay or plastic material.
5. Proper drainage systems should be installed around the landfills so as to check water accumulation.
6. Monitoring wells to evaluate potential groundwater pollution should be installed.

Advantages

1. No open burning
2. Less odour
3. Less cost
4. Accommodates huge amount of wastes
5. Landfills can be used for other purposes

Disadvantages

1. More noise pollution
2. More dust

3. Space problem
4. Emits greenhouse gases (methane)
5. Ground water pollution
6. Slow decomposition

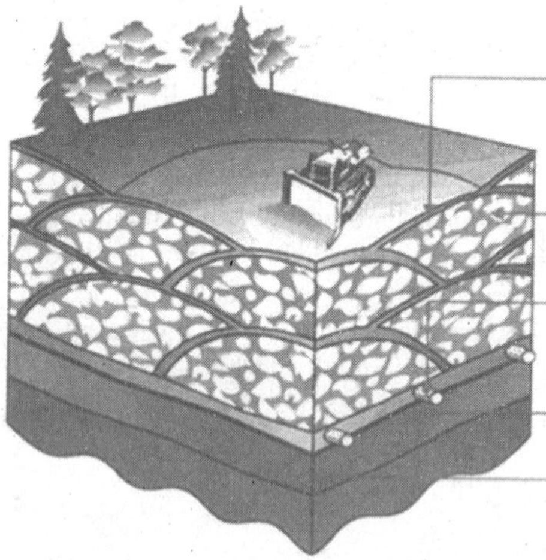

Fig. 17.3 Sanitary landfills

Ocean Dumping

Oceans are the biggest reservoirs. Every year approximately 25000 metric tonnes of solid wastes including packaging materials, bottles, cans, containers, etc are dumped into the sea. This careless dumping of wastes into the oceans, has many disadvantages which are as follows:

- It has littered many coastal areas with non-biodegradable wastes.
- Reduction of the dissolved oxygen content of marine environment.
- Eutrophication caused by nutrient rich wastes especially in shallow waters, such as estuaries.
- Habitat change of the organisms adversely affecting the marine ecosystem.

Drainage

It is one of the most common practices in which the wastes are drained directly into the water bodies like rivers, lakes, ponds, oceans, etc. Although it is economical, it pollutes the water making it unfit for use. This practice is now avoided in most of the affluent countries and other safe disposal methods are being implemented.

Incineration

As space for sanitary landfills is decreasing day-by-day, burning is the most modern method of solid waste disposal. This process is called incineration, in which the hazardous wastes are destroyed through high temperature burning (1300°C). The plant in which the wastes are burnt is called **incinerator** (Fig. 17.4). In some incinerators, the resultant steam is generated which is used to drive a turbine to generate electricity. In advanced countries especially designed municipal incinerators are capable of burning thousands of tones of wastes per day. There are two types of incinerators:

i. **Refuse derived:** The wastes are segregated to remove uncombustible materials before combustion.

ii. **Mass burn:** In which all the solid wastes are burnt without sorting.

During the process of incineration, the volume of garbage is reduced by 80 to 90% and the weight by 75%, thus making disposal an easier task. In India, incinerators are not successful because of the huge costs involved and also because the wastes produced in the Indian cities contain less percentage of combustible matter.

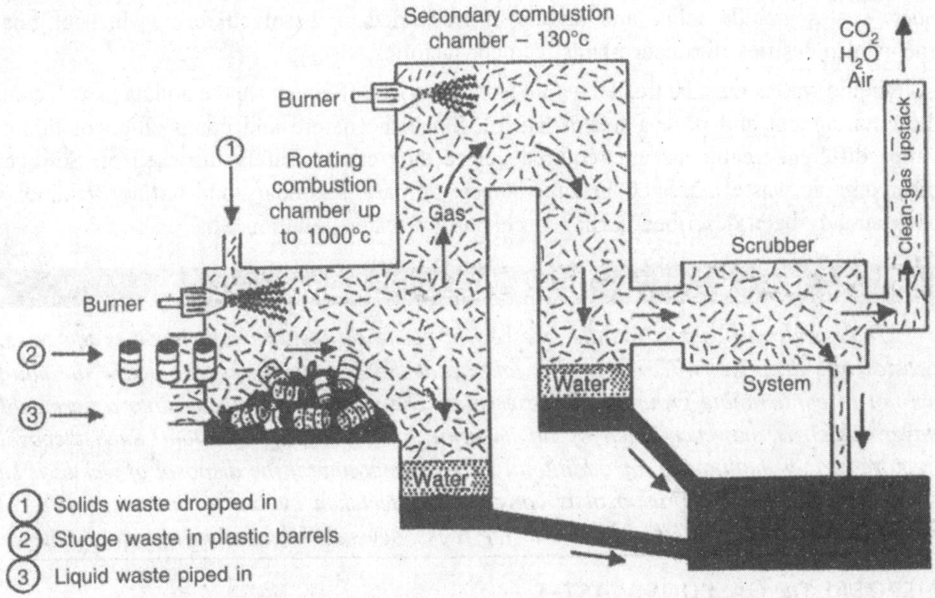

Fig. 17.4 Generalized diagram of an incineration system

Composting

Composting is a biochemical process in which organic materials are decomposed to form humus. It involves partial decomposition of moist solid organic wastes under aerobic conditions by the action of microorganisms. The organic material present in the wastes are not buried or burnt but are thus converted to useful manures.

During the process of composting the organic wastes like food products, leaves, paper, saw dust, etc are deposited in a trench, layer by layer. When the trench is filled, the pile of wastes is dampened and covered by a layer of soil. By the action of microorganisms, some amount of heat is generated, which helps in the decomposition of the wastes. After a time of three to four months, the wastes are converted into nutrient rich compost.

The benefits of composting include:

1. Compost acts as manure.
2. Conversion of household organic wastes into manure.
3. Nutrient-rich compost increases the water holding capacity of soil and improves crop yields.
4. It is an easy, inexpensive and eco-friendly fertiliser.

DISPOSAL OF LIQUID WASTES

The liquid wastes can be broadly divided into two groups:

- Biological wastes
- Non-biological wastes

The biological liquid wastes, for example, municipal sewage, wastes from tanneries, paper mills, sugar mills, distilleries, food manufacturing uits, etc contain organic matters that are degradable in nature.

The non-biological liquid wastes comprise of inorganic substances that are non-degradable. Such wastes include solids and liquids in suspended or dissolved form, which originate from chemical industries, fertiliser plants, cement factories, etc.

Liquid wastes must be treated before being disposed of otherwise the pollutants will contaminate the environment and pose a health hazard. Since the nature and composition of liquid wastes varies different treatments are required before they can be safely disposed of. Sewage or any other organic waste is treated in three stages: primary, secondary and tertiary treatments which have already been described earlier in chapter on water pollution.

DEEP WELL DISPOSAL

Deep well disposal is another method of waste disposal, which involves the injection of wastes into deep wells. Care must be taken that the deep well must be below the aquifers so as not to contaminate groundwater reservoirs. The wastes are injected into a permeable rock layer, which is, however, lined by an impermeable layer at the bottom. Such disposal wells require further monitoring, by additional well, to determine if the disposal of wastes is secured. Such type of deep well disposal is constantly undertaken by the oil industry where oil field brine (salt water) is injected back into the rocks, several thousand meters below the surface.

DISPOSAL OF GASEOUS WASTES

The gaseous wastes are the main source of air pollution. As air pollution is most dangerous to human health, the gaseous wastes must be properly treated before they are disposed of into the atmosphere. The following techniques are undertaken for the management and disposal of gaseous wastes.

Filters

Filters are employed to remove the particulate matter in the gaseous wastes. The filters remove the particles by trapping them into porous bags which allow only air to pass through. Such filters are also called **bag filters.** The bags are prepared from cotton, polyester, spun glass fibers, etc. The bags are 10-15 m long and 2-3 m wide. When the gaseous wastes are blown into the bottom of the bag, gases escape through the pores and the particulate matter is trapped. After regular intervals the bags are cleaned to remove the accumulated particles. Such filters are commonly used in textile, glass and asbestos industries.

Scrubbers

Scrubbers are devices that are installed to separate particulate matter as well as gaseous pollutants. It consists of a device in which the gaseous wastes are passed through a space containing wet packing. Therefore, these are also called **wet scrubbers** (Fig. 17.5). Water is most commonly used for wet packing, while some other liquids may also be used to remove the pollutants. The solid particles as well as the gaseous pollutants are adsorbed on the wet packing. Though, scrubbers are highly efficient as they remove 80 to 90% of the hazardous wastes, they are quite costly as they consume water and elecricity. A cost effective method is to install **spray towers** where the gaseous pollutants are sprayed with water to remove the gaseous and particulate contaminants.

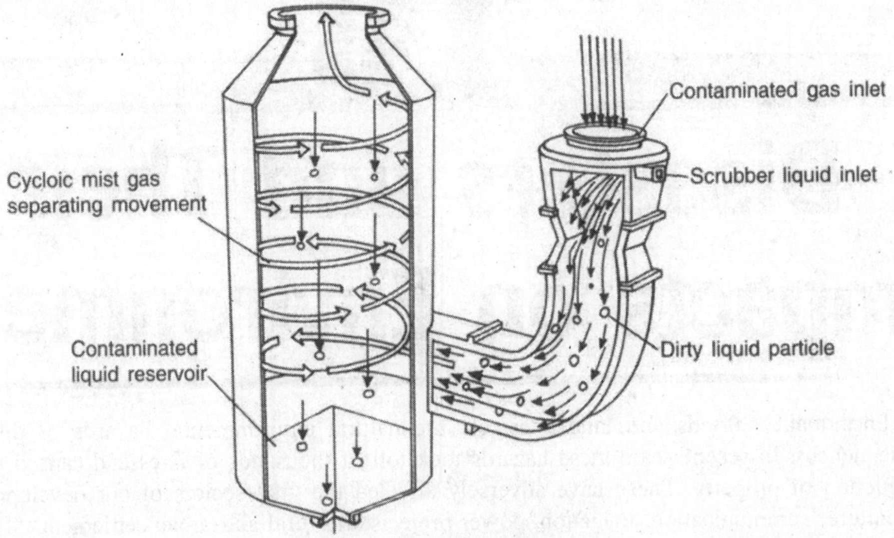

Fig. 17.5 Scrubber

Electrostatic Precipitators

Electrostatic precipitators (ESP) are electrical devices for controlling particulate matter from smoke and other gaseous pollutants. The gaseous wastes are passed through a collection plates at high voltage (30,000 to 1,00,000 volts) whereby negative electrical charge is imparted to the suspended particles. These negatively charged particles migrate towards the collection plates which are positively charged. During collection, the charge is neutralized and the particles are removed by washing or by gravity action. Electronic precipitators are extremely efficient (99%) as they can remove micron size particles and are also simple to maintain. However, these precipitators consume large amount of electricity.

Catalytic Converters

Air pollution from the automobiles is quite common in the urban areas. Automobiles emit carbon monoxide, hydrocarbons, etc that are a constant health hazard. The control of these gases is of considerable importance, not only for human health, but also for controlling the greenhouse effect and ozone layer depletion. Catalytic converter is the most common device installed in the automobiles that convert carbon monoxide to carbon dioxide and hydrocarbons to carbon dioxide and water.

QUESTIONS

1. Define waste. Describe the sources of different types of waste.
2. What do you mean by biodegradable and non-biodegradable wastes? Describe briefly the impacts of waste accumulation on the environment.
3. What is meant by Waste management? Throw light on the different steps of Waste management.
4. Write short notes on :

   (*i*) Hazardous wastes

   (*ii*) Composting of wastes

18

Disasters and Their Impacts on Environment

Earthquakes, floods and landslides, etc are natural environmental hazards of disastrous consequences. In recent years these hazards took toll of thousands of lives and caused massive destruction of property. These have adversely affected the vital sectors of our development as agriculture, communication, irrigation, power projects and rural and urban settlements. The time and cost overrun in some cases have been enormous but their indirect impact on our economy has never been calculated. India is among the world's most disaster prone areas and a large part of the country is exposed to natural hazards, which often turn into disasters causing loss of life and property. The unique geo-climatic conditions have exposed this country to natural catastrophes.

Disasters can be defined as a **sudden, accidental event of great magnitude that causes considerable damage to life and property.** They are sudden, drastic and normally occur without any alarm or warning. Some disasters may be short lived such as earthquakes and some other may be of long duration, such as floods. However, irrespective of the duration of a disaster, the damage in the form of deaths, injuries and losses of property is immense. The magnitude of the disasters can be judged by the fact that only during the past two decades, occurrences of floods, earthquakes, landslides, cyclones, etc have killed several million people.

Most of the disasters have a natural origin, however, some disasters are man made as well. On this basis, disasters can be broadly classified into two groups:

Natural disasters: When disasters occur due to natural forces they are called natural disasters, over which man has hardly any control. Some common natural disasters are earthquakes, landslides, floods, droughts, cyclones, etc. Tsunamis, volcanic eruptions and wildfires are also included under natural disasters. These disasters cause enormous loss to life and property.

Man-made disasters: When the disasters are due to carelessness of human or mishandling of dangerous equipments they are called man-made disasters. Common examples of these disasters are train accidents, aereplane crashes, collapse of buildings, bridges, mines, tunnels, etc.

NATURAL DISASTERS

Some of the common natural disasters, their impact on environment, and their prevention, control and mitigation are discussed below:

Earthquakes

An earthquake is the shaking of the earth's surface caused by rapid movement of the earth's crust or outer layer. Ever since it came into existence 4.6 billion years ago, the earth has been a dynamic, evolving system. The position of the different continents and oceans that we see today, has changed a number of times in the earth's history. The earth is primarily composed of three layers:

1. the outer **crust**,
2. the middle **mantle**, and
3. the inner **core**

The Earth's outer layer or crust is made up of a number of zig-saw pieces like structures that interlock into one another. These pieces are called **tectonic plates**, These plates are in continuous motion over the mantle, which is known as **tectonic movements.** These tectonic processes are also responsible for the mountain building processes. The plates that are moving past over one another are slowed by friction along their boundaries. Due to this, the rocks are under strain. When the stress on the rocks exceeds certain limits, the rocks rupture and form a **fault** along which the rocks are displaced during tectonic movements. This sudden rupture of the rocks releases energy in the form of earthquake waves (Fig. 18.1).

Thus earthquake is a form of energy, which is transmitted to the surface of the earth in the form of waves called **seismic waves.** The study of earthquakes and the waves they create is called **seismology** (from the Greek *seismos,* "to shake"). Scientists who study earthquakes are called **seismologists.** The instrument that records the seismic waves is called **seismograph.**

The exact spot under the earth surface at which an earthquake originates is called the **focus** or **hypocentre.** The point on the earth surface above the focus is called the **epicentre.** The **Richter scale** is used to measure the intensity of earthquakes. The intensity is measured on a scale of 0 to 8 and above (Table 18.1).

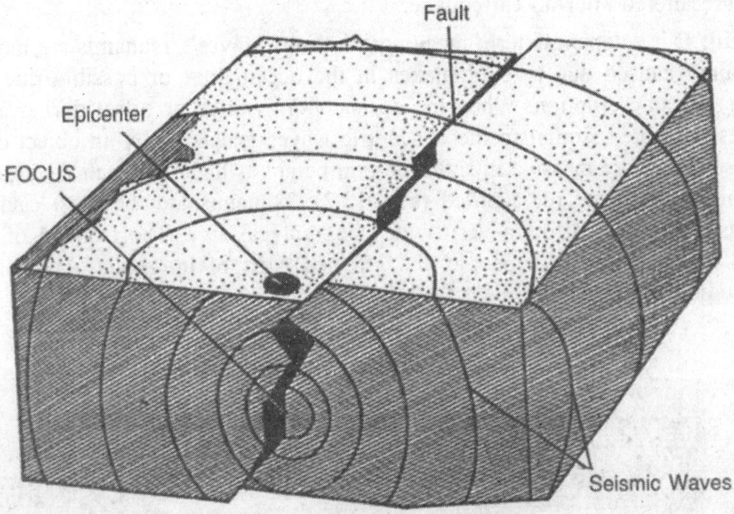

Fig. 18.1 Elements of earthquake.

Table 18.1 Richter scale

Grade	Effects
2.5	Generally not felt, but recorded on seismographs
3.5	Felt by many people
4.5	Some local damage may occur
6.0	A destructive earthquake
7.0	A major earthquake
8.0 and above	Massive earthquakes

Impact of Earthquake on the Environment

The destruction, an earthquake causes, depends on its magnitude and duration or the amount of shaking that occurs. In the last 500 years, earthquakes around the world have killed several million people. Earthquake is one of the most catastrophic natural disasters. Massive loss of life and property occurs due to collapse of buildings. Besides, roads, bridges, canals, electric poles, etc are severely damaged. Certain regions of the earth are more prone to earthquakes. These are places located in the unstable regions of the earth crust, which are subjected to tectonic activities. Countries like Japan, parts of Southeast Asia, Turkey, Iran, Mexico, etc are affected by severe earthquakes. In India, the entire Himalayan region, parts of the Gangetic Plain, Kutch and Andaman and Nicobar islands are in the earthquake hazard zone (Table 18.2). The major impacts of earthquakes are as follows:

Shaking of the ground and surface rupture: This is the main cause of destruction in which buildings, bridges, roads, canals and other structures are damaged.

Liquefaction: Earthquakes make sands and silts to transform from a solid to liquid state. This also results in building collapse.

Landslides: Earthquakes of high intensity often trigger many landslides in the hilly regions.

Fires: It is a major hazard associated with earthquakes. The shakings of the ground and building damage often break the gas pipes and electric lines that cause fires.

Changes in the land elevation: The surface topography of a region and groundwater conditions are altered after an earthquake.

Tsunami: It is a Japanese term meaning **'harbour waves'.** Tsunamis are massive sea waves that are mainly caused due to earthquakes in the ocean floor or possibly due to an undersea landslide or volcanic eruption. When the ocean floor is tilted or offset during an earthquake, a set of waves is created similar to the concentric waves generated by an object dropped into the water. These waves are massive in size and gain height as they approach the seashore. Tsunamis upto the height of 30 m are recorded (Fig. 18.2). Tsunamis are the most catastrophic among natural disasters as they affect a very wide geographical area. The tsunami of 26th December, 2004 killed around three lakh people and affected parts of Indonesia, Andaman and Nicobar Islands in India, Sri Lanka and even Somalia.

Fig. 18.2 Tsunami waves

Prevention and Mitigation

Despite the advances made by modern science, the exact time and place where an earthquake may strike cannot be predicted. Hence, the occurrence of an earthquake cannot be prevented.

However, there are certain regions that are earthquakes prone and so the administration must work before hand to minimize the damages due to occurrence of earthquakes in such areas.

The control and mitigation measures in earthquake prone regions include hazard reduction programmes, development of critical facilities and proper landuse planning.

Hazard reduction programmes: These include the following:

- Earthquake education and evacuation plans
- Use of proper construction material that is not injurious even if the structures collapse
- Construction of quake resistant buildings having proper structural design

Development of critical facilities: These include the following:

- Establishment of earthquake regulatory agencies for fast relief
- Establishment of specific health care units for treating earthquake injuries

Proper landuse planning: These include the following:

- Mapping of faults and weak zones in earthquake prone areas
- Buildings such as schools, hospitals, offices, etc should be in areas away from active faults

HUMAN INDUCED EARTHQUAKES

Though the earthquakes are exclusively a natural phenomenon, certain human activities also cause or induce earthquakes. Human induced earthquakes may be because of the following reasons:

- *Huge dams or reservoirs*
- *Underground nuclear explosions*

The Koyna earthquake was due to the construction of a dam. Underground nuclear explosions generate seismic waves similar to those of earthquakes, which are known to have damaged buildings.

Table 18.2 A brief list of recent major Indian earthquakes

Date	Details
October, 2005	Jammu and Kashmir, intensity 7.4, about 40,000 people died
26th January, 2001	Gujarat, intensity 7.9, about 20,000 people died.
29th March, 1999	Chamoli, Uttaranchal, intensity 6.8, about 1000 people died
22nd May, 1997	Jabalpur and Mandla, M.P., about 50 people died
30th September, 1993	Latur and Osmanabad, Maharashtra, about 10,000 people died
20th October, 1991	Uttarkashi, Uttaranchal, intensity 6.6, about 1000 people died

Floods

Floods refer to the 'inundation of large parts of land which otherwise remain dry by water for some duration of time'. Floods are one of the most common natural disasters occurring in many parts of the world every year. Floods occur due to heavy rainfall within a short duration of time in a particular region which cause the rivers and streams to overflow. Since most of the precipitation occurs within span of two to three months during the rainy season, most floods occur during that time. The floods in the mountainous regions due to cloudbursts or damming of streams are referred to as **flash-floods.** In flash-floods, the water drains away quickly but only after causing extensive damage. The plain areas of a region which are drained by a number of rivers, are the places most affected by floods.

In India, states like Assam, Bihar and parts of Gangetic Uttar Pradesh are quite prone to floods during the rainy season (Fig. 18.3). The Ganga and Brahmaputra rivers and their tributaries are most susceptible to floods. However, heavy rains cause occasional floods in parts of Gujarat, Maharastra, Karnataka and Tamil Nadu. Flooding, in India, is a major problem and some part or the other is affected by the fury of floods usually during the months from July to September.

Floods cause untold miseries to the affected regions in the form of huge losses of life and property. There is great damage to agriculture and livestock. Flood affected areas face acute shortages of food and drinking water. Besides, floods cause a number of water borne diseases such as diarrhoea, gastroenteritis, jaundice, malaria, etc.

Fig. 18.3 Flood affected area

Impact on the Environment

Though the lives lost in floods may not be as high as in case of earthquakes or cyclones, the damage to the environment is immense. The problem is further aggravated if the floods last for a longer duration of time. Floods not only damage property and endanger if lives of humans and animals, but have other effects as well, such as:

1. Floods cause the spread of many epidemic diseases.
2. Rapid runoff causes soil erosion.
3. Wildlife habitat and forests are often destroyed.
4. Man made structures like buildings, bridges, roads, sewer lines, power lines, etc are damaged.
5. Floods cause widespread damage to the standing crops and degrade the agricultural land.
6. Flood affected areas are faced with acute shortage of food and drinking water.

Prevention, Control and Mitigation

Though floods are a natural hazard, it is sometimes intensified due to undesirable human activities. The measures that can be taken to control the extent of flood damage include landuse planning, building of physical barriers, preventing human encroachment and use of technology for relief.

Landuse planning: Proper landuse planning in flood prone areas includes:

1. Demarcation of the floodprone areas that are first inundated during floods.
2. Construction work and concentration of human population should be avoided in the floodplains.
3. Afforestation on the upper reaches of the river (catchment areas) to control soil erosion and excessive runoff.

Building of physical barriers: Flood can be prevented by building certain structures, such as:

1. Embankments along the banks of rivers in densely populated areas.
2. Building of reservoirs to collect excess water during floods.
3. The construction of channels that divert floodwater.

Preventing human encroachment: Human encroachment should be avoided in the following areas:

1. Floodplains and catchment areas.
2. This would control deforestation and soil erosion which would prevent excessive runoff.

Use of technology for relief: Advanced technology can be used in the following ways:

1. Advanced communication techniques for flood forecasting and warning.
2. Fast evacuation of people.
3. To provide relief in temporary shelters.
4. Immediate supply of medicines, drinking water, food and clothes.
5. Epidemic diseases must be controlled through spraying, vaccination, etc.

Drought

Drought is a condition of abnormally dry weather within a geographic region. Drought refers to the lack or insufficiency of rain for an extended period of time in a specific region. During droughts, rainfall is less than normal causing a water imbalance and resultant water shortage. It occurs when the rate of evaporation and transpiration exceeds precipitation for a considerable period. Drought should not be confused with dry climate, as in the Sahara or Thar Desert. It is marked by an unusual scarcity of water and food for the humans as well as animals.

Certain regions of the world, such as parts of Central Africa, are characterized by low amount of rainfall resulting in perennial drought-like conditions. Some part of India is often affected by drought even during the rainy season. As India is primarily an agricultural country, droughts cause untold miseries to the common people. Many Indian farmers are still totally dependent on rainfall for irrigation and because of abnormally dry spells there is extensive crop damage. The main drought prone areas of the country are parts of Rajasthan, Maharastra, Karnataka, Orissa, Tamil Nadu and Chhattisgarh. However, sometimes drought-like conditions also prevail in the Gangetic Plain also.

Impact on the Environment

The severity of the drought is gauged by the degree of moisture deficiency, its duration, and the size of the area affected. If the drought is brief, it is known as a dry spell or partial drought. Drought causes serious environmental imbalances, which are summarized below:

1. Water-supply reservoirs become empty, wells dry up and there is acute water shortage.
2. Groundwater level is also depleted because of less recharge.
3. Soil degradation and erosion occurs. Soil cracks because of shrinkage during desiccation (Fig. 18.4).

4. There is extensive crop damage.
5. People become impoverished and there are diseases due to malnutrition.
6. Widespread damage to flora and fauna air including domestic animals.

Prevention, control and mitigation

Rains are caused by a number of natural factors like air currents, wind direction, etc. Thus, droughts are a natural phenomenon, beyond human control and prevention. Though, global warming may have changed the pattern of rainfall in the recent times. In modern times, by the use of satellites, we can predict the weather pattern over a particular area. Drought-like conditions can be overcome by better water harvesting techniques. Certain precautions can be taken in drought prone areas, which relate to management of water resources, proper agricultural techniques and relief by different agencies.

Management of water resources: These include the following:

1. Conservation of water through rainwater harvesting, building check dams, bunds, etc.
2. Construction of reservoirs to hold emergency water supplies.

Proper agricultural techniques: These include the following:

1. Increased use of drought resistant crops.
2. Proper irrigation techniques, such as drip and trickle irrigation that minimise the use of water.
3. Over-cropping and overgrazing should be avoided.

Fig. 18.4 Drought affected land with cracked soil.

Relief measures: Immediate relief to the drought-affected people should be provided in the form of:

1. Employment generation programmes, like 'food for work' in the drought affected areas.
2. To provide fodder for domestic animals.

Cyclones

Cyclone is an area of low atmospheric pressure surrounded by a wind system blowing in anti-clockwise direction, formed in the northern hemisphere. In common terms, cyclone can be described as a giant circular storm system. In a cyclone, the wind speed must be more than 119 km/hr. Cyclones generate in the seas and oceans and move with a very high speed towards the land. Cyclones form when moisture evaporates from the warm oceans during the hot season. The air rises, condenses and gathers momentum as it moves over the ocean. Due to the extreme low pressure in the centre, more and more air rushes inwards and it grows to a considerable size and intensity. It strikes the land with a devastating force and gradually withers off on land when they are cut from their source of ocean moisture. Cyclones are named variously depending on their source of origin. They are called **hurricanes** in the Atlantic, **typhoons** in the Pacific, **cyclones** in the Indian Ocean and **willy-willies** around Australia.

Impact on the Environment

Cyclones are quite common in the Bay of Bengal and often cause much damage in Bangladesh and coastal areas of West Bengal, Orissa, Andhra Pradesh and Tamil Nadu. Bangladesh has been devastated by cyclones a number of times. In November 1970, a severe cyclone caused a 6 m rise in sea-level and the consequent flooding killed approximately three lakh people. Another cyclone in 1971 killed more than one lakh people. The cyclone that hit Orissa in 2001, is the worst recorded natural disaster in India. Even an advanced country like America recorded more than 10,000 deaths and huge financial losses when New Orleans was hit by a hurricane named Katrina, during August 2005.

Cyclones cause devastation when they hit the landmass in the form of very strong winds, heavy rains and storm tides. The impact on the environment is severe, some of which are as under (Fig. 18.5):

Fig. 18.5 Cyclone affected area

1. The coastal low lying areas are most affected.
2. The affected areas are inundated both with rainfall and the surge of seawater.
3. Devastation is also increased due to the accompanying high velocity winds.
4. Widespread damage in the form of uprooted trees, blown-off roof tops, standing crops, injuries and death to humans and animals.
5. Many shipwrecks occur during cyclonic storms.
6. The affected areas are impoverished and are followed by spread of epidemic and diseases.

Prevention, Control and Mitigation

The occurrence of cyclones is a natural phenomenon, over which humans have no control, hence it cannot be prevented. However, some scientists have speculated that rise in global warming may cause an increased occurrences of cyclones. The devastating effects of cyclones can only be controlled and mitigated through some effective policies such as use of advanced technology, hazard reduction initiatives and relief measures.

Use of Advanced Technology

1. Satellites can easily forecast the origin of cyclones in advance.
2. Satellite images can track the movement and intensity of cyclones.
3. Installation of early warning systems in the coastal areas.

Hazard reduction initiatives

1. Increasing public awareness regarding cyclones.
2. Increasing the public response to cyclone warnings through training.
3. Development of underground shelter belts in the cyclone prone areas.

Relief measures

1. Rushing relief to the affected areas in the form of medicines, food, clothes, etc.
2. Checking the spread of epidemic water borne diseases as cyclones are generally accompanied by flooding.

Landslides

Landslides refer to a rapid down-slope movement of rocks or soil mass under the force of gravity. It is also known as **slope failure and mass wasting.** Landslides may be typed as **mudflow** where there is down-slope movement of soil and **debris flow,** which is the down-slope movement of coarse material and rocks. Landslides may occur when water from rain and melting snow, seeps through the earth on a sloppy surface and encounters a layer of loose, unstable material such as clay. Landslides mostly occur on unstable hillsides by the action of rain or snow that seep through the soils and rocks (Figs. 18.6 and 18.7). This results in the sliding of earth and rock masses down the hill slopes. These are further triggered due to deforestation and human encroachment on unstable slopes. All the hilly regions of our country are prone to landslides. The important factors responsible for landslide occurrence are as follows:

1. Stability of slopes
2. The type of earth and rock material
3. The type of vegetation
4. The role of ground water conditions and precipitation
5. Presence of streams, etc.

Avalanche: It is a type of landslide involving a large mass of snow, ice and rock debris that slides and fall rapidly down a mountainside. Avalanches are initiated when a mass of snow and ice begins to rapidly move downhill because of the overload caused due to a large volume of new snowfall. This result in internal changes of the snow pack, producing zones of weakness along which fissure occurs.

Impact on the Environment

Landslides, though local in nature, occur quite often in many parts of the world. Landslides occur in the hilly regions; the Himalayan region in India is particularly prone to landslides. Every year landslides occur, especially during the monsoon season and cause much damage to life and property. For example, Malpa landslide in 1999 in the Kumaon hills, took the lives of many pilgrims who were going to Mansarovar in Tibet. The impact on the environment is manifested in the form of:

1. Uprooted trees and degraded soil
2. Buried building and settlements
3. Damage to crops and plantation
4. Frequent roadblocks in the hilly areas
5. Injuries and death to humans and animals

Prevention, Control and Mitigation

Though landslides are a natural phenomenon and may occur without human interference, in certain cases human activities like deforestation, mining, etc. can also induce landslides. Landslides can be controlled, to some extent, by adopting initiatives, such as providing slope support and minimising human encroachment.

Fig. 18.6 Landslide

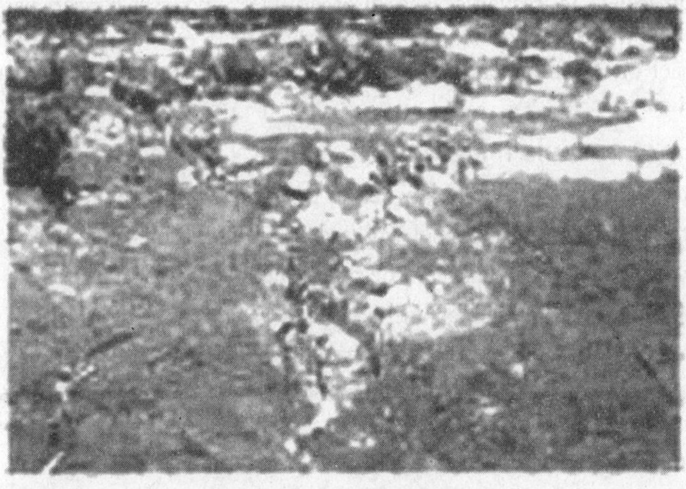

Fig. 18.7 Malpa landslide

Providing slope support

- By building retaining walls made of concrete, gabions (stone filled wire blocks) and wooden and steel beams, etc.
- By providing drainage control measures so that water may not infiltrate into the slope

Minimising human encroachment

- Mining activities should be monitored in the hilly, unstable regions.
- Plantation of trees should be undertaken on the unstable hilly slopes.
- By preventing human encroachment in the form of buildings, roads, agriculture, grazing, etc. on unstable slopes.

MAN-MADE DISASTERS

Man-made disasters are the result of carelessness or human errors during technological and industrial use. The disasters are in the form of accidents, which occur all of a sudden and take a huge toll on life and property. Mostly such disasters cause injuries, diseases and casualties where they occur.

Man-made disasters are mainly of two types:

Local disasters

These are small-scale disasters such as train accidents, plane crashes and shipwrecks.

Industrial and technological disasters

These are much larger in scale and are the result of technology failures or industrial accidents. Such disasters affect both local population and may even cover a much larger area.

Industrial disasters result due to accidental leakage of water or air pollutants. Many of the chemicals are extremely toxic and carcinogenic which affect the human population in an adverse way. Some people die instantly while others are crippled for whole life in the form of blindness, paralysis and many other chronic diseases.

Impact on the environment

Leakage of toxic chemicals from the industries and accidents in the nuclear reactors have short-term and long-term effects on the environment and human health. Short-term effects on human health relate to casualties and diseases like blindness, cancer, paralysis, heart trouble, gastric and respiratory abnormalities. Long-term effects include genetic imbalances in humans and its impact on the future generations. Soil and water sources also remain polluted for long durations of time.

Prevention, control and mitigation

Man-made disasters can be minimised to a large extent by adopting the following measures:

1. Proper training of personnel working in the hazardous industries
2. Proper maintenance and care of safety measures
3. Removing human encroachments around hazardous industries
4. Making the people aware about the first-aid methods in case of accidents
5. Applying wet cloth over the mouth and nose in case of gas leakages minimises the health hazards
6. Remaining indoors in case of radioactive accidents
7. Providing the people with proper medical care, in some cases throughout their life
8. Providing adequate compensation to the affected people by way of money and employment.

Bhopal Gas Tragedy (BGT)

The most serious industrial disaster occurred on December 3, 1984 at Bhopal, India, which is known as the Bhopal Gas Tragedy (BGT). The Bhopal gas tragedy occurred due to leakage of methyl isocyanide (MIC) gas from the factory of Union Carbide of India Ltd. MIC gas is used as an ingredient in pesticides. It leaked from the factory and formed the deadly cloud over Bhopal. People living in slums in the vicinity of the factory were the most affected and more than 5000 people were killed, half of them due to direct exposure and other half due to after affects. MIC is a colourless gas which causes severe irritation, violent coughing, swelling of the lungs, bleeding and death due to direct inhalation. It also caused loss of eye-sight in more than 1000 people. More than 50,000 people were affected with respiratory, eye, gastric, neurological and gynaecological problems (Figs. 18.8 and 18.9).

Another technological disaster is due to the potential damages of nuclear fallout.

An example is the **Chernobyl Nuclear Disaster.**

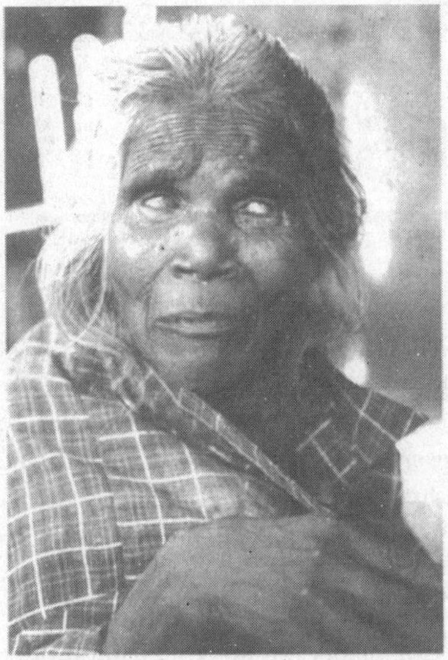

Fig. 18.8 People with damaged eyes due to BGT

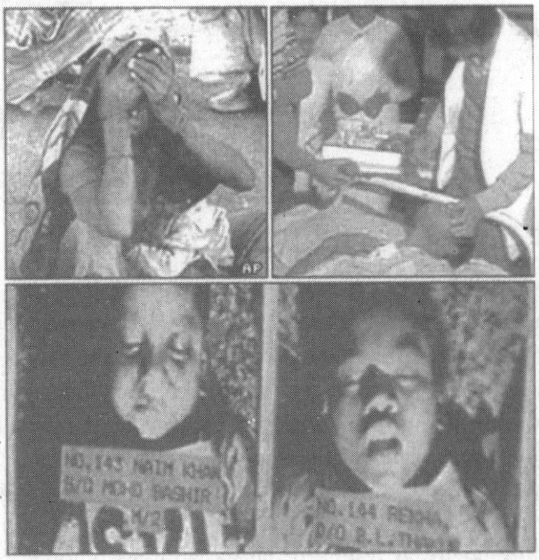

Fig. 18.9 Effects of BGT

Chernobyl Nuclear Disaster

This nuclear disaster occurred at the Chernobyl Nuclear Power Plant, which was one of the largest power plants in the Ukrainian Republic of erstwhile USSR, on April 26, 1986. It is the worst nuclear disaster recorded in a nuclear power plant. This nuclear power plant had four reactors of 1000 megawatt each for electricity generation. A sudden power surge resulted in two explosions, which destroyed the reactor core and blasted a large hole in the roof of the reactor building. The Radioactive debris moved up through that hole to heights of 1 km. Approximately 100 to 150 million curies of radiation (radioactive isotopes of iodine and caesium) escaped into the atmosphere. To reduce emissions, the rescue team bombarded the reactor with 5,000 metric tonnes of shielding material consisting of lead, boron, sand and clay. Soviet officials placed the toll of human lives to 31. However, according to western estimates, 2000 people were killed. Large areas of the Ukrainian, Byelorussia Republics of the USSR and even parts of Poland, Denmark and Sweden were contaminated. Around 200,000 people had to be evacuated and resettled. The after affects lasted for many years and a rise in the incidence of thyroid and blood cancer has been observed in a wide group of people. Other affects on the human health included skin diseases, hair loss, nausea, anaemia, respiratory and reproductive diseases.

QUESTIONS

1. Define disasters. Discuss briefly one important natural disaster.
2. Discuss the impacts of floods on environment and mention the control measures for floods.
3. What are the impacts of cyclones on environment? How modern methods have helped in controlling the impacts of cyclones ?
4. Write brief notes on the following :

   (*i*) Bhopal gas Tragedy

   (*ii*) Nuclear disaster

19

Biodiversity

CONCEPT OF BIODIVERSITY

It has been estimated that more than 50 million species of plants, animals and micro-organisms are existing in the world. Out of these, about 1.4 million species have been identified so far. Each species is adapted to live in specific environment, from mountain peaks to the depth of seas, from polar ice caps to tropical rain forests and deserts. All this diversity of life is confined to only about one kilometer thick layer of lithosphere hydrosphere and atmosphere which form **biosphere.** Though the study of environment and ecology is quite old, the term **biodiversity** has been introduced by **Walter Rosen** in 1986. Biological diversity or Biodiversity is defined as **the variety and variability among the living organisms and the ecological complexes in which they occur.** It refers to the variabilities among species of plants, animals and microorganisms; ecosystems; ecosystem including terrestrial, aerial, marine and other aquatic system and ecological complexes of which they are part. In simpler terms, biodiversity is the assemblage of different life forms (Fig. 19.1). If reflects the number of different organisms and their relative frequencies in an ecological system. It includes the organisation of organisms at many levels ranging from complete ecosystems to the chemical components that form the molecular basis of heredity. Thus, biodiversity is sum of all the genes, varieties, species, populations in different ecosystems and their relative abundance.Scientists are aware of the immense potentials of various life-forms existing on the earth. Our planet's requirements and services depend mainly on the biological resources. Biological resources not only provide us nourishment, clothing, housing, fuel and medicine but also meet our several other requirements. Therefore the knowledge of biodiversity is of immense utility in planning sustainable livelihood and conserving the natural resources.

Significance of Biodiversity

Biodiversity, besides its ecological significance provides a socio-economic and monetary asset to the nation. Human society depends on biological resources, their diversity and the ecosystems that sustain them to provide essential goods and services. Values related to biodiversity can be grouped into three categories as below :

Productive use. This is assigned to the products that are commercially harvested for exchange in formal markets and is, therefore, the only value of biological resources that is concerned in national income. Biodiversity provides us many products, such as fuel, timber, fish, fodder, skin, fruits, cereals and medicines. In 1994-95 the income from agriculture, forestry and fisheries in India was nearly 30 per cent or 736.88 billion rupees.

Consumptive use. Consumption value is related to natural products that are consumed directly, *i.e.,* the goods which do not come under normal circulation of trade. For example, a significant number of such non-timber forest products as soft broom grass and cane come under this category.

Indirect use. Indirect use of biodiversity is of much significance because this value is related primarily with functions of ecosystem and is concerned with national accounting systems. They may provide us indirect benefits as non-consumptive values. Maintenance of ecological balance, conservation of natural resources and prevention of soil erosion may be considered as the examples of indirect use of biodiversity.

Fig. 19.1 The different species of mammals

TYPES OF BIODIVERSITY

Biodiversity is of three types:

1. Species diversity
2. Genetic diversity
3. Ecological diversity

Species Diversity

According to Biological Species Conepts (BSC), species is a basic unit of classification and is defined as a group of similar organisms that interbreed with one another and produce offsprings and share a common lineage. Species diversity refers to biodiversity at the most basic level and is the 'variety and abundance of different types of individuals of a species in a given area'. It includes all the species on Earth, ranging from plants such as bacteria, viruses, fungi, algae, bryophytes, pteridophytes, gymnosperms, angiosperms and all the species of animals including unicellular protozoans to mammals.

Certain regions support a more diverse populations than others. Regions that are rich in nutrients and have well balanced climatic factors, such as moderate temperature, proper light and adequate rainfall, show high degree of diversity in their life forms. The tropical areas support more diverse plant and animal communities than the desert and polar areas, as for examples, tropical forest has a higher species diversity as compared to a timber plantation. The regions that are rich in species diversity are called hotspots of biodiversity.

GENETIC DIVERSITY

'Genetic diversity pertains to the range of diversity in the genetic resources of the organisms'. Every individual member of a plant or animal species differs from other individuals in its genetic constitution. Each individual has specific characters, which is due to the genetic makeup or code. The genes present in the organisms can form infinite number of combinations that causes genetic variability. Thus, we find that each human, who is representative of the same species, i.e. *Homo sapiens,* is distinct from another. Similarly, there are many varieties within the same species such as rice, wheat, apples, mangoes, etc. that differ from one another in shape, size, colour of flowers and taste of fruits and seeds due to the variations at the genetic level.

The term **'gene pool'** has been used to indicate the genetic diversity in the different species (Fig. 19.2). This also includes the diversity in the wild species, which through intermixing in nature over millions of years have given rise to newer varieties. The domesticated varieties of agricultural crops and animals have also evolved from the wild gene pool. The genetic variability is essential for healthy breeding population, the reduction in genetic variability among breeding individuals leads to inbreeding which in turns can lead to extinction of species. In the recent decades, a new science named '**biotechnology**' has emerged. It manipulates the genetic materials of different species through various genetic re-combinations to evolve better varieties of crops and domestic animals.

Fig. 19.2 Genetic diversity in squirrels

Ecological/Ecosystem Diversity

Each ecosystem consists of organisms from many different species, living together in a region connected by the flow of energy and nutrients. The Sun is the ultimate source of energy for all the ecosystems. The Sun's radiant energy is converted to chemical energy by plants. This energy flows through the different systems when animals eat the plants and then are eaten, in turn, by other animals. Fungi and bacteria derive energy from the decomposing dead organisms, releasing nutrients back into the soil as they do so. An ecosystem, therefore, is a collection of living components, like microbes, plants, animals, fungi, etc. and non-living components, like climate, matter and energy that are connected by energy flow. Ecological diversity refers to the 'variability among the species of plants and animals living together and connected by flow of energy and cycling of nutrients in different ecosystems or ecological complexes'. It also includes variability within the same species and variability among the different species of plants, animals and microorgnisms of an ecosystem. Thus, it pertains to the richness of flora, fauna and microorganisms with in an ecosystem or biotic community.

The richness of the biosphere in terms of varied life forms is due to the variations in the ecosystems. The earth has a number of ecosystems like grasslands, forests, semi arid deserts, marine, freshwater, wetland, swamp, marshlands (Fig. 19.3) etc. each one having its distinct floral, faunal and microbial assemblages. Ecological diversity represents an intricate network of different species present in local ecosystems and the dynamic interaction among them. The ecological diversity is of great significance that has developed and evolved over millions of years through interactions among the various species within an ecosystem.

Fig. 19.3a Desert

Fig. 19.3b Forest

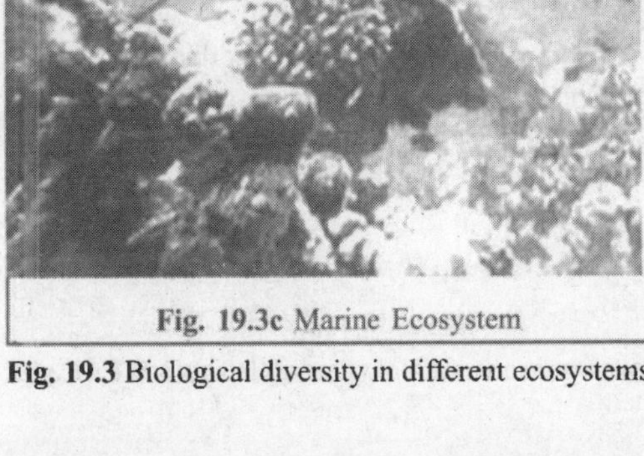

Fig. 19.3c Marine Ecosystem

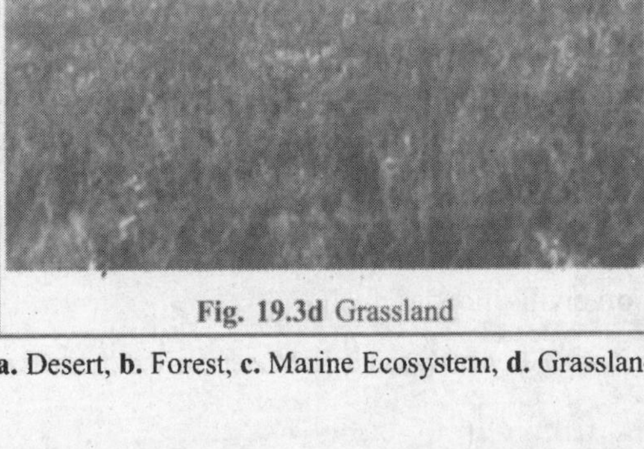

Fig. 19.3d Grassland

Fig. 19.3 Biological diversity in different ecosystems, **a.** Desert, **b.** Forest, **c.** Marine Ecosystem, **d.** Grassland

MEASURING BIODIVERSITY

There are various mathematical ways of measuring biodiversity, which calculate the number of species diversity in different regions. The measure of diversity of species is also known as **species richness.** These are as follows:

Alpha diversity

This is the diversity in species, i.e. the number of species within a community. This depends on the interaction between the biotic and abiotic factors and also takes into account immigration from other locations.

Beta diversity

This is the change in the composition of the species with reference to the changes in the environment.

Gamma diversity

This refers to the overall diversity and is applied to larger areas in which both alpha and beta diversity are measured.

VALUE OF BIODIVERSITY

Biodiversity is the most precious gift of nature the mankind is blessed with. The uniqueness of our planet Earth is due to the presence of life manifested through the diversity in flora and fauna. As all the organisms in an ecosystem are interlinked and interdependent, the value of biodiversity in the life of all the organisms including humans is enormous. Besides its ecological and environmental value, biodiversity has significant socio-economic values as well. The value of biodiversity can be grouped under the following heads:

Environmental Value

The diverse group of organisms found in a particular environment together with the physical and biological factors that affect them, constitute an **ecosystem.** Healthy ecosystems are vital to life. The natural environment is responsible for the production of oxygen, maintenance of water-cycle and other biogeochemical cycles. The more a region is rich in terms of biodiversity, the better are the different cycles regulated. For example, forests regulate the amount of carbon dioxide in the air by releasing oxygen as a by-product during photosynthesis, and control rainfall and soil erosion. As you are well aware, deforestation would further increase carbon dioxide in the earth's atmosphere leading to greenhouse effect and global warming. This will cause irreparable damage to all organisms including mankind. Ecosystems depend on the health and vitality of the individual organisms that compose them. As all the organisms in an ecosystem are interdependent, removing just one species can prevent the ecosystem from operating normally.

Consumptive Value

This is related to natural products that are used directly for food, fodder, timber, fuel wood, etc. Humans use at least 40,000 species of plants and animals on a daily basis. Many people around the world still depend on wild species for most of their needs like food, shelter and clothing (Fig. 19.4 and Table 19.1). The tribal people are completely dependent on the forests for their daily needs. Similarly, fishermen in the coastal areas are dependent on the marine resources. The wood derived from the forests has been used from the birth of civilisation as fuel.

Table 19.1 Plant parts used directly by the indigenous people

Name of the plant	Part used	Use
Tulsi	Leaves	Cold and cough
Jambul	Bark	Sore throat, asthma
Neem	All parts	Skin diseases, antiseptic
Rauwolfia (Sarpgandha)	Root	Blood pressure and heart diseases
Cinchona	Bark	Malaria
Ashwagandha	Root	Strength
Amla	Fruit	Rich source of vitamin C and iron
Vasaka	Leaf extract	Cough
Ishabgul	Seeds	Stomach disorder
Ginger	Stem	Sore throat and cold
Cotton	Fruit	Textiles
Sal, Teak	Woody stem	Timber
Wheat	Fruit and stem	Cereal and fodder
Sugarcane	Stem	Cane sugar
Mango	Fruit	Food
Pulses	Seeds	Food and green manuring
Bamboo	Stem	Making huts, etc

Fig. 19.4a. Firewood

Fig. 19.4b. Food

Fig. 19.4c. Timber

Fig. 19.4 Consumptive value of biodiversity a. Firewood, b. Food, c. Timber

Productive Use Value

This is assigned to products that are commercially harvested and marketed. Almost all the present day agricultural crops have originated from wild varieties. The biotechnologists continuously use the wild species of plants for developing new, better yielding and disease resistant varieties. Biodiversity represents the original stock from which new varieties are being developed. Similarly, all our domesticated animals came from wild-living ancestral species. Through scientific breeding techniques animals giving better yield of milk, meat, etc are being developed. The commonly used animal products used by the modern society come from the advances made in the fields of poultry farming, pisciculture, silviculture, dairy farming, etc.

Even the fossil fuels like coal and petroleum are the products of biodiversity from the geological past. Most of the drugs and medicines used in the present times are extracted from different plant parts. The commonly used drugs derived from plants are given in Table 19.2 and Fig. 19.5.

Table 19.2 Common modern drugs obtained from plants

Drug	Plant	Use
Penicillin	*Penicillium* (Fungus)	Antibiotic and antibacterial
Quinine	Cinchona	Anti-malarial
Atropine	Belladonna	Painkiller
Caffeine	Tea, coffee	Stimulant
Cocaine	Cocoa	Analgesic and anaesthetic - painkiller
Morphine	Opium	Analgesic - painkiller
Menthol	Mint	Rubefacient (reduces pain on local application)
Papain	Papaya	Highly digestive
Tetracycline	Bacterium	Antibiotic
Resepirine	*Rauwolfia serpentina*	Controls blood pressure

Fig. 19.5a. Aloe

Fig. 19.5b. Cinchona

Fig. 19.5c. Ginseng

Fig. 19.5d. Tulsi

Fig. 19.5 Plants of medicinal values a. Aloe, b. Cinchona, c. Ginseng, d. Tulsi

Social Value

The lifestyle of the ancient people was closely interwoven with their surroundings. The life of the indigenous people in many parts of the world still revolves around the forests and environment, even in the modern times. Many of them still live in the forests and meet their daily requirements from their surroundings. Due to modernisation, their habitats are being encroached upon and their very survival is at stake. It is ironic that the societies, whose whole life is intricately associated with the forests, are now not able to use the natural resources for their sustenance.

The biodiversity in different parts of the world has been largely preserved by the traditional societies. Since the indigenous people always protect the forests for their own benefit, the Government should formulate plans to involve such people for environmental protection.

In ancient times, especially in India, the environment in totality i.e. flora, fauna, etc were held in high esteem. Trees like Peepal, Banyan and Tulsi are still worshipped. Ladies offering water to Tulsi daily is considered good and there are festivals when ladies tie sacred threads around Peepal and Banyan trees and pray for the welfare of their families. Similarly, certain animals and birds were represented as vehicles of Gods and were duly respected. Thus, the different facets of biodiversity were closely linked to the social values in many regions.

Ethical and Moral Values

It is based on the principle of 'live and let others live'. Morality and ethics teach us to preserve all forms of life and not to harm any organism unnecessarily. Some people take pleasure in the hunting of animals. People also sometimes degrade and pollute the environment by their unethical actions. Through proper education and awareness, the people's conscience against such practices must be raised. We may not be deriving direct benefits from many plants and animals, but should they be harmed because of this? Each species has its own utility in the world of biodiversity and has every right to live.

Aesthetic Value

The beauty of our planet is because of biodiversity, which otherwise would have resembled other barren planets dotted around the universe. Biological diversity adds to the quality of life and provides some of the most beautiful aspects of our existence. Biodiversity is responsible for the beauty of a landscape. Humans are also attracted towards the biologically rich regions and nobody likes to live or visit a barren place. People go to far off places to enjoy the natural surroundings and wildlife. This type of tourism is referred to as **eco-tourism,** which has now become a major source of income in many countries (Fig. 19.6). Eco-tourism includes visiting wildlife sanctuaries, national parks, coral reefs, exotic islands, safaris and trekking in the mountainous and forested areas. In some countries like Nepal, Bhutan, Kenya, Rwanda, eco-tourism has now become the major source of foreign currency income.

In many societies, the diversity of flora and fauna has become a part of the traditions and culture of the region and has added to the aesthetic values of the place. For example, in India, the richness of flora and fauna are depicted in many paintings; animals are represented as vehicles of Gods; the dances and festivals are intricately associated with nature.

Fig. 19.6 Biodiversity promotes eco-tourism

Optional Value

This refers to the value of biodiversity that is yet unknown, but needs to be explored for future possibilities and use. Scientists have discovered and named about 1.75 million species, which is of utmost importance. We should preserve all the world's biodiversity that can be used by the future generations.

Economic Potential of Biodiversity

As discussed above, the economic potential of biodiversity is immense in terms of food, fodder, medicinal, ethical and social values. Biodiversity forms the major resource for different industries which govern the world economy. The salient features regarding the economic potential of biodiversity are given below:

1. The major fuel sources of the world including wood and fossil fuels have their origin due to biodiversity.
2. It is the source of food for all animals and humans.
3. Biodiversity constitutes the gene pool of the world. Agricultural production of many crops depends on the introduction of fresh genetic material form wild 'varieties to create new hybrids.
4. By introducing fresh genetic characters from the gene pool, new hybrid varieties of both plants and animals are being developed that are high yielding and disease resistant.
5. Biodiversity has the potential for developing new crops or commercial products from wild plants and animals.
6. Many important chemicals have their origin from the diverse flora and fauna, which are used in various industries.
7. Most of the modern medicines are derived from the treasures of biodiversity. Many marine organisms are also being tested for use in the pharmaceutical industries.
8. Diverse groups of animals like rodents, rabbits, monkeys, etc are used for medical research during the testing of new drugs (Fig. 19.7).
9. The richness of diversity of vegetation is an important factor in regulating the world's climate. If the earth were devoid of the green cover it would lead to global warming, which would have disastrous effect on the world's agricultural economy.
10. Biodiversity provides excellent pollution control. Plants remove toxic substances from the air, water and soils. As different species have different capabilities for removal of toxins, a diversity of species can provide a better range of pollution control.
11. Biological diversity is of great importance to many indigenous people, especially in poor countries. For them, diversity in forests and wildlife provides food, wood, shelter, material for clothing and medicine. A reduction in the biodiversity will increase the poverty of these people.
12. Biodiversity provides excellent potential for eco-tourism and thereby generates revenue.

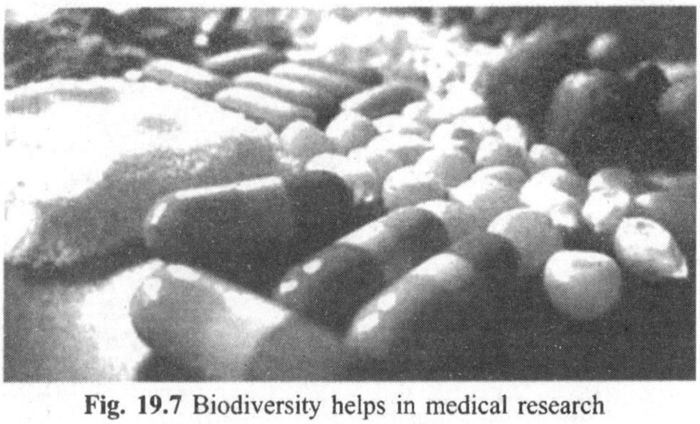

Fig. 19.7 Biodiversity helps in medical research

ECONOMIC VALUE OF ANIMALS

- *A Kenyan elephant in its life span earns about 1 million dollars from tourists.*
- *Coral Reefs in Australia generate about 2 billion dollars annually through eco-tourism.*
- *Eco-tourism is also becoming popular in Andaman and Nicobar Islands and Lakshadweep*
- *Wildlife Sanctuaries and National Parks like Gir, Kanha, Bandhavgarh, Corbett, Kaziranga, etc are quite popular and help the Government and local people in generating employment and revenue.*

ECOLOGICAL ROLE OF BIODIVERSITY

Ecological diversity is the intricate network of different species present in different ecosystems and the dynamic interaction between them. An ecosystem consists of interacting organisms of many different species living together in a region that are connected by the flow of energy and nutrients. The radiant energy of the Sun provides the ultimate source of energy in nearly all the ecosystems. The Sun's radiant energy is converted to chemical energy by the green plants through the process of photosynthesis. This energy flows from the producers to the consumers and lastly to the decomposers.

The following points characterise the ecological role of biodiversity:

(*i*) **Species diversity:** Biodiversity reflects the natural assemblage of large number of plant and animal species in a given area. It is part of a larger ecosystem in which biotic and abiotic components interact and bring about circulation, transformation and accumulation of energy and matter. A distinct area with uniform habitat conditions and supporting characteristic type of flora and fauna is termed **biotope.** Each species of a community has got a definite range of tolerance towards the physical and biological environmental conditions of the habitat. The range of environment that a species can tolerate is called its **ecological amplitude.**

(*ii*) **Coexistence:** The species occurring in a particular habitat do not live in isolation but coexist with mutual adjustment. The coexisting populations are interrelated and they show some sort of interaction. The interactions between two coexisting species are of the following types:

1. **Exploitation:** Where one species lives at the expense of another.
2. **Mutualism:** In this, the coexisting species benefit from the relationship.
3. **Competition:** In this, the coexisting species compete for the same resources.
4. **Neutralism:** In this, the coexisting species are independent of one another.

(*iii*) **Interdependency:** Here, the different species within a community live under similar environmental conditions and are interdependent on each other.

(*iv*) **Species dominance:** Not all the species of a community are found in abundance. Only a few species are abundant, either in number or in biomass, while the majority of the species are rare. The species that are most common and abundant and contain maximum biomass are known as **dominants.**

(*v*) **Stratification:** In a plant community, the different plant species are represented by trees, shrubs, herbs, etc. These plants form, more or less, distinct strata or layers on vertical as well as horizontal plains known as stratification.

(*vi*) **Succession:** The interacting species within an ecosystem are characterized by death and replacement, which are continuous processes. In this way, the composition and shape of an ecosystem remains dynamic. This is known as succession. The changes go on taking place until a complete balance is established between the species and the environment.

Biodiversity for Sustenance of Mankind

The importance of biodiversity for mankind is immense which have been discussed already in this chapter under 'Value of Biodiversity'. Mankind is totally dependent on the biotic reserves for day-to-day sustenance. The cereals that we eat, the fruits that we relish, meat, poultry and various other products, we consume are all part of the biological diverse wealth.

Broadly speaking, biodiversity caters to human needs in two different ways – **direct** and **indirect.** As already discussed, biodiversity not only helps in increasing agricultural productivity but also in developing disease resistant varieties. This was evident in the early 1970s when an epidemic called 'Grassy stunt virus' destroyed thousands of hectares of rice fields in Asia. It was eventually controlled from a single species of wild rice - *Oryza nivara* from Central India, which was found to be the only known genetic source of resistance to the virus. Similarly, many genes from wild rice are now used in rice improvement programmes.

Besides food, the richness of biodiversity is being used in human health programmes. All the common drugs owe their origin to plants or animals. Newer drugs are being manufactured from plants for treating deadly diseases. For example, scientists have developed a drug called Vincristine, from a plant of the tropical forests *Vinca rosea,* which is being used to treat leukaemia. Thus, the mass extinction of certain species could cause a serious setback in the fight against many diseases.

Indirect benefits include the trapping of the Sun's radiant energy, maintenance of different bio-geochemical cycles, soil formation and protection, absorption and breakdown of pollutants, provision of recreation, aesthetic, scientific, social and other benefits. Scientists have estimated that there are more than 50 million species of plants, animals and microorganisms constituting the richness in biodiversity. Of these, less than 2 million species have only been identified so far. There is every possibility that as our knowledge increases; any one of the unidentified species may turn out to be of immense value to mankind. Therefore, biodiversity provides tremendous potential to benefit mankind in the context of advances being made in science and technology.

HOTSPOTS OF BIODIVERSITY

Hotspots are the areas that are biologically rich and exhibit high species diversity most of which are **endemic** in nature. **Endemic** species are those that are restricted to a specific area. Hotspots are characterised by richness in floral and faunal wealth. Though the area covered by the hotspots represent only two per cent of the world's land area, it preserves about 50% of the total terrestrial biodiversity. This term was introduced by Myers in 1988, who has presently recognised twenty-five such hotspots of biodiversity on a global level. These hotspots include 49955 endemic species which account for about 20% of world's total plant species, Mittermeir and Werner (1990) on the basis of species richness put forth the concept of megadiversity centres. The recognised 12 countries/regions are megadiversity centres. These are Mexico, Columbia, Peru, Equador, Brazil, Zair, Medagascar China, India, Malaysia, Indonesia and Australia.

List of Global Hotspots of Biodiversity

1. Tropical Andes
2. Madagascar
3. Brazil's Atlantic forest region
4. The Philippines
5. Meso-American forests
6. South Africa's Cape Floristic region
7. Wallacea (Western Indonesia)
8. Western Sunda (Indonesia, Malaysia and Brunei)

9. Brazil's Cerrado
10. Polynesia and Micronesian island complex
11. The Darien and Choco of Panama, Colombia, Western Equador
12. The Eastern Mediterranean region
13. **The Western Ghats of India and the Island of Sri Lanka**
14. **Indo-Burma Eastern Himalayas**
15. The Guinean forest of West Africa
16. New Caledonia
17. South-eastern Australia and Tasmania
18. Caribbean
19. Central Chile
20. California Floristic Province
21. Eastern Arc and Coastal forest of Tanzania / Kenya
22. Succulent Karoo
23. Caucasus
24. South Central China
25. Newzealand

BIODIVERSITY IN INDIA

Varying physiographic features at different places have resulted in almost all sorts of ecological conditions in India. Indian flora and fauna president very rich biodiversity. Varied physiographic and climatic conditions in different parts of India have resulted in a high level of biodiversity in the country from sea level to the highest mountain ranges in the world, the long coastal line (7500 km long with a shelf zone of over 43 m and exclusive economic zone of over 210 m), hot arid conditions in the west, cold desert conditions in the north and humid tropical conditions in other parts of the country. Thus, India, with a geographical area of 329 million hectare has almost all the ecological zones found in the world. By virtue of being a meeting ground for three major global biogeographic region : (*i*) Indo Malayan (*ii*) Eurasian and (*iii*) Afrotropical, India is one of the twelve major or mega diversity centres in the world. It also has two of the 25 recognized biodiversity "hot spots" of the world—Eastern Himalayas and Western Ghats. "**Hot spots**" are the areas that are usually very rich in species, most of which are endemic and are under constant threat of being overexploited. Their floral wealth is very rich.

India is very rich in all aspects of biodiversity and is one of the most significant biodiversity centres of the world. About 48,000 species of plants which represent 11 per cent of the total world flora and 81,000 species of animals representing 6.4 per cent of the world's fauna are found in India. India's floral diversity includes over 64 species of Gymnosperms; 1,022 species of Pteridophytes; 2843 spp. of Bryophytes; 1940 species of lichens; 13,000 species of fungi and 12281 species of Algae. Similarly, the faunal diversity of our country includes 5,000 species of Molluscs; 57,000 species of insects; 2546 species of fishes; 204 species of Amphibians; 420 species of Reptiles; 1228 species of Aves (birds) and 372 species of Mammals. According to an estimate 35 per cent of the flowering plants and 18 per cent of the total flora are endemic to India. The endemism among the animals is high and it may be as high as 62 per cent in case of Amphibia. Biologically rich areas of India are either unexplored or under-explored. Thus recorded biological diversity of India may be only a part of the actually existing biological wealth. Besides, the present knowledge about the diversity in micro-organisms, wetland flora and fauna and marine life is scanty and the major gap in our knowledge still persists. The knowledge of diversity in plant and animal genetic resources will provide a sound basis for generating additional resources for conservation and sustainable use.

The Indian subcontinent commonly known as "Hindustan Centre" is one of the Twelve centres of origin and diversity of crop plants which alone has given to the world nearly 167 species of domesticated agri-horticultural crop plants. Over 320 species which are wild progenitors of these domesticated crop plants are native to this country. The important crop plants are: rice, sugarcane, millets, Brassica, Asiatic vignas, egg plant, banana, citrus, mango, jack fruit, jute, black pepper, cardamom, turmeric, ginger, several umbellifers, cucurbits, colocasia and many herbal drugs as jasmine, rhododendron, some orchids, betel nut and bamboos.

The rich germplasm reserve of Indian subcontinent includes 51 species of cereals and millets, 104 species of fruits, 27 species of spices and condiments, 55 species of vegetables and pulses, 24 species of fibre crops, 12 species of oilseeds and various strains of wild tea, coffee, tobacco and sugarcane. Besides above, 1,000 species of plants are of ethnobotanical interest (Plants used by tribal populations) and 2,000 species are used in various traditional systems of medicine. Further diversity has been increased through genetic engineering and biotechnology by transferring genes in the species which enhance agricultural productivity.

India's indigenous livestock diversity includes about 27 breeds of cattles, 40 breeds of sheep, 22 of goats, 8 of camels, 6 of horses, 2 of donkeys and 18 types of poultry. Eight indigenous breeds of buffalo in India represent the entire range of its genetic diversity on the earth. These prized plants and animals together with their ancestral wild forms are used by plant and animal breeders for the development of improved varieties.

Hotspots of Biodiversity in India

Of the twenty-five hotspots of biodiversity, recognised in the world, two are found in India, which extend into the neighbouring countries: (i) the Indo-Burma region covering the Eastern Himalayas and (ii) the Western Ghats – Shri Lankan region. The hotspots are rich in floral wealth, reptiles, amphibians, mammals and also in their endemism. The botanical hotspots of India include (1) *Western Ghats*, (2) *North-East India*, (3) *Himalayas* and (4) *Andman and Nicobar Islands* (Table 19.3).

Eastern Himalayas: As compared to the Western Himalayas that are colder and drier, the Eastern Himalayan ranges are much wetter with suitable climatic conditions. This supports endemism and biodiversity. The Eastern Himalayas comprise of .parts of Nepal, Bhutan, Sikkim, Arunanchal Pradesh and extends upto Burma. The forest vegetation ranges from tropical rain forests to temperate alpine forests. *Rhododendron* is the dominant tree of this region.

The topography in the Eastern Himalayan region is quite varied that helps biodiversity and endemism. The rugged mountains and valleys support the virgin forests which are rich in many endemic plant species. In the Indian part of the Eastern Himalayan hotspot, about 5800 plant species are found of which around 2000 are endemic. Sikkim is one of the most blessed Indian states so far as endemism is concerned; as of the 4250 plant species found there, 60% are endemic to the region.

Palaeobotanists consider the Indo-Burma region to be one of the centres of origin for the flowering plants (angiosperms). Many primitive angiosperm families occurring there, such as Magnoliaceae, Winteraceae, etc support this contention. The Eastern Himalayas are home to around 8000 species of angiosperms of which nearly 40% are endemic. The characteristic floras are *Rhododendron, Alnus, Betula, Magnolia,* etc. The animals include members of the goat family, antelopes, musk deer, snow leopard, brown bear, black bear and the red panda. Due to continuous habitat destruction and hunting, many of these species are highly endangered.

Andaman and Nicobar Islands are rich in littoral and Island type floras. About 2500 species of flowering plants have been recorded from there, of which 250 species are endemic.

The important hotspots of Andman & Nicobar Islands are North Andaman, spike Island, Table Island, South reef Island, Little and Great Nicobars.

Western Himalayas are well-known for Alpine flora which abounds in Gymnosperms. Some 5000 species of flowering plants are known to occur in this region of which 800 species are endemic. The valley of flowers, Pithoragarh, Gori Valley, Mandal Chopta Valley, Karakoram and Laddakh are some of the major hotspots of this region.

Western Ghats: The Western Ghats run parallel to the west coast of India and constitute more than 16,000 km strip of forests in the states of Maharashtra, Goa, Karnataka, Tamil Nadu and Kerạla. Locally they are also known as the *Sahyadris*. The Western Ghats are characterised by low hills; however, it achieves the highest elevation of 2675 m at Annamalai. Upto an elevation of 500 m the forests are evergreen in nature, while the forests occurring between 500-1500 m altitude are of semi-evergreen type.

The Western Ghats by virtue of having a humid tropical climate and geological stability supports one of the most biodiversity rich areas of the country. According to an estimate, of the 17,000 flowering plants species reported from India, more than 4,500 occur in the Western Ghats region.The Botanical Survey of India has listed 518 endangered species endemic to Peninsular India, the majority of which occur in the Western Ghats.

The dominating plant families are Acanthaceae, Graminae, Orchidaceae, Rubiaceae, Labiatae, Compositeae and Leguminoseae. More than 200 species of rare orchids, many of them are endemic, are found in Western Ghats. Many economically important plants such as banana, rice, black pepper, ginger, etc have spread to other parts of the country from here.

The Western Ghats are home to a rich variety of fauna which are unique and many are endemic to this region. Nilgiri Langur, Lion-tailed Macaque, tiger, leopard, elephant, rare species of tortoise and other amphibians represent the faunal wealth. Many varieties of birds and fishes are also found here.

Sacred Forests. These real hotspots of biodiversity are protected and worshiped by tribals due to religious sanctity. These are known as **Devaskadu** in Karnataka **Devarahati** in Maharashtra and **Lakyntok** in Meghalaya. These forests contain many rare endangered and endemic species.

Mangroves. Mangroves are salt tolerant forest ecosystems found in Saline habitat, tropical and subtropical intertidal zones near estuaries. The total area under mangrove vegetation in India is estimated to be 6740 km^2 and of this 4200 km^2 is covered by mangrove forest of sundarban alone and the second being in the Andaman & Nicobar Islands.

Table 19.3 The major hotspots of endemic and genetic diversity of India

S.No.	Hotspots	Bio-geographic Zone
1.	Karakoram and Ladakh	Trans Himalaya
2.	Kumaon and Garhwal Himalaya	Western Himalayas
3.	Siwaliks	Himalaya
4.	Sikkim Himalaya	Eastern Himalaya
5.	Arunanchal Pradesh	Eastern Himalaya
6.	Lushai Hills	North-eastern India
7.	Tura-Khasi Hills	Meghalaya
8.	Aravallis	Semi-arid zone
9.	Bundelkhand	Central India Plateau
10.	Chota-Nagpur Plateau	Deccan Plateau
11.	Panchmarhi - Satpura ranges	Deccan Plateau
12.	Simlipal and Jeypore Hills of Orissa	Deccan (Eastern Ghats)

13.	Bastar and Koraput Hills	Deccan Plateau
14.	Vishakhapatnam Hills and Araku Valley	Eastern Ghats
15.	Tirupati - Cuddappa Hills	Eastern Ghats
16.	Marathwada Hills	Deccan Plateau
17.	Saurashtra Kutch	Deccan Plateau
18.	Mahabaleshwar - Khandala ranges	Western Ghats
19.	Agumbe - Phonda ranges	Western Ghats
20.	Ratnagiri and Kolaba ranges	Western Ghats
21.	Nilgiris	Western Ghats
22.	Silent valley and Wynaad	Western Ghats
23.	Annamalai	Western Ghats
24.	Idduki - Sabarigiri	Western Ghats
25.	Kalakad and Agastyamalai Hills	Western Ghats
26.	Andaman and Nicobar	Islands

Source : *Diversity of Indian flora, R. R. Rao (1997).*

MEGA DIVERSITY OF INDIA

Varying physiographic features at different places have resulted in the presence of almost all sorts of ecological conditions in India. The varied physiographic and climatic conditions, in the different parts of India, have resulted in a high level of biodiversity in the country. As already mentioned, India possesses the highest mountain range in the world, a long coast line, an arid desert in the west, a cold desert in the north and humid tropical conditions in some other parts. Thus India is blessed with almost all the ecological niches of the world. India is also a meeting ground of the following three major global bio-geographic regions:

(*i*) Indo-Malayan

(*ii*) Eurasian

(*iii*) Afro-tropical

All this has helped India to become a megadiversity nation and it is recognised as one of the twelve mega diversity centres of the world. India has about 48,000 species of plants and 81,000 species of animals, representing 11% and 6.5% of the world's flora and fauna respectively (Table 19.4 & 19.5) According to an estimate, 35% of the flowering plants and 18% of the total flora are endemic to India. The endemism among the animals is even higher and in some cases it may be as high as 62%, as in the amphibians.

Table 19.4 Number of species in major groups of flora and fauna in India

Plant	Number	Animal	Number
Bacteria	800	Molluscs	5,000
Fungi	13,000	Insects	57,000
Algae	12,281	Fishes	2,546
Lichen	1,940	Amphibians	204
Bryophytes	2,843	Reptiles	420
Pteridophytes	1,022	Birds	1228
Gymnosperms	64	Mammals	372
Angiosperms	17,000	Lower groups	Around 10,000

Table 19.5 India's world ranking different groups of organisms

Group	India's world ranking	No. of species in India
Angiosperms	15th	17,000
Birds	8th	1228
Reptiles	5th	420
Amphibians	15th	204
Mammals	8th	372

The biodiversity of India can be observed in the following:

Wetlands

India, due to its varied geological terrain and climate, has a variety of wetlands that range from the cold lakes of Ladakh to saline water lakes of the arid region. A variety of wetlands are found in India, such as the fresh water lakes in the hilly regions; coastal lagoons; the fresh water marshes of the Gangetic Plain and Brahmaputra flood plains; swamps in the north-eastern region; saline lakes of the arid region and the mangroves and delta wetlands. Examples of wetlands are Chilka Lake, Ghana Bird Sanctuary, Sunderban Delta, Pullicat Lagoon, etc.

The wetlands are a complete ecosystem and support a multiple variety of life forms. Wetlands are the habitat of many endemic and endangered species of the world. The inland fresh water wetlands are the breeding ground of many indigenous and migratory birds. The coastal wetlands are the breeding grounds of many threatened species like Olive Ridley turtles, Green sea turtles, dugongs and dolphins. Many unique plant species are found in the mangroves, such as *Avicennia* and *Rhizophora.* The mangroves are also home to a number of fish species, crustaceans, monkeys, otters, fishing cats, snakes, wild pigs and tigers.

Forests

The diversity and vastness of the Indian terrain supports diverse types of forests. The main types of forests present in India include evergreen tropical rainforests, semi-evergreen rainforests, deciduous monsoon forests, sub-tropical dry deciduous forests, sub-tropical conifer forests and dry alpine scrubby forests.

The evergreen tropical rainforests are found in Andaman and Nicobar islands, parts of Western Ghats and north-eastern India. The rainforests are characterized by significant species diversity and density of vegetation. Sub-tropical dry deciduous forests occur in major parts of India, in the central regions, and Deccan Plateau. Sub-tropical conifer forests are present throughout the length of the Himalayas from 1000 m to 3000 m elevation. Beyond this height, the plants are represented by alpine scrubby vegetation.

Forests cover around 22 per cent of India's total land area which is however unevenly distributed. Industrial timber species include teak, deodar, pine, sheesham and sal. Products such as charcoal, fruits and nuts, fibres, oils, gums and resins are among the most valuable commodities from India's forests. The wildlife of India is wholly concentrated in the forested areas and are represented by a wide variety of deer, bear, tiger, elephant, rhinoceros, wild buffaloes, monkeys, birds, reptiles, amphibians, etc.

Deserts

The Indian desert is present in the western region of the country and is characterised by low rainfall, low humidity, hot winds and sandy terrain. Though, the desert presents a barren landscape, the flora of the Indian desert is represented by a wide variety of plants. Important plants are *Acacia, Phoenix, Ephedra, Euphorbia, Amaranthus, Agave,* etc. The faunal diversity is also rich, as all the major groups of animals are found in the Indian deserts. More than 1700 species of fauna exist in the Indian desert, of which birds, reptiles and mammals are the major groups. Some of the important animals are desert fox, camel, gazelle, black buck, Great Indian bustard, wild ass, etc.

Besides the Great Indian Desert in the western part of the country, desertic conditions also prevail in the Trans-Himalayan rain shadow zone. These are called cold deserts, found in Ladakh (J&K) and Lahaul-Spiti (H.P.). The vegetation mainly comprise of shrubs or stunted trees. The common trees are fir, kail, bhojpatra, oak, juniper, populus, etc. Due to high elevation and accompanying arid conditions, trees are rarely found whereas grasslands are quite common. Many medicinal plants occur in this region. Though, the fauna is sparse, it is represented by unique and endangered species, such as snow leopard, Tibetan wild ass, wild yak, mountain goats, etc.

Coastal Marine Environment

India has a coastline of more than 7000 km and the coastal marine environment has a diversity of flora and fauna, some of which are important sources of revenue for the local people. Lakhs of people along the coastline are dependant on fishing for their daily sustenance and as a major source of income. Sardines, hilsa, pomfret, crabs, prawns and oysters are the important catches.

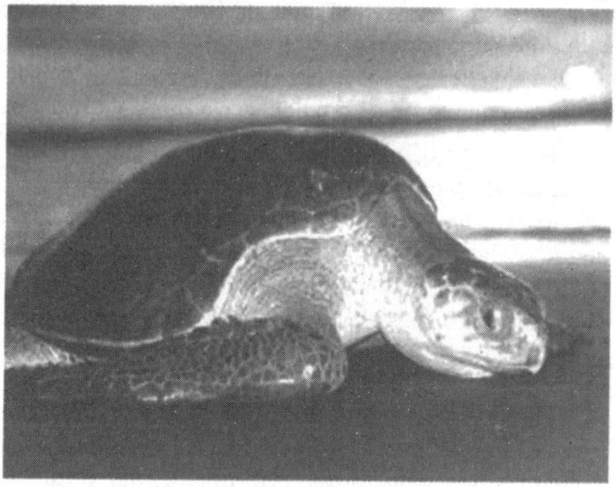

Fig. 19.8 Olive Ridley turtle

Coral reefs are one of the most biodiverse ecosystems of the world and these occur in Andaman, Nicobar and Lakshadweep islands and in isolated patches in the Gulf of Kutch and in southern India. Coral reefs that are of commercial value also contribute towards the economy of the local people. Sea grasses and seaweeds are used for decorative purposes in the aquaria and for commercial purposes in the harvesting of prawns. A number of turtle species like Green-turtle, Olive-Ridley turtle (Fig. 19.8), Leatherback turtle are quite common in the Indian marine waters.

Agriculture

India has been one of the major centres of origin and diversity of crop plants. Its contribution to agro-diversity has been quite impressive, as it has given nearly 167 species of domesticated agri-horticultural crop plants to the world. Over 320 species of wild progenitors, from where these domestic crops have originated, are native to India.

The rich genetic reserve of the Indian subcontinent includes 51 species of cereals and millets, 104 species of fruits, 27 species of spices, 55 species of vegetables and pulses, 24 species of fibre crops, 12 species of oil seeds and various strains of tea, coffee, tobacco and sugarcane. Besides the above, over 2000 species of different plant parts are used by the indigenous people for medicinal purposes.

Livestock

India has played a major role in the field of animal husbandry as it has contributed such useful animal species like water buffalo and camel. India's indigenous livestock includes 27 breeds of cattle, 22 breeds of goats, 40 breeds of sheep, 8 breeds each, of camels and buffaloes 6 breeds of horses, 2 breeds of donkeys and 18 breeds of poultry. The 8 indigenous breeds of buffaloes in India, represent the entire range of its genetic diversity on earth. Further, diversity has been increasing through genetic engineering and biotechnology to enhance their productivity.

The biodiversity rich areas of India are mostly unexplored or under-explored. Thus, the recorded biological diversity of India may only be a part of the actually existing biological wealth. Major gaps in our knowledge still persist, especially about the diversity in microorganisms, tropical flora and fauna and marine life. Persistent efforts are required to enhance our knowledge regarding our biological wealth. The knowledge of diversity in plant and animal genetic resources will provide a sound basis for further developments in the fields of biotechnology and genetic engineering, which will be helpful to mankind.

EFFECT OF HUMAN ACTIVITIES ON GLOBAL BIODIVERSITY

Human Population
(Size and resource use)

↓

Human Activities
(Agriculture, Industry, production, consumption and recreation)

↓

Direct Effects

1. Destruction or degradation of natural ecosystems
2. Changes in number of species
3. Alteration of natural and chemical cycles
4. Energy flows
5. Pollution of air, water and soil

↓

Indirect Effects

1. Climatic changes
2. Loss of Biodiversity

LOSS OF BIODIVERSITY

Biodiversity is under tremendous stress and is undergoing a crisis as many species are threatened or endangered and some have already become extinct.

Threatened Species

Threatened species are those organisms or species whose existence is in danger due to various pressures. The threatened species include the organisms that are decreasing due to habitat destruction, over-exploitation, environmental degradation and pollution. It also includes the species whose numbers are seriously depleted and are under constant threat due to adverse factors in their habitats. If the pressure persists, the threatened species are liable to become endangered and extinct. Thus, on the basis of severity of threat, the species are further categorised as endangered, vulnerable and rare.

1. **Endangered.** Category includes the species that are in danger of being extinct if the factors that threaten them continue to operate.

2. **Vulnerable** species are those that may move into the endangered category if the pressure on their existence and environment continues unabated.
3. **Rare** species are characterised by organisms with low populations that are not presently endangered or vulnerable but are at danger of being threatened. For example, some species like the whooping crane are naturally rare. Thus their rarity may become the likely cause for their extinction.

Endangered Species

The endangered species are those that are in danger of being extinct and whose very existence is at stake in the near future. This category includes the species whose numbers have been reduced critically and whose habitats have been drastically damaged so that they are finding difficulty in regenerating themselves. Such species are on the verge of extinction.

The species in danger of extinction are divided into several categories. A number of agencies, both international and national, study and maintain the lists of endangered species. They also suggest measures to protect and preserve their natural habitats and to promote programmes for the recovery and re-establishment of these species. The main international agency working in this field is the 'International Union for the Conservation of Nature' (IUCN) based in Switzerland. It provides a list and publishes information on the endangered species worldwide in the form of a book called the **Red Data Book.** Another agency that looks into the endangered species is the 'Secretariat for the Convention on International Trade in Endangered Species of Wild Fauna and Flora'.

The list of endangered species provided by the 'IUCN Red Data Book of Threatened Animals' is quite alarming. The 1996 Red Book reported that one-fourth of all known species of mammals, 11% of the birds, 20% of the reptiles, 25% of the amphibians and 34% of fresh water fishes are under threat of extinction (Table 19.6 and 19.7 and Fig. 19.9). Similarly, the 1997 'IUCN Red Data Book of Threatened Plants' estimates that more than 33,000 species of higher plants have already become extinct or endangered. Over 8,700 tree species, nearly 9% of the world's tree flora, are listed to be globally threatened by the IUCN.

The percentage of actually threatened species may be much more because many areas are still unexplored or have remained underexplored. As already pointed out, only a small per cent — age of the existing species have so far been identified.

Table 19.6 Endangered and threatened animals worldwide (IUCN)

S.No.	Group	Endangered	Vulnerable	Rare	Indeterminate	Total
1.	Mammals	177	199	89	276	741
2.	Birds	188	241	257	284	970
3.	Reptiles	47	88	79	102	316
4.	Amphibians	32	32	55	50	169
5.	Fishes	158	225	246	349	979
6.	Molluscs	309	422	139	312	1182
7.	Annelids	6	145	6	2	159
8.	Spiders	3	2	4	9	18
9.	Crustacea	10	6	28	114	158
10.	Insects	252	119	241	572	1184
	Total	**1182**	**1479**	**1144**	**2070**	**5876**

Table 19.7 Endangered Species of Birds, Reptiles and Mammals in India

Birds	Reptiles	Mammals
Great Indian Bustard	Gharial	Black Buck
Great Pied Hornbill	Green Sea Turtle	Musk Deer
Himalayan Pheasant	Leatherback Turtle	Lion-tailed Macaque
Jerdon's Courser	Tortoise	Wild Ass
Black necked Crane	Crocodile	Sloth Bear
Pelican	Monitor Lizard	Snow leopard
Siberian White Crane	Pythons	Rhinoceros
Vulture		Tiger
		Asiatic Lion
		Indian Wild Dog

Fig. 19.9 Some of the endangered animal species of India.
A. Great Indian bustard **B.** Pelican, **C.** Tortoise, **D.** Asiatic lion **E.** Wild ass **F.** Black Buck

ENDANGERED PLANTS OF INDIA

Indian flora is very rich in diversity due to varied topography ranging from sea level to highest mountain, the long coastal line in peninsular India, desert in the west, cold desert in the north and humid tropical conditions in the eastern part of the Indian subcontinent. In the recent past the forest wealth of India is known to have depleted to a large extent both in terms of quality and quantity ultimately leading to various kinds of environmental hazards which not only deplete the biomass reserve but also degrade the land, affect the climate thus disturb ecosystem of the area. Frequent occurrence of floods and droughts are the resultant effects of deforestation which ultimately influences the human and animal populations to a great extent. Scenic beauty of mountains is fading due to destruction of forest. According to Ministry of Agriculture (1980) about 175 mha area is suffering from environmental degradation due to deforestation alone. About 90 million hectare area is badly degraded and has become completely unproductive. According to S.K. Jain (1987) out of about 15,000 species of Indian flowering plants, nearly 2,500 species are under various degrees of threat. The National Remote Sensing Agency (NRSA, 1983) has reported 16.52 per cent loss of total forest area in India during 1972 to 1982. In U.P. alone vegetation cover was 17.41 per cent in 1982 of the total geographical area.

CAUSES OF THREAT

The depletion of forest wealth is mainly due to heavy and uncontrolled biotic interference which has greatly affected natural forest environment thereby adversely influencing the wild life. Many species of tropical flora and fauna have reached near extinction and some are known to have even become extinct. For better survival of wild animals in future, it is necessary that they must be allowed to live undisturbed in their own habitat.

The causes of threats to plant species may be of two types :

1. Natural
2. Artificial (Man-made)

The natural causes responsible for endangering plants are land slides, droughts, floods, storm, earthquakes, diseases, avalanches etc. Other threats include invasion of exotic and other aggressive weeds, air, water pollution and lack of pollinators.

Man-made threats include grazing, commercial exploitation, industrialization, forestry, urbanisation, scientific and educational research, roads, dams, increasing townships, internal tourism, mining and pressure of introduced plants. Introduction and growth of weeds and pests result into reduction in vegetation of natural flora. Over exploitation of plants for commercial purpose leads to destruction of their habitat. Supply of raw material to the forest based industries is also a major cause of forest destruction in the hills. The other cause of forest destruction is due to increase in cattle population.

Conservation or Protection of Endangered Plants

Throughout the world, there is growing awareness for conservation and protection of endangered flora and fauna. *Red Data Books* on endangered animals, birds and flowering plants have been prepared by the survival service commission of the International Union for conservation of Nature and Natural Resources (IUCN). It has been estimated that out of 18,000 to 20,000 species of flowering plants in India, about 1,000 species fall into the conservation category (rare, endangered, vulnerable or depleted) and need adequate conservation. In fact, the 10th General Assembly of the IUCN held in New Delhi in 1969 had focussed attention on the urgent need to protect endangered species in Indian flora and urged the government to bring these endangered plants into effective cultivation. Thus some of our notable plants like *Rauvolfia serpentina* (Sarpgandha), *Dioscorea deltoides, Cypripedium cordigerum* (Lady slipper orchid), *Nepenthes khasiana* (Pitcher plant) and *Rheum nobile* have been decimated and are being threatened with extinction in their original habitat. Of many valuable use of botanic gardens and their ancillary research establishment, the conservational ones are the most important at least for the future as human population pressure and resultant degradation of the biosphere increase. Therefore, it is proposed to set up a conservation unit to cater the needs of endemic and vanishing plant taxa in Indian flora and to prepare a baseline upon which future action plans can be prepared.

There has been some consciousness about the depletion of forests mainly due to alarming situation. The forest cover of the country especially that of U.P. is comparatively much less than recommended by the National Forest Policy, Govt. of India (1952). Accordingly, vigorous afforestation and social forestry programmes have been launched to increase the tree cover. One of the important aspects of these programmes is the selection of suitable land which can be acquired with care without changing their present land use which may otherwise lead to socio-economic problems. Therefore, the best location for afforestation programme would be the degraded or waste lands. This would not only increase the productivity of such land but would also lead to qualitative improvement of such degraded lands.

During the last decade, however, various efforts have been made to rehabilitate the forest cover through launching of afforestation programme. These can be implemented well only when the problems related to these are fully understood. It is possible to identify and map such lands where social forestry and afforestation programme can be undertaken more effectively. This would not only increase the forest area to meet ever increasing demand of timber, fuel wood etc. but would simultaneously improve their lands. The more important aspect of these programmes is the continuous monitoring of the afforested area.

Botanical gardens and sanctuaries play a major role in the conservation and protection of endangered flora and fauna. Exchange of information on plant material would serve in preserving germplasm. Cultivation of endangered plants in botanical gardens would serve as a germplasm collection for better genetic resources for future generation. Popularization of rare and endangered plants in the agricultural fields may be used to protect them from extinction. Detailed data on the Morphology and Autecology of endangered species are of immense importance in monitoring populations for successful reintroduction. Studies on germination, growth, flowering, fruiting and acclimatization of a species are also of great importance for successful cultivation. In various centres of India plant tissue culture techniques are now being adopted to preserve highly endangered plants. For reintroduction of plants, knowledge of seed germination, survival, transplanting, plantation, transportation and documentation is essential. In order to get better results of reintroduced plants for the establishment in the future, help from forest department and voluntary organizations should be taken. These steps will create and preserve endangered plants and help in the evolution of new and more useful cultivars suitable for present ecological conditions.

ENDANGERED PLANTS

In India, amongst the commercially exploited plants are many species of medicinal and economic importance belonging to various genera such as *Aconitum, Nardostachys, Podophyllum, Commiphora, Coptis, Coscinium, Vanilla* etc., and plants of horticultural importance such as *Cycads,* tree-ferns, *Orchids* (*Vanda, Dendrobium*). Over exploitation of even some very widely growing species can also result in their rarity; examples of which are species of *Dioscorea, Rauvolfia* and *Colchicum.* Over exploitation of biologically important species such as primitive and insectivorous species also results in their rarity or loss. Examples of this category are *Gnetum, Ephedra, Psilotum, Ophioglossum* and *Pinus.*

A list of some economically and medicinally important endangered plants from Indian subcontinent is given below along with their outline information :

Rare, Endangered and Threatened Economic Plants of India

1. *Aconitum deinorrhizum* Stapf. (Ranunculaceae)
   Herb. Important aconite of medicinal value. Rare in occurrence, exploited for trade.
2. *Acorus calamus* Linn. (Araceae)
   Rhizome carminative, emetic, antispasmodic used in Dyspepsia and in chronic diarrhoea.
3. *Allium strachevi* Baker (Liliaccae)
   Leaves are medicinal and used for flavouring also.

4. *Angelica glauca* Edgew (Apiaceae)
   Roots used for confusions and as blistering agent, seeds purgative.
5. *Arundina gramnifolia* (D. Don) Hochr. (Orchidaceae)
   Flowers are ornamental. Rare in occurrence.
6. *Atropa acuminata* Royle (Solanaceae)
   Used in reducing hypertension.
7. *Betula utilis,* D. Don (Betulaceae)
   Timber used for commercial purposes. Bark of the tree is sacred.
8. *Balanophora involucrata* Hk. f. (Balanophoraceae)
   Used as illumination in java.
9. *Castanopsis tribuloides* A. Dc. (Fagaceae)
   Used as timber and fodder.
10. *Colchicum luteum* Baker (Liliaceae)
    Carminative, luxative and used in Rheumatism. Rare occurrence, exploited in trade for its valuable alkaloid colchicine.
11. *Commiphora wightii* (Arn.) Bhandari (Burperacae)
    Small tree, exploited for its gum, resin from the bark. Scarce.
12. *Coptis teeta* Wall. (Ranunculaceae)
    Rhizomes are medicinal, rare occurrence, exploited in trade for its high medicinal value.
13. *Carduns nutana* Linn. (Asteraceae)
    Used as a blood purifier.
14. *Cyperipedium cordigerum* D. Don. (Orchidaceae)
    Highly ornamental, rare, threatened due to trade.
15. *C. elegans* Reichb. b. (Orchidaceae)
    Ornamental, rare occurrence, threatened due to trade.
16. *Dendrobium densiflorum* Wall. ex Lindl. (Orchidaceae)
    Highly ornamental, exploited in trade.
17. *Dioscorea deltoidea* Wall. ex Kunth (Dioscoreaceae)
    Used in washing silk, wool, dyeing etc.
18. *D. prazeri* Prain & Burkil (Dioscoreaceae)
    Used as fish poison, also to kill mouses.
19. *Dischidia rafflesiana* Wall. (Asclepediaceae)
    Roots chewed with betel to cure cough.
20. *Drosera peltata* Sm. (Droseraceae)
    Used in preparation of Gold Bhasm.
21. *Gentiana kurroo* Royle (Gentianaceae)
    Bitter tonic for improving appetite, urinary trouble. Threatened due to trade.
22. *Grewia sapida* Roxb. Dc. (Tiliaceae)
    Economically important for fibres, fodder and fuel, Exploited in trade.
23. *Lagerstroemia hypoleuca* Linn. f. (Lythraceae)
24. *Mahonia jaunserensis* Ahrendt. (Berberidaceae)
    Medicinal plant, threatened due to trade.
25. *Meconopsis betonicifolia* Franchet (Papaveraceae)
    Due to medicinal value exploited in trade.
26. *Nardostachys jatamansi* DC (Valerianaceae)
    Used as diuretic, luxative, and in hysteria.

27. *Nepenthes khasiana* Hook. f. (Nepenthaceae)
Used in urinary trouble and itching of eyes.
28. *Nymphaea pygmaea* Ait. (Nymphaeaceae)
Root used medicinally.
29. *Papliopedilum harsutissimum* (Lindl). Pfitz. (Orchidaceae)
Flowers are ornamental, extremely rare due to indiscriminate collections from the wild state.
30. *P. venustum* (Lindl). Pfitz. (Orchidaceae)
Flowers ornamental, rare due to over exploitation from the wild state.
31. *P. villosa.* (Lindl.) Pfitz. (Orchidaceae)
Flowers ornamental, rare and threatened due to over collections.
32. *Podophyllum hexandrum* Royle (Podophyllaceae)
Purgative, as source of medicinal resins.
33. *Primula sikkimensis* Hook. f. (Primulaceae)
Used for perfumery purposes.
34. *Rauwolfia serpentina* (Linn.) Benth ex Kurx. (Apocynaceae) Used in antihypertensive and in nervous diseases.
35. *Rheum emodi* Wall ex Neissr. (Polygonaceae)
Powdered roots used as tooth poweder and paste, used against swelling.
36. *R. nobile* Hook. f. & Thomas. (Polygonaceae)
Dry beans are used as substitute of tobacco.
37. *Rhododendron arizelum Balb.* & Farr. (Ericaceae)
Ornamental flowers.
38. *R. edgeworthii* Hook. f. (Ericaceae)
Leaves are used in fish poisoning.
39. *Santalum album* Linn. (Santalaceae)
Timber & heart wood have economic value.
40. *Saussurea lappa* clarke (Asteraceae)
Used as perfumary, cosmetics and antiseptic.

List of Rare and Endangered Himalayan Plants

TREES

1. *Aglaia perviridis* Hiern. (Meliaceae)
Local name—Thaidubriki-buphang (Assami)
Locality—Cultivated in India.
Uses—Medicinally used.
2. *Abies delavayi* Franch (Pinaceae)
Locality—Balipara Frontier Tract Mountain.
3. *Castanopsis tribuloidis* A.D. (Fagaceae)
Local name—Chongom, Bakhalai.
Locality—Himalayas.
Uses—Fruits edible.

4. *Elaeocarpus prunifolius Wall.* (Elaeocarpaceae)
   Local name—Dieng-la-khmar, Diengruin (Khasi).
   Locality—Sylhet, Khasi, Jaintia Hills.
   Uses—Fruits edible.
5. *Euonymus echinatus Wall.* (Celastraceae)
   Local name—Laduli.
   Locality—In Himalayan region (Khasi & Jaintla Hill).
6. *Grewia sapida* Roxb. ex DC. (Tiliaceae)
   Local name—Phuhura (Assami).
   Locality—Assam.
   Uses—Fodder & edible fruits.
7. *Homalium schlichii* Kurz. (Somyaceae)
   Local name—Dieng-soh-mara (Khasi).
   Locality—Khasi Hills.
8. *Larix griffithiana,* Gord, Larix griffithi, Hook. f. & Thoms.
   Local name—Saar.
   Locality—Himalayas, Eastern Nepal to Bhutan.
   Uses—Non-medicinal.
9. *Mahonia jaunsarensis* Ahrendt. (Berberidaceae)
10. *Magnolia griffithii* Hook. f. & Thoms. (Magnoliaceae)
    Local name—Gauri-sopa, Gahorisopa (Assami).
    Locality—Common in Lakhimpur, Sibsagar and Darrang Distt.
    Uses—Not used for timber.
11. *M. gustavigustavi* (King)
    Local name—Khorokia-sopa (Assami).
    Locality—Makum Forest Lakhimpur.
12. *Magnolia pealiana* King (Magnoliaceae)
    Local name—Gahori-Sopa (Ass.).
    Locality—Makum Forest, Lakhimpur.
    Uses—Wood is white soft and light, fairly strong.
13. *M. pterocarpa* Roxb. (Magnoliaceae)
    Local name—Patpate.
    Locality—Himalayas to Assam.
    Uses—Wood used as fuel and tea boxes.
14. *Malus baccata* Desf. (Rosaceae), *Pyrus baccata* Linn
    Local name—Banmehal (Hindi).
    Locality—Siberia to Manchuria & Japan, Southwards to Himalayas.
    Uses—Ornamental, fruits edible.
15. *Michelia lanugiorosa Wall.* (Magnoliaceae)
    Local name—Gogichamp (Nepali).
    Locality—South Asia to China.
    Uses—In timber, fuel and making charcoal.

16. *Palaquium polyanthum* Engl. (Sapotaceae)
Local name—Tali (Bengali).
Locality—In Garo Hills.
Uses—In carpentary work, timber.
17. *Pinus gerardiana Wall.* (Coniferae)
Local name—Chilgoza.
Locality—Upper Assam.
Uses—Edible proteinaceous kernel; commercial timber.
18. *Populus gamblei* Haines (Salicaceae)
Local name—Bahan, Bhan.
Locality—Punjab plains, Himalayas.
Uses—Plywood, boats, shoe-heels.
19. *Rhododendron arboreum* Sm. (Eriaceae)
Local name—Burans (Nepali).
Locality—Himalayas, from Kashmir to Bhutan.
Uses—Tender leaves are used as vegetable and to relieve headache.
20. *Populus euphratica* Oliv (Salicaceae)
Local name—Bahan, Bhan.
Locality—Punjab plains, Himalayas.
Uses—Plywood, boats, shoe-heels.
21. *Rhododendron edgeworthii* Hook. f. (Eriaceae)
Locality—Sikkim Himalayas (7,000–9,000 ft).
22. *Rhododendron dalhousiea* Hook. f. (Eriaceae)
Local name—Sikkim Himalayas (6,000–9,000 ft).
23. *Zanthoxylum khasianum* Hook. f. (Rutaceae)
Locality—Soh-mrit.
Locality—Khasi Hills (5,000–6,400 ft).
24. *Picea brachytila* Masters (Pinaceae)
Locality—N.E. Assam.
Uses—As timber.

HERBS

1. *Acorus calamus* Linn. (Araceae)
Local name—Bach (Hindi, Bengali).
Locality—Throughout India, 6,000 ft in the Himalayas, Kashmir & Sirmoor, Sikkim.
Uses—Rhizome carminative, emetic, Antispasmodic use in Dyspepsia and in chronic Diarrhoea.
2. *Alpinia galanga* S-W. (Zingiberaceae)
Local name—Kulanjan.
Locality—Throughout India, from the foot of Himalaya to Ceylon.

3. *Allium stracheui* Baker (Liliaceae)

   Locality—Western Himalaya, from Kashmir at altitude 9,000 ft to Kumaon10,000–12,000 ft.

4. *Angelica glauca* Edgew. (Umbelliferae)

   Local name—Chora.

   Locality—From Kashmir to Simla, alt. 8000–10,000 ft.

5. *Atropa acuminata* Royle (Solanaceae)

   Local name—Belladonna.

   Locality—Western Himalayas, alt. 6,000–11,000 ft from Kashmir to Simla.

   Uses—To reduce hypertension.

6. *Allium clarkei* Hk. f. (Liliaceae)

   Locality—Kashmir to Skrado, alt. 7,000–11,000 ft.

7. *A. consanguineum* Kunth (Liliaceae)

   Locality—Western Himalayas, Kashmir, alt. 8,000–10,000 ft.

8. *Allium thomsonii* Baker (Liliaceae).

   Locality—Kashmir, alt. 12,000 ft.

9. *Anemone tetrasepala* Royle (Ranunculaceae)

   Locality—Western Himalaya, Marri, Kashmir and Kistwar, alt. 8,000–11,000 ft.

10. *Aquilegia nivalis* Folc. ex Jackson (Ranunculaceae)

11. *Artemisia amygdalina* Dcne. (Compositae)

    Locality—Kashmir, Pir Panjal.

12. *Botrychium virginianum* (Sw.) (Ophioglossaceae)

    Local name—Clark's Fern.

    Uses—In dysentery and Healing of wounds.

13. *Balanophora involucrata* H.k. f. (Balanophoraceae)

    Local name—Gajpipal.

    Locality—Tropical Himalaya, Simla, alt. 6,000 ft Sikkim, 7,000–10,000 ft.

    Uses—Used as illumination in Java.

14. *B. dioica* Brown (Balanophoraceae)

    Locality—Tropical and subtropical Himalaya, From Nepal to Mishmi, alt. 3,000–7,000 ft Khasia Hills 4,000–6,000 ft.

15. *Coptis teeta* Wall.

    Local name—Mamira.

    Locality—Assam.

    Uses—As Antibiotics.

16. *Colchicum luteum* Baker (Liliaceae)

    Local name—Hiran tutiya.

    Locality—Western, Temperate Himalaya Kashmir, altitude 4,000–7,000 ft.

    Uses—Carminative, Laxative, Rheumatism.

17. *Cyathea gigantia* Halltt. (Cyatheaceae), a tree fern.

18. *Cardus nutans* Linn. (Compositae)
 Local name—Kanchari.
 Locality—Western Himalaya.
 Uses—Blood purifier.
19. *Dioscorea prazeri* Prain & Burkill
 Locality—Assam.
 Uses—To kill mouses and as fish poison.
20. *D. deltoidea* Wall ex Kunth (Dioscoreaceae)
 Locality—Assam.
 Uses—To kill mouses and fish posion.
21. *Drosera peltata* Sm. (Droseraceae)
 Local name—Mukhjali.
 Locality—Throughout India.
 Uses—In preparation of Gold Bhasm.
22. *Dischidia rafflesiana* Wall. (Asclepiadaceae)
 Local name—Hankhao jharmona (Assami).
 Uses—Roots chewed with betel to cure cough.
23. *Dianthus cachemiricus* Edge (Caryophyllaceae)
 Locality—Kashmir.
24. *Dicentra roylei* Hook. f. & Thoms. (Fumariaceae)
 Locality—Temperate Himalaya, alt. 5,000–6,000 ft. Simla, Khasi Hills.
25. *Gentiana kurroo* Royle (Gentianaceae)
 Local name—Kutki.
 Locality—Kashmir & N-W. Himalaya.
 Uses—Bitter tonic for improving appetite, Urinary trouble.
26. *Helminthostachys zeylanica* Linn. Fern. (Ophioglossaceae)
 Local name—Kamraj.
 Locality—Eastern part of India.
 Uses—In toxication, in Sciatica & in cough.
27. *Iris aurea* Lindl (Iridaceae)
 Locality—Western Himalaya.
28. *I. gilgitensis* Baker (Iridaceae)
 Locality—Western Tibet, alt. 12,000 ft.
29. *Impatiens acuminata* Benth (Balsaminaceae)
 Locality—On Wet rocks in the Khasia, alt. 4,000–5,000 ft.
30. *I. bella Hk.* f. & Thoms. (Balsaminaceae)
 Locality—In marshes in the Khasia, 4,6000 ft.
31. *I. fimbriata* Hook. (Balsaminaceae)
 Locality—In marshes in the Khasia, 4,6000 ft.
32. *I. paludosa* Hk. f. (Balsaminaceae)
 Locality—Khasia, 5,000–6,000 ft.

33. *Meconopsis betonicifolia* Franch (Papaveraceae)
Locality—Himalayan region.
34. *Nardostachys jatamansi* DC. (Valerianaceae)
Local name—Jatamansi.
Locality—Eastern & Western Himalaya.
Uses—Spasmodic, Diuretic, Luxative, in hysteria.
35. *Nepenthes khasiana* Hk. f. (Nepenthaceae)
Local name—Ghatparni.
Locality—Khasia, Garo Hills.
Uses—In Urinary trouble, etching of eyes.
36. *Nymphaea pygmaea* Ait (Nymphaeaceae).
37. *Osmunda regalis* Linn. (Osmundaceae)
Locality—Khasi Hills, Eastern Ghats.
Uses—In Intestinal grip, Rheumatism, Dysentery.
38. *Podophyllum hexandrum* Royle (Podophyllaceae)
Local name—Bhavanbakara.
Locality—Himalayan, from Kashmir to Sikkim.
Uses—Purgative, Source of Medicinal Resin.
39. *Potentilla clarkei* Hk. f. (Rosaceae)
Locality—Kashmir, Srinagar, alt. 7,000 ft.
40. *P. curviseta* Hk. f. (Rosaceae)
Locality—Kashmir, alt. 12,000 ft.
41. *Primula sikkimensis* Hk. f. & P. Repenta Hk. f. (Primulaceae)
Locality—Temperate region of Himalaya.
Uses—Delicious perfume.
42. *Rauwolfia serpentina* Benth & Kurz. (Apocynaceae)
Local name—Sharpagandha, Chandrika.
Locality—In Himalayas, Sikkim.
Uses—Antihypertensive, and in nervous disorder.
43. *R. tetraphylla* Linn. (Apocynaceae)
Local name—Bara chandrika.
Locality—Western India, U.P.
Uses—Poisonous, given in mouth pain and burning sensation of Stomach.
44. *Rheum nobile* Hk. f. & Thoms. (Polygonaceae)
Local name—Tchuka.
Locality—Himalaya, from Nepal to Bhutan.
Uses—Dry beans are used as substitute of tobacco.
45. *R. emodi* Wall. ex Meissn. (Polygonaceae)
Local name—Gandhini (Sankrit).
Locality—In Himalaya region.
Uses—Powdered roots as tooth powder.

46. *Saussurea obvallata* Wall. (Compositae)
    Local name—Brahma Kamal.
    Locality—Himalaya from Kashmir.
47. *S. bracteata* Done (Compositae)
    Locality—Eastern Tibet, North Kumaon.
48. *Saussurea gynaphaloides* Sch.–Bip (Compositae) (Royle, Sch. Bip).
49. *S. clarkei* Hook. f. (Compositae)
    Locality—Kashmir.
50. *S. elliptica* Ck. (Compositae)
    Locality—Eastern Tibet, Karakoram, alt. 14,000 ft.
51. *S. glandulifera* Schultz. (Compositae)
    Locality—Eastern, Tibet, Ladakh, alt. 14,000–16,000 ft.
52. *S. chultzii* Hook. f. (Compositae)
    Locality—Western Tibet from North Kashmir to Karakoram, alt. 14,000–17,000 ft.
53. *S. lappa* C.B. Clarke (Compositaeae)
    Local name—Kuthkur, Pachak.
    Locality—Valley of Kashmir.
    Uses—In Perfumery, Cosmetic, Antiseptic.
54. *Swertia thomsonii* Cl. (Gentianaceae)
    Locality—Eastern Himalaya, Kashmir.
55. *Vaccinium nuttallii* Sleumer nuttallii (C.B. Clarke) Sleum.
56. *Viola falconeri* Hook. f. & Thoms. (Violaceae).
    Locality—Kashmir.
57. *Senecio jocquemontianus* Benth (Compositae)
    Local name—Hatermul Poshkar.
    Locality—Western Himalaya.
    Uses—Roots in skin diseases.
58. *Salvia hians* Royle
    Locality—Kashmir to Himalaya region.
    Uses—Essential oils is extracted.
59. *Psilotum nudum* Beauv. (Psilotaceae)
    Locality—In moist shady places throughout India.
    Uses—Purgative, in diarrhoea, antibiotic.

EXTINCT SPECIES

Extinct species are those that no longer exist. The primary causes of species extinction are habitat destruction, commercial exploitation such as collecting, hunting and trading in animal parts and environmental pollution. Of the above, direct habitat destruction is the main cause of extinction. Extinction is of two types: **Local extinction** - when a species disappears from a part of its range but persists elsewhere, and **Global extinction** - when a species does not exist at all in the world.

Extinction is a normal process in the course of evolution. Species have slowly evolved and disappeared throughout the geological time because of climatic changes and their inability to adapt to the changed environment. The rate of extinction has varied greatly over geological time. Over much of the earth's history, the rate of evolution of new species equalled or slightly exceeded the rate of extinction. However, there have been episodes in the earth's history, where the fossil records suggest periods of catastrophic losses of species and periods of rapid evolution of new species. The catastrophic losses of species within a short time are referred as **mass extinctions.** There have been five major episodes of mass extinctions in the geological past. One such most interesting mass extinction episode occurred 65 million years ago, when the dinosaurs became extinct.

Since the 18th century, the rate of extinction has accelerated rapidly because of human population growth and resource consumption. Today, most of the world's habitats are changing faster than most species can adapt to such changes through evolution or natural selection. The current global extinction rate is estimated to be around 10,000 species per year. Many scientists believe that we are in the middle of the greatest mass extinction episode, which this time is however triggered by humans.

Causes of Extinction

The main causes of extinction can be summarized below:

Population risk: Variations in population risk can cause a rare species to become extinct.

Environmental risk: The population size can be affected by the changes in the physical or biological environment, competition and predator-prey relation.

Genetic risk: Detrimental changes in the genetic constitution can cause a species to loose its vitality.

Natural catastrophe: Sudden change in the environment like fires, violent storms, earthquakes volcanic eruptions, etc. can also bring about extinction.

Fig. 19.10 The cheetah is no longer found in India

Anthropogenic causes:

1. Habitat destruction
2. Hunting and commercial exploitation
3. Introduction of pests, parasites and predators
4. Pollution of the habitat

EXTINCT ANIMALS

In the recent past, the Mauritian bird Dodo, the North American Passenger Pigeon, the Elephan Bird of Madagascar, Painted Vulture of Florida, The Cape Lion are some of the fauna that became extinct. So far 23 species have been reported to be extinct from India, the most important of which are: cheetah (Fig. 19.10), pink-headed duck and mountain quail.

Causes of Biodiversity Loss

Increasing population pressure and over-exploitation of the biotic resources are taking their toll on biodiversity leading to its loss. The main factors causing loss of biodiversity are summarised as under:

Habitat degradation and destruction: There is tremendous pressure on the limited land resources for food, fuel, fodder, timber etc. Thus, the increase in population and developmental activities ultimately results in the degradation of habitat of the flora and fauna. Many species that are endemic to a specific ecosystem cannot cope with the changed environment, leading to loss of biodiversity.

Over-exploitation of biotic resources: Sometimes, the increasing demands of forest products and medicinal plants severely erode the biotic resources. Though, these resources are renewable in nature, due to over-exploitation they do not get sufficient time for rejuvenation, leading to their loss.

Changing pattern of landuse: The changing pattern of landuse affects the forests adversely as large-scale deforestation is done for agricultural, mining, industrial and other urban uses.

Overgrazing: The feeding of the world's livestock is a major problem as fodder is not available in plenty throughout the year, in many areas. The poor people allow their livestock to graze the forests and grasslands which also causes biodiversity loss.

Poaching and hunting: Organized poaching of animals for fur, food, bones, horns, etc has driven many animals to the verge of extinction. Many animals are killed, as there is a misconception that medicines prepared from their body parts increase vitality. For example, tigers and rhinoceros are ruthlessly hunted for this purpose.

Shifting cultivation: In many parts, as in Northeastern India, the practice of shifting cultivation is the major cause of habitat degradation in the region, as it leads to large-scale deforestation.

Fishing: Modern fishing techniques like the use of huge fishing nets and bottom vacuuming techniques that remove everything in their paths including fish and other marine fauna of no commercial use. This puts a large strain on the biodiversity, as the species are not able to recoup due to heavy losses.

Natural calamities: Catastrophic events like floods, droughts, cyclones, volcanoes, fires, etc cause severe biodiversity losses from time to time.

Introduction of exotic species: Human beings, by introducing exotic species (species belonging to some other place) whether intentionally or accidentally, have created ecological crisis in many regions. Sometimes, the exotic species disrupt local ecosystems and in some cases even drive the native species to extinction.

Developmental activities: The impact of developmental activities in the form of mining, construction of roads, canals, dams, etc are responsible for large-scale degradation of the biotic resources.

Environmental causes: At the global level the impacts of environment pollution, particularly the depletion of ozone layer, acid rains, global warming, etc is bound to affect biodiversity adversely. The rapid destruction of forest vegetation is the main reason for biodiversity loss on land.

STRATEGIES FOR CONSERVATION OF BIODIVERSITY

Humans have been directly or indirectly dependent on biodiversity for sustenance to a considerable extent. However, increasing population pressure and developmental activities have led to large-scale depletion of the natural resources. There is an urgent need not only to manage and conserve the biotic wealth but also restore the degraded ecosystems. Conservation of biodiversity and genetic resources is essential for food security, medicinal benefits and ecological balance in nature.

Conservation of biodiversity can be achieved by the following two methods:

1. In-situ conservation
2. Ex-situ conservation

Fig. 19.11 In-situ conservation is done in the natural habitats.

In-situ conservation aims at:

1. Consolidating the network of protected areas for wildlife to ensure the conservation of ecosystems and bio-geographic units.
2. Establishing new protected areas based on utility, distinctiveness and endangerment of species.
3. Coordinating new and existing protected areas to facilitate migration among populations and to ensure proper representation of species and habitats.
4. Ensuring conservation of biodiversity and rich ecosystems outside the network of protected areas.
5. Minimising or banning activities like over-exploitation, pollution, poisoning and introduction of exotic species leading to loss of biodiversity and habitat destruction.
6. Enhancing the ecological and social value of protected areas and providing incentives for biodiversity conservation on adjacent private lands.
7. Initiating regional cooperation for conservation for ecosystems and species.
8. Conducting periodic reviews of the protected areas and plan for assessing present and future needs.
9. Encouraging public participation in planning and management of protected areas.

Advantages of In-Situ Conservation

- The flora and fauna live in natural habitats without human interference (Fig. 19.11).
- The life cycles of the organisms and their evolution progresses in a natural way.
- In-situ conservation provides the required green cover and its associated benefits to our environment.
- It is less expensive and easy to manage.
- The interests of the indigenous people are also protected.

Ex-situ Conservation

This refers to the preservation of biotic wealth **away from their natural habitat.** Under ex-situ conservation, the whole organisms are conserved as in botanical gardens, zoos, aquaria, etc or some parts of the organisms such as pollen, seed, sperm, etc are preserved under artificial conditions.

Botanical Gardens

Initially, botanical gardens were created for aesthetic reasons where different plants were grown for pleasure. However, over a period, the botanical gardens have become centres for scientific research for the propagation of many local and exotic plants. The first botanical garden was established at Calcutta in 1787. Today, there are more than 70 botanical gardens spread across the country, contributing immensely in the conservation of biodiversity.

Zoological Parks (Zoos)

Zoos are places where animals are kept in captivity. Many such animals are rare or endangered and in many cases, extinct in their natural habitat. The zoos play an important role in the breeding of rare and endangered species. Sometimes, after successful breeding, many species are released in their natural environment. Thus, the zoos are a major contributor in checking the extinction of the endangered species. The zoos also significantly contribute in creating awareness and interest in wildlife, especially among the children. The first modern zoo in India opened in Madras in 1855. Today, there are more than 100 zoological parks in the country.

Gene Banks/Seed Banks

These are cold storages where seeds are kept under controlled temperature and humidity for storage. This is an important way of preserving the genetic resources. Seeds preserved under controlled conditions (minus temperature), remain viable for long durations of time. Besides storage, gene banks are also well equipped for scientific researches related to seed processing, seed drying and testing.

Tissue Banks

There are many plant species that do not propagate through seeds but by vegetative means, for example, potato, garlic, sugarcane, etc. The tissue banks conserve such vegetatively propagated materials. Other plant parts are also preserved in tissue banks for carrying out research related to tissue culture and biotechnology. This forms the basis for the generation of high yielding and disease resistant varieties of crops and vegetables.

Cryopreservation

This is the newest application of technology for preservation of biotic parts. Here, the organism or their parts are stored in liquid nitrogen (–196°C). The metabolic activities of the organisms are suspended under low temperature which are later used for research purposes.

The Advantages of Ex-Situ Preservation

1. It is useful for declining population of species.
2. Endangered animals on the verge of extinction are successfully breeded.
3. Threatened species are breeded in captivity and then released in the natural habitats.
4. Ex-situ centres offer the possibilities of observing wild animals, which is otherwise not possible.
5. It is extremely useful for conducting research and scientific work on different species.

Proposed Mechanism for Monitoring Conservation and Utilization of Biological Resources

A model for monitoring conservation and utilization of biological resources have been proposed. In this model it is suggested to set up conservation and utilization board which should have the powers to regulate a National Biodiversity fund. The biodiversity monitoring and conservation efforts should be initiated at the grass-root level and involve people, students, teachers and scientists. The information generated may be stored and documented at several levels involving villages, talukas, districts and states. Finally the information may be passed on to a

National Biodiversity resources and information data bank. The resources and information can be assessed by multinational companies and national industrial houses for a fee, which becomes a part of the National Biodiversity Fund. However, biological resources and information should be made available to the National Research and Development (R and D) establishments free of charge with an understanding that any outcome or product of these resources and information will be National property.

Present Information on Biodiversity

At present regional, National and International institutes are working on collection and compilation of information and research activities on biodiversity. Botanical Survey of India (BSI), Zoological Survey of India (ZSI), Forest Survey of India (FSI), Indian Council of Forestry Research and Education (ICFRE), Indian Council of Agricultural Research (ICAR), Council of Scientific and Industrial Research (CSIR) laboratories, such as National Botanical Research Institute (NBRI), Lucknow, Central Drug Research Institute (CDRI), Lucknow, Central Institute of Medicinal and Aromatic Plants (CIMAP), Lucknow and various regional research laboratories are prominent. Other centres are G.B. Pant Institute of Himalayan Environment and Development (GBPIHED), Wildlife Institute of India (WII), World Widelife Fund for Nature (WWFN), Institute of Ethnobotany (IOE) and various educational institutes and universities. These institutes are engaged on different lines as data collection, survey, documentation of floral and faunal diversity, identification of endangered plants and animals, habitats survey of national resources, use of Remote Sensing Techniques for mapping and assessment of bioresources and monitoring of changes in biodiversity of different ecosystems. More attention is needed for transfer of informations to the agencies implementing conservation of genetic resources.

India has already established a network of protected areas measuring 1,40,675 sq km which account for 4 per cent of the geographic area of the land. The existing protected area consists of 95 National Parks, 421 Sanctuaries. These are to be further expanded to 148 National Parks and 503 Sanctuaries. Besides these, 21 wetlands, 15 Mangroves and 4 Coral Reef areas have also been identified for conservation.

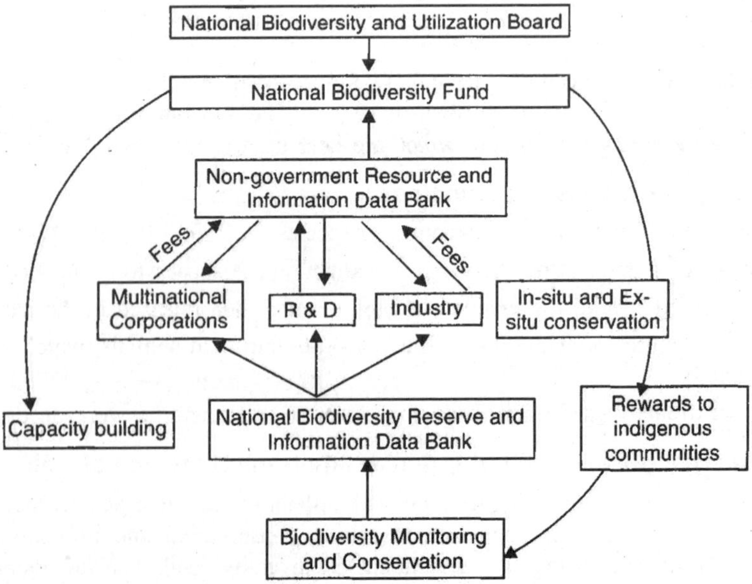

Fig. 19.12. Proposed Model for sustainable utilization of Biological Resources

QUESTIONS

1. What is Biodiversity? Describe different types of Biodiversity.
2. What are different causes of loss (depletion) of Biodiversity? In what ways it can be checked?
3. Write brief notes on the following :
   (*i*) Ex-situ and In-situ conservation.
   (*ii*) Genetic and Ecological Biodiversity.
   (*iii*) Present information on Biodiversity.
4. What do you mean by "hot spots "? Mention different genetic and endemic hot spots of India.
5. Write short notes on:
   (*a*) Red Data Book
   (*b*) Conservation and protection of endangered plants.
   (*c*) Chipko movement (Chipko Andolan)
6. Discuss in brief the causes of biodiversity loss.
7. Discuss in brief the economic potential of Biodiversity.

20

Ethnobotany

The term *ethnobotany* was coined by J.W. Harshberger in 1895 to include the study of plants used by the primitive and aboriginal people. Though this discipline has existed for ages, ethnobotany emerged as a distinct academic branch of natural science in twentieth century.

DEFINITIONS AND CONCEPTS

Several definitions have been assigned to ethnobotany but the basic concept has not changed much and the scope of ethnobotany is increasing day by day.

According to Schultes (1962), ethnobotany is defined as the study of the relationships between the people of a primitive society and plants.

Alcom (1984) states that ethnobotany is the study of contextualized plant use.

Jain (1987) elaborated it as the total natural and traditional relationship and interaction between man and his surrounding plant wealth.

Recently, Wickens (1990) defined ethnobotany as the study of useful plants prior to commercial exploitation and eventful domestication. Infact, ethnobotany is the first knowledge on plants which primitive and aboriginal people had aquired by sheer necessity, intuition, observation and experimentation in the forests. It is now almost universally recognised as the total direct or natural relationship between man and plants and it includes the use of plants by both tribals and non-tribals without any implication of primitive or developed societies.

Ethnobotany is an interdisciplinary science and has relevance to sociology, anthropology, taxonomy, phytochemistry, archaeology, ecology, agriculture, medicine, linguistics etc. Today, ethnobotany has become an important and crucial area of research and development in resource management, sustainable utilization and conservation of biodiversity and socioeconomic development. Now the botanists, social scientists, anthropologists, the practitioners of indigenous medicines all over the world are engaged in the study of man-plant interactions in natural environment.

HISTORY OF DEVELOPMENT OF ETHNOBOTANY

Prior to coining of the term Ethnobotany usage of plants by human being for various purposes found place in the ancient Sanskrit, Greek, Arabic literatures, ethnographics, travelogues etc. Some compilation works of repute are *Foods and Drinks through Ages* (2500 BC to 1937) (Anonymous, 1937), *Arabian medicines and influences on the middle ages* by Campbell (1926), *Indo-Europian and Greek Legends by* Halliday (1932), *Foods and Drinks in ancient India* by Ray (1933), *The Economic plants of Bible* by Moldenke (1954), *Glossary of vegetable Drugs in Vaagbhatta by Godbole et al* (1966), *Plants used against cancer* by Hartwelt (1967–1971), *Glossary of vegetable drugs in Vrihattrayi* by Singh and Chunekar (1972), *Economic plants of ancient North China as mentioned in Shih Ching* by Keng (1974), *Flora and Fauna in Sanskrit literature* by Banerjee (1980), *Plants of Quran* by Farooqui (1989).

In the early nineteenth century the isolation of alkaloids such as morphine, quinine etc. as effective drugs from the herbs heralded new era in the use of plant products in modern medicines. The last thirty years or so witnessed a resurgence of interest in the traditional medicines, ethnomedicines and drugs derived from plants all over the world.

The twentieth century has witnessed the emergence of ethnobotany as a distinct academic branch of natural science. There has been an increasing interest in the scientific study of ethnic plants used by various indigenous peoples commonly designated as aboriginal natives, first people, original settlers, aadivasees, Vanya jaati (forest castes), Aadim Jaati (primitive castes), Jan Jaati (folk communities) and tribals. More than a dozen books and a thousand research publications on ethnobotany have appeared in the world.

DIFFERENT ASPECTS OF ETHNOBOTANY

Due to interdisciplinary nature and socio-economic impacts, the linkages of Ethnobotany have proliferated. Ethnobotany thus has relevance with problems of nutrition, health care and life support system, social customs, mythological association or faith in plants, cottage industries, economic upliftment, conservation of biodiversity and sustainable use of plant resources.

Ethnic Groups in India

There are over 200 tribes, subtribes, clans and some other ethnic groups in India. The following list provides statewise brief general information on some selected groups :

Andhra Pradesh

Andh tribe, Bagata (Bagatha), Chenchu, Hill Reddis, Jatapau, Kanmara (Konda), Khond (Konod), Kolam, Kollavaru, Konda-Dhora, Konda-Kapus, Konda-Reddis, Koya, Lambadis (Sugalis), Pardhan, Paroja, Saora, Valmiki, Yenadis, Yerukulas.

Bihar

Asur, Bathudi, Bedia (Beria), Birhor, Gond, HO, Karmali, Kharia, Kherwar, Kora, Kurmi, Santhal, Sauriya Pahariya.

Jharkhand

Baiga, Korwa, Lohra, Munda, Mahto, Mallar, Parhaya, Porja, Sauria Pahariya.

Chhatisgarh

Kol (Munda), Majhi, Majhwar, Muria, Gond, Nagesia, Pao, Abujhmaria.

Gujarat

Chaudhri (Chandhra), Ravalia, Damor, Dhodia, Dubla, Gamit, Kathodia, Katkuri, Kokna, Kunbi, Paradhi, Patelia, Rathawa, Vasavas.

Himachal Pradesh

Bhat, Gaddi (Gadi), Johari, Kanaura.

Jammu & Kashmir

Amchi (Laddakh), Bakarwala (Gujjars), Gujjar.

Karnataka

Badaga (Nilgiri Hills), Koli (Dhor-Koli), Koya, Marati, Mullukurumban, Wynaadan-chetty, Yerava.

Kerala

Panyan, Pathiyar, Pulayan, Uridavan-Gowdatu, Adiyan, Hill Pulaya, Ilava, Irula, Kadai, Kanikkar, Karimpalan, Kundu Vadian, Kurichya (Kurichchan), Kuruman, Kurumba, Malaarayan, Malakkaran, Malapantaram, Malavettivam, Malayar, Muthuvan.

Madhya Pradesh

Panika Gond tribe, Abujhmaria, Agariya, Baiga (Panda), Bhaina, Bharias (Bhumia), Bhatra, Bhil, Bhumiya, Binjhal, Binghwar, Birhor, Dhurwa, Gudaba, Gond, Halwa, Hillmaria, Kamar (Kanwar), Khairwar, Khariya, Khond, Kol (Col), Korku, Mahto, Manjhi, Majhwar, Munda, Maria, Mina, Saharias, Saora (Sawara), Pao, Pardhan, Pardhi, Paroja, Nagesia, Oraon.

Maharashtra

Bhil, Dhanka, Dhanwar, Dhodia, Dubla, Halba (Holwa), Kokna, Kolam, Kolimahadev, Koli Malhar, Korku, Maria, Pardhan, Pardhi, Pathawa, Thakur, Varli.

Orissa

Bagata, Banjara, Bathudi, Bhatra, Bhuiya, Bhumia, Bhumij, Binjhal, Buijhwar, Gadaba, Gondaru, HO, Juang, Kharia, Kharivar, Kisan (Kuda), Kol, Kolha, Konda, Dhora, Kora, Koya, Lodha, Mallar, Mirdha, Munda, Mundari, Porja Santal Sounti.

Rajasthan

Bhil, Garasia, Kathodia, Katkari, Mina, Saharias.

Tamil Nadu

Irula, Kadar, Kota, Kuruman, Kurumla, Malasar, Malayali, Mullukurumban, Panyan, Toda, Urali, Kurumba.

Uttar Pradesh

Dusadh, Bhoxa (Bhoksa), Kol, Bhil, Kharwal, Tharu (Tippera), Bhotia, etc.

Uttaranchal

Bhotia, Gangwal, Jodh, Jaunsari, Khaseas (different from Khasia, found in Kumaon), Tharu.

West Bengal

Asur, Bathudi, Bhotia, Bhumij, Bishor, Chick Barak, Ho (Munda), Kharwar, Kora, Korwa, Lepcha, Lodha, Lohra, Mahali (Mahli), Male (Maler), Parhaya, Santhals, Sherpa, Toto.

Assam

Bodo or Boro-Kachari, Chakma, Deori, Dimasa-Kachari, Garo, Hojong, Hmar, Hojai, Kachari including Sonwal, Khasi and Jaintia, Kuki, Barmons, Lalung (Bodorace), Mech, Mikir (Karbi) Mishings, Hill miris.

Arunachal Pradesh

Abor (Adi), Aka (Hrusso), Apatani, Bagung, Bangni, Deuri, Digaru-Mishmi, Hill-miri, Idu-Mishmi, Kangbo, Khampti, Meyer, Miji, Miju, Mishing, Mishmi, Monpa, Na, Nishi, Nocte, Sherdukpen, Simpha, Sulung, Tagin, Tangsa, Tangkhul, Wanehu, Yabin.

Manipur

Amol, Anal, Angami, Chira, Chothe, Gangte, Hmar, Kubui, Kacha Naga, Koirao, Koireng, Kom, Lamgang, Mao, Maram, Maring, Mizo, Monsang, Moyon, Paite, Purum, Ralte, Sabte, Sema, Simte, Tangkhul, Thodou, Vaiphei and Zou.

Meghalaya

The Garo, Khasi and Jaintia, Banai, Baro, Bhoi, Biate, Dalu, Hajong, Koch, Lyngam, Man, Rabha and War Jaintia.

Mizoram

Miza (Hmar, Lushai, Paite, Pawi, Ralti), Lakheri, Kuki, Thode, Chakma, Chowngthu, Abzia (Mora), (Chawhto, Nagenta, Khanlthting, Khaingte, Pautu, Rawite, Renthlet, Tlau, Vongachhia, Zawngte).

Nagaland

Nagas (16 scheduled tribes) including Angami, AO, Chakhesang, Chang, Chirr, Konyak, Khei, Mnungan, Lotha, Makware, Phom, Rengma, Sangtam, Sema, Tikhir, Yimchungrel, Zeliang.

Sikkim

Bhotia and Lepcha.

Andman & Nicobar Island

Andmanese, a small tribe of Negrito race of a few dozen persons, Jarawa (Jorawa), Nicobarese, Onge Sentinelese, Shompen (Great Nicobar).

Importance of Ethnobotany

Ethnobotanical knowledge is very ancient. It provides information regarding the traditional uses of plant wealth which can be utilized in integrated tribal development. The ethnobotanical studies throw light on certain unknown useful plants and new uses of many known plants which can be exploited for developing new sources for some plant products and agrobased industries such as, food processing, fibres and floss, cordage and basketry, extraction of edible and non-elible oils, gum, resins, tannin, dye extraction for the upliftment of tribal communities.

With the opening of new vistas of ethnobotany, the scope of this natural science has now greatly increased both in terms of its theoretical contribution towards the understanding of plant-human relationship and the practical knowledge of tribal people in medicine, agriculture, health and industry.

The tribals depend mostly on forest flora for meeting their day to day needs and primary health care. They collect and utilize many wild plants for food, medicines, fibres, oils, gums, tannins and dyes from the ambient vegetation of their localities. The agricultural practices are not technologically advanced and most tribal groups in north-east part of India resort to shifting cultivation widely known as '*Jhum*'. There are community land, clan land and individual land and the individual users of land have right of possessing inherited land. Jhuming or shifting cultivation involves felling of forest trees, clearing of shrubs and undergrowth in limited area and turning of soil for sowing crops. The cleared land is used for agriculture for short time and when the fertility of the soil decreases, the area is abandoned by tribals and then they shift to new area and clear the forest vegetation for agricultural purpose. Jhuming has caused deforestation and soil erosion. Thus the existing land use and its unscientific management have aggravated the problems of soil erosion, floods, depletion of water resources and overall productivity. Therefore, it has now become essential for botanists, agricultural scientists, anthropologists and government agencies to take stock of the problems of the tribal communities and make coordinated efforts in this direction. It requires proper management of settlements of tribal population because the migration of tribal communities compels them for shifting cultivation. Tribals have knowledge about the utility of many plant species which are unknown to modern society. It is for the botanists to collect information regarding the traditional uses of such wild plants from them. Anthropologists have to deal with the cultural aspects of the life of tribals. The tribal culture is changing rapidly due to urbanization, rapid industrialization and changing pattern of life.

The medicinal plants are almost the exclusive source of drugs for the majority of world population even today. Plant products constitute approximately 25% of the total prescribed medicines even in most advanced countries like U.S.A. The record of use of herbal medicines in India is very old. India with diverse ethnic groups and rich biodiversity has a century old heritage of medicinal phytotherapy for the treatment of various diseases and promotion of health. The branch of ethnobotany which deals with the traditional systems of medicine or folk lore medicines is called **medico-ethnobotany.**

Tribal communities utilize ethnomedicinal plants for the treatment of diseases and disorders like diarrhoea, dysentery, fever, headache, skin diseases, boils and blisters, rheumatism and gout, piles, jaundice, ophthalmic diseases, toothache, bone fracture, snake bite, helminthic or worm infection, cuts and wounds, cough, cold, asthma, leprosy etc. The ethnomedicinal data will serve as a useful source of information for the chemists, pharmacologists and practitioners of herbal medicines for detection and isolation of bioactive compounds used in modern medicines.

During the last few decades, a succession of the so called *Wonder drugs e.g.* reserpine, quinine, ephedrine, cocaine, emetine, khelline, colchicine, digitoxin, artemisin, podophyllotoxin, gugulipid, taxol etc. have been discovered from plants with rich ethnobotanical lore in tribal societies. The isolation of these alkaloids from plants heralded a new era in the use of plant products in modern medicines. The last thirty years or so witnessed a resurgence of interest in the traditional medicines and drugs obtained from plants all over the world. The tranquilizers rescinnamine and reserpine have been isolated from the roots of *Rauvolffia serpentina* which has been in use in India for more than thousand years in folk medicine for the treatment of snake bite, insanity, epilepsy, fever and high blood pressure. The native of Medagascar used *Catharanthus roseus* (Sadaa bahar) as an oral hypoglycemic agent. It yielded two important drugs; Vinblastine and Vincristine which are known to be effective in leukaemia (blood cancer). Out of 120 active compounds currently isolated from higher plants and used in modern medicines 75% show a positive correlation between their modern therapeutic use and the traditional use of plants from which they are derived.

The efficacies of a number of phytopharmaceuticals derived from plants such as atropine (pupil dialator), berberin (used in gastrointestinal disorders), caffeine (a stimulant), digitoxin (a cardiac tonic), emetine (antiamoebic), ephedrine (anti asthmatic), morphine (analgesic), papain (protein digestant and anthelmintic), quinine (antimalarial), reserpine (tranquilizer), Vinblastine and Vincristine (antileukaemic), Camptothecin, (antitumour), forskoline (hypotensive and antispasmodic) with rich folk-lore have been discovered.

The recent discovery of certain bioactive compounds such as artemisin (antimalarial drug from *Artemisia annura*), taxol (anticancer drug obtained from *Taxus brevifolia*), hypericine (antiviral from *Hypericum perforatum*), Gossypol (a male contraceptive from *Gossypium* spp), yue chukene (antifertility agent from *Murraya penniculata* triggered the interest in medicinal plants all over the world. Further efforts are being made with a view to discover potent herbal medicines based on ethnotherapeutics prevalent in tribal and aboriginal societies.

The primitive societies in India have been dependent on herbal medicines from the time immemorial. In fact, all traditional systems of medicine had their roots and origin in folk medicines or ethnomedicines. The knowledge of ethnobotany plays a vital role in the primary health care and economy of the tribals and aboriginal populations of our country and has potential for the discovery of new herbal drugs and new sources of nutraceuticals etc. The study of ethnobotany provides valuable informations to the scientists, planners and administrators for the preparation of action plan for the economic emancipation of tribals and ecodevelopment of tribal areas.

Ethnobotany and Conservation of Natural Resources

Ethnobotany has relevance with natural resources which provides raw materials for agroforestry. Ethnobotanical studies have helped in revealing numerous germplasm-stocks of cultivated plants and vegetables. The tribals living in interior forests cultivate numerous vegetable crops. Those crops are in cultivation for generations and thus they represent distinct genetic stocks adapted to local conditions. Some examples of such crops are *Piper peepuloides* which is used as condiment in Khasi Hills, *Parkia roxburgii* which is grown in country yards for fruit vegetable in North-Eastern states of India, *Moghania vestita*, a tuber crop of Khasi, Jaintia Hills, *Hodgsonia heteroclita*, a cucurbit which is grown by Mizotribals for oil seeds and edible Karnel.

The worship of nature is prevalent in India and all forms of life are regarded as sacred in Hindu mythology. The most important patches of some rich and diverse forests are called *sacred groves*. The sacred groves are found mainly in Megalaya and Sikkim where they are locally known as *Lawkyntang* and Lungchuk respectively. Khasis believe that groves are the abode of deities who look after the welfare of the people. Every thing within these groves is under the protection of a reigning deity and removal of any thing, even dead wood or twig is taboo. The Khasis maintain these sacred groves in a comparatively undisturbed condition, protected for reasons of traditional beliefs and customs. These sacred groves form a repository of wild germplasm and biological diversity and help in conservation of indigenous flora of the area including many endangered species of orchids and medicinal plants.

The habitat and environment where the tribals have learnt useful folklore through generations are changing due to deforestation and unplanned management of environment. It is, therefore, important that the oral folk-lores about the plants must be preserved and properly documented before it is lost for ever.

Plants of Ethnobotanical Importance

The tribals and natural populations living in different parts of India use plant species of forest floras for food, fodder, fibres house building, fuels, medicines, beverages, oils, gums, resins, dyes, basketry, timber and wood works, musical instruments, fish poisons, religious ceremonies, narcotics etc. About 5000 plant species have been recorded so far which are used by tribals and aboriginal communities in different states. Some of the plants are used for particular purposes in more than one state while the others are used only in one state. The ethnic plant wealth may be divided into the following groups on the basis of their uses :

(1) Wild edible plants (food plants)
(2) Ethnomedicinal plants
(3) Fibre and floss yielding plants
(4) Oil yielding plants
(5) Dye yielding plants
(6) Gum and resin yielding plants
(7) Ethnotoxic and fish poison plants
(8) Timber and wood work
(9) Tannin yielding plants
(10) Plants used in basketry and brooms. Some less known ethnic plants which are used by tribals for food, ethnomedicine and narcotic purposes are listed here.

Tribal Food and Food Plants

Tribals grow a number of cereals, millets, pulses, fruits, vegetables etc. They are largely dependent on wild leafy vegetables, tubers, fruits, seeds etc. Rice is the staple food for the majority of tribals. They also eat a variety of meat of wild and domestic animals, the most important being the meat of cow, wild buffalloes (called Mithuns in Nagaland), pigs, fowls, fishes, dogs and some other wild animals as well as snakes and birds. The tribals of Arunanchal Pradesh, Meghalaya and Nagaland take meat of dog on special occasions. They are fond of hunting and fishing. For fishing they have developed various types of fish traps made up of bamboo and canes, rods and hooks. Sometimes they put some poisonous plants in the rivers and streams to make the fishes unconscious. They use sticky seeds of some plants *e.g. Loranthus scurulla* to trap the birds. Tribals in North-east and Southern parts of India are fond of crabs. Some plants are used for making arrow poison.

Rice beer or millet beer is an integral part of tribal's life. The beverages, tea and local beer account for one quarter to one half of the daily calorie intake of the inhabitants. Taking beer is essential on all occasions. It is offered in wooden pots, bowls or bamboo tumblers. Chewing betelnut is very common.

The ethnobotanical studies conducted in various regions of India have brought to light hundreds of less known plants of food value which are utilized by tribals. Some of the important and less known edible plants used by tribals are listed here as under (Table 20.1) :

Table 20.1. Some plants of ethnobotanical importance

Plant species	Family	Plant parts used	Tribal areas
Vegetables			
Acetophila excelsa	Euphorbiaceae	leaves	
Actinidia callosa	Actinidaceae	flowers and fruits	
Alpinia nigra	Zingiberaceae	rhizomes	
Amaranthus spinosus	Amaranthaceae	young shoots	
A. Viridis	,,	,, ,,	
Amorphophalus paeonifobius	Aracaceae	corm and leaves	
Angiopteris erecta	Angiopteridaceae	rhizomes	
Antidesma acidum	Euphorbiaceae	leaves	
A. acuminatum	,,	,,	
A. bunius	,,	,,	
A. ghaesembilla	,,	,,	
Arisaema conseinguineum	Araceae	rhizomes and stems	
Arundina graminifolia	Orchidaceae	tender shoots and leaves	
Astilbe revularis	Saxifragaceae	tender leaves and flowers	Sikkim, Meghalaya
Asparagus racemosus	Liliaceae	Tuberous roots and tender shoots	North Bengal
Bambusa khasiana	(Poaceae)	young shoots	Meghalaya & Manipur
B. tulda	,,	,,	,, ,,
B. vulgaris	,,	,,	,, ,,
Bauhinia purpurea	Caesalpiniaceae	flower buds	Sikkim, Darjeling
Begonia josephi	Begoniaceae	leaves	N-E India.
B. palmata	,, ,,	tender shoots and leaves	,, ,,
Boerhaavia diffusa	Nyctaginaceae	Leaves	
Bombax ceiba	Bombacaceae	flower buds & fleshy calyx	Assam, Manipur, Meghalaya.
Buddleja asiatica	Loganiaceae	leaves (eaten raw)	Arunachal Pradesh
Bupleurum falcatum	Apiaceae	boiled roots	Assam & Sikkim.
Calamus erectus and C.	Arecaceae	young shoots	Arunachal, Assam
tenuis	,,	,,	,, ,,
Caseurea gravedems	Flacourtiaceae	Tender shoots and leaves	Assam & Meghalaya
Catimbium malaceanse (Alpinia malaeceanse)	Zingiberaceae	young stem	Arunachal & Assam
Chenopodium album	Chenopodiaceae	young shoot India	N.E. India, Central
Chlorophytum arundinaceum	Liliaceae	whole plant	N.E. India.
Clerodendron spp	Verbenaceae	tender shoot & leaves	N.E. India.

(Contd..)

Plant species	Family	Plant parts used	Tribal areas
Colocasia esculenta	Araceae	tender leaves & corm.	All over India.
Commelina bengalensis	Commelinaceae	young leaves	N.E. India, U.P., Bihar, M.P.
Costus spaciosus	Zingiberaceae	flowers & rhizome	Sikkim, Manipur
Cyathea gigantea (*Tree fern*)	Cyathiaceae	Pith	Nishi & Shulung of Arunachal Pradesh.
Debregeasia longifolia	Urticacae	leaves	Manipur, Arunachal Pradesh
Dillenia India	Dillaniaceae	clayx	Meghalaya and Manipur.
D. Pentagyna	,, ,,	flower buds cooked with fishes	,, ,,
Dioscoria spp.	Dioscoriaceae	Tubers and Bulbils	N-E, Uttaranchal, U.P.
Diplazium esculentum	Athyriaceae (fern)	young fronds	N-E. India.
Duchesnea indica	(Rosaceae)	leaves	Assam, Meghalaya, Manipur.
Elatostema platiphyllum	Urticaceae	leaves	Assam, Arunanchal states
Eryngium foetidum	Apiaceae	leaves (in chatani)	Megalaya, Sikkim, Manipur.
Eriosema himalaicum	Fabaceae	tuberous roots	Sikkim
Fagopyrum dibotrys (*Syn. F. cymosum*)	Polygonaceae	leaves	N.E. India, Uttaranchal, J & K, Shimla.
F. esculentum	,,	Leaves and seeds	,, ,,
Gardenia companulata	Rubiaceae	leaves & fruits	Assam & Meghalaya
Girardinia palmata	Urticaceae	tender shoots	Sikkim, Darjeeling
Gnetum gnemone	Gnetaceae	leaves & fruits	Assam, Manipur
Hovenia acerba	Rhamnaceae	Inflorescence stalk	Sikkim, Meghalaya
Indigofera atropurpuria	Fabaceae	flowers	Sikkim
Ipomoea aquatica	Convolvulaceae	tender shoots	N-E. India, Uttaranchal, U.P., M.P., Bihar.
Lassia spinosa	Araceae	rhizomes	Assam, Nagaland
Ludargia ascendens (*Syn. Jussiaea repens*)	Onagraceae	tender shoots	Manipur.
Maesa chisia	Myrsinaceae	young shoots	Sikkim, Darjeeling
Meliosma pinnata	Salriaceae	young leaves & shoots	Nepalese in Sikkim
Meyneca spinosa	Rubiaceae	”	Manipur, Meghalaya
Moringa oleifera	Moringaceae	young leaves and fruits	All tribals in India use it
Musa balbisiana	Musaceae	young fruits and inflorescence	Sikkim, Assam, Darjeeling.
Natsiatum herpeticum	Icacinaceae	leaves & young shoots	Sikkim, Assam, Arunachal Pradesh
Oenothera javanica	Apiaceae	young shoots	Sikkim, Arunachal
Olax acuminata	Olacaceae	young shoots	Meghalaya, Assam
Oxalis cornienlata	Oxalidaceae	whole plant	Manipur, Arunachal.

(Contd..)

Plant species	Family	Plant parts used	Tribal areas
Pegia nitida (*Syn-Tapirira hirsuta*)	Anacardiaceae	young leaves	Manipur, Sikkim, Arunachal.
Phytolacca acinosa	Phytolaccaceae	tender shoots	Arunachal, Sikkim
Plantago erosa (*Syn. P. major*)	Plantaginaceae	,, ,,	Arunachal, Manipur.
Polygonum molle	Polygonaceae	Tender shoots	Arunachal, Sikkim
P. nepalensis	,,	,, ,,	Arunachal, Meghalaya
Portulaca oleracea	Portulacaceae	leaves & shoots	all over India
Pouzolzia hirta.	Urticaceae	leaves	Khasi Hills
P. sanguinea	,,	young shoots	Assam, Meghalaya
Premna latifolia	Verbenaceae	tender shoots, leaves & Bark	Assam, Meghalya
Pereraria tuberosa	Fabaceae	tubers & young shoots	Sikkim
Rhus javanicus	Anacardiaceae	tender shoots Manipur.	Sikkim, Darjeeling, Manipur
Rosippa indica	Brassicaceae	leaves	Arunachal, Sikkim Manipur.
Rumex hastata	Polygonaceae	leaves	N-E. India, Bihar, U.P., Uttaranchal.
R. nepalensis	,,	leaves & young shoots	N-E. India, U.P., Meghalaya.
Solanum crassipetalum	Solanaceae	leaves	Sikkim
S. erianthum (Syn. *S. verbascifolium*)	,,	fruits	Sikkim
S. indicum	,,	fruits	N-E. India,
S. kurzii	,,	,,	,, ,,
Sterculia indica	Sterculiaceae	tender shoots	,,
Urtica ardens	Urticaceae	young leaves & shoots	Sikkim, Manipur
Vaccinum vacciniaceum	Vacciniaceae	leaves & flower	Manipur, Sikkim
Zanthoxylum acanthopodium	Rubiaceae	tender shoots & leaves	Manipur, Meghalaya, Manipur, Arunachal Pradesh
Fruits and Seeds			
Aegle marmelos	Rutaceae	Pulp of ripe fruits	N-E. India, Bihar, U.P., Uttaranchal, M.P., Orissa, Bihar
Aglaria edulis	Meliaceae	Aril	Sikkim, Darjeeling
Ampelocissus barbata	Vitaceae	ripe fruits	Sikkim, Manipur
Antidesma acuminatum	Euphorbiaceae	,, ,,	,, ,,
Aporusa octandra	,, ,,	,, ,,	Sikkim, Manipur, Arunachal, Meghalaya.
Aphania ruba	Sapindaceae	,, ,,	Sikkim, Nagaland, Manipur.
Aristida macrocarpa	Myrsinaceae	ripe fruits	

(Contd..)

Plant species	Family	Plant parts used	Tribal areas
A. solanacea	,,	,, ,,	,, ,,
Artocarpus chama	Moraceae	,, ,,	N-E. India, U.P., Bihar, Uttaranchal.
Baceaurea ramiflora (Syn. *B. sapinda*)	Euphorbiaceae	,, ,,	Manipur
Bauhinia purpurea	Caesalpiniaceae	seeds	Sikkim, Darjeeling
B. vahlii	,,	seeds (roasted)	,, ,,
Castanopsis indica	Fagaceae	fruits & seeds	Assam, Meghalaya
C. tribuloides	,,	,, ,,	Assam, Manipur
Clausena dentata	Rutaceae	ripe fruits	Sikkim
Cornus capitata	Cornaceae	,, ,,	Sikkim, Meghalaya
Cythocalyx martabanicus	Annonaceae	,, ,,	Arunachal Pradesh
Daphniphyllum himalayense	Daphniphyllaceae	,, ,,	Arunachal Pradesh
Dandrocalamus hamiltonii	Poaceae	seeds used as rice	Arunachal, Assam, Sikkim, Darjeeling.
Diploknema butyracea	Spotaceae	fruits seeds used for edible fat	Sikkim, Arunachal
Docynia indica	Rosaceae	fruits	Meghalaya
Duchesnea indica (Syn. *Fragaria indica*)	,,	,, ,,	Khasia, Manipur, Arunachal.
Echinochloa coloum	Poaceae	Grains	Arunachal, Meghalaya.
Ehratia acuminata	Boraginaceae	Ripe & unripe fruits	Sikkim, Manipur
Elaeagnus caudata	Elaeagnaceae	Ripe fruits	N-E. India.
E. pyriformis	,,	,, ,,	,, ,, ,,
Elaeocarpus floribundus	,,	fruits ,,	Manipur, Meghalaya
E. lancifolius	,,	,, ,,	Sikkim, Darjeeling.
Emblica officinalis	Euphorbiaceae	fruits	N.E. India, U.P., Bihar, M.P.
Entada pursaetha (Syn. *E. phaseoloides*)	Fabaceae	seeds	Assam, Arunachal, Manipur.
Erioglossum rubiginisum	Sapindaceae	ripe fruit	Sikkim, Manipur.
Euryale ferox	Euryalaceae	fruits	Assam, Manipur.
Evodia fraxinifolia	Rutaceae	fruits	Sikkim, Manipur.
Fagopyrum esculentum	Polygonaceae	seeds	N.E.India,
F. dibotrys	,,	(made into flouer)	Uttaranchal.
F. tataricum	,, ,,	,, ,,	Himachal Pradesh, J & K.
Ficus hispida	Moraceae	ripe fruits	Sikkim, Manipur, Assam.
Flacourtia jangomas	Flacourtiaceae	ripe fruits	Sikkim, Manipur, Assam
Garcinia anomala	Clusiaceae	fruits	Manipur
G. cowa	,,	,,	Manipur and Meghalaya.
G. kydia	,,	,,	Arunachal, Assam
G. pedumeulata	,,	,,	Assam, Manipur

(Contd..)

Plant species	Family	Plant parts used	Tribal areas
Gnetum gnemon	Gnetaceae	Seeds	Assam, Manipur
Gynocardia odorata	Flacourtiaceae	fruit pulp and seeds	Sikkim, Meghalaya.
Hodgsonia macrocarpa	Cucurbitaceae	Roasted seeds	N.E. India.
Knema lincifolia	Myristicaceae	fruits	Manipur, Meghalaya.
Lepisanthes senegalensis	Sapindaceae	ripe fruits	Sikkim, Meghalaya
Litsea cubeba	Lauraceae	fruits	Sikkim, Darjeeling.
Mahonia nepalensis	Barberidaceae	ripe fruits	Sikkim, Arunachal
Mangifera sylvatica	Anacardiaceae	fruits	Sikkim, Darjeeling
Melia dubia	Meliaceae	fruits	Sikkim, Darjeeling
Microcos paniculata	Teliaceae	fruits	Manipur, Maghalaya, Uttaranchal
Morus australis	Moraceae	fruits	N.E. India, Uttaranchal.
Musa balbisiana (*Syn. M. sikkimensis*)	Musaceae	fruits	Sikkim, Darjeeling, Assam.
Myrica esculenta	Myricaceae	fruits	Sikkim, Darjeeling.
Parthenocissus himalayana	Vitaceae	fruits	Sikkim, Darjeeling.
Persea fructifera	Lauraceae	fruits	Sikkim, Darjeeling.
Phoenix acaulis	Aracaceae	ripe fruits	Sikkim
Prunus cerasoides	Rosaceae	,,	Arunachal, Manipur, Uttaranchal, H.P.
P. nepalensis	,,	fruits	Sikkim, Manipur, Darjeeling.
Pereraria phaseoloides	Fabaceae	fruits	Meghalaya, Arunachal
Pyrus pashia	Rosaceae	fruits	Meghalaya, Manipur
Pyrularia edulis	Santalaceae	fruits	Sikkim, Manipur
Rhus javanica	Anacardiaceae	fruits	Sikkim, Darjeeling, Manipur.
Rubus biflorus	Rosaceae	fruits	Sikkim, Manipur.
R. ellipticus	,,	,,	Sikkim, Manipur.
R. niveus	,,	,,	Meghalaya.
R. rugosus	Rosaceae	Fruits	Manipur, Meghalaya
Saurauria nepalensis	Saurauriaceae	fruits	Arunachal, Sikkim, Meghalaya.
Sterculia villosa	Sterculiaceae	roasted seeds	N.E. India.
Stixis suaveolens	Capparidaceae	ripe fruits	Arunachal, Meghalaya and Sikkim.
Syzygium claviflorum	Myrtaceae	ripe fruits	Sikkim, Darjeeling
S. formosa	,,	ripe fruits	Sikkim, Arunachal
S. kurzii	,,	ripe fruits	Sikkim, Darjeeling
Tetrastigma bracteolatum	Vitaceae	ripe fruits	Sikkim, Manipur
Viburnum cotinifolium	Caprifoliaceae	ripe fruits	Sikkim
Zizyphus apetata	Rhamnaceae	ripe fruits	Sikkim
Z. rugosa	,,	ripe fruits	Sikkim, Darjeeling

(Contd..)

ETHNOMEDICINAL PLANTS

Acorus calamus	Araceae	Root and rhizome paste used as antiseptic, paste is used in snake bite, Leprosy.	Tharu in Kheri district Nagaland, Arunachal, Sikkim.
Aconitum ferox	Ranunculaceae	corm and root powder is given to animals to cure sickness	Nagaland, Arunachal.
Alpinia galanga	Zingiberaceae	rhizome powder is given orally for rheumatism, piles and respiratory troubles in children	Manipur Nagaland, Arunachal.
Amomum aromaticum	"	paste of rhizome and seed given for abortion and as purgative.	Arunachal, Meghalaya.
Begonia palmata	Begoniaceae	Plant extract is cure for stomachache and diarrhoea.	Nagaland
Berginia ciliata	Saxifragaceae	Plant extract cures cough and cold, paste of plant stops bleeding	Arunachal, Sikkim.
Clerodendron colebrookianum	Verbenaceae	leaf paste is applied in rheumatism and gout, plant decoction is a cure for Malaria	Assam, Nagaland.
Coptis teeta	Ranunculaceae	Plant decoction is used in cough, cold and fever, backach and Malaria.	Arunachal
Costus speciosus	Zingiberaceae	Rhizome decoction given in kidney stone, burning pain during urination	Nagaland, Manipur.
Croton roxburghii	Euphorbiaceae	Seeds are purgative, seed oil is insecticide, plant juice is antidote to snake poison.	Nagalland, Manipur, Arunachal.
Curcuma angustifolia	Zingiberaceae	Rhizome juice is rubbed on swollen parts and paste applied on bone fracture	Nagaland
Eclipta prostrata	Asteraceae	Aqueous extract of plant given in body swelling	Tharus of U.P.
Eryngium foetidum	Apiaceae	Plant juice is used for headache, fever and as tonic	Nagaland
Euphorbia acaulis	Euphorbiaceae	Paste of root stock boiled with *linseed* oil is applied on rheumatism.	Tharus of U.P.
Geranium nepalense	Geraniaceae	Powder of whole plant is given in stomach disorders.	Nagaland, Sikkim, Darjeeling.
Helictres isora (Vsn. Marorphali)	Sterculiaceae	Seed extract is given in dysentery	Tharu of U.P.
Hedychium spp.	Zingiberaceae	Decoction of rhizome is cure for bronchitis, tonsilitis, throat and stomach trouble.	Arunachal, Manipur

(Contd..)

Hydnocarpus kurzii	Flacourtiaceae	Seed oil is a cure for leprosy	Nagaland, Manipur
Helminthostachys zeylanica (Kamraj)	Ophioglossaceae	rhizome decoction is given in impotency and leaf juice cures tongue blisters.	U.P. (Tharu)
Piclorrhiza kurroa (Kutki)	Scrophulariaceae	Root decoction is given in diarrhoea, cough, influenza, fever. It is effective in liver and spleen disorders.	Arunachal, Sikkim, Dharchula in Pithoragarh (Uttaranchal).
Piper spp.	Piperaceae	Paste of stem and black pepper is given for sterilization of women, and leaf juice cures eye troubles.	Assam, Darjeeling, Sikkim, Meghalaya.
Podophyllum hexandrum	Podophyllaceae	Decoction of roots and rhizome is given for tumor and skin diseases and as purgative	Arunachal.
Rubia cardifolia	Rubiaceae	Root and stem decoction is given for stomach ailment, chest trouble, Jaundice and irregular menstruation and cancer.	Meghalaya, Darjeeling, Nagaland, Pithoragarh (Uttaranchal)
Solanum viarum	Solanaceae	Seed powder mixed with mustard oil is inhaled to relieve headache, cold, blocked nose and insanity.	Assam, Meghalaya, Nagaland, U. P. (by Tharu)
S. torvum	Solanaceae	Leaf decoction is given in snake bite, and fruits given in cough and tonsil complaints.	Khasi, Jantia, Manipur
Stephania glabra	Manispermaceae	Tuber powder along with honey is given in Asthma and stomach tumor. Juice of tuber is dropped in eyes to cure eye diseases.	Nagaland
Swertia chirayeta	Gentianaceae	Plant powder is given in stomachache and root decoction in fever and influenza.	Nagaland, Sikkim, Uttaranchal
Taraxacum officianalis	Asteraceae	Given in malaria and urinary complaints	Arunachal

NARCOTIC PLANTS

Primitive man has been using plants and plant products for chewing and smoking to alleviate sufferings, to get relief from fatigues and mental tension, as well as for pleasure sake during religious ceremonies and social functions. Chewing of tobacco (*Nicotiana tabacum*), betel leaf (*Piper betel*), betel nut (*Areca catechu*) and kolanuts, (*Cola nitida*), smoking of tobacco, opium (madak and chandu), Ganja, charas or hasish (*Cannabis sativa*), Junson weed or thorn apple (*Datura stramonium*) Hentane and drinking of beverages prepared from Bhang (*Cannabis sativa*), opium (*Papaver somniferum*), kava-kava (*Piper methysticum*) are very common among the tribal communities.

QUESTIONS

1. What is ethnobotany? Discuss the importance of Ethnobotany.
2. Discuss the development of ethnobotany as a branch of natural science.
3. Give a brief account of some less known plants which are used as food plants by tribal people in India.
4. Discuss giving suitable examples of plants how ethnomedicine has helped in the development of modern medicines.
5. Write brief notes on :
   (*i*) Religious groves
   (*ii*) Plant narcotics used by tribal communities
   (*iii*) Tribals of Uttar Pradesh and Madhya Pradesh.

SOIL SCIENCE

21

The Soil

The word soil is derived from a Latin word *'solum'* meaning earthy material in which plants grow. The study of soil is known as *Soil science* or *Pedology* (pedos = earth) or *Edaphology* (edaphos = soil). The study of soil is important in many respects. Soil is natural habitat for plants and animals. It provides water and nutrients to the living organisms. Knowledge of soil is helpful in agricultural practices, such as cultivation, irrigation, artificial drainage and use of fertilizers. It is also important from geological, petrological, mineralogical and paleobotanical points of view.

The earth is, more or less, spherical and its surface is highly irregular, marked by deep oceans, high mountain ranges and plains in between. The internal composition of earth is not known exactly. According to a generally accepted interpretation, the earth has three zones, viz., core, mantle and crust (Fig. 21.1). The core is the central fluid or vapourised sphere having diameter of about 2,500 kms from the centre and is possibly composed of nickel-iron (Urey, 1952). The mantle extends about 2,900 kms above the core. This is in molten state. The outermost solid zone of the earth is called crust which is about 8 to 40 kms above the mantle. The crust is very complex. We live on its surface. Soil developed from the pre-existing rocks, the regolith of the crust. Here soil means the *loose, friable, unconsolidated top layer of the earth crust.* The soil is differentiated into several layers or horizons which can be distinguished from one another by their colours, textures and other characteristics. It contains water, gases, complex minerals, organic substances and micro-organisms. The dead remains of plants and animals are degraded by micro-organisms and after degradation a number of organic substances, generally called humus, are contributed to the soil. The mineral component of the soil is derived from the rocks.

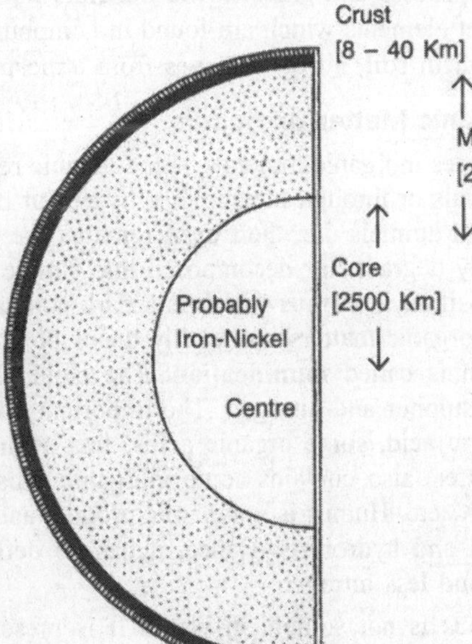

Fig. 21.1. Diagram of the supposed cross section of the earth.

DEFINITION OF SOIL

Soil may be defined as "the part of earth crust in which humus is present".

According to R.F. Daubenmire, "soil is the upper part of earth crust in which plants are anchored." He defines soil as weathered superficial layer of earth crust with which are mingled living organisms and products of their decay.

According to Hilgard, 1917 (American school), "it is, more or less, loose, friable material in which, by means of their roots, plants may or do find a foothold, nourishment as well as other conditions of growth."

According to Raman, 1928 (German school), "soil is the upper weathering layer (*i.e.,* layer subjected to physical and chemical changes) of the solid earth crust."

Joffe and Marbut, two well-known American soil scientists, have defined soil in the following way : "soil is a natural body developed by natural forces acting on natural materials. It is usually differentiated into horizons of minerals and organic constituents of variable depths which differ from the parent materials in morphology, physical constitutions, chemical properties, composition and biological characteristics."

According to Russian school, "soil is natural body differentiated into horizons of usually unconsolidated minerals and organic constituents of variable depths."

According to Wadia (1945), "soil is the topmost layer of earth crust capping the rock." It is natural body of variable thickness, composed of disintegrated rock materials together with variable proportions of organic matters, generally differentiated into zones or layers and mostly unconsolidated.

In brief, soil can be defined as that region on the earth surface where geology and biology meet each other.

COMPONENTS OF SOIL

The soil is made up of the following components :

(1) Mineral particles,

(2) Dead organic matter or humus,

(3) Soil atmosphere;

(4) Soil water, and

(5) Biological system or soil micro-organisms.

1. Mineral Components

The mineral constituents of the soil are derived from the parental rocks or regolith. They may be found in the form of particles of different sizes; from clay (.0002 mm or less in diam) to large pebbles and gravels. The minerals represent about 90% of the total weight of the soil. Important elements which are found in compound state are Oxygen, Si, Fe, Al, N, P, K, Ca, Mg, C, H, etc. In soil, nitrogen comes from atmosphere in the form of nitrogen salts.

2. Organic Matter or Humus

Besides inorganic minerals, some organic residues derived either from dead remains of plants and animals or through metabolic activities of living organisms are present in the soil. When the plants and animals die, their dead remains are acted upon by a number of microorganisms and are finally degraded or decomposed into simple organic compounds. A product of this microbial decomposition is **humus** which is a dark coloured, jelly-like amorphous substance composed of residual organic matters not readily decomposed by soil microorganisms. The process of humus formation is called **humification**. The chief elements found in humus are carbon, hydrogen, oxygen, sulphur and nitrogen. The important compounds found in it (humus) are carbohydrates, phosphoric acid, some organic acids, fats, resins, urea, etc. Tree litter (very little decomposed dead matter) also contains some inorganic substances as lime, potash, Mn, Mg, silica, Cu, Al, Ca, Na, K, etc. Humus is a dynamic product and is constantly changing because of its oxidation, reduction and hydrolysis. Hence, it has no definite chemical composition. It has much carbon content and less nitrogen.

Humus is not soluble in water. It is present in soil in the form of organic colloids. The amounts of humus in different soils vary greatly. Humus percentage in the soil is affected by climatic and biological factors. It is less in arid soils and very high in humid soils. In the top layer of the soil, humus quantity is greater than in the deep layers. In dark humid areas which are thickly covered with vegetation, the humus may be found in the following three stages of degradation :

(*i*) The top floor is covered with dead organic parts showing low degree of decomposition. These poorly decayed dead parts of plants form *litter.* (Fig. 21.2).

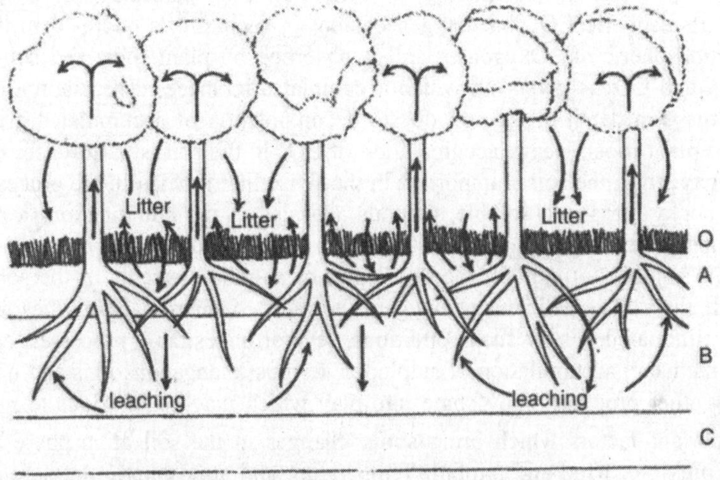

Fig. 21.2. Litter covering the floor of the forest.

(*ii*) Below the litter may be found a layer of partially decomposed organic matter which is known as *duff layer.*

(*iii*) When the duff is decomposed completely into organic substances, the decomposition products, generally called *leaf moulds,* are accumulated below duff layer.

Sometimes under anaerobic conditions, the dead remains are not at all acted upon by the microorganisms. Accumulation of such undecomposed organic remains is termed as *peat.*

Humus plays many important roles in the soil, such as :

(*a*) It makes the soil fertile.

(*b*) It provides nutrients to the plants and microorganisms.

(*c*) On complete decomposition, it forms several organic acids which serve as solvents for soil materials. Thus humus increases the availability of minerals in dissolved state to plants.

(*d*) Because it is porus, it has got high capacity for retaining water.

(*e*) Humus makes the soil porous, thus increases the aeration and percolation which make the soil more suitable for the plant growth.

(*f*) It also acts as weak cement thus binds the sand particles.

(*g*) Presence of humus in the soil increases the rate of absorption in plants.

The factors which influence the rate of humifications are outlined below :

(*i*) Nature of plants, animals or soil organisms.

(*ii*) Rate of decomposition.

(*iii*) Temperature (increase in temperature upto a certain limit increases the rate of humification).

(*iv*) Aeration and moisture. These *increase* the rate of humification.

3. Soil Atmosphere

Gases found in soil profiles are said to form the soil atmosphere which is one of the most important components of the soil. The spaces between soil particles and soil organisms are called *pore spaces.* These are filled with moisture and air in varying quantities which account for approximately half of the total volume of soil. In dry soils, percentage of moisture is lesser than that in wet soils.

The soil atmosphere contains three main gases, namely oxygen, carbon dioxide and nitrogen. In soil atmosphere, oxygen is 20%, nitrogen is approximately 79 per cent and carbon dioxide is 0.15 to 0.65 per cent by volume. In the cultivated land, percentage of CO_2 is much higher than that of atmospheric CO_2, but oxygen content in such soil is poorer than the percentage of oxygen in atmospheric air. Oxygen of soil is absorbed by plant roots and soil microorganisms in respiration and CO_2 is given out which accumulates in spaces. The amount of CO_2 increases with the increase in depth of the soil due to decomposition of accumulated organic matter and abundance of plant roots. Heavy accumulation of CO_2 in the soil is harmful for the plant growth. Presence of oxygen in the soil is important in the sense that it helps in the process of break down of insoluble rocky mass into soluble minerals and also in the humification (a process in which insoluble minerals and organic nutrients locked up in the dead remains of plants and animals are converted into soluble forms). The accumulation of soluble nutrients in the soil makes it more productive. If the soil is deficient in oxygen, the rates of microbial activities are slowed down and may be eliminated. Under such conditions, several undesirable processes, such as evolution of nitrogen, methane, accumulation of sulphides, ferrous, manganous ions and organic inhibitors, and so many other processes may come into play which may be injurious to plants.

The important factors which bring about changes in the soil atmosphere are temperature, atmospheric pressure, wind and rainfall. Temperature and atmospheric pressure cause expansion and contraction of the soil air. Wind helps the soil in sucking the air in and rain water displaces the soil air. Any considerable change in the soil atmosphere affects the size and function of microflora and other biological populations.

4. Soil Water

Soil water plays very important role in the plant growth. Plants absorb a small quantity of rain water and dew directly from their surfaces but most of water absorbed by them comes from the soil.

Soil water maintains the soil texture, arrangement and compactness of soil particles. It is good solvent for minerals and it makes the concentration of nutrients low so that nutrients may be absorbed by plants easily. Water affects the plant growth and other physiological activities. In plant growth, water forms a major part of the plant itself. It is essential for the process of photosynthesis, it maintains the turgidity of the plants and acts as a medium by which mineral salts essential for plant growth enter the plants from the soil. In brief, water regulates the physical, chemical and biological activities in the soil.

Water in the soil comes mainly through infiltration of precipitated water (rain, sleet, snow and hail) and irrigation whereas it is lost from the soil chiefly through evaporation, percolation, stream flow and transpiration.

The quantity of water available in the soil varies from place to place. The amount also depends upon the quality of soil. In loamy, silty and clay soils, the amount of water is greater than that in coarse sandy soil. Water is held in the soil in the following forms :

(*i*) Gravitation water,
(*ii*) Capillary water,
(*iii*) Hygroscopic water,
(*iv*) Water vapour, and
(*v*) Combined water.

(*i*) **Gravitational water.** After complete water saturation of soils the excess water displaces air from the pore spaces between soil particles and percolates downwardly under gravitation influence and finally it is accumulated in the pore spaces. This excess water is called *gravitational water.* The amount of water held in the soil, when all pores are filled and when drainage is restricted, is *maximum water holding capacity.*

When the gravitational water percolates down and reaches to the level of parental rock it is called *ground water.*

(*ii*) **Capillary water.** The amount of water present around the soil particles at saturation stage, when gravitational water has drained away through capillaries or channels, is called *capillary capacity* or *field capacity* and the water which is held by surface tension and attraction force of water molecules as thin film around soil particles in the capillary spaces is called *capillary water*. It moves in the direction where capillary tension is more.

(*iii*) **Hygroscopic water.** Water which is adsorbed on the soil particles and held on the surface of soil particles by forces of attraction and cohesion of its molecules is called hygroscopic water.

(*iv*) **Water vapour.** This is the water present in the soil atmosphere in the vapour form.

(*v*) **Combined water.** It is water of chemical compounds held by chemical forces of molecules (as for example, $CuSO_4.5H_2O$). It can be driven out from the compounds only at bright red heating.

Water requirement of plants varies from individual to individual. Some absorb large quantity, while some others require very small quantities of water for their normal growth. A major fraction of total water absorbed from the soil is transpired by the plants and only a small quantity of it enters the composition of protoplast. The availability of soil water to plants depends primarily on its diffusion pressure deficit, often termed the *soil moisture stress*. The total of all the forces which the plants must overcome to take up water from soil is called *soil moisture stress*.

Water lost from the soil surface by evaporation and through absorption by plants is replaced by rise of capillary water from root zone. The continuous loss of water may finally result in a stage at which water content of the soil becomes so poor as it (soil) cannot supply water to growing plants rapidly enough to maintain them turgid. Under such conditions, permanent wilting occurs in the plants. At permanent wilting stage, the percentage of moisture in the soil is termed as *wilting coefficient* or *permanent wilting percentage* (Fig. 21.3 C, G).

The difference between field capacity and wilting coefficient is termed as **maximum available water** and the water content of the soil at any time over and above the wilting coefficient is referred to as *available water.* The amount of water to be added to a soil at the wilting point to reach the field capacity is called the *available water capacity.* It need not be emphasized here that the actual moisture content of a soil has little meaning in respect to plant growth unless (*i*) the field capacity and (*ii*) the permanent wilting percentage of the soil are also known. Water contents above field capacity displace so much of the soil air that the plant roots usually suffer from inadequate aeration and serve to be detrimental. Although plants usually continue to absorb water in the soil drier than at permanent wilting stage, absorption is too slow to replace water losses and the resulting water deficit causes cessation of growth and finally results in death from dehydration.

The moisture at the field capacity is held with a force of one-third atmosphere and that at permanent wilting stage is held with a force of 1.5 atmosphere.

5. Biological System of the Soil or Soil Microorganisms

Organisms present in the soils are called soil organisms. Important group of soil organisms are given below (Fig. 21.4).

Soil Flora	Soil Fauna
1. Bacteria	1. Protozoa
2. Soil fungi	2. Nematodes
3. Soil actinomycetes	3. Insects and mites
4. Algae	4. Rodents and earthworms
5. Root, rhizoid and rhizome of higher plants.	5. Burrowing vertebrates

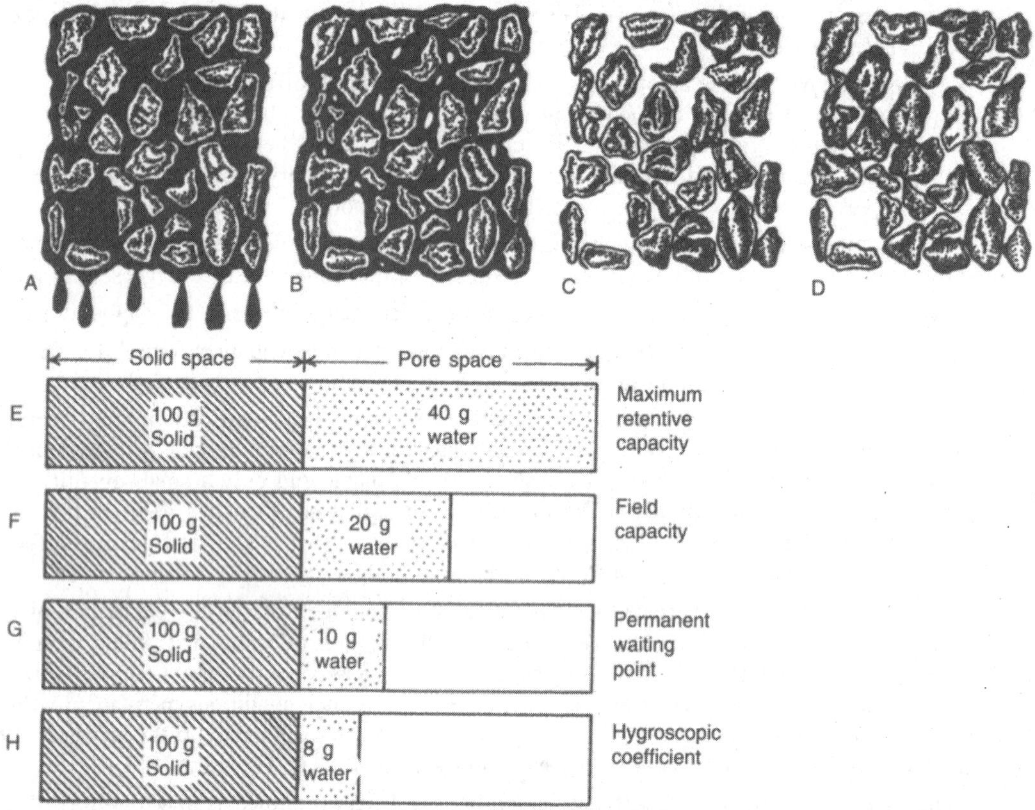

Fig. 21.3. Illustrations of the water contents of a soil show it at saturation (A, E), at field capacity (B, F), at its permanent wilting point (C, G), and in a state of extreme desiccation, the hygroscopic coefficient (D, H).

Many of these soil organisms are stable, many are mobile, but some are held in the colloidal films of the soil particles. Protozoa, mites and insects are example of moving organisms. They move in or on the surface of soil in search of food. Earthworms by the burrowing habit make the soil loose and fertile. They are found in abundance. In some forests their number may reach up to 10,000 per square foot. These soil organisms feed on the organic matter of the soil.

The majority of soil fungi are found in acidic soils. Actinomycetes prefer saline soils and soil bacteria grow fairly well in the neutral soils richly supplied with organic nutrients. These microorganisms are found in the soil at variable depths.

Algae are found in the top layer of soil under the conditions of constant shade and moisture.

It is estimated that in soil microflora bacteria form about 90 per cent of the total microbe population. Fungi and algae together represent only one per cent and actinomycetes cover 9 per cent.

Density of microbial population is actually governed and influenced by climatic conditions, physical and chemical nature of soil and vegetation cover. The greatest amount of microbes (10,00,000 per cubic cm) is found in the top layer of soil at a depth of 5 to 15 cm. In deeper layer (1.5 to 5 m) individual microbes are found. However, they have been discovered at a depth of 17.5 m in coal, oil and artesian water. It has been calculated that in the ploughed layer of cultivated soil over an area of one hectare there may be from 5 to 6 tons of microbial mass and one gram of ploughed soil contains 1-10 thousand million bacteria.

Role of Soil Organisms

Soil organisms take part in a number of processes in the soils. Some of their important roles are as follows :

(1) decompose the dead organic matter and increase plant nutrients in available forms,
(2) production of toxins,
(3) production of growth stimulating substances,
(4) nitrogen fixation in the soil,
(5) mixing of soil,
(6) improvement in soil aeration,

Fig. 21.4. Certain most common living organisms (biota) inhabiting the soil.

(7) improvement in the aggregation of soil particles or soil binding, and

(8) cause injury to the plants.

1. Decomposition of dead organic matter. A number of soil microbes attack the dead remains of plants and animals and cause decomposition. In the process of decomposition, complex organic matters are converted into simple organic compounds. Compounds like sugars, starch and proteins are decomposed first in the decomposition process and then cellulose, fatty substances and lastly lignin and woody substances are degraded. Proteins when acted upon by microbes are converted into amino acids, ammonium salts, nitrates and nitrites. Humus, an intermediate product of decomposition process, is formed by micro-organism in optimum physical conditions. In the decomposition process, a number of complex mineral compounds are also converted into simpler and soluble compounds. Organic acids and carbon dioxide that are released by decomposition make insoluble phosphates and other unavailable compounds more easily available to plants.

Decomposition of dead organic matter primarily helps in the feeding and growth process of these micro-oganisms and secondly, increases the nutrient contents of the soil. Bacteria and soil fungi are main agents which bring about the process of decomposition in the soil.

2. Production of toxins. In the absence of oxygen some soil microbes secrete chemicals, such as, aldehydes, organic acids, etc. which may show toxic effects on many plants. Examples of toxin secreting organisms may be found in fungi, bacteria and algae. *Fusarium lini,* which causes wilt of flax (Alasi) secretes HCN, a deadly poisonous substance and *Fusarium udum,* a fungus causing wilt of pigeon pea (Arhar) secretes fusaric acid in the roots of the host plants. These toxic chemicals secreted by fungi may be responsible for causing wilt in the flax and arhar (*Cajanus cajan*).

3. Production of growth stimulating substances. Many soil organisms including soil fungi and bacteria produce growth stimulating substances such as 3-Indol acetic acid, Gibberellins and Gibberellic acid in the soil. *Fusarium* species too have been found to secrete Gibberellin and Gibberellic acid ($C_{19}H_{22}O_6$).

4. Nitrogen fixation. Many bacteria inhabiting the root nodules of leguminous plants (*Rhizobium*), nitrogen-fixing bacteria living free in the soil (*Azotobacter, Clostridium pasteurianum*), actinomycetes, fungi, purple bacteria and a number of blue-green algae are known to fix free atmospheric nitrogen gas into nitrogenous compounds, such as nitrates and nitrites, etc., and thereby increase the fertility of the soil. It has been established that in each hectare of ordinary soil every year 25—50 kg of nitrogen are fixed and in cultivated soil and in soil containing legume plants 35 to 60 kg and 100 to 400 kg of nitrogen are fixed respectively. *Anabaena, Nostoc, Microcystis* are important nitrogen fixing blue-green algae. De and Fritsch (1938) have found that certain blue-green algae are able to fix 20 lbs of atmospheric nitrogen per acre in a rice field. They increase the yield of rice from 15 per cent to 25 per cent. Singh, R.N. and Relwani and others have also shown experimentally that some blue-green algae fix nitrogen in the paddy soils.

5. Soil mixing. Many organisms by their mechanical activities help in mixing and weathering of soil. Roots of the higher plants take active part in the disintegration of rocky mass and also make the compact soil loose. Many rodents, insects and earthworms turn over the soil and sometimes also expose the rock surface for physical and chemical weathering. Burrowing animals, such as rodents, bring soil from deeper regions to the surface. The excreta of soil animal is deposited on the surface of soil in the form of casts which increases the fertility of the soil.

6. Soil aeration improvement. Soil micro-organisms improve aeration of soil. Burrowing worms are also helpful in improving the aeration and percolation.

7. Improvement in aggregation of soil particles. Bacteria, blue-green algae, and some other micro-organisms secrete mucilaginous substances which bind the soil particles into soil aggregates.

8. Injury to plants. Not all the soil organisms are beneficial in their properties and behaviour. Some microbes become parasites of higher plants and cause considerable damage. Nematodes are important animals which cause a number of diseases in plants. Besides these, many soil bacteria and fungi cause many diseases, such as damping off, seedling blight, root rot, mildew diseases in a number of crops.

QUESTIONS

1. Define the soil. Discuss briefly the different components of soil.
2. What is humus? Give and account of composition and properties of humus and mention its important roles in the soil.
3. Discuss the biological system of soil in detail and mention the role of micro-organisms in soil.

22

Origin and Formation of Soil

Soils are derived from parent rocks by the process called weathering. In the weathering process, rocks are disintegrated and decomposed into smaller pieces by physical, chemical and biological agencies. The products of weathering are called regoliths which are small particles of rock materials. These particles are basic materials which under the influence of pedogenic or soil forming process finally develop into mature soil. C.F. Marbut (1926) states that "a mature soil is one that has assumed the profile features characteristic of predominant soils on the smooth uplands within the general climatic and botanic regions in which it is found."

The processes of soil formation from the rocks can be presented as follows :

Consolidated parent rocks	——————→ (Weathering)	small particles of parental material or
Mature soil with profile differentiation.	←—————— (Pedogenesis)	regolith

Weathering is the initial stage and the soil forming processes or pedogenic stages come afterwards. Practically no definite line can be drawn where weathering process terminates and pedogenesis takes its way. In fact, both these processes run simultaneously. **Weathering processes** lead to disintegration or degradation of complex mineral substances that are locked up in the rocks and eventually simple compounds are formed. The **pedogenesis** or **soil forming process**, on the other hand, are constructive and are of bio-geochemical nature in which biological influences play important role.

Rock is a chemical mixture in which a number of minerals are locked up in very complex forms. The chemical and physical properties of minerals composing the rock directly govern the properties of the rock. The principal minerals which occur in the earth crust are listed in Table 22.1. In addition to these, there are a number of other minerals also, such as Tourmaline (Borosilicates of aluminium with alkali metals and iron and magnesium), Rutile (Titanium oxide), Zircon (Zirconium oxide), Gauconite (hydrated silicates of iron and potassium). These minerals are found in very small quantities in the soil.

According to Clark (1924), the major minerals in the earth's crust occur in the following approximate proportion :

Feldspars 58%; amphiboles and pyroxenes 16%; quartz 13%; micas 4%; and remaining portion consists of accessory minerals. (Ref : *Data of Geochemistry.* U.S. Geological Survey Bull. 770.1924).

Table 22.1. Chemical composition of some common soil minerals

Minerals	Chemical Constituents
A. Sand and silt minerals	
1. Quartz or silica	SiO_2
2. Feldspars	
(*a*) Orthoclase	$K_2Al_2Si_6O_{16}$
(*b*) Plagioclase	$NaAlSi_2O_8$
(*c*) Calcium feldspar	$CaAl_2Si_2O_8$
3. Micas	
(*a*) Muscovite	$KAl_2\ (AlSi_3O)_{10}(OH)_2$
(*b*) Biotite	K, Mg, Fe, Al silicates
4. Pyroxenes	(Mg, Fe) SiO_3
5. Amphiboles	(Mg, Fe), $(Si_4O_{11})_2\ (OH)_2$
6. Olivine and serpentine	$(Mg, Fe)_2\ SiO_4$
7. Calcite; magnesite; and dolomite	$CaCO_3$; $MgCO_3$; and ($CaCO_3$, $MgCO_3$)
8. Iron oxides	
(*a*) Haematite	Fe_2O_3
(*b*) Magnetite	Fe_3O_4
(c) Limonite	FeO (OH), xH_2O
B. Clay minerals	
1. Kaolin	Al_2O_3, $2SiO_2$, $2H_2O$
2. Montmorillonite	(Ca, MgO) Al_2O_3, $5SiO_3$, $5H_2O$

Rocks are generally classified into three types on the basis of their origin and structure. These are as under :

(1) igneous rocks,
(2) sedimentary rocks,
(3) metamorphic rocks.

1. Igneous rocks. These are formed by cooling and solidification of molten magma or lava which escapes through fissures from the interior of the earth. The solidification may occur at great depths resulting in the *plutonic rocks,* or at moderate depth in the earth forming the so-called *intrusive rocks,* or on the surface of the earth forming the *extrusive* or *effusive rocks.* Extrusive rocks are formed as a result of volcanic activities.

Important igneous rocks involved in the weathering process are Granite, Basalt, Gabbro, Diabase, Syenite, Andesite etc. They are composed of primary minerals such as quartz, feldspars and biotite, augite and hornblende.

2. **Sedimentary rocks.** These develop due to gradual accumulation and consolidation of *weathering product* or *mineral particles* brought by wind or water on the surface of earth. Such rocks are characterised by the presence of distinct sediments or layers in them. Some important examples of sedimentary rocks are limestone, shale, conglomerate, sandstone, siltstone, etc. Calcite, clays, quartz and dolomite are the common dominant minerals in them. Sedimentary rocks do not weather as rapidly as the igneous rocks.

3. **Metamorphic rocks.** These are formed after transformation of igneous and sedimentary rocks when they are subjected to intense heat and pressure and are influenced chemically by active liquids and gases. Important metamorphic rocks which take part in weathering process are slate (formed from shale), marble (formed from limestone), schist (formed from shale) and quartzite (from sandstone). Dominant minerals are quartz, clays and calcite.

Wallwork (1970) has summarised the details of the principal soil forming rocks in Table 22.2.

Table 22.2. Classification, characteristics and mineral composition of the principal soil-forming rocks (Wallwork, 1970)

Class of rock	Mode of formation of rock	Type	Description of rock-types	Principal minerals of soil-forming rocks
A. Igneous rocks	Formed by cooling of molten megma (lava)	1. Granite	Usually light in colour, coarse to medium grain	Quarts, feldspar, mica, amphibole, iron oxides
		2. Diorite	Grey to dark in colour, coarse to medium grain	Feldspar, amphibole, iron oxide, biotite
		3. Basalt	Dark to black in colour, dense to fine grain	Feldspar, pyroxene, iron oxide, biotite
B. Sedimentary rocks	Formed by deposition of weathered minerals which are derived from igneous rocks	1. Shales	Light to dark in colour, thinly laminated	Clay minerals, quartz
		2. Sand stone	Light to red in colour, granular and porous	Quartz, clay minerals, iron oxides, calcium carbonate
		3. Limestone	Light grey, red, brown, or black in colour, fine grained and compact	Calcite, dolomite, iron oxide, clay minerals
C. Metamorphic rocks	Formed by change of pre-existing rock (*i.e.*, igneous or sedimentary) through heat and pressure	1. Gneiss	With light and dark bands	Much as in granite from which it is formed
		2. Schist	Foliated structure	Much as in basalt or shale from which it is formed
		3. Slate	Grey to black in colour, compact and uniform texture	Composition similar to shale from which it is formed
		4. Quartzite	Light to brown in colour, compact and uniform texture	Similar in composition to sandstone from which it is formed
		5. Marble	Light, red, green or black in colour, compact, fine to coarse texture	Calcite, dolomite formed from limestone

On the basis of their silica content, rocks have also been classified into acidic, basic and intermediate rocks.

Acidic rocks	Basic rocks	Intermediate rocks
Silica content 65-75% Examples : sand-stone, granite etc.	Silica content 40-55% Examples : gabbro, basalt, limestone etc.	Silica content 55-65% Examples; syenite, diorite, andesite etc.

WEATHERING PROCESS

It is a process in which the massive consolidated rocks are broken down into smaller particles and eventually into the individual minerals of which they are composed. As a result of weathering the rock fragments and the minerals are changed to new minerals either by alteration or by complete chemical changes.

Weathering processes are distinguished into the following three types on the basis of nature of agencies which bring about weathering :

(1) Physical weathering,

(2) Chemical weathering, and

(3) Biological weathering or Biogeochemical weathering.

1. Physical Weathering

Physical weathering of rocks is a mechanical process which is brought about by a number of factors, such as :

(A) Temperature,

(B) Water,

(C) Wind.

(A) **Temperature.** It causes breakdown of rocks in the following ways:

(*i*) **Differential expansion and contraction of materials.** Minerals composing the rock show different degrees of expansion (coefficient of expansion). These minerals expand in the high temperature of day and contract when the temperature falls. The differential expansion and contraction of different minerals set up internal tension and produce cracks in the rocks and thus the rocks weather into finer and finer particles.

(*ii*) **Exfoliation.** The arrangement of layers in rock is called stratification. Layer differentiation is not common in all types of rocks. The upper layer of rocks expand and contract faster than those of deeper region. The temperature changes bring about separation and disintegration of the layers of rocks. This process is known as *exfoliation.*

(*iii*) **Frost action.** Sometimes, temperature of rocks reaches below freezing point. This causes accumulation and freezing of water in the crevices and rock joints. In freezing, water expands to about 9 per cent of its original volume and exerts a pressure of approximately 150 tons per square feet which is more than enough to break the rocks.

(B) **Water.** Water causes weathering of rocks in the following ways:

(*i*) **Rain.** Natural water falling either in the form of rain drops or as hail storm on the surface of rocks with beating effect bring about abrasion of massive rocks into smaller particles.

(*ii*) **Running water.** Rapidly flowing water rolls the heavy rock masses (rock boulders) along the bottom of stream and grinds them into finer particles.

(*iii*) **Wave action.** It is most active in sea shores. The water waves striking with great force on the rock surface break and grind the rock into pieces.

(*iv*) **Glacier formation.** At mountain tops, ice formation takes place in the winter season. When the summer approaches, ice starts melting and glaciers (huge sliding masses of ice) move downwardly on the slopes. In the glacier movement, the rocks are corroded and finally broken into sand particles (Fig. 22.1).

(C) **Wind.** Rapid stormy wind carrying suspended sand particles causes the abrasion of exposed rock. The Fig. 22.2 shows rocks which were subjected to wind erosion.

2. Chemical Weathering

Chemical weathering brings about disappearance of original rock minerals either completely or partly. In this process secondary products may be formed from parent materials. This process is also known as chemical transformation. Presence of moisture and air is very essential in the chemical weathering. This is why chemical weathering is not so effective in desert.

The chemical weathering takes place in the following ways :

(*i*) **Solution.** Solvent action of water helps in the weathering of rocks. It dissolves soluble minerals of rocks. Solution helps in the removal of weathered materials but total loss is negligible. Solvent action is increased in presence of CO_2 and organic acids formed by decomposing dead organic remains of plants and animals. Sodium, potassium, calcium and magnesium are easily removed from rocks in dissolved state.

Fig. 22.1. Moving glaciers cause cutting and crushing of the bed rocks.

Fig. 22.2. Effect of wind on the rocks.

(*ii*) **Hydrolysis.** It is essentially an exchange of constituent parts between water and rock minerals. When water reacts with strong base it produces hydroxides. The soluble products of hydrolysis are usually removed by water. Sometimes soluble products may react with insoluble ones and form clays. Hydroxides in presence of CO_2 change to carbonates and bicarbonates. Water in ionised state acts as a weak acid on silicious matter, *e.g.*,

$$KAlSi_3O_8 + HOH \rightarrow HAlSi_3O_8 + KOH$$

$$2HAlSi_3O_8 + 8HOH \rightarrow Al_2O_3 . 3H_2O + 6H_2SiO_3$$

(*iii*) **Oxidation.** It means addition of oxygen to mineral compounds. The reaction produces oxides which when dissolve in water weaken the rock and bring about weathering. Iron, aluminium, manganese oxides and sulphides are easily oxidised.

$$\underset{\text{(Ferrous oxide)}}{4FeO} + O_2 \rightarrow \underset{\text{(Ferric oxide)}}{2Fe_2O_3}$$

(*iv*) **Reduction.** It means removal of oxygen from minerals, *e.g.*,

$$2Fe_2O_3 \rightarrow 4FeO + O_2$$

Reduction takes place in the deep zone where oxygen is not available.

(*v*) **Carbonation.** It occurs simultaneously with hydrolysis. In this process, CO_2 unites with water to produce carbonic acid which is a weak acid.

$$CO_2 + H_2O \rightarrow H_2CO_3 \text{ (Carbonic acid)}$$

The carbonic acid reacts with hydroxides of soil forming minerals and forms insoluble carbonates. Sometimes it dissolves minerals and thus weakens the rock promoting thereby the weathering. Carbonation of hydroxides results in the formation of carbonates and bicarbonates, *e.g.*,

$$Ca(OH)_2 + CO_2 \rightarrow CaCO_3 + H_2O \xrightarrow{CO_2} Ca(HCO_3)_2$$

(*vi*) **Hydration.** In this process, water molecules become chemically attached to particular rock material, for example :

$$\underset{\text{(Hematite)}}{2Fe_2O_3} + 3H_2O \rightarrow \underset{\text{(Limonite)}}{2Fe_2O_3.3H_2O}.$$

Soil forming minerals in the rock do not contain any water. They undergo hydration when they come in contact with water. In this process, the volume of the parental material increases and hydrated materials become soft and more readily weatherable.

3. Biological Weathering

Many organisms play important roles in the weathering of rocks through physical and chemical means. Important organisms concerned with the decomposition of rocks are lichens, bacteria, fungi, higher plants, nematodes and other soil microbes. Lichens and some other organisms in presence of moisture secrete carbonic acid which corrodes the rock.

The presence of roots on the surface of rock exerts a considerable pressure by which rocks are broken into smaller fragments. The root exudates also weaken the rocks and weather them to a small extent.

Joffe (1949) states that there is no biogeochemical weathering. According to him, it is either physical or chemical weathering by biological agencies.

SOIL FORMING PROCESS

The surface rocks of the earth are weathered and as a result of weathering, small particles of parental materials are formed. The weathering process continues until all the essential elements locked up in the rocks become available to all sorts of organisms. Simultaneously with weathering process, pedogenesis or soil forming process also runs which leads weathered materials to develop into mature soil with well differentiated soil profiles and biological system.

SOIL PROFILE

Weathering of parental rocks results in the development of several loose layers or horizons of weathered materials. Biological system, addition of organic matter (humus) and interaction between organic and mineral compounds make the horizons more distinct. The complete succession of horizons down to the level of undifferentiated parent materials is called ***soil profile***. In other words, it is vertical section of earth crust showing different layers or **horizons** of soil.

The soil profile may be divided into the following three zones or horizons (Fig. 22.3) :

(1) A horizon (top soil or the zone of extraction),

(2) B horizon (subsoil or subsurface zone), and

(3) C horizon (Regolith).

Some workers have recognised D horizon below C horizon.

1. **A horizon.** It is the top soil which consists of mineral matter and organic residues. The soluble chemicals are leached away by water.

Upper zone or A horizon is very loose and it supports vegetation. Because it is full of organic elements and minerals, the top soil is very fertile. The thickness of A horizon varies from one to ten feet. A horizon is differentiated into several sublayers which are Aoo, Ao, A_1, A_2, A_3. These can be distinguished by their distinct colours and textures.

2. **B horizon.** A zone of soil present below the top soil or A horizon is called B horizon. It is also known as subsoil. In this zone, the weathered substances or minerals are deposited and organic residues are present in very small quantities. This horizon is *lighter in colour* than A horizon. B horizon can also be differentiated into several subzones, as B_1, B_2, B_3 and so on, on the basis of colour and texture. Upper subzone of B horizon, which is just below the top soil (A horizon), is looser than lower subzones.

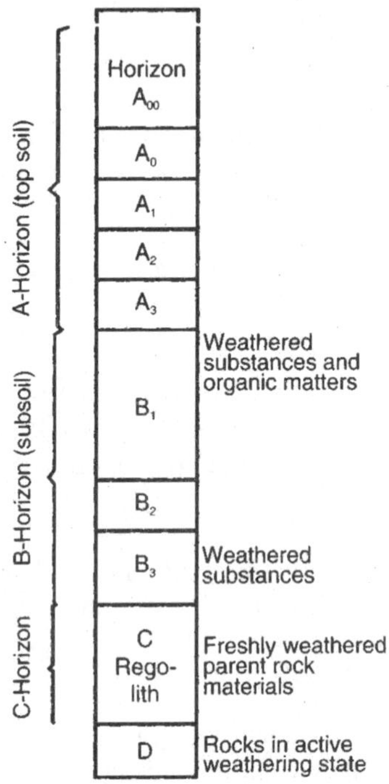

Fig. 22.3 Soil profile

3. **C horizon (regolith).** Below B horizon and above the surface of weathered rock, there is a third horizon, the C horizon or zone of regolith. It is formed of freshly weathered parental materials.

4. **D horizon**. It is also called R. horizon (rock in active weathering state). Below the C horizon, there may be found rock in active weathering state. This is recognized as a fourth zone or **D horizon** by some workers.

Morphological Characteristics of Soil Profile

The following morphological characteristics are recorded in respect of each of the horizons :

Colour, texture, structure, consistence, concretions, calcareousness, roots, etc.

Colour. Colour is one of the most noticeable characteristics of a soil. As we examine a soil profile, we are likely to see definite changes in colour from the surface through the subsoil and underlying parent material. The colours of the surface and subsoil may be classified as follows:

Surface soil—Dark, moderately dark, light, very light.

Subsoil—Dull, mottled, bright.

As a rule, the surface soil is darker than the subsoil since it contains more organic matter. Subsoil colour is less affected by organic matter than surface soil colour.

Texture. The soil particles which resulted from weathering vary greatly in size. They are classified on the basis of size into gravel, sand, silt, and clay.

Most soils are mixtures of the various sized particles except that many contain no gravel and some very little sand. If a soil contains considerable quantities of atleast two sizes of particles, it is known as loam. It may be called a gravelly loam, sandy loam, silty loam, or clay loam, depending on the size of particle that predominates in the mixture. If however, a mixture consists mostly of one size of particle, it is not called as loam.

Structure. Structure means the aggregation or arrangement of the soil particles into clusters or shape of various sizes. Structure does not change soil texture. The structure of soil particles may be crumb, granular, angular, sub-angular, blocky, prismatic, platy and columnar type. Details are given in the next chapter.

Consistence. Consistence is a term expressing the degree of cohesion of the soils and the resistance opposed to forces tending to deform or rupture the aggregates.

Consistence of soil materials is determined at dry, moist and wet conditions. The terms used are :

Dry — Loose, soft, slightly hard, hard, very hard, extremely hard.

Moist — Loose, very friable, firm, very firm, extremely firm.

Wet — Non-plastic, slightly plastic, plastic, very plastic, non-sticky, slightly sticky, sticky, very sticky.

Concretions. Iron and maganese concretions alone have no definite meaning, although their presence usually indicates that conditions of alternate oxidation and reduction have taken place. Presence of calcium carbonate concretions throughout the profile and on the surface may be the result of resistance action. These concretions at deeper layers may indicate the usual depth of penetration of rainwater.

Calcareousness. Soil horizons are tested with dilute HCl and based upon the briskness of the effervescence, they are classified as weakly, moderately and strongly calcareous. Calcareousness has an effect on the availability of plant nutrients.

Roots. The abundance of roots in each horizon is normally recorded in terms of few, common, abundant etc.

Other features like presence of clay, pores etc. within each horizon are also recorded.

SOIL-FORMING PROCESSES

The soil profile is formed by interaction of various pedogenic factors under a special set of conditions.

The following fundamental processes namely ***humification*** and ***eluviation*** **and** ***illuviation*** develop a profile. These are described below :

(*i*) **Humification.** Top layer of the soil, called A horizon, contains abundant dead remains of plants, animals and other products of microbial metabolism. These products undergo decomposition which may produce some soluble organic compounds and some amorphous incompletely decomposed black coloured organic residue, the ***humus***. The process of humus formation is called ***humification***. The humus and organic compounds become mixed with the fine particles of weathered rock. Some soluble organic compounds are removed from the top soil by water which percolates downwardly through humus mixed soil.

(*ii*) **Eluviation and illuviation.** Water is important agent which helps in the development of soil profile. While percolating downwardly through mineral and organic substances of A horizon or top soil, it removes a number of soluble chemicals from the top soil. This process of washing away of soil constituents is termed as ***eluviation*** (meaning wash out) and the surface layer from which components are lost is called ***eluvial layer*** or A horizon. The eluviated substances move downwardly and are deposited in the lower zone or B horizon which is called ***illuvial layer*** (meaning 'wash in'). The process of accumulation of eluviated materials is called ***illuviation***.

The fundamental soil-forming processes involved in the development of soil profile are described below :

(*i*) **Gleization.** This process takes place in wet and cold Tundra regions where saline conditions do not exist. In this process there develops a compact structureless and sticky surface layer. This layer is blue-green in colour, poorly aerated and has reduced content of iron compounds. It favours surface accumulation of peat materials and undergoes a series of chemical, physical and biological changes which produce a characteristic soil.

(*ii*) **Podzolisation.** In temperate zones where climate is cold and moist, the rate of decomposition is slow and in the decomposition process acids are produced which make the soil acidic. The acidic unproductive soil is called **podzol** and the process is called **podzolisation**. In the process of podzolisation humus and some minerals including dissolved Si, Fe and Al salts from A horizon move downwardly with percolating water and accumulate in the lower horizons. This process is more effective in sandy base-poor parent materials under intense leaching and thick vegetational cover.

(*iii*) **Laterisation.** In tropical and subtropical regions, when rainfall occurs the organic matter and minerals are leached away and hydroxides of aluminium and iron are precipitated in the form of residue which is called **laterite**. This process is termed as ***laterisation***.

In this process, silica is completely removed. Laterites usually do not show well differentiated horizons.

Podzol and laterites are collectively described as **pedalfer** group' (iron accumulating group).

(*iv*) **Calcification.** In subhumid and dry regions, due to lack of excessive moisture the soil accumulates considerable amount of soluble materials, and carbonates of calcium and magnesium are deposited in B horizon. The soil having such features is called ***pedocal*** (calcium accumulating soil). Pedocals are not acidic and calcium content in them is very high. This process is common in grasslands.

Hydromorphic profile development. Such soil-forming processes result in swamp, bog, marsh and peat soils and occur under conditions when percolation is restricted and certain horizons become saturated with water. These two conditions result in the development of anaerobic conditions, as well as beginning of more chemical reduction processes.

FACTORS AFFECTING SOIL FORMATION

These are outlined as under :

(1) Climate, (Cl.)

(2) Biosphere or Organism (O)

(3) Parent material (P)

(4) Relief (r)

(5) Time (t)

All these factors are interrelated and complementary to one another. Some of them may be more influential under a particular set of conditions in determining the nature of soil while the others may be less influencing. Any soil property(s) is a collective function or the effects of all pedogenic factors (Jenny, 1991):

S = f(cl, o, r, p.t.) where f = function dependent upon a set of conditions.

Joffe (1949) creates the following two categories in the soil forming factors:

(*i*) **Active factors.** These include agents which supply energy for soil formation. Climate and biosphere are the two important active factors.

(*ii*) **Passive factors.** These include the sources of soil forming mass and the conditions affecting it. Parent material, relief and time are the three passive factors.

1. **Climate.** It shows the following effects on pedogenesis :

(*i*) Heavy rainfall and high relative humidity favour the soil formation or pedogenesis. Low rainfall and low relative humidity show retarding effects on the process.

(*ii*) Severe erosion of materials by wind also affects the process adversely.

(*iii*) Horizons tend to develop faster under cool and humid forest conditions.

2. **Biological factors influencing the pedogenesis.** Biological agents affect the soil formation in a number of ways :

(*i*) Large number of burrowing animals upturn the deep soils and thus disturb the soil profile development.

(*ii*) Man's activities, such as manuring, ploughing, drainage, irrigation, cropping system, reclamation also affect the pedogenesis to a great extent.

(*iii*) Addition of organic residues and mixing them with minerals promote the differentiation of soil profile.

(*iv*) Vegetation exerts its main influence on pedogenesis through the amount and nature of organic matter which it adds to the soil and by checking the soil erosion.

3. **Nature of soil forming parental material mass.** This affects pedogenesis in the following ways :

(*i*) High percentage of clay in the weathered materials has retarding effect on pedogenesis.

(*ii*) Resistant parental materials, such as granite if present in parent material mass, do not promote the soil formation.

(*iii*) Excessive sandyness of parent materials reduces the rate of soil formation.

(*iv*) Materials with low silt and clay percentages do not differentiate into soil profile quickly.

(*v*) Heavy accumulation of soil materials also has retarding effects.

(*vi*) High lime content in the parent material checks profile differentiation.

Time. Length of time required for soil formation depends upon many interrelated factors, such as climate, biosphere, nature of parent materials, etc.

Relief. Some of the important effects of relief in the pedogenesis are as follows :

Topography determines the drainage condition and groundwater level in the soil. High water table shows influence in the horizon differentiation. Very steep slopes do not promote pedogenesis because of the fact that deposition of parent materials is affected by downwardly flowing water. If anyhow soil is formed, water content in such a soil is very poor.

Soil materials on moderate slopes promote deeper profile development and more luxuriant vegetation than in soil on steep slopes.

In soil survey Manual of U.S. Department of Agriculture (USDA) the interrelationship of pedogenic factors has been concluded as follows :

"Soil characteristics in any one place result from the combined influence of climate and living matter, upon the parent rock materials, as conditioned by relief (topography) over periods of time including the effects of the cultural environment and man's use of soil."

QUESTIONS

1. What are different minerals composing the rocks? Give an account of different weathering processes which bring about breakdown of mineral rocks.
2. Give an account of soil forming process and discuss the effects of various factors on pedogenesis.
3. Write short notes on :

   (*i*) Humification. (*ii*) Eluviation and illuviation.

   (*iii*) Soil profile.

23

Properties of Soils

A. PHYSICAL PROPERTIES OF SOILS

Physical properties of the soil can be discussed under the following heads :

(1) Soil separates and texture,

(2) Structure of soil,

(3) Weight and soil density,

(4) Porosity of soil,

(5) Permeability of soil,

(6) Soil colour,

(7) Temperature of soil, and

(8) Soil Plasticity, Compressibility and Erodibility.

1. Soil Separates and Soil Textures

Mineral fraction of soil consists of particles of various sizes. According to their size, soil particles are grouped into the following types (Table 23.1).

Table 23.1. Soil separates

	Name of soil particle	Diameter range of soil particle in millimetre
1.	Coarse particles or gravels	More than 2.00
2.	Coarse sands	2.00—0.20
3.	Fine sands	0.20—0.02
4.	Silts	0.02—0.002
5.	Clays	Below 0.002

The particle sizes of above groups are suggested by International Society of Soil Science. In India, international system of particle differentiation is commonly followed. The particle types are generally called 'soil separates' or 'soil fractions'.

Amount of soil separates is determined by a process known as mechanical analysis. In this process, soil sample is crushed and screened through a 2 mm round hole sieve. The screened soil is then homogeneously dispersed in water and allowed to settle.

In suspension, particles of largest dimensions will settle first and those of smaller dimensions will settle afterwards. Individual soil separates are identified on the basis of their respective diameter ranges.

Soil separates (sand, silt and clay) differ not only in their sizes but also in their bearing on some of the important factors affecting plant growth, such as, soil aeration, workability, movement and availability of water and nutrients. Important characteristics of different soil separates are as follows:

Sand. This fraction of soil consists of loose and friable particles of 2.203—.02 mm diameter. Sand particles can be seen by unaided eye. These particles, although inactive, constitute the framework of the soil. They play less important role in physicochemical activities. When coated with clay, these sand particles take very active part in chemical reactions. Sands increase the size of pore spaces between soil particles and thus, facilitate the movement of air and water in the soil.

Silt. It consists of soil particles of intermediate sizes between sand and clay (diam range .02—.002 mm). Silt, when wet, feels plastic but in dry state feels like flour or talcum. Coarse silt shows little physicochemical activities but finer grades play important role in some chemical processes. Silty soil has got larger exposed surface area than the sandy soil. Silty soils contain sufficient quantities of nutrients, both organic and inorganic. That is why they are very fertile. Soils rich in silt possess high water holding capacity. Such soils are good for agriculture.

Clay. This soil fraction contains smaller particles than silt (below .002 mm diameter) which exhibit plasticity and smoothness when wet and hardness when dry. Owing to their smallest size and colloidal nature, the clay particles expose extremely large surface area. They take very active part in physicochemical reactions of the soil. Clay soils have fine pores, poor drainage and aeration and thus they have highest water holding capacity. The clay acts as store house for water and nutrients.

Some soils are fine, while others are coarse. It is so because of the fact that the relative percentage of sand, silt and clay differ from soil to soil. The relative percentage of soil separates of a given soil is referred to as soil texture. Texture of soil for a given horizon is almost a permanent character, because it remains unchanged over a long period of time. The relative percentages of soil separates of average samples are almost infinite in possible combinations. It is, therefore, necessary to establish limits of variations among soil fractions so as to group them into textural classes. The common textural classes, as recognised by USDA (U.S. Department of Agriculture) are given in the following table. These classes are recognised on the basis of relative percentage of separates; sand, silt and clay (Table 23.2).

Table 23.2. Soil textural classes

Soil classes or Textural name	Range in relative percentages of soil separates		
	Sand	**Silt**	**Clay**
Sandy soil	85—100	0—15	0—10
Loamy sand	70—90	0—30	0—15
Sandy loam	43—80	0—50	0—20
Loam	23—52	28—50	7—27
Silt loam	0—50	50—88	0—27
Silt	0—20	88—100	0—12
Sandy clay loam	45—80	0—28	20—35
Clay loam	20—45	15—53	27—40
Silty clay loam	0—20	40—73	27—40
Sandy clay	45—65	0—20	35—45
Silty clay	0—20	40—60	40—60
Clay	0—45	0—40	40—100

This chart is adapted from fraction system of U.S.D.A.

If relative percentages of soil separates are known, the soil can be given textural name. For this purpose equilateral triangles are used. The most widely used Equilateral triangles are international equilateral triangle and the one used by USDA (Figs. 23.1, 23.2). These consist of three angles and its area is divided into twelve groups representing twelve different textural classes. Each group covers definite range of percentages of sand, silt, and clay. In the triangles, left side line represents the clay %, right side line represents percentage of silt and base represents percentage of sand. Each arm of the triangle is divided into ten divisions representing soil separate's percentage. These divisions are further divided into ten small divisions; each small division represents one per cent of soil separate. The percentages of sand, silt, and clay obtained after mechanical analysis of the given soil are read on the equilateral triangle. In using the diagram as indicated the percentages of silt and clay should be located on silt and clay lines respectively. The line in case of silt is then projected inward parallel to clay side of the triangle and in case of clay it should be projected parallel to the sand side. The three lines; one representing sand percentage, other representing silt percentage and the third clay percentage meet at a point in the triangle. The compartment in which the point falls indicates textural name for the given soil sample.

The knowledge of soil texture is of great help in the classification of soil and in determination of degree of weathering of rock.

2. Structure of Soil

Sand, silt and clay are found in aggregated form. Arrangement of these soil particles on certain defined patterns is called *soil structure*. The natural aggregates of soil particles are called *peds*, whereas an artificially formed soil mass is called *clod*. Ped differs from *fragment* because the latter refers to the broken ped. Ped differs from concretion in the sense that the latter is formed in the soil by precipitation of salts dissolved in percolating water. Soil structure also reveals the colour, texture and chemical composition of soil aggregates. Soil structure is influenced by air, moisture, organic matter, micro-organisms and root growth. When many particles or peds are aggregated into cluster, a compound particle is formed.

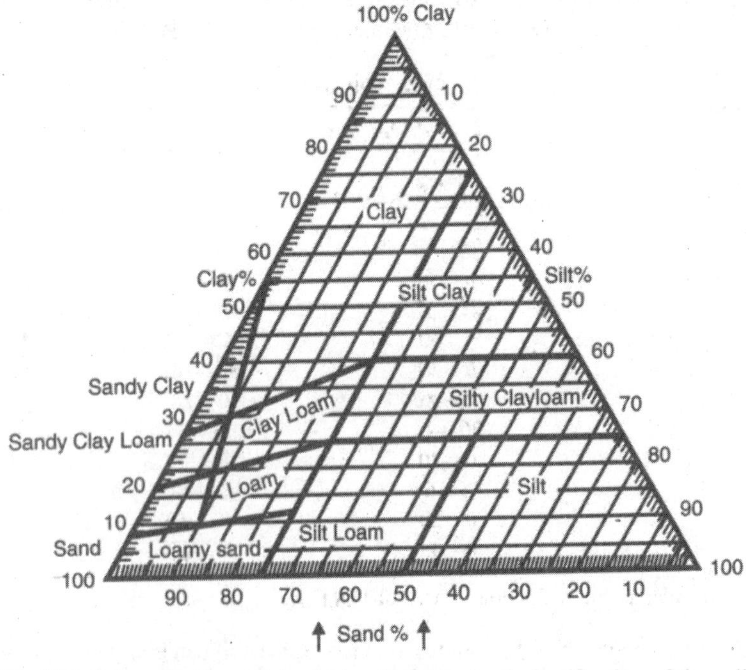

Fig. 23.1. International equilateral triangle for textural classification of the soil.

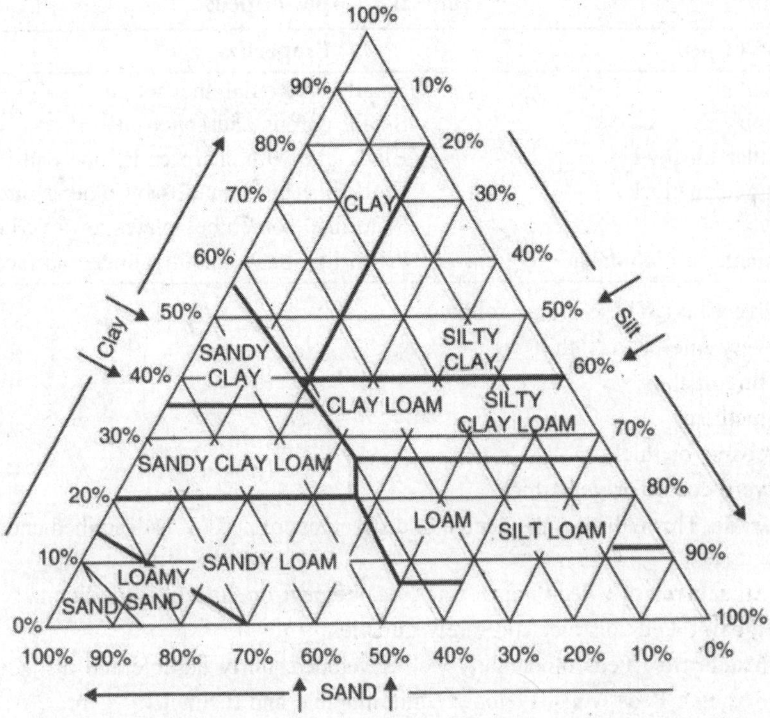

Fig. 23.2. Equilateral triangle used by USDA for textural classes of soil.

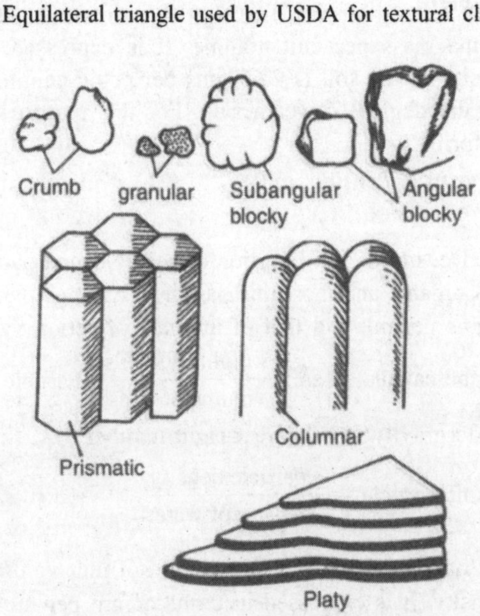

Fig. 23.3. Soil structure.

Soil structure is described under the following three categories :

A. Type. This indicates the shapes or forms and arrangement of peds. Peds may be of various shapes, such as granular, crumb, angular blocky, subangular blocky; platy and prismatic (Fig. 23.3). Different types of peds and their properties are described in (Table 23.3).

Table 23.3. Types of peds

Type of ped	Properties
(1) Granular	Small, spheroidal and nonporous.
(2) Crumb	Small, porous and spheroidal.
(3) Angular blocky	Block-like with sharp ends, one end may be pointed.
(4) Subangular blocky	Block-like but bounded by other aggregates.
(5) Platy	Plate like, sometimes plates are overlapped.
(6) Prismatic or Columnar	Prism like but without rounded surface.

B. Size Class. These are as follows:

(*i*) very fine or very thin

(*ii*) fine or thin

(*iii*) medium

(*iv*) coarse or thick

(*v*) very coarse or very thick

C. Grade. This indicates the degree of distinctness of peds. It is described under the following four categories :

(*i*) **Structureless :** Peds not distinct, *i.e.,* cement or sand like condition.

(*ii*) **Weak :** Peds distinct and rarely durable.

(*iii*) **Moderate :** Peds moderately well developed, fairly durable and distinct.

(*iv*) **Strong :** Peds well developed, quite durable and distinct.

3. Density and Soil Weight

Density of soil is the mass per unit volume. It is expressed in terms of gm per cubic centimeter. Average density of the soil is 2.65 gms per cubic centimeter. Density of soil varies greatly depending upon the degree of weathering. For this reason soil density is expressed in two generally accepted forms :

(*i*) particle density or true density; and

(*ii*) bulk density.

(*i*) **Particle density.** Density of solid portion of soil is called particle density. It is sum total of densities of individual organic and inorganic particles. Average particle density of organic soil varies from 1.2 to 1.7 gms per ml. and that of inorganic fraction varies from 2.6 to 2.78 gms/ml. Particle density may be calculated as : $\frac{\text{weight of solids}}{\text{volume of soils}}$. Particle density divided by density of water gives the specific gravity or relative weight number.

$$\text{Specific gravity of soil particles} = \frac{\text{particle density}}{\text{density of water}}.$$

(*ii*) **Bulk density or apparent density.** Dry weight of unit volume of soil inclusive of pore spaces is called bulk density. It is expressed in terms of gm per ml or lbs per cubic foot. It is lesser than the particle density of the soil. Bulk density of soil may be calculated as: $\frac{\text{weight of soil}}{\text{volume of soil}}$. Bulk density of the soil divided by density of water gives volume weight or apparent specific gravity of soil. Bulk density of soil changes with the change in total pore space present in the soil and it gives a good estimate of the porosity of soil. Average density of soil in bulk is 1.5 gm/ml. Organic soils have low bulk density as compared to mineral soils.

Soil weight varies in relation to textural classes. Average weight of loam or sandy soil is 80—110 pounds/cubic foot but that of clay ranges between 70 and 100 pounds/cubic foot.

4. Porosity of Soil

The spaces occupied by air and water between particles in a given volume of soil are called pore spaces. The percentage of soil volume occupied by pore space or by the interstitial spaces is called porosity of the soil. It depends upon the texture, structure, compactness and organic content of the soil. Porosity of the soil increases with the increase in the percentage of organic matter in the soil. Porosity of soil also decreases as the soil particles become much smaller in their dimension because of decrease in pore spaces. It also decreases with depth of the soil. The pore spaces are responsible for better plant growth because they contain enough air and moisture. Percentage of solids in soils can be determined by comparing bulk density and particle density and multiplying by hundred.

$$\text{Percentage of solids} = \frac{\text{bulk density}}{\text{particle density}} \times 100$$

This percentage of solids subtracted from total volume (100%) will give the percentage of pore space. Hence, the formula :

$$100\% - \left[\frac{\text{bulk density}}{\text{particle density}} \times 100\right] = \text{percentage of pore space.}$$

Depending upon the size pore spaces fall into two categories. These are :

(1) Micropore spaces (capillary pore spaces)

(2) Macropore spaces (non-capillary pore spaces)

Capillary pore spaces can hold more water and restrict the free movement of water and air in soil to a considerable extent, whereas macropore spaces have little water holding capacity and allow free movement of moisture and air in the soil under normal conditions.

5. Permeability of Soil

The characteristic of soil that determines the movement of water through pore spaces is referred to as soil permeability.

Soil permeability, because it is directly dependent on the pore size, will be higher for the soil with large number of macropore spaces than that for compact soil with a large number of micropore spaces (capillary spaces). Permeability of soil also varies with moisture status and usually decreases with the gradual desiccation of soil. In the arid regions, groundwater moves upwardly through capillary action and bring sodium, potassium and calcium salts with it in dissolved state on the surface of soil. The water evaporates and inorganic salts precipitate on the surface of the soil. As a result of this, the soil becomes less permeable and the productive capacity of soil is reduced.

6. Soil Colour

Soils exhibit a variety of colours. Soil colour may be inherited from the parental material (*i.e., lithochromic*) or sometimes it may be due to soil forming processes (***acquired*** **or** ***genetic colour***). The variations in the soil colour are due to organic substances, iron compounds, silica, lime and other inorganic compounds.

The organic substances impart black or dark greyish-black colour to the soil. Iron compounds are responsible for brown, red and yellow colours of soils. Iron oxides in combination with organic substances impart brown colour which is most common soil colour. Silica, lime and some other inorganic compounds give light white and grey tinges to the soil.

Soil colour influences greatly the soil temperature. The dark coloured soils absorb heat more readily than light coloured soils. The work of Ramdas, L.A. and David, R.K. (1936) at Poona showed that black cotton soil absorbed 86% of the total solar radiations falling on the soil surface as against 40% by the grey alluvial soil. Soil colour is used as an important criterion for description and classification of soil. Many soils are named after their prominent colours, such as black cotton soil, red-yellow latosol, grey hydromorphic soils and so on.

7. Soil Temperature

The chief sources of soil heat are solar radiations and heat generated in the decomposition of dead organic matters in the soil and heat formed in the interior of earth. The soil temperature greatly affects the physico-chemical and biological processes of the soil. Temperature of soil depends upon the temperature of atmospheric air and on moisture content. It is controlled by climate, colour of soil, slope, and altitude of the land and also by vegetational cover of the soil. The average annual temperature of soil is generally higher than that of its surrounding atmosphere. Ramdas and M.S. Katti (1934) recorded surface temperature of black cotton soil as high as 165°F at Poona (Maharashtra), India. Surface temperature of soil shows considerable fluctuations but soil temperature below certain depth remains more or less constant and is not affected by diurnal or regional temperature changes.

Studies made by Leather at Pusa Research Institute in Bihar (India) showed that diurnal temperature difference at the level 12 inches below the soil surface was only 1°C and at a depth of 24′′ it seldom exceeded 0.1°C. At the depth of 3 or 4 feet, the temperature remains almost constant.

8. Soil Plasticity, Compressibility and Erodibility

Soil plasticity is a property that enables the moist soil to change shape when some force is applied over it and to retain this shape even after the removal of the force from it. The plasticity of soil depends on the cohesion and adhesion of soil materials. Cohesion refers to the attraction of substances of like characteristics, such as, that of one water molecule for another. Adhesion refers to the attraction of substances of unlike characteristics. Soil consistency depends on the texture and amount of inorganic and organic colloids, structure and moisture contents of soil.

Compressibility. It refers to the tendency of soil to consolidate or decrease in volume. The compressibility is partly a function of elastic nature of soil particles and is directly related to settlement of structures. With the decrease in the moisture contents soils gradually tend to become less sticky and less plastic and finally they become hard and coherent. Plastic soils have great cohesion force. It is only because of cohesion property the moist clay soils frequently develop cracks when they become dried (Fig. 23.4). Coarse materials such as gravels and sands have low compressibility and the settlement is considerably less in these materials as compared to highly compressible fine grained organic soils.

Fig. 23.4. Diagram showing soil cracking due to shrinkage.

Erodibility. It refers to the ease with which soil materials can be removed by wind or water. Easily eroded materials include unprotected silt, sand and other loosely consolidated materials, Cohesive soils (with more than 20% clay) and naturally cemented soils are not easily removed from its place by wind or water and, therefore, have a low erosion factor.

B. CHEMICAL PROPERTIES OF SOILS

Chemical properties of soils can be described under the following heads :

(1) Inorganic matters of soil,

(2) Organic matters in soil,

(3) Colloidal properties of soil particles, and

(4) Soil reactions and Buffering action,

(*i*) Acidic soils,

(*ii*) Basic soils,

1. Inorganic Matters of Soil

From the accounts given in the description of weathering process it is clear that compounds of aluminium, silicon, calcium, magnesium, iron, potassium and sodium are chief inorganic constituents of soils. Besides these, the soils also contain small quantities of several other inorganic compounds, such as those of boron, magnesium, copper, zinc, molybdenum, cobalt, iodine, fluorine etc. The amounts of these chemicals vary in soils of different places. Chemical composition of soil of one horizon differs greatly from the composition of soil in the other horizon.

2. Organic Matters in Soil

Organic component of the soil consists of substances of organic origin; living and dead. In sandy soil of arid zone, it is found in very poor quantity (one or less than one per cent) but in peaty soil, it may be as high as 90%. When the plants and animals die, their dead remains are subjected to decomposition. As a result of decomposition a number of different organic products or compounds are formed from the original residues. In the course of decomposition, the original materials are converted into dark coloured organic complexes, called *humus*. Sometimes living micro-organisms add sufficient amount of organic matters in soil in the form of metabolic wastes. Chemists have been attempting to unravel the details of humus composition since the earliest days of soil science, and have got much success but more is yet to be discovered. In terms of specific elements, the organic component of soil contains compounds of carbon, hydrogen, oxygen, phosphorus, nitrogen, sulphur and small amount of other elements also. Only small fraction of total organic matter is soluble in water but majority of them are soluble in alkali solution. Chemically humus contains the following organic molecules :

A. Amino Acids

(*i*) Glutamic acid

(*ii*) Alanine

(*iii*) Valine

(*iv*) Proline

(*v*) Cystine

(*vi*) Phenyl alanine

B. Proteins

(*I*) *Purines*

(*i*) Guanine

(*ii*) Adenine

(*II*) *Pyramidines*

(*i*) Cytosine

(*ii*) Thymine

(*iii*) Uracil

C. Aromatic Molecules

D. Uronic Acids

(*i*) Glucuronic acid

(*ii*) Galacturonic acid

(*iii*) Lactic acid

E. Aliphatic Acids

(*i*) Acetic acid

(*ii*) Formic acid

(*iii*) Succinic acid

F. Aminosugars

(*i*) Glucosamine

(*ii*) N. Acetylglucosamine

G. Pentose Sugars

(*i*) Xylose

(*ii*) Arabinose

(*iii*) Ribose

H. Hexose Sugars

(*i*) Glucose

(*ii*) Galactose

(*iii*) Manose

I. Sugar Alcohols

(*i*) Inosital

(*ii*) Mannitol

J. Methyl Sugars

(*i*) Rhamnose

(*ii*) Fucose

(*iii*) 2-O, methyl D-xylose

(*iv*) 2-O, methyl D-arabinose

Besides these compounds locked up in the humus fraction, the soil also contains fats, oils, waxes, resins, tannin, lignin and some pigments.

Waksman and Stevans have proposed the following method for separation of different organic compounds present in the soil :

Organic matter

- Soluble fraction (fats, oils and waxes) 0.54-7%
- Insoluble fraction
  - Insoluble fraction + alcohol
    - Soluble fraction Resins 0.5-3%
    - Insoluble part — add hot 2% HCl
      - Soluble fraction Hemicelluloses 5-12% (Pentosans and Hexosans)
      - Insoluble part — Add cold H_2SO_4
        - Cellulose 3-5%
        - Lignin 35-50%
        - Protein 30-55%

Another modified method for separation of the various organic compounds from the soil is as follows :

Soil organic matter

- Dark coloured matter (Humus) — Dissolve in alkali solution
  - Insoluble in cold alkali (Humin)
  - Soluble in cold alkali — add acid and adjust pH at 12
    - Insoluble in dilute acid (humic acid or humus) — add alcohol
      - Soluble in alcohol (Humic acid)
      - Insoluble in alcohol (Hymatomelonic acid)
    - Soluble in dil. acid (Fulvic acid) — add NaOH and adjust pH at 4.8
      - Soluble (Fulvic acid)
      - Insoluble (B-Humus)
- Undecomposed nonhumic matter

The fractions are not pure chemical compounds but are in the form of mixtures of several substances. They are found in colloidal state in the soil.

3. Colloidal Properties of Soil Particles

There are two types of substances namely crystalloids and colloids.

Crystalloids are those crystalline solid substances which form true solution on being mixed with other substances. In true solution, crystal particles cannot be seen with the help of microscope.

The word colloid first coined by Grahm (1849) is derived from Greek words *kolla* meaning glue and *eoids* meaning appearance, *i.e.*, glue like in appearance. Colloid is really speaking amorphous state of the substances which do not form true solution if mixed with other substances. The particles of colloidal substances float in the solvent in suspension state but do not tend to settle at the bottom. Colloids are not found in ionic or molecular form but are found in aggregates of atoms or molecules. Colloidal system or suspension contains two phases which are :

(*i*) dispersion phase, *i.e.*, medium in which the particles are suspended, and

(*ii*) dispersed phase, *i.e.*, suspended particles.

Colloidal suspension may be of different kinds, such as

(1) Suspension of liquid, in liquid, as milk (fats in water).

(2) Suspension of solid in liquid as India ink (or clay suspension in water).

(3) Suspension of solid in gas, as smoke (coal particles suspended in air).

(4) Suspension of liquid in gas, *e.g.*, cloud and fogs in atmosphere.

The commonest colloids are those which remain suspended in a liquid medium.

If the colloidal suspension exhibits properties of fluid, it is called *sol,* but sometimes sols exhibit solid like behaviour and form solid or nearly so. This condition is called *gel.* Some sols form reversible gel while the others form irreversible gel. Some of the important properties of colloids in general are as follows :

1. **Particle size.** Crystalloids and colloids differ from each other in their size range. Particles of crystalloid in true solution are 0.2 to 1 mμ (millimicron) while those of colloids in suspension are 1 to 200 mμ.

2. **Adsorption.** Because the colloidal particles of dispersed phase are very small, they have got large exposed surface areas. Owing to their large exposed surface areas, these colloidal particles show great adsorptive capacity. In adsorption, particles of particular substances come to lie on the surface of colloids and they do not enter deep in the colloidal particles.

3. **Electrical properties.** The electrically charged colloids are termed as *micelles.* Colloids have some electrical charge on them. They may be charged either positively or negatively. Colloidal particles of one electrical charge have tendency to attract colloids of opposite charge. In the soil, clay particles are negatively charged, thus they attract cations (+ charged ions).

Colloid particles differ from electrolytes in the fact that when electric current is passed in the colloidal suspension, all the colloidal particles are attracted towards one electrode or the other depending upon the nature of charge they carry on them. This phenomenon is called *electrophoresis.* The electrolytes, when dissolved in solvent dissociate into two types of ions among which half will bear positive charge (cations) and remaining half will bear negative charge (anions). When electric current is passed in the solution of electrolyte all the positively charged ions will accumulate on negative pole and remaining negatively charged ions will collect on positive pole.

4. **Coagulation or flocculation of colloidal particles.** Colloidal particles in the suspension can be coagulated either by heating or by adding some substances which contain opposite charged ions. When substances carrying positive ions are added in suspension containing negatively charged colloid particles, ions will move and accumulate on the surface of colloids carrying opposite charge. Finally a stage comes when colloidal particles cannot attract more opposite charged ions. This is called *isoelectric point.* As a result of ion accumulation on their surface, the colloids

first become large and heavier and finally they tend to settle at the bottom in floccules. This process is known as ***flocculation***.

5. **Tyndal phenomenon.** Colloidal particles in suspension can be seen when a strong beam of light is passed through suspension and observer looks it from the place at right angle to the path of light. The colloidal particles become visible as strongly illuminated particles and they appear bigger than normal size. This phenomenon is known as "Tyndal effect".

6. **Brownian movement.** Colloidal particles when suspended in dispersion medium show a characteristic continuous zig-zag motion, called ***Brownian movement***. This type of movement was first observed by English botanist Robert Brown, hence it is called Brownian movement. The movement is exhibited because of characteristic collision of one particle with others. This prevents the particles from settling down.

7. **Dialysis.** Because colloidal particles in suspension are larger than the particles of crystalloid in true solution and are larger than the diameter of pores of porous membranes, *e.g.*, parchment membrane, they are not allowed to be filtered down and are retained by the membrane. Thus, they can be separated in pure state from the mixture of crystalloids and colloids by filtration process. This separation process is known as *dialysis*.

COLLOIDAL FRACTION OF SOIL

There are two types of colloids in the soil. These are

(1) mineral colloids or clay colloids, and

(2) organic or humus colloids.

These two colloidal fractions of soil are very intimate to each other and it is very difficult to separate them. The inorganic colloids occur as very fine particles and organic colloids occur in the form of humus particles. The soil colloid particles show almost all the characteristics of typical colloidal system, *i.e.*, adsorption, Tyndal effect, Brownian movement, coagulation, electrophoresis, dialysis etc.

1. **Clay colloids.** As regards the size, clay fraction of soil contains both non-colloidal and colloidal particles. Some clay particles may be as large as .002 mm in diameter but some may be smaller than normal colloid size (normal size of colloid particle is from 1 to 200 mμ). The clay particles are formed mainly of silica, alumina, iron and combined water. Colloidal clay may also contain rich accumulation of plant nutrients. Early researchers of soil science have described clay colloids as spherical particles and their sizes were mentioned in terms of their diameters, but recent electron micrographs reveal that particles occur in layers or plates and each clay particle appears as if it is composed of a large number of plate-like units. These units or flakes of clay are held together by a force of attraction. The plate-like clay particles expose large surface area on which moisture and cations (+ ions) are held. The finer the clay particles the greater will be the percentage of hygroscopic moisture.

If clay is suspended in distilled water, shaken, and then a little NH_4OH is added to suspension and allowed to settle, after a few minutes large particles settle down but finer particles remain in suspended state. When a little limewater is added to suspension, fine suspended particles increase in size and form small floccules which have a tendency to settle down. Thus, finer clay particles show flocculation property. Now if some acid is added, the floccules are broken and the clay particles will return to their normal size. This process is known as ***deflocculation***.

The clay particles are negatively charged, hence they can hold thousands of positively charged ions of mineral nutrients on their surfaces.

The clay colloids are lyophilic (water loving). So, they are important from the standpoint of the adsorption of large quantity of water (perhaps, 5-10 layers of water molecules are held on the surface of clay colloid).

2. **Organic colloids or humus colloids.** Organic colloids in the soil are chiefly due to presence of humus. The humus contains 8% each of lignin, protein, polyuronides (sugars and uronic acid complex). Organic colloids may be present in appreciable proportion in the soils. In sandy soil, it forms minor part of colloids. In peaty soil, organic colloids may be more than 50%. These colloids show adsorptive capacity many times greater than clay colloids. Organic colloids are negatively charged like clay colloids. Addition of organic colloids to the sandy soil increases temporarily its moisture and nutrient retaining capacity.

CATION EXCHANGE

Since the soil colloids (clay and organic colloids) have negative charges on them, they attract and hold positive ions (cations). When cations are added to the soils such as Ca^{++} in the form of lime, K^+ ions in the form of potassium fertilizer, and NH_4^+ in the form of ammonium fertilizer, the adsorption of cations will take place on the surface of colloid micelle and this will be accompanied by release of one or more ions held by colloid micelle. This is known as *cationic exchange.* For example, suppose that colloid micelle has one half of its capacity satisfied with Ca^{++} ions, one quarter with K^+ ion and remaining one quarter with H^+ ions. Now the colloid is treated with KCl solution. The K^+ ions will first replace Ca^{++} ions and then H^+ ions. The Ca^{++} and H^+ ions will combine with Cl^- ions of KCl and will form $CaCl_2$ and HCl respectively.

The cation exchange in the soil may take place between :

(1) cations present in the soil solutions and those already present on surface of soil colloids,

(2) cations released by plant roots and those present on the surface of soil colloids, and

(3) cations present on the surface of two clay crystals either two organic colloids or an organic colloid and a clay colloid.

Exchange reaction is very quick and reversible and the exchange of ions continues till equilibrium is attained. All cations are not adsorbed with equal ease. Some are easily adsorbed while others are replaced with difficulty. Divalent cations are more effective than the monovalent ones. Hydrogen is exception because it is held by colloids most tenaciously and it is most powerful replacer of cations. Replacing capacities of some cations are compared here : $H^+ > Ca^{++} > Mg^{++}, K^+ > Na^+$

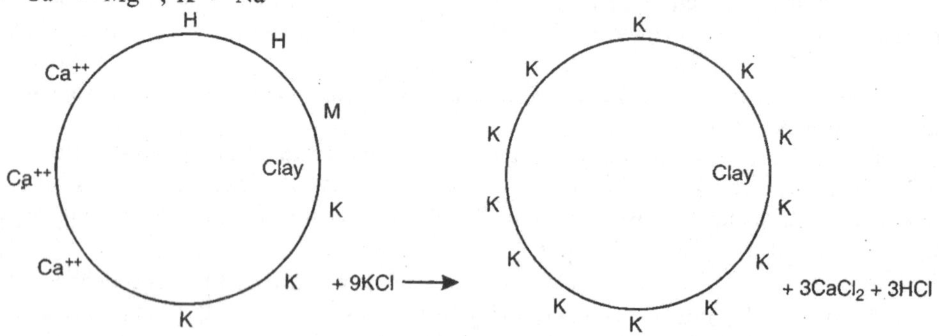

Fig. 23.5. Cation exchange

The number of cations adsorbed per unit weight of one hundred grams dry soil is called **cation exchange capacity**. More scientifically, cation exchange capacity of soil is the sum total of exchangeable cations adsorbed per unit weight of one hundred gms of dry soil.

Factors which are responsible for cation exchange or base exchange are as follows :

(1) relative concentration and number of cations present in the soil,

(2) replacing capacity of the ions, and

(3) number of charges on the ions.

Anion Exchange

Soils rich in organic colloids show anion exchange also. In this process, negatively charged ions held by colloids are replaced by OH^-, $H_2PO_4^-$, SO_4^{--}, and NO_3^- ions. The relative order of exchange is

$$OH^- > H_2PO_4^- > SO_4^{--} > NO_3^-$$

Among these anions, exchange of PO_4^{---} ions is most important. SO_4^{--} and NO_3^- are not retained in the soil for long period of time, hence not available for anionic exchange. Laterite soils have high adsorptive and fixation capacity for PO_4^{---} than black soils.

Anionic and cationic exchange reactions are important in agriculture. Several soil scientists have shown that the capacity of soil to exchange cations is the best index of soil fertility. The predominance of desirable ions in the exchange complex brings about good physical cations and favourably influences the microbial activities in the soil, such as ammonification, nitrification, etc. The knowledge of cation and anion exchange is of great help in reclaiming acidic and saline or alkaline soils.

4. Soil Reaction

Many chemical properties of soils centre round the soil reaction. As regards their nature, some soils are neutral, some are acidic and some basic. The acidity, alkalinity and neutrality of soils are described in terms of hydrogen ion concentrations or pH values. In order to understand soil reaction, the knowledge of pH is very necessary. It can be understood in the following ways :

Water dissociates into H^+ ion and OH^- ion. Hence the ionic constant of water can be represented as follows :

$$\text{Ionic constant of water} = \frac{[H^+][OH^-]}{[H_2O]}$$

But the rate of dissociation of water is so slow that ionization constant of water can be expressed simply as product of concentration of H^+ and OH^- ions, thus ionization constant of water $Kw=[H^+][OH^-]$. Concentration of H^+ and OH^- ions are expressed in terms of equivalents per litre. Only one molecule in ten million water molecules is in dissociated condition.

At neutrality, H^+ concentration is 0.0000001 or 10^{-7} gm of hydrogen per litre solution. The ionization constant of water is 10^{-14} at 25°C and thus in any aqueous system products of H^+ and OH^- ion concentration is 10^{-14}. Now, the above equation can be written as :

$$10^{-14} = [H^+][OH^-]$$

This can also be represented in the following way by dividing both sides in one and taking logarithms.

$$\text{Hence,} \quad \log\frac{1}{[Kw]} = \log\frac{1}{[OH^-]} + \log\frac{1}{[H^+]} = 14$$

The value of $\log\frac{1}{[H^+]}$ and $\log\frac{1}{[OH^-]}$ are generally called pH and pOH respectively. These pH and pOH are indices of the acidity and alkalinity respectively. Thus, pH can be defined as negative logarithms of the H^+ ion concentration. When the system is neutral, pH will be equal to pOH and when Kw is 10^{-14}, the value of pH and pOH at neutral point will be 7 for each. When pH value is less than 7, it is acidic. The pH value above 7 indicates alkalinity. If in a system, hydrogen ion concentration is 1/.000001 or 000001 gm/litre, the pH value will be 6. It is more clear from the following calculation :

$$pH = \log\frac{1}{[H^+]} = \log\frac{1}{0.000001} \text{ or } \log 10{,}00{,}000 = 6.$$

Thus at pH value of 6, the H^+ ion concentration is increased 10 times than the H^+ concentration at pH value of 7. Like this, at every lower point H^+ ion concentration will increase by a multiple of 10.

As the product of H^+ ion and OH^- ion concentrations at neutrality is always 10^{-14}, the pOH can be determined from pH value. Suppose, pH value of a solution is 6, the pOH value will be 8 ($10^{-6} + 10^{-8} = 10^{-14}$). pH scale is divided into 14 divisions or pH units from 1 to 14. Soil with pH value of 7 is neutral, that below pH 7 is acidic and that with pH value above 7 is alkaline.

From the pH value intensity of acidity in the soil is expressed but these values are not the measure of total acidity because they do not indicate the reserved acidity or relative acidity. For example, there are two soil samples which have similar pH values but they require different quantities of lime for neutralization. It means that the quantities of acids are different in the given weight of above two soils.

Buffer Action

It refers to the resistance to change in pH of a system. Such solutions as are reasonably permanent in pH value even after addition of some alkali or acid to them are called solution with reserved acidity or alkalinity or more often "buffer solutions". Suppose, a certain amount of acid is added to distilled water, the resulting solution will show acidic reaction and that will have a pH below 7, but if the same quantity of acid is added to a neutral soil suspension there would be very minor change in pH. This property of soil to resist a change in pH is called "buffer action".

Buffer solutions are usually formed of a mixture of salt of weak acid and acid itself in various proportions, as for example, a mixture of sodium acetate and acetic acid if added to water will result in a buffer solution. In the mixture solution we have mainly sodium ions and acetate ions. In water there will be some H^+ and OH^- ions. In buffer solution, acetate ions are in excess, owing to presence of well ionised sodium acetate. If H^+ ions are added to this solutions they will combine with acetate ions to give acetic acid of low ionisation power.

$$H^+ + \text{acetate}^- \rightleftharpoons \text{acetic acid}$$

$$K = \frac{[H^+][\text{acetate}^-]}{[\text{acetic acid}]}$$

Hence, there will be a little increase in pH. The addition of acid to buffer solution then makes little difference in the pH value.

Buffering in Soil

Soils should have good buffering capacity. Therefore, it is necessary to add considerably large amount of acids or alkalies in order to bring about any change in the original pH of soil. Buffering action is due to presence of large quantity of weak acids and their salts in the soil. Phosphates, carbonates, bicarbonates and other salts of weak inorganic acids and corresponding acids themselves are important buffering agents in the soils. Besides these, colloids associated with cations are important buffering agents. The buffering action of soil is directly governed by the amount and nature of clay and organic or humus colloids present in it.

Buffering action of soil is important in agriculture in the following respects :

(1) ***Stabilization of pH.*** This protects the higher plants and micro-organisms form direct adverse and injurious effects of sudden change in soil reaction.

(2) ***Amount of amendments necessary to correct the soil reaction.*** The greater is the buffering capacity of soil the smaller will be the amount of the amendments required such as lime, sulphur etc. to correct the acidity or alkalinity.

ACID SOILS

Soils with pH values below 7 are acid soils. In the regions of high rainfall, soils are acidic in their reaction because of the facts that soluble basic salts such as those of Ca, Mg, K, Na are leached away by drainage water and insoluble acidic residues composed chiefly of oxides and silicates of iron, silicon, aluminium are left which accumulate in pretty high amount. These salts are acidic in reaction, hence the soils are acidic. Besides that reason, there may be other causes also which produce acidity in the soil. Important factors which produce acidity in soil are as follows :

(1) Continuous removal of lime and other base elements by crops and accumulation of acids contained in the manures.

(2) Application of acid forming fertilizers in the soil.

(3) Microbial action.

(4) Formation of soil on the acidic rocks.

In India, acid soils occur in the high rainfall areas covering about 25 million hectares of land with a pH below 5.5 and 23 million hectares of land with a pH between 5.6 and 6.5. These estimates are calculated by Bhaumik, H.D. and Donahue, Roy, L., 1964 (Reference: Soil acidity and the use of lime in India. Farm Information unit, Directorate of Extension, Ministry of Food and Agriculture, Government of India). In India, acid soils occur in Assam, Meghalaya, Arunachal Pradesh, Mizoram, Nagaland, NEFA, Manipur, Tripura, West Bengal, Bihar, Uttar Pradesh, Himachal Pradesh, Jammu and Kashmir, M.P., Maharashtra, Kerala, Karnataka, Tamil Nadu and Andhra Pradesh. Punjab, Haryana, Rajasthan and Gujarat are the only states in India where acid soils do not occur.

Very few plants can grow well in strong acid soils. Soil acidity below pH value 5.5 is generally injurious to plants. Plant roots are badly affected if the pH value exceed limits of tolerance for particular crops. High degree of soil acidity (pH 5 to 6.5) decreases the availability of plant nutrients particularly phosphorus, calcium, magnesium, molybdenum, iron, manganese, potassium, sulphur, nitrogen, boron, copper and zinc. It also affects adversely the important microbiological processes, such as nitrogen fixation by *Azotobacter, Clostridium* and nodule inhabiting bacteria (*Rhizobia*) of leguminous plants.

ORIGIN OF ACID SOILS

Several factors are responsible for the origin of acid soils. Generally climate, hydrologic cycle, vegetation, parent rocks and human interference play important roles in the origin and development of acid soils. Acid soils occur generally in humid regions where the rainfall is regular and very heavy. Dry regions are devoid of acid soils.

Climate. In humid regions where evaporation is less than precipitation, chances for the development of acid soils are good. For the development of acid soils it is also necessary that water percolating down the soil profiles must reach the water table. In India, it is believed that the regions with acid soils must receive more than 750 mm annual rainfall. The regions with annual rainfall 1350 mm may have acid soils with pH value 5.0 or even less than that. In temperate regions the acid soils can develop even if the rainfall is scanty. In hilly regions where the loss of water through evaporation is very slow due to very low temperature the conditions for the development of acid soils are very favourable, although the rainfall is scanty there.

Vegetation cover. In temperate regions or hilly areas covered with conifers the acid soils can develop easily. According to Bloomfield (1953), the foliage leaves of conifers lack alkali elements and their mineralization process is very slow. When the leaf-litter on the ground is degraded, organic acids are released which gradually make the soils acidic. Plants found in the coastal regions and marshy places after death and decay produce acids which render the soils acidic.

Parental rocks. Though the development of acid soils is possible on all types of rocks and parental rock materials in presence of favourable climate and vegetation, yet the development of acid soils on alkaline rocks take longer time as compared to the acid soils developing on the acidic parental rocks. Acid soils develop more quickly from parental rock materials with simple composition than from the parental rock materials of complex composition. It is so on account of presence of less adsorbed cations, poor buffering capacity and quick percolation of water through them.

Topography. Slopy places with good drainage conditions are supposed to be good for the development of acid soils. On hill slopes, the development of acid soils is easy. Acid soils do not develop generally in river basins. The plains with good drainage may also develop acid soils in due course of time.

Human interference. Continuous efforts by man for developing permanently submerged areas into cultivable land, or for improving drainage in submerged or saline lands, regular use of nitrogen fertilizers like ammonium sulphate which cause acidity in the soils are responsible for decrease of soil pH.

In urban areas, industrial wastes containing sulphur or sulphur dioxide also contribute much in the development of acid soils.

Classification of Acid Soils

According to the intensity of acidity, the acid soils are of the following five types :

(1) Slight acidic (pH range 6.6 to 6.1)

(2) Medium acidic (pH 6.0 to 5.6)

(3) Strong acidic (pH 5.5 to 5.1)

(4) Very strong acidic (pH 5.0 to 4.6)

(5) Extremely strong acidic (pH 4.5 or lower)

Acid soils occurring in different climatic regions are classified as follows :

(*i*) Acid soils of temperate climate including podzol, brown podzol, grey brown podzol, brown forest soils, and grey forest soils.

(*ii*) Acid soil of tropical and subtropical climates including yellow podzolic soil, lateritic soil and latosols.

(*iii*) Acid soils of other great soil groups including wet soils (hydromorphic soils), washed peaty soils, mucky, cat-clay (acid sulphate) soils. Cat-clay or acid sulphate soils with pH 3.5 or lower are the soils that abound in organic material as well as H_2SO_4.

According to soil classification system (1970) developed by U.S. Soil Scientists, soils of the world have been classified into 10 soil orders. Among these Aridisols, Vertisols and Mollisols are nonacid soils and the remaining 7 orders contain acid soils. But acid soils occur mainly in three orders : Oxisols, Alfisols and Histosols.

In modern soil classification soil orders have been divided into suborders. Suborders *Humox, Humod, Aqualf* and *Udalf* include acid soils.

In modern system of soil classification, acid sulphate soils have been assigned separate position and these soils have been placed in a group called *sulpha-aquepts.* This group includes soils in which the top horizon contains sulphuric horizon at some level or the other in top 25 cm thick layer. This is mineral or organic sublayer with yellow colouration due to xarocite. Acid sulphate soils of tropics possibly belong to *Typic Sulpha-aquepts* and those of temperate regions are mainly *Typic Sulpha-aquepts* and *Hapla aquepts.* Mineral and organic soils rich in sulphur which remain regularly submerged are referred to as *sulphidic soils.* Since such soils develop under the influence of saline water which is rich in sulphur, they are placed in *Halic* subgroup.

Mandal (1974) has classified acid soils of India into the following seven groups :

(1) Laterite soils

(2) Lateritic and laterite red soils

(3) Mixed yellow red soils

(4) Ferruginous red soils

(5) Podzolic soils

(6) Terai soils

(7) Peaty soils.

In this classification, acid sulphate soils and degraded alkali soils have not been assigned proper places. Nevertheless, it would be appropriate if they are classed with acid soils as the pH levels of such soils indicate that they are acidic in nature.

Recently Mishra, S.G. (1976) has suggested that the acid soils should be classified into the following two categories on the basis of organic contents in them :

(1) Acid mineral soils (organic matter less than 20%)

(2) Acid organic soils (organic matter 20% or more)

1. **Acid mineral soils.** Such soils are further classified into the following three subgroups :

(*i*) **Acid mineral soils rich in organic matter in upper layer.** Such soils are commonly found in temperate and sub-temperate regions and develop by podzolisation process. Since these regions are covered with thick forest vegetation, the surfaces of such soils are covered with decomposing organic matter. The degradation of organic matter results in the production of various organic acids, such as citric acid, acetic acid, oxalic acid and so on. Microbial degradation of organic matter also produces CO_2 which combines with water to form carbonic acid (H_2CO_3). These acids along with rain water percolate down through soil profiles. Along with these acids and rain water sesquioxides are also leached out from the upper horizons and become deposited in lower horizons. This depletion of sesquioxides from the top layer and their accumulation in lower sublayers is referred to as *Podzolisation* and such soils are called Podzols.

(*ii*) **Acid mineral soils devoid of organic layer.** Acid mineral soils found in plains usually do not possess organic layer. Such soils originate mainly as a result of *laterisation* and partly due to podzolisation process. Laterite soils, Red soils, and hydromorphic acid soils found in India belong to this category. Such soils originate in the following ways :

(*a*) CO_2 of atmosphere as well as of soil dissolves in water to form carbonic acid (H_2CO_3) which, when percolates down the soil profiles, degrades carbonates and primary minerals present in the soils and makes the soil acidic.

(*b*) In tropics, at high temperatures maximum degradation of silica takes place and in top soil layer the quantity of sesquioxides increases. This process is referred to as *laterisation.* Laterite and Red loam soils found in India have probably originated through this process.

At low temperatures Yellow red Podzolic and Grey Podzolic soils originate which are less acidic.

(*iii*) **Degraded alkali soils.** The top layer of some alkali soils shows a pH value less than 7 due to desalinisation or dealkalization. Such soils are referred to as degraded alkali soils. In this process. The alkali salts are washed by irrigation or rain water and exchangeable Na^+ ions of soils are displaced by H^+ ions of water.

2. **Acid organic soils.** According to the amount of organic matter, acid organic soils can be classified into the following two types :

(*i*) Peaty Soils

(*ii*) Mucky Soils

(*i*) *Peaty soils.* Peaty soil are characterised by presence of poorly degraded organic matter. In India, peaty soils occur in Kashmir, Himachal Pradesh, Assam and other Hill states.

(*ii*) *Mucky soils.* Such soils contain highly degraded organic matter. They have relatively higher pH values than the peaty soils. Thus they are less acidic. Mucky soils are also found in Kashmir, Himachal Pradesh, Assam, and some other states.

EFFECTS OF SOIL ACIDITY ON PLANTS

Soil acidity affects the plants both directly and indirectly. These effects are briefly mentioned below :

1. **Direct influences.** These are as follows :

(*a*) Toxic effects of low H^+ ion concentrations on root tissues.

(*b*) Influence of soil acidity on the permeability of the plasma membrane for cations.

(*c*) Disturbance in the balance between basic and acid constituents through roots.

(*d*) Affects enzymatic processes since enzymes are particularly sensitive to pH changes. Different crop plants have their specific optimum pH requirement. Rice, oat and linseed can endure a fairly acidic reaction (pH = 5.0) while barley, sugar-beet, lucern etc. can tolerate a fairly alkaline reaction (pH = 8.0)

2. **Indirect effects.** These are listed below :

(*a*) Availability of various nutrients, *e.g.*, phosphorous, copper, and zinc.

(*b*) High solubility and availability of elements like aluminium, manganese and iron in toxic amount due to high acidity in the soil.

(*c*) Deficiency of some nutrients such as calcium and potassium due to soil acidity.

(*d*) Prevalence of plant diseases.

(*e*) Beneficial activities of soil microbes are adversely affected.

RECLAMATION OF ACID SOILS OR CORRECTION OF SOIL ACIDITY

Acidity of soil is due to predominance of H^+ ions over OH^- ions, the bulk of H^+ ions being held in close association with clay-organic colloid complex. Strong acid soils are not much productive. The soils which are less productive owing to high degree of acidity can be made more productive by liming (application lime).

When lime is added to moist soil, the soil solution becomes charged with cations and the exchangeable hydrogen and aluminium ions on clay-organic colloid complex as well as the H^+ ions in soil solution are displaced by calcium ions. Hydrogen combines with OH^- to form neutral water or with CO_3^- or HCO_3^- to form unstable H_2CO_3 which readily dissociates to form CO_2 and water.

Acidity of soil can also be corrected by adding exchangeable Mg^{++} to exchange complex. But addition of Ca^{++} or Mg^{++} or both to the soil will not necessarily solve the problem of soil acidity. The important points to be considered in liming are : (*i*) the salts of these elements which are going to supply these ions (Ca^{++} or Mg^{++}) and (*ii*) the overall reactions of salts in the soils. Salts of strong acids as gypsum ($CaSO_4$) or calcium chloride ($CaCl_2$) can be applied to supply calcium ions to the soils but it is worth considering what will be the effects of these salts on soil acidity. The application of these salts will indeed increase the acidity in the soil, instead of decreasing it. Therefore, it is suggested that calcium salts of strong acids must not be applied for correcting the acidity of soils.

Liming materials. More than 90 per cent of the lime used in agriculture for reclamation of acid soils is generally in the form of calcium carbonate, some in calcium and magnesium carbonates, and much smaller quantity in the form of calcium oxide or calcium hydroxide. To a chemist lime is calcium oxide but to a farmer, agronomist or soil scientist lime usually means calcium carbonate or calcium carbonate equivalents.

The common liming materials used for reclamation of acid soils are as follows :

(1) Calcic limestone ($CaCO_3$) which is ground limestone.

(2) Dolomite ($CaCO_3.MgCO_3$).

(3) Quick lime (CaO) which is burnt limestone.

(4) Hydrated (slaked) lime [$Ca(OH)_2$].

(5) Coral shell lime.

(6) Marl or chalk ($CaCO_3$).

(7) *Slags.* Obtained as by-products from iron and steel plants, slags are used in agriculture for reclaiming acid soils. The slags are of three types : (*i*) blast furnace slags, (*ii*) basic slag and (*iii*) electric furnace slag. These slags are rich in phosphorus and mixture of CaO and $Ca(OH)_2$. Besides, Ca, Mg, Al, silicates are also present in them.

(8) *Press-mud.* It is obtained from carbonation plants of sugar mills. Press mud and some other matters containing calcium are used to decrease acidity in the soils.

(9) *Miscellaneous sources of lime,* such as, wood ash, ground oyster shells, by-product lime resulting from paper mills, tanneries, water softening plants, and by product $CaCO_3$ form fertilizer factories using gypsum process (such as Sindri Fertilizer Factory, Bihar, India).

The rate of lime application should always be determined after soil testing. When excessive amount of lime is applied to sandy soils low in humus, injury to plants may be caused which may be attributed to one or more of reasons listed below :

(1) Boron deficiency.

(2) Iron, manganese and zinc deficiency.

(3) Reduced availability of phosphorus to a critically low level.

(4) Reduced potassium uptake.

Such injurious effects may be reduced by application of large amount of compost manure, green manure crops, phosphorus fertilizers, boron or a mixture of minor elements.

While applying liming agents of acid soils, the following points must be taken into consideration.

(*i*) The liming agents should be used in highly powdered state. The smaller the particles of liming agents the greater will be their effectiveness in correcting the soil acidity.

(*ii*) Liming materials should be in direct contact with clay organic exchange complex so that H^+ ions of exchange complex may be easily displaced by Ca^{++} ions.

(*iii*) Liming agents should be applied to soils at least one month before sowing the crops or they should be applied thorougly mixed with soils just after harvesting the crops.

Important Roles of Liming Agents in Soils

(1) Liming agents reduce soil acidity and stabilize pH of the soils.

$$\text{Acid-clay} + Ca(OH)_2 \rightarrow \text{Ca-clay} + H_2O$$

(2) Lime makes phosphorus easily available. This is true mainly because in acid soils phosphorus is fixed by soluble iron and aluminium. Liming reduces the solubility of iron and aluminium and therefore less phosphorus is held in these insoluble and unavailable forms.

(3) Lime makes potassium more efficient in plant nutrition. When K is in sufficient amount in soil, plants absorb more potassium than is actually needed but at the same time when lime is available in plenty, plants take up more calcium and less potassium. Economically liming is more desirable because plants absorb more cheap Ca than expensive potassium.

(4) Lime enhances the decomposition of organic matter, thereby increases the availability of nitrogen and other nutrients locked up in complex forms to plants.

(5) Lime promotes beneficial activities of soil bacteria.

(6) Liming programme extended over a period of years improves the physical conditions of the soil by causing granulation of soil particles, decreasing its bulk density, and increasing its infiltration rate.

(7) Ca and Mg found in liming agents, particularly in Dolomite act as essential elements in the nutrition of plants.

(8) Lime converts toxic elements such as aluminium, Mn, Fe of the soil in insoluble and harmless compounds.

Acid Tolerance in Crops

Some plants are adversely affected and they suffer injuries when grown in acid soils. Familiar crops which can endure fairly acidic soil conditions are oat, rice and linseed and those which are not adapted to acid soils are wheat, barley, cabbage, *sorghum* (Jowar), tobacco, lettuce, spinach, onion, eggplant or brinjal.

SALINE AND ALKALI SOILS

Saline soils represent a group of soils in which percentages of soluble salts, usually chlorides and sulphates of the alkali bases are very high. The pH of saline soils are always high. In India saline soils occur in many provinces, as U.P., West Bengal, Punjab, Bihar, Orissa, Maharashtra, Tamil Nadu, M.P., Andhra Pradesh, Gujarat, Delhi and Rajasthan covering an area of about 7 million hectares.

There are 4 major tracts in India where salinity problem is acute. These are (*a*) the arid tract of Rajasthan and Gujarat, (*b*) semi-arid alluvial tracts of Punjab, Haryana and Uttar Pradesh, (*c*) the arid and semi-arid tracts of Southern States, and (*d*) the coastal alluvium.

In U.P., the saline (usar) soils are distributed in Kanpur, Lucknow, Hardoi, Unnao, Allahabad, Rai Bareli, Azamgarh and many other districts covering about 1.29 million hectares. In Punjab alone saline soil covers about 2 million hectare area. In U.P. and Punjab the saline soils are gradually increasing in area. These lands are known by a variety of names in local agricultural parlance. By far the most common of them is *usar* derived from the Sanskrit word *Ushtra* meaning sterile or barren. Other terms like *reh,* char, *lone, thur* or *shora* are also popular. The word *alkali* is of Arabic origin meaning ash-like and is used to designate hard and intractable soils generally known by the names *rakkar, kallar, bara* and *bari.*

The salty soils are of three types :

(*i*) **Saline or solonchak or white alkali soils.** In these, salinity is caused by soluble salts other than alkali salts. They have high soluble salts and low exchangeable sodium.

(*ii*) **Alkali or sodic or solonetz or black alkali soils.** These are formed by accumulation of alkalies, such as Na, K etc. in excess. Such soils have low salt content but high exchangeable sodium.

(*iii*) **Saline-alkali soils.** In these, alkali and other soluble salts have combined effects. They are also called *saline sodic* as they have high salt content and high exchangeable sodium. The United States Salinity Laboratory recently designated these soils scientifically on the basis of soil analysis following the ideas of Sigmond and Gedroiz (1954) (Table 23.4).

Table 23.4. Three classes of salty soil

Soil Property	Saline soil	Alkali soil	Saline-alkali soil
1. Conductivity of saturated extract (ECe)	more than 4 m mhos/cm at 25°C	less than 4 m mhos/cm at 25°C	greater than 4 m mhos/cm at 25°C
2. Exchangeable sodium percentage (ESP)	less than 15	greater than 15	greater than 15
3. pH range	usually below 8.5	between 8.5 and 10	above 8.6

Soils in which salinity is mainly due to accumulation of alkali salts are called alkali soils or usar soils. High alkalinity in the soil adversely affects the plant growth, thereby reduces the crop yield. Such sterile or unproductive soils are called barren soils. The main salts present in the alkali soils are Na_2SO_4, K_2SO_4, NaCl, and KCl.

Types of alkali soils. These soils are of two types :

(*i*) Black alkali soils—In these soils, Na_2CO_3 is found in excess.

(*ii*) White alkali soils—In this group, NaCl is present in excess.

Russians call such soils as *solonchack.* Bertholet suggested that Na_2CO_3 was formed in black alkali soil by interaction of NaCl and $CaCO_3$.

$$2NaCl + CaCO_3 = Na_2CO_3 + CaCl_2$$

Factors Which Make the Soils Alkaline

These are :

1. Poor drainage in arid region,
2. Rapid evaporation of alkaline soil solution, and
3. Excess uptake of alkaline salts and little percolation.

In arid and semi-arid regions, the rainfall is too low to leach or remove the saline matter from the top soils. Besides this, water along with dissolved alkali salts moves upward by capillary action which on reaching to the soil surface evaporates and the salts accumulate in the form of a hard layer or pan in the subsoil. This hard layer is responsible for impermeability of such soils.

Miller is of the opinion that many plants absorb excess acidic ions, *e.g.,* $NO^-{}_3$, than the basic ions. This excessive removal of acidic ions results in the accumulation of basic ions which make the soil alkaline.

According to a chemical hypothesis, alkali soils may result in the following steps :

(*a*) reaction between NaCl or KCl and soil (S) :

$$NaCl + S(Soil) \rightarrow Na\ (S) + Cl^- \text{ ion}$$

(*b*) then the soluble products are leached away from the soil surface by drainage water, and

(*c*) finally, reaction between insoluble Na (S) complex and carbonates:

$$2\ Na\ (S) + CaCO_3 \rightarrow \underset{\text{(alkaline)}}{Ca\ (S)} + \underset{\text{(alkaline)}}{Na_2CO_3}$$

Effects of alkali salts on vegetation. The alkali salts show the following effects on plants:

(*i*) Due to excessive accumulation of salts, concentration of soil solution, becomes high. This decreases absorption of nutrients by plants and causes plasmolysis of cell cytoplasm in the plants which may be fatal sometimes. All these effects are responsible for stunted growth of plants.

(*ii*) If sodium is absorbed by the plants in excess, it shows toxic effects. Chloride salt of alkaline elements cause the death of trees. $BaCO_3$ and BaCl are toxic to all plants.

(*iii*) Presence of excess salts in the soil retards the germination of seeds and growth of seedlings. Plants die before bearing fruits.

(*iv*) Alkali salts in the soil also affect the plant growth by reducing the size of leaves. In alkali soil, plant roots remain superficial, bark of stem turns brown or black, green tissues are less developed.

Alkali tolerance in plant. Some plants are resistant to alkali salts. Barley, wheat, oats *sorghum*, sugar-beet, berseem are best suited to grow in alkaline soil. Cotton and grapes are also alkali tolerants. Uppal *et al* (1961) prepared a list of crops that can be grown at various stages of reclamation. They can be categorized as high, medium and low salt tolerant crops. *High salt tolerant crops* are *Dhaincha,* paddy, sugarcane in *kharif* and oat, berseem, lucerne, sanji (*Trigonella* spp), and barley in rabi. *Medium salt tolerant crops* to be tried during second stage of reclamation are castor, cotton, jowar, bajra, and maize for *kharif* and mustard and wheat for *rabi.*

Low salt tolerant crops are sesamum, moong, urd, arhar, and sannhemp in *kharif* and gram, peas, linseed during *rabi.*

Uppal *et al,* also listed babul, dhak, jhand, khair, chokra, neem, Lasora, Sisham, siris, bahera and reonjha as trees that can be planted on saline alkali soils. The alkali tolerance of plants depends upon :

(*i*) Physiological constitution of cell cytoplasm of the plant.

(*ii*) Length of roots. Shallow roots are more affected by alkalinity than the deeper roots.

(*iii*) Alkali salts reduce and correct the soil acidity and improve the physical conditions of the soils.

(*iv*) Calcium salts provide calcium to the plants.

(*v*) Many alkali salts change toxic elements, such as aluminium and Mn into their harmless compounds.

RECLAMATION OF ALKALI, SALINE AND SALINE—ALKALI SOILS

The excessive accumulation of alkali salts in the soils is injurious for plants growth. It is necessary, therefore, to reduce the percentage of salts to optimum or normal level so that plants may grow luxuriantly in such soils. There are several methods of reclamation which can be grouped as follows :

(A) Chemical method in which some chemicals are added to the soil in order to bring the alkalinity to desired level.

(B) Mechanical practices such as improving drainage and leaching, mechanical shattering of clay pans, and scrapping.

(C) Cultural method (growing salt tolerant plants).

Since fundamental causes in various groups of salty soils are different, their reclaiming techniques are different. Hence, these are discussed separately.

1. Reclamation of Alkali Soils

Alkali soils are best reclaimed by the following methods :

(A) Chemical method. (1) By cationic exchange (replacement of alkali from soil colloids by calcium ions). Application of calcium sulphate (gypsum) in the soil reduces alkalinity to a great extent and makes the soil fertile. The reaction proceeds in the following way :

$$2\text{Na-clay} + CaSO_4 \rightarrow Na_2SO_4 + \text{Calcium-clay}$$

$$Na_2CO_3 + CaSO_4 \rightarrow CaCO_3 + Na_2SO_4$$

Good drainage leaches away Na_2SO_4.

(2) Alkali salt percentage can also be reduced in the soil by the use of acid forming chemical amendments such as sulphur, ferrous sulphate and limestone. Sulphur, when applied to the soil, oxidises and forms sulphuric acid which converts carbonates of sodium and potassium to Na_2SO_4 and K_2SO_4 respectively that may be removed from top soil by drainage water.

The amount of gypsum and sulphur required to reclaim the alkali soils will be different depending upon the degree of alkalinity, drainage and buffering capacity of soils.

The types of reaction which occur when an amendment is applied to an alkali soil are given below:

(1) **With Sulphur**

(*i*) $2S + 3O_2 = 2SO_3$

(*ii*) $SO_3 + H_2O = H_2SO_4$

In the next step, if soil is calcareous—

(*iii*) $H_2SO_4 + CaCO_3 = CaSO_4 + CO_2 + H_2O$

(*iv*) $2Na\text{–}Clay + CaSO_4 = Ca\text{–}Clay + Na_2SO_4$

But if the soil is non-calcareous—

(*iii*) $2Na\text{–}Clay + H_2SO_4 = 2H\text{–}clay + Na_2SO_4$

(2) **With lime-sulphur**

(*i*) $CaS_5 + 8O_2 + 4H_2O = CaSO_4 + 4H_2SO_4$

Now if the soil is calcareous—

(*ii*) $H_2SO_4 + CaCO_3 = CaSO_4 + CO_2 + H_2O$

(*iii*) $2Na\text{–}Clay + CaSO_4 = Ca\text{–}Clay + Na_2SO_4$

But if the soil is non-calcareous—

(*ii*) $10Na\text{–}Clay + 4H_2SO_4 + CaSO_4 = 8H\text{–}Clay + Ca\text{–}Clay + 5\ Na_2SO_4$

(3) **With ferrous sulphate**

(*i*) $FeSO_4 + H_2O = H_2SO_4 + FeO$

Now if the soil is calcareous—

(*ii*) $H_2SO_4 + CaCO_3 = CaSO_4 + CO_2 + H_2O$

(*iii*) $2Na\text{–}Clay + CaSO_4 = Ca\text{–}Clay + Na_2SO_4$

But if the soil is non-calcareous—

(*ii*) $2Na\text{–}Clay + H_2SO_4 = 2H\text{–}Clay + Na_2SO_4$

(4) **With limestone on non-calcareous soils**

(*i*) $Na\text{–}Clay + H_2O = H\text{–}Clay + NaOH$

(*ii*) $2H\text{–}Clay + 2NaOH + CaCO_3 = Ca\text{–}Clay + Na_2CO_3 + 2H_2O$

(5) **With any H-Clay**

(*i*) $2H\text{–}Clay + CaCO_3 = Ca\text{–}Clay + CO_2 + H_2O$

The use of pyrite (FeS_2) as an amendment is a recent development in the chemical amelioration and reclamation of alkali soils. In presence of moisture and air, pyrite is converted into sulphuric acid which then replaces exchangeable sodium by hydrogen or calcium released from insoluble calcium present in the soil. In addition it is said to correct iron deficiency and lime induced iron chlorosis in alkali soils. It is important to mention that the formation of H_2SO_4 in the soil by the application of pyrite may take place through chemical and microbiological actions. Pyrite is oxidised according to the following equation suggested by Bloomfield (1973).

$FeS_2 + 2Fe^{+3} = 3Fe^{+2} + 2S$ (Chemical)

Sulphur thus formed could be the substrate for *thioxidants* which convert it into H_2SO_4.

$S + 3(O) + H_2O = H_2SO_4$

Temple and Kochler (1954) explained the action of *ferroxidans* on the formation of H_2SO_4 as follows :

$FeS_2 + H_2O + 7O = FeSO_4 + H_2SO_4$

$2FeSO_4 + O + H_2SO_4 = Fe_2(SO_4)_3 + H_2O$

$FeSO_4$ formed in the above reaction may be converted into H_2SO_4 by hydrolysis.

$FeSO_4 + H_2O \rightarrow H_2SO_4 + FeO$

In brief, the pyrite is oxidized in soils to ferrous sulphate and sulphuric acid as depicted in the following equation:

$2FeS_2 + 2H_2O + 7O_2 = 2FeSO_4 + 2H_2SO_4$

Both sulphuric acid and ferrous sulphate help in reclamation of calcareous as well as non-calcareous salt affected soils by lowering the pH and solubilising free calcium from calcium carbonate present. The reactions are given below:

In salt affected calcareous soils:

Ia. $CaCO_3 + H_2SO_4 = CaSO_4 + CO_2 + H_2O$

Ib. $2Na\text{–}Clay + CaSO_4 = Ca\text{–}Clay + Na_2SO_4$

II. $FeSO_4 + H_2O \rightarrow FeO + H_2SO_4$

H_2SO_4 formed in reaction II reacts as per equations Ia and Ib

III. H_2SO_4 also neutralizes $NaHCO_3$ and Na_2CO_3 present in these soils.

$2NaHCO_3 + H_2SO_4 \rightarrow Na_2SO_4 + 2H_2O + 2CO_2$

$Na_2CO_3 + H_2SO_4 \rightarrow Na_2SO_4 + H_2O + CO_2$

But if the soil is non-calcareous :

(I) $2Na\text{–}Clay + H_2SO_4 = 2H\text{–}Clay + Na_2SO_4$

(II) $FeSO_4 + H_2O = FeO + H_2SO_4$

H_2SO_4 formed in reaction II acts in similar manner as in reaction I.

(3) *Dhar's method.* In India, Dr. Neel Ratan Dhar (1935) succeeded in reducing the alkalinity and salinity of the soil by the use of molasses (Hindi—raab) and press-mud. For one acre land he recommended the mixture of the following substances :

(*i*) 2 tons of molasses, (*ii*) 1-2 tons of press-mud (a waste product of sugar industry), and (*iii*) 50-100 pounds of P_2O_5 in the form of basic slag.

The molasses is fermented by soil microbes and as a result of fermentation organic acids are produced which lower the alkalinity and increase the availability of phosphates. The press-mud contains Ca which forms calcium salts that reduce the content of exchangeable sodium. Phosphate helps in the microbial fixation of nitrogen into nitrogenous compounds in the soil.

(B) Mechanical methods. The alkali salts are removed by

(1) scraper or by rapidly moving streams of water,

(2) deep ploughing of the land which reduces the alkalinity and makes the soil more permeable.

(3) application of green manures of Dhaincha, guar, jantar (*Sesbania aculeata*) has been found most successful in reclamation of alkali and saline soils.

(4) spreading of straw and dried grasses and leaves on the alkaline soil.

(C) Cultural method. Growing of alkali tolerant crops and plants, such as sugar-beet, rice, patsann (*Hibiscus cannabinus*), wild indigo and babul in such soils successfully reduces alkalinity. Rice is commonly the first crop grown on salty lands to be reclaimed. In Punjab, the usual practice of reclamation of salty lands involves growing of paddy after first initial leaching, followed by berseem or senji which has higher water requirement than *Dhaincha* as green manure, which is followed by sugarcane and then wheat or cotton. Introduction of leguminous crops helps in building up of nitrogen supply and opens the soils. *Dhaincha*–paddy–berseem rotation has been found to be the best cropping pattern on mild type of alkali soils in Punjab region. In U.P. also, paddy or *dhaincha*-paddy are the usual crops taken during first stage of reclamation of salty soils. This is followed by berseem or barley in winter. Pulse crops like gram or peas show poor performance.

II. Reclamation of saline soil

Saline soil can be reclaimed by the following methods :

(1) By lowering the water table 5-6 feet below the surface. In slopy area, it can be done by making network of 5-6 feet deep trenches at right angles to the slopes. In course of 2 or 3 successive leaching, harmful salts are removed. A deep ploughing is also helpful in reclamation of saline soil. This also makes the soil loose and thus facilitates the downward movement of salty water in the soil.

(2) Salt tolerant crops, *e.g.*, rice, sugar cane, barley and castor gradually remove salts from the soil.

(3) In case of saline soils which do not contain calcium salts, the addition of $CaSO_4$ (gypsum) is beneficial. Supply of calcium in the soil can indirectly be maintained by addition of organic matters which on decomposition produce CO_2. The CO_2 gas, so produced, combines with insoluble calcium carbonate in moist condition to form soluble calcium bicarbonate. This also reduces alkalinity.

(4) Application of green manure, organic manures, organic residues, acids or acid formers is yet another good way to reduce salinity.

III. Reclamation of saline-alkaline soil

Here the problem of reclamation is two-fold because of

(*a*) heavy accumulation of different types of salts,

(*b*) poor percolation due to the presence of hard clay pan and highly dispersed sodium clay.

Such soil can be reclaimed by :

(*i*) *Mechanical shattering of clay pans.* This helps in downward movement of water.

(*ii*) *Application of gypsum in the soil.* This is followed by flushing with plenty of water.

(*iii*) *Green manuring* with Dhaincha (*Sesbenia aculeata*).

(*iv*) *Growing of salt tolerant plants, e.g.*, paddy in kharif and oat and barley in rabi seasons are recommended for such soils.

Schoonover (1959) worked on the soils of India and enlisted the following technical requirements for reclamation of saline and alkaline soils:

(1) necessity of good drainage.

(2) Availability of sufficient water to wash the excess salts from the top soils.

(3) Good soil management including land levelling, good bunding for irrigation and recent and advanced agronomic practices.

(4) Protection of soil from erosions.

(5) Good quality of irrigation water.

QUESTIONS

1. Discuss in brief chemical properties of soil.
2. Give a brief account of physical properties of soil.
3. Write short notes on : (*i*) soil pH, (*ii*) ionic exchanges in the soil, (*iii*) Buffer action.
4. Discuss the origin, classification and effects of acids soils and mention the methods of reclamation of acid soils.
5. What are the different types of salty soils and how do they originate? Discuss the various methods of reclamation of alkali soils, saline soils and saline-alkaline soils.
6. Write brief description about the following :

   (*a*) Usar (*b*) Cation exchange (*c*) Tyndal phenomenon (*d*) Brownian Movement (*e*) Flocculation.

24

Principles of Soil Classification, Soil Surveys and Chief Soil Types of India

CLASSIFICATION OF SOILS

Soils are of various types as regard their physical, chemical and biological properties. Soil samples collected from distant places and from different climatic regions may show considerable resemblances and differences in their physical, chemical and biological properties. In order to make the study of soils convenient, soils are classified into well defined categories. The classification of soils, in brief, aims mainly at the orderly and systematic arrangement of soils on the basis of resemblances and differences in their physical, chemical and biological properties. As in biological systems plants and animals are classified into classes, orders, suborders, families, etc., the soils are also classified into well defined classificatory units as orders, suborders, great soil groups and subgroups. Lower classificatory units in the soil classification are families, series, types and phases.

Soil Orders

These are the largest categories. They are described into the following groups:

(*a*) **Zonal soil.** This is characterised by well differentiated profile. Soil develops on well drained parental rock where environmental factors are little active.

(*b*) **Intrazonal soil.** In this group, soils show well differentiated profiles which may be influenced by climatic and biological factors.

(*c*) **Azonal soil.** It is characterised by the absence of profile. Soil may be in the formation stage and it may be either too immature or may be subjected to extreme environmental influences. Such soils are formed on rocky slopes or on steeps in the form of alluvial deposits, coarse sand deposits, etc.

Suborders of Soils

These are recognised on the basis of factors that influence soil formation and the distinguishing features of soil groups itself, as for example, cold and arid soils are direct products of climatic conditions. Similarly, grassland, transitional land and forest soils are products of combined effects of climates divided into great soil groups.

Soil Series

It consists of a group of soils with similar profile differentiation and similar horizon characteristics, such as colour, thickness, structure, reaction and approximate organic contents. Textures of surface soils are not taken into consideration in the recognition of soil series. Soil series are usually named after town, river or other geographic places near which they are first identified, for example, Miami series which was first recognized in Miami river shades in Ohio (U.S.A.).

Soil Types

Soil types are subdivisions of soil series. They are recognised on the basis of soil texture. The names of soil types are given after series names and the terms describing texture of the surface soils, for example, Indogangetic alluvial soil in which Indogangetic plains are the place of occurrence and alluvium is the texture of surface soil.

Soil Phase

Soil types are sometimes further divided into phases on the basis of slopes, stoniness, degree of erosion and surface characteristics.

In classification of soil, the following descending sequence is followed:

Order
Suborder
Great soil group
Series
Soil type
Soil phase

SOME SOIL CLASSIFICATIONS

Middelburg, H.A., proposed a tentative scheme of classification for tropical and subtropical soils (Reference : Transactions; Formal International Congress of Soil Science, Part 4 : 139-147) The scheme is presented in Table 24.1.

Table 24.1. Soil classification

Order	Suborders	Great soil groups
1	2	3
Zonal	(1) Humid regions	(1) Podzolic yellow soils with black humic top soils
		(2) Red yellow podzolic soils
		(3) Non-laterised red soils
		(4) Laterised red soils
		(5) Degraded grey and black clay soils
		(6) Illuvial soils
	(2) Arid region (Desert soils)	(1) Red desert soil
		(2) Dark desert soil
Intrazonal	(1) Calcimorphic Soils (Calcareous soils)	(1) Red soil (developed over lime stone)
		(2) Black soil (developed over lime stone)
		(3) Red soil (developed over marls)
		(4) Black soil (developed over marls)

	(2) Halomorphic	(1) Saline soil
	(3) Hydromorphic soil	(1) Ground water lateritic soil
		(2) Planosols
		(3) Organic soil
Azonal	Lithosols	
	Regosols	
	Alluvial soil	

On the basis of properties of soils, recently a new system of soil classification has been proposed by E. Joseph Larsen (Soil Conservation, 30, No. 5, Dec. 1964). The scheme has now become official Soil classification system for United States Department of Agriculture (USDA) from January 1, 1965. Brief outline of this classification is given in Table 24.2.

Table 24.2. Comprehensive soil classification of U.S.D.A.

	Soil order	Important characteristics	Old equivalent type
1.	Entisols	No stratification	Azonal soils
2.	Vertisols	Excess of clay	Grummusols
3.	Inceptisols	Usually moist, incomplete weathering, no clay, Fe and Al oxide saturated layers absent	Humic clay soils, brown forest soils
4.	Aridisols	Low organic matter, more than six months dryness in a year	Desert soils, sierozens, solonchak
5.	Mollisols	Almost black, organic upper horizon, high base content	Chestnut, chernozemes, rendzinas, prairie soils, brown forest soils, solonetz
6.	Spodosols	Amorphous substances excessively present in middle horizons	Podzol, brown podzols, ground water podzol
7.	Alfisols	Brown upper layer, high base content, excess clay in middle layers, usually moist	Brown soil, chernozem, half bog soil, planosols, grey-brown podzols, noncalcic brown soils
8.	Ultisols	Normally moist, low base content, clay content high	Reddish yellow podzol, reddish brown laterite
9.	Oxisols	Fully weathered, excess of Fe and Al hydroxides, and clay in middle layers	Laterite, latosols
10.	Histosols	Excess of organic matter	Bog soils

SOIL SURVEY

Soil survey is the study and mapping of soils in the fields. It is starting point for all soil researches. The following are the main objectives of soil survey:

(1) To describe and classify soils giving uniform system of classification with uniform nomenclature in order to correlate the soils of different area.

(2) To show distribution of different soils in the field (soil mapping).

(3) To provide data for making interpretations as to the adaptability of particular soils for agricultural purpose and also for many other purposes (as in soil management). Soil surveys are of the following three types :

(1) Detailed

(2) Reconnaissance

(3) Detailed reconnaissance

1. Detailed Soil Survey

In this, soil boundaries are plotted accurately on maps on the basis of observations made throughout the surveyed area. In this, geographical distribution of soil is also described. Detailed soil surveys are important in the sense that they provide informations needed for planning land use and management and formulating agricultural research and extension programmes.

2. Reconnaissance Survey

In this, soil boundaries are plotted from the observations made at intervals.

3. Detailed Reconnaissance Survey

In this, a part of surveyed area is plotted on the map by detailed method and remainder by reconnaissance method. In mapping of soils either aerial photographs are taken or good topographic maps are made.

In mapping of cultivated areas, generally a scale of four inches to a mile is used but detailed or special maps are made on scales of six inches, eight inches or twelve inches to a mile. In India, soil surveys are being carried out by central and provincial agencies. Some surveys of limited areas have been conducted for specific objectives, such as fertility surveys, surveys for soil classification, survey from geological point of view, for physicochemical properties of surface sample and surveys for genetic classification. Pre and post-irrigation surveys have also been used for limited areas under command of big irrigation projects to determine the nature, quality and the concentration of soluble salts in different horizons of soil. In India, a few generalized soil maps have been prepared under the auspices of International Society of Soil Science, Geological Survey of India and I.A.R.I. (Indian Agricultural Research Institute), New Delhi. The latest soil map of India is given in Fig. 24.1.

Aerial Photo-interpretation for Soil Surveys

Normally the scales recommended for detailed survey are 1 : 10,000 or 1 : 25,000. In such surveys, the air photos serve as an ideal base for plotting soil boundaries. This is on account of the wealth of pictorial detail available in the air photo, thus rendering orientation and navigation very easy. Normally a limited use of pocket stereoscope is made for delineating boundaries based on relief and other features. The final soil map in such surveys as well as the base photo are prepared on large scale.

Medium and Small Scale Map

Aerial photo-interpretation is of greater utility in this case. The degree of utility increases from medium to small scale. The procedure for photo-interpretation is based mainly on the principle that physiographic units have unique soil patterns. The scale recommended for semi-detailed reconnaissance surveys are 1 : 25,000 to 1 : 1,00,000 and 1 : 2,50,000. In similarity to the techniques used by geologists and foresters, the soil scientists also make use of such elements as relief, slope, geological features, vegetation, tone, textures, pattern etc.

Systematic Photo-interpretation

The most favoured method adopted by soil scientists involves delineation of the soil boundaries by air photos which are sub-divisions of physiographic units. In the next stage, they select sample areas that represent all the delineations of the photo-interpretation. These consist of long, narrow areas which cut across as many sub-divisions as possible. Detailed soil observations are taken up in each of the delineation of the interpretation. Based upon the number of such observations, the soils of each unit are classified and related to the photo-interpretation unit. Thus soil composition of each unit is established.

With the finding of the sample areas as basic, the remaining aerial photographs are interpreted by the principle of extrapolation of knowledge gained in the sample area studies. It may be mentioned that this "adjusted photo-interpretation" is very often a partial revision of the pre-field interpretation done in the office.

Final phase comprises a "selective ground check". This achieves the testing of the validity of the adjusted photo-interpretation. The resulting field sheet namely air photos contain the correct soil boundaries.

CHIEF SOIL TYPES OF INDIA

Depending upon the mode of origin, soils can be classified into two main groups :

(*i*) Sedimentary or residual soils

(*ii*) Transported soils

(*i*) **Residual soils.** These soils are formed from the disintegrated rocky material and remain at the same places where they are formed.

(*ii*) **Transported soils.** These are shifted soils and are brought from one place to another by various agencies. They are of different types such as :

(1) *Illuvial soils* :— Soils that move due to gravitational influence.

(2) *Alluvial soils* :— Soil is brought by running water.

(3) *Glacial soils* :— Soil is brought or transported to a particular place by glaciers.

(4) *Aeolian soils* :— Wind moved soils, *e.g.*,

(1) Dune sands of variable fineness,

(2) Volcanic ash,

(3) Loess, silt like material.

Geographically soils of India are divided into three groups :

(*i*) Mature soils of Peninsular India (include red soils, black soils, lateritic soil).

(*ii*) Alluvial soils of Indogangetic plain.

(*iii*) Scanty soils of Himalayas.

According to Indian Council of Agricultural Research, the soils of India are classified into the following main groups :

(1) Laterite or Lateritic soils,

(2) Black cotton soils or 'regur',

(3) Red soils including red loam, yellow earths etc.

(4) Alluvial soils including deltaic, coastal and inland alluvium,

(5) Alkali soils and saline soils,

(6) Peaty and other organic soils of forests,

(7) Desert soils or arid soils,

(8) Scanty soils of mountains and hills.

1. Laterite or Lateritic Soils or Latosols

Laterite is generally reddish or yellowish red in colour which turns black on exposure to sun. This group of soil occupies belts of various widths around the peninsular zone. Such soils occur on the plateau of Malwa, Madhya Pradesh, Central India, Bihar, Orissa, Tamil Nadu, Eastern and Western Ghats, Assam, Bengal, Hyderabad covering a total area of about 2,48,000 square kilometres.

Origin of lateritic soils is still a matter of dispute. Important rocks which produce such soils are *greisze,* sand stone, basaltic rocks and granite. These are residual or sedimentary rocks. Laterites are formed in the region of high rainfall and with alternating wet and dry seasons. When the alkali and silica are leached away the residual compounds rich in aluminium and iron oxides form such soils. Lateritic soils are compact and composed of hydrated oxides of aluminium and iron with small amount of manganese and titanium oxides. When soil is broken and carried to lower levels by water it again becomes cemented into compact indurated honey-combed mass.

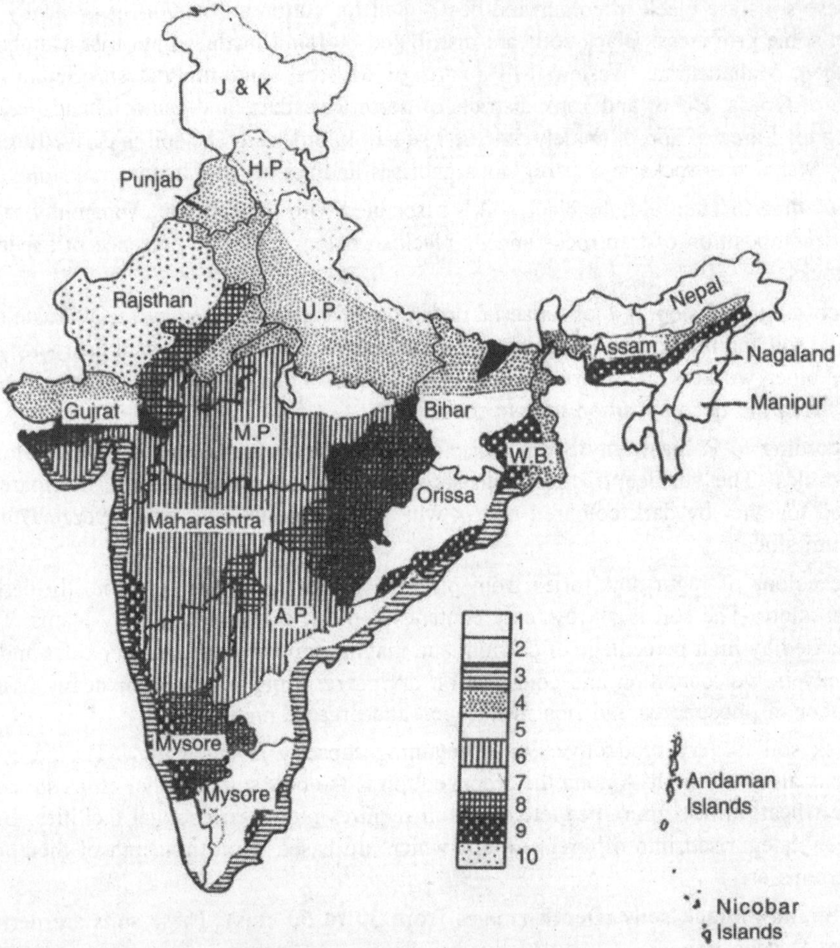

Fig. 24.1 Important soil types of India

1. Red
2. Red and yellow soils } Red soils
3. Coastal alluvial soils
4. Alluvial soil
5. Desert soil } Alluvial soils
6. Shallow black soil
7. Black soil } Black soils
8. Peat soil
9. Laterite soil
10. Mountain or Scanty soil.

Lateritic soil is usually poor in lime, magnesia, nitrogen, phosphate, and potash. Humus is present in abundance in this soil. Base exchange capacity is low. Soil is good for agriculture purpose. Valley soils can produce good crops of rice and sugar cane whereas at higher elevations, such soils favour the growth of coffee, rubber, tea, cinchona etc.

2. Black Cotton Soils or Regur

(*Regada* = *Black*)

These soils are black in colour and best suited for cotton cultivation. It is locally known as regur in some provinces. Black soils are distributed in Tamil Nadu, South-east Mumbai, Eastern Hyderabad, Maharashtra, Western M.P., parts of Mysore, Andhra Pradesh, Gujarat, Southern districts of Orissa, Bundi and Tonk districts of north Rajasthan, and Bundelkhand region of U.P. covering total area of approximately 5,46,000 square kilometres. The soil is derived from Deccan and Raj Mahal trap rocks and ferruginous gneisses and schists in Chennai.

According to Theofold, the black soil is associated with Deccan trap. Apparently, it is derived by the decomposition of trap rocks and its blackish colour is due to presence of superficial iron in the rocks.

According to Hislop, it is of subaerial origin or it has been formed by the alternate deposition of organic soil and the primary deposition of weathering products of almost all the rocks. According to some other workers, it is produced by assimilation of humus of different argillaceous soils derived from the decomposition of trap rocks.

According to Ramaswamy Sivan, it is formed due to accumulation of black coloured light soil particles. The surface fraction consists of small transparent or semi-transparent grains cemented together by dark coloured matrix which is a mixture of double hydrated ferrous and aluminium silicate.

The colour of regur soil varies from place to place. Regur soils are mostly dark grey to black in colour. The soil is clayey (clay content 40–60%) or loamy or sandy loams. The soil is characterised by high percentage of calcium and magnesium carbonates, iron oxides and alumina. It is sticky in wet condition and contracts on drying resulting in cracks or heavy fissures. The soil is poor in phosphorus and rich in nitrogen and organic matter.

Black soil is very productive. Water retaining capacity is much high. The base exchange capacity is also very high. Among the crops, cotton is the best suited. Other crops suited to black soils are wheat, millet, gram, pea, etc. The soil requires proper irrigational facilities. Black soils have been categorised into different groups which are based upon the depth of their formation. These groups are :

(*i*) **Shallow black soils (Depth ranges from 30 to 50 cms).** These soils are derived from basalts of Deccan trap.

(*ii*) **Medium black soils (Depth ranges from 50 to 120 cms).** They are derived from a variety of rocks including basaltic trap, Dharwar schists, basic granites, gneisses, hornblende and chloride schists.

(*iii*) **Deep black soils (Depth ranges from 120 to 200 cms).** These are derived from basaltic traps in Deccan plateau and are commonly identified as "regur soils". Distribution of irregular nodules of lime (Kankar) throughout is regular feature of this subgroup of black soil.

3. Red Soils Including Red Loam and Yellow Earth

These soils are distributed in India mainly in the peninsular regions. The soils occur mainly in M.P., S.E. Mumbai, Mysore, Andhra Pradesh, West Bengal, Assam, Orissa, Chhota Nagpur covering a total area of about 3,50,000 square kilometres. Red soils are also found in some parts of Bihar and in Mirzapur, Hamirpur, Banda and Jhansi districts of U.P. and territory of East Aravali hills in Rajasthan.

The red soils develop from metamorphic sediments of granite gneisses and schists.

The red soils of valleys and plains are dark and fertile. These are composed of silica and aluminium with free quartz or sands. Soils are low in bases and base exchange capacity. These soils are characterised by absence of lime nodules (Kankars). Percentages of humus and other organic nutrients are very poor in red soils. The colour of red soil is due to high proportion of iron compounds. Ferrous oxide is precipitated after evaporation of water from the soil. They are neutral to acidic in their reaction. Soils are generally rich in potash but nitrogen content is low. The red soils are light, friable and porous. Their water-absorbing capacity is very low. Cultivation on such soils depends upon irrigation and rainfall.

4. Alluvial Soils including Deltaic Coastal and Inland Alluvium or Alluvial Soils of Indogangetic Plains and River Basins

It is the largest soil group in India covering a total area of about 15,00,000 square kilometres.

The soils are immature and can be divided into two geological groups namely old and new alluvial sub-divisions. The old alluvial soils, also called **'Bangar'**, are the soils of uplands and are composed of massive clay or compact silt. These are mixed with sands and kankars or lime nodules and are pale brown or dark in colour. The old alluvial soils are in the process of denudation. The new alluvial soil group, also known as **Khadar**, is sandy with less kankars. It is light in colour.

The soil is best suited for agriculture provided rains occur regularly. Alluvial soils are suitable for cultivation of wheat, rice, cotton, maize, millet, oils, sugarcane, groundnut, vegetables and jute.

The various kinds of alluvial soils have same origin. They are formed from the materials transported by rivers over long period of time. These soils are found in Rajasthan, Punjab, Haryana, the valleys of Brahmaputra in Assam; Narmada and Tapti rivers in M.P.; Godawari, Krishna and Kaveri in the south, Yamuna, Ganga and their tributaries; Gandak, Gomati, Ghagra and others in U.P., Bihar, West Bengal and Jharkhand.

The soils of dry regions are rich in K, P and alkalies. CaO and FeO are present in varying amounts. In Brahmaputra and Ganga valleys and Assam plains, the great *Khadar* is low in lime, K, and Mg contents. In dry regions of India, alluvial soils contain KNO_3 and alkali in excess. Scanty rainfall, high temperature and natural drainage are chief conditions which result in the accumulation of salts in the form of underground obstructions and impervious clay pans which inhibit the leaching. These factors are also responsible for the development of barren land (usar) which is useless for cultivation.

The deltaic soils of Ganga and Brahmaputra are rich in humus and are loamy, and hence they are fertile. But by floods, surface soils are removed and also replaced by fresh silty soils. This induces the soil fertility. Alluvial soils of peninsular regions including those of Narmada and Tapti valleys are very fertile. In eastern ghats, the deltaic soils of Godawari, Krishna and Kaveri rivers are black, humid and sandy. The soils of Orissa are sandy and swampy with fine texture. They are rich in potash.

Soils of Chennai are transported by rivers and consist of alternate layers of sands and silts. The alluvial soils of Gujarat, locally known as *Goradu,* consists of brown clay and *kankars.* Soils are poor in organic content and nitrogen but rich in potash. Such soils are very common in north Gujarat and Ahmedabad.

Yellow and red alluvial soils of M.P. are found in the basin of Mahanadi river covering Balaghat, Durgapur, Raipur and Bilaspur districts. These soils are grouped as *Bhata, Matasi, Dorsa,* and *Kanhar* as these are found at different altitudes.

Comparative Account of Soils in India (ICAR)

	Type of soil	Formation of soil	Distribution	Composition	Characteristic features	Crops grown
1.	Alluvial soils	By silt deposition brought by rivers	River valleys deltas of rivers and coastal strips of Peninsular India. Punjab, Haryana, Uttar Pradesh, Bihar, Jharkhand and West Bengal.	Rich in potash and lime but deficient in nitrogen, phosphoric acid and humus. (The Ganga delta is rich in humus).	In texture, it is coarse in upper region, medium in middle region and fine in the lower region.	Large variety of *rabi* and *kharif* crops: wheat, rice, sugarcane, cotton, jute and oil seeds.
2.	Black soils or Regur	By weathering of lava-flow rocks	Deccan Plateau, valleys of Krishna and Godavari, Maharashtra, Gujarat, Andhra Pradesh, M.P., Chhattisgarh, and Tamil Nadu.	Rich in iron, lime, aluminium potash, magnesium, calcium. It lacks in nitrogen, phosphorous and humus	When wet it is fine grained, moisture retentive and sticky. Cracks and crevices develop when it is dry.	Sugarcane, cotton, jowar, tobacco, wheat, citrus fruits and vegetables and oil-seeds.
3.	Red soils including red loam and yellow earth	By decomposition of old crystalline or metamorphic rocks.	Eastern parts of Deccan Plateau. Tamil Nadu, Goa, South Karnataka, Orissa and Meghalaya.	Rich in iron and potash and deficient in lime, nitrogen, phosphorous and humus.	Acidic, porous and coarse. Becomes fertile with use of fertilizers.	Wheat, rice, cotton, sugarcane and pulses (with the use of fertilizers).
4.	Laterite or Latosols	By weathering of rocks under monsoon-climatic conditions such as heavy rainfall and high temperature with alternating wet and dry periods	In patches in the Eastern and Western Ghats and Assam hills. Andhra Pradesh, Tamil Nadu, Karnataka, West Bengal and Orissa.	Rich in iron, and poor in silica, lime, nitrogen, phosphoric acid, potash and humus.	Highly acidic, porous, crumbly and coarse. Red in colour due to presence of iron oxide.	Tea, coffee, rubber, cashew, tapioca and millets.
5.	Desert soils or Arid soils	By weathering of rocks due to high diurnal range of temperatures and dryness	West and north-west of India. Rajasthan, north Gujarat and southern Punjab.	Rich in salts and deficient in organic matter.	Loose, porous and coarse.	Generally, unsuitable for cultivation but with irrigation, useful for wheat, gram, bajra, melon, ec.
6.	Scanty soils of Mountain and Hills	By weathering of sedimentary rocks	Hills of Jammu & Kashmir 5 slopes of mountains in Uttarakhand, Assam hills and West Bengal.	Rich in iron and deficient in lime.	Acidic, coarse porous and thin.	With fertilizers tea plantation, fruits and medicinal plants.
7.	Saline and Alkaline soils	By the accumulation of tidal water.	Orier parts of Bihar, Jharkhand, Uttar Pradesh. Haryana, Punjab, Rajasthan, and Maharashtra.	Many salts such as of sodium, magnesium and calcium	Infertile and render.	Unfit for agriculture.
8.	Peaty and Marshy soils	Originate in humid regions as a result of organic matter.	Kerala, coasal regions of Orissa and Tamil Nadu and Sunderbans of West Bengal.	Rich in iron and organic matter. Lack in potash and phosphorus.	Black and heavy.	Useful for jute and rice cultivation.

Bhata is barren soil consisting of red gravels and sands. This is found on uplands.

Matasi is yellow loamy clay soil.

Dorsa is dark coloured clay soil of slopes.

Kanhar is darker and heavier than *Matasi* and *Dorsa* and is deposited in low lying areas.

The alluvial soils of U.P. are divided into the following four regions:

(*a*) Alluvial soil of West and North U.P. (light texture)

(*b*) Alluvial soil of East U.P. (heavy texture)

(*c*) Alluvial soil of Central region consisting of old alluvium which is neither too light nor too heavy, *i.e.*, this is intermediate in texture.

(*d*) Alluvial soil of north-east U.P which contains $CaCO_3$ and other salts and shows alkaline reaction.

5. Saline Soils and Alkaline Soils

In these soils, percentage of salts is very high. These soils are distributed in Kanpur, Lucknow, Hardoi, Unnao, Raibareli, Azamgarh and many other districts of U.P., Punjab, West Bengal, Bihar, Orissa, Maharashtra, Tamil Nadu, M.P., A.P., Gujarat and Delhi covering an area of about 70,00,000 hectares. Some soils are salty mainly due to presence of alkali salts in them. They are called alkali soils. In some soils salts other than alkali are present. These are commonly referred to as saline soils or non-alkali saline soils. Alkali soils are barren and are also called usars in Hindi.

The detailed accounts on saline and alkali soils have already been given in the chemical properties of soils.

6. Peaty and Marshy Soils of Forests

Peaty soils are also known as peaty saline soils. At some places it is called *Kari*. The peaty soils are peculiar blue coloured soils. The waterlogged peaty soils appear in the low lying submerged areas. They are also termed as marshy soils. The peculiar colouration is due to presence of free aluminium and ferrous compounds. Organic matters are accumulated in poorly decomposed conditions on account of the lack of oxidation and poor drainage conditions. Peaty soils are distributed in coastal tracts of Orissa, West Bengal (in Sunderban), North Bihar, Almora in Uttaranchal, S.E. Tamil Nadu and Kerala covering an area of about 150 square kilometres. Soils are rich in humus and are acidic. Sometimes water soluble alkali salts are also accumulated in the soils. During monsoon period these soils are flooded with water.

7. Desert Soils or Arid Soils

Desert soils are commonly found in arid and semi-arid zones in India, *e.g.*, North-Western Rajasthan and South Punjab (between Indus river in west and Aravali hills in the east) and semi-arid area of Rann of Kutch covering a total area of about 1,42,000 square kilometres. In arid area, rainfall is very poor (upto 62.5 cm) and temperature goes very high during the summers. The soils in these areas are coarse textured regosols and wind moved sand and sand-dunes. Soils are salty and they contain good amount of alkalies and humus content. Their water holding capacity is very low. At several places due to poor rainfall soils become barren but they can be reclaimed to normal conditions provided irrigational facilities are available. The scarcity of water and excessive accumulation of salts make the soils unfit for cultivation. This is why vegetation is very poor, forests are totally absent and desert is spreading progressively eastward in Rajasthan.

Considerable attention is being given these days for checking desertification and for converting the desert land into fertile area. Recently Central Arid Zone Research Institute, Jodhpur, has recommended the stabilization of sand by growing thick vegetation in the area. Central Arid Zone Research Institute has further recommended the cultivation of many indigenous and exotic varieties of trees, grasses and salt tolerant species which grow in desert conditions. Important plants well suited to desert land are *Calligonum polygonoides, Leptadenia, Crotolaria,* some grasses and herbs like *Citrulus, Rhynchosia,* etc. Some trees of deserts are *Prosopis, Acacia, Albizzia lebbeck, Dalburgia sissoo,* etc.

8. Scanty Soils of Hills and Mountains

Mountain or hill soils are distributed in India from north-west to eastern part of the country up to 12,000 feet altitude. At higher altitudes, entire rocky surface is covered with snow. Mountain and hill soils are loose and are young or immature. In these soils, A horizon is formed on the surface of rocks directly. Soil particles are loosely aggregated in soft sandy beds. The scanty soils may be of two types:

(1) acidic soils with acidic humus.

(2) basic soils having high base status.

In foothill areas of Kumaon and Garhwal districts, soils are coarse and gravel type mixed with clay and iron salts. In these areas at the altitude of 5,000 to 11,000 feet sal forests and pine forests are common. Important crops grown on such soils are wheat, maize and tea. The soils of hilly regions of Assam and Darjeeling contain high degree of organic matter, nitrogen and minerals. These soils are acidic in reaction.

The Himalayan region of U.P. consists of soils known as Bhabar, Terai and Plains.

The soils of Himachal Pradesh may be divided into low, mid and high hill soils and mountain soils.

The Himalayan soils can further be divided into the following soil groups on the basis of distribution and types of vegetation.

(*i*) **Terai soil.** It is distributed near the foothills of Simla. Soil is clayey in composition and has got high water retaining capacity and humus content. Hence, it is very fertile.

(*ii*) **Tea soils.** These are found in Assam, Dehradun, and Darjeeling. Soils are light loamy with low lime contents. Rainfall is extremely high. Soils are well suited for the plantation of tea.

(*iii*) **Igneous soil.** This soil is derived from granite rocks. It is fertile and good for cultivation purpose.

(*iv*) **Soils of older rocks.** Such soils are found in Nainital. They are rich in clay and iron contents thus suitable for cultivation of certain crops and vegetables.

(*v*) **Lime stone soil.** It occurs near Mussoorie. Rice is the main crop of this hill soil.

Important tree species suitable for planting in different soil groups are given in Table 24.3.

Table 24.3. Important tree species suitable for planting in soils of different climatic regions

	1 *Species*	2 *Sites*	3 *Soil Groups*	4 *Uses*	5 *Remarks*
A.	**Temperate Region** (*i*) Himalayan				
1.	*Alnus nepalensis* (units)	River banks, ravines & newly formed soils	Brown hill soils (on sand stones & shales) Sandy	Timber & fuel	Large tree, grown by direct sowing or transplanting
2.	*Alnus nitida*	-do-	-do-	-do-	-do-
	(*iii*) Dry Areas of Rajasthan & Peninsular India (Dry tropical)				
	1. *Acacia auriculiformis* 2. *Acacia nilotica* 3. *Ailanthus excelsa* 4. *Albizzia lebbeck* 5. *Azadirachta-indica* 6. *Cassia siamia* 7. *Eucalyptus hybrid* 8. *Leuceina leucocephala* 9. *Madhuca lalifolica* 10. *Prosopis Juliflora*	Trees suitable for afforestation in arid regions, grown by entire transplanting, other details mentioned earlier.			

QUESTIONS

1. Give brief accounts of soil classification and soil surveys.
2. Describe the main soil groups found in India and metion the distribution and properties of each group.
3. Discuss the distribution and properties of (*i*) arid soils, and (*ii*) scanty soils of hills and mountains.
4. What is soil survey? Discuss the methods through which soil is mapped.
5. Write short notes on :
   (*a*) Laterite Soil
   (*b*) Regur (Black cotton soil)
   (*c*) Aerial Photo-interpretation for Soil Surveys
   (*d*) Alkali Soil.

25

Soil Erosion and Soil Conservation

Soil is a store house for organic and inorganic plant nutrients and water. Some soils are rich in organic and humus contents and are more productive while others are less productive and have very poor percentage of organic contents. The soil is subjected to a continuous and simultaneous depletion or loss and addition of soil resources. The various soil components are being removed by living organisms and are returning to the soil by way of death and decay of organisms on it. If the rate of removal or loss of components is greater than the rate of addition, the soil will naturally become less fertile. The various factors which make the soils less fertile are :

(1) Leaching,

(2) Biological agencies, and

(3) Soil erosion.

Leaching is an important factor which makes the soil poor in its resources. In this process, minerals and organic substances are removed from the top layer of soil by rain water.

Biological agencies are also active in causing loss of resources from the soil. Cultivation of crops regularly year after year makes the soil less productive. In cultivation there are little chances for the compensation of lost nutrients of the soil by way of death and decay of the vegetation which grows on it. Leguminous plants, however, compensate the loss of nitrogenous compounds because bacteria inhabiting in their root nodules fix considerable amount of free atmospheric nitrogen into its compounds, such as nitrites, nitrates, ammonia and so on.

SOIL EROSION

The word 'erosion' literally means *to wearing away.* In soil erosion, fertile soil surfaces are detached and removed from their original places and are deposited at some other places. According to Fox (1950), "the term soil erosion covers a wide range of physical and chemical actions, such as removal of soluble matters, chemical changes, disintegration by frost or by rapid changes of temperature, attrition by dust charged wind, scouring by silt laden currents, alternate impact and succession by storm waves, landslides and so on". Thus, soil erosion is the removal of soil from its upper part.

There are two main types of erosions, namely (1) *normal erosion* and (2) *accelerated erosion.*

1. Normal Erosion

When the top soils are gradually removed under normal conditions of physical, biotic and hydrological equilibria it is called **normal erosion**. Sometimes, it is also called **geologic erosion**. It is very slow process in which complete equilibrium is maintained between soil removing and soil forming processes. The normal erosion tends to produce wavy or undulating land surface with alternating ridges and depressions. This is accomplished chiefly by means of slow migration of soil particles from soil surface in successive rains. In arid region, wind during the long dry season is an important factor for normal erosion.

2. Accelerated Soil Erosion

When the removal of soil does not keep harmony with the soil formation and it is much faster than the latter, it is called **accelerated soil erosion**.

AGENCIES CAUSING SOIL EROSION

Soil erosion is caused by the following two agencies :

(A) Climatic.

(B) Biotic.

A. Climatic Agencies Causing Erosion of Soil

These are water and wind.

1. **Water.** Water is an important factor in soil erosion. Snow and melting ice also remove the top soil to considerable extents. Soil is directly affected by heavy rainfall, rapidly running water and by wave action.

Erosion caused by water may be of the following types :

(*i*) **Sheet erosion.** Uniform removal of a thin layer of soil from large area is called sheet erosion. It is affected by run-off effect of rain water.

(*ii*) **Rill erosion.** In this type of soil erosion, heavy rainfall and rapidly running water produce finger-shaped grooves or rills over the entire field (Fig. 25.1).

(*iii*) **Gully erosion.** It is more prominent type of erosion in which heavy rainfall, rapidly running water and transporting water may result in deeper cavities or grooves called gullies.

Gullies may be 'V' shaped or 'U' shaped. Gullies cut the fields into small fragments and make them uncultivable (Fig. 25.2). Continuous flow of water through gullies further deepens the grooves and may ultimately result in ravines. Ravines are 15 to 30 m deep and with steep vertical sides.

(*iv*) **Landslides or slip erosion.** This type of soil erosion is caused by heavy rainfall and it occurs in slopy lands, such as mountains and hills. In this type of erosion when the running water percolates through the crevices of rocks great masses of soils and loose rocks lying on the steep slopes slip downward.

(*v*) **Stream bank erosion.** On the banks of swollen rivers it is most active. During the rainy season when fast running water streams take turn in some other directions, they cut the soil and make caves in the banks. As a result of this, quite often large masses of soils become detached and washed away from the banks and are deposited at places in course of streams.

2. **Wind erosion.** Removal of soil by wind is called wind erosion. Stormy winds carry the soil particles to distant places and sometimes form sand-dunes. Wind currents usually remove the top soil which is fertile and full of humus and minerals. Wind causes the following three types of soil movements, viz., (*i*) saltation; (*ii*) suspension; and (*iii*) surface creep.

(*i*) **Saltation.** Under the influence of direct pressure of stormy wind small soil particles of 1 to 1.5 mm diameters move up from the soil surface, generally in vertical direction. Major part of wind carried soil is moved in a series of bounces, called saltation.

Fig. 25.1. Rill formation.

Fig. 25.2. Gullies.

(*ii*) **Suspension.** In this, fine soil particles (diameter less than 1 mm) are suspended in air. These suspended particles are kicked up when particles of saltation strike on the soil. The soil particles are deposited at distant places.

(*iii*) **Surface creep.** In this, there are larger particles ranging from 5 to 10 mm in diameter. Because they are too heavy to move in saltation, they creep on surface of soil.

B. Biotic Agencies Causing Soil Erosion

Excessive grazing, deforestation, undesirable forest biota, and mechanical practices by man are important factors which cause soil erosion.

Deforestation is the commonest factor which is responsible for soil erosion. Ruthless felling has exposed soils to direct effects of rain, snow and drought and has set in soil deterioration of the gully and sheet types over extensive areas in Kashmir (J.S. Singh and M.K. Wali, 1961).

Fig. 25.3 Overgrazing by sheep and other animals

Grazing is yet another destructive biological factor for the soil erosion. Cattle and sheep during the summer graze the forest vegetation and make the soil bare (Fig. 25.3). Whyte (1957) while commenting on the problem of soil erosion of Kashmir Himalayas, mentioned, "control of fluctuating grazing was made the responsibility of forest department in 1939 but this appears to have little effect and the problem is as acute as ever".

SOME OTHER CAUSES OF SOIL EROSION

1. Uneven distribution of rainfall.

2. **Shifting cultivations.** Shifting cultivations are usually noted in the mountains which are geographically young and degraded into soil easily and the whole of the land is covered with a thick mantle of tropical forest vegetation. The removal of the forest or bush cover by felling and burning for shifting cultivation and the resulting exposure of the bare soil to rains and sun, cause enormous soil losses especially on hill slopes. Both surface layer of the soil and large quantity of plant nutrients are washed away under the influence of intense rainfall. Shifting cultivation is a major problem in the hilly areas of Assam, Manipur, Tripura, Arunachal Pradesh, Nagaland and Orissa. It is reported that about 207,267 hectares in Assam, 46,963 hectares in Tripura, 21,862 hectares in Manipur and 308,502 hectares in Orissa are under shifting cultivation. Shifting cultivation is practised sporadically in Andhra Pradesh, Madhya Pradesh, Maharashtra, Kerala, Karnataka and Tamil Nadu also. The process of shifting the area of cultivation is called by various names : *Jhum* or *Jum* in Assam, *podu, dahi* or *kamana* in Orissa, *penda* in Madhya Pradesh and so on. In English it is described as slash, burn or *swidden* or simply as *shifting cultivation.*

3. Fields on steep slopes are cultivated and top soil is washed away by rains. The loss of soil is too much and the fields become uncultivable.

4. Forest fires are responsible for burning down forest trees on huge scale.

5. Faulty agricultural methods—Sometimes farmers do not care towards levelling and terracing of their upland fields. Rainfall washes away the top soil and results in erosion.

6. Over-grazing by cattle causes removal of vegetational cover of the soil.

Factors Affecting Soil Erosion

There are several factors which affect the soil erosion in a particular area. These factors are given below :

1. Nature of soil.
2. Distribution, nature and amount of precipitation.
3. General slope of the soil.
4. Vegetational cover.
5. Soil management.
6. Land use practices.

Consequences of Soil Erosion

There are several serious effects of soil erosion which are as follows:

1. Due to uprooting of trees the shortage of timber and fuel wood results.
2. Loss of soil stability and fertility.
3. Shortage of fodder.
4. Destruction of land in plains.
5. Formation of sand dunes.
6. Greater frequency of floods and threat to communication channels.
7. Silting of river bed, lakes and dams.
8. Higher temperature and scanty rainfall.

SOIL CONSERVATION

The main aims of soil conservation are as follows :

(1) To protect the soil from erosion.

(2) To maintain the productive capacity of the soil.

Soil erosion is a natural phenomenon by which soil is removed from rocks. Accelerated erosion due to misuse of resources of land, water and soil is today one of the most difficult and pressing problems before man. Both engineering and biological methods have been used to check the soil erosion but it is still without a plausible check. Vast tracts of fertile land are rendered useless on account of industrialisation and development. Erosion has posed a serious challenge in the United States, South and North Africa, Japan, Mesopotamia, North China, India, Pakistan and in a number of other countries. In India, we are aware of advancing deserts of Rajasthan and erosional losses, floods etc. in other parts of the country. The problem has received the attention of forest ecologists, soil scientists and engineers only recently.

Practical methods of soil conservation are broadly grouped as follows:

(A) Biological measures,

(B) Mechanical or Engineering methods.

A. Biological Measures

The following are the biological methods which are helpful in checking the soil erosion :

1. Agronomic practices;
2. Agrostological methods;
3. Dry farming practices.

1. **Agronomic practices.** The important agricultural practices which contribute to the conservation and productivity of cultivated lands are referred to as conservational farmings or advanced agronomical methods. These are listed as under :

(*i*) Contour farming:

(*ii*) Tillage and keeping the land fallow

(*iii*) Crop rotation, sowing of leguminous crops and mixed cropping

(*iv*) Mulching

(*v*) Strip cropping

(*i*) **Contour farming.** It is practised in the hilly regions or on the slopes. In such areas, the rain water is absorbed in very little amount because of its quick downward movement on the slopes. If these slopy areas are ploughed up and down the slope, the heavy rainfall may cause

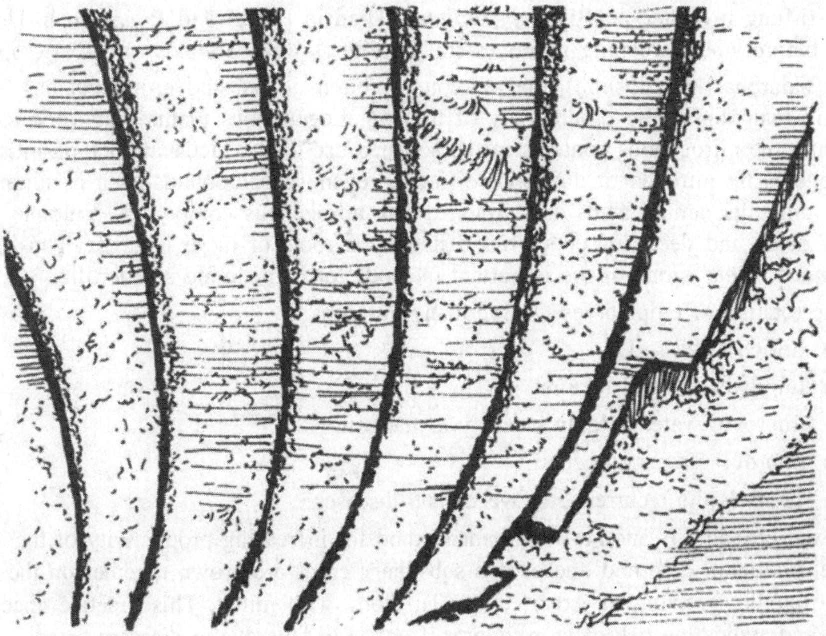

Fig. 25.4. Contour strip cropping for controlling soil erosion and strip cropping for controlling soil and moisture losses on uniform slopes.

gully development. Taking into consideration this defect, the slopy areas are ploughed and seeded against the slope, *i.e.,* in circular furrows around the slopes. This process is termed as *contour farming.* The contours (circular or peripheral furrows) catch the downwardly moving water until it is absorbed in the soil. The ridges reduce the flow of water. The circular rows of plants across the slopes check the soil erosion. Thus, contour farming reduces run off, saves more water for crops, reduces soil erosion and increases the yield of crops (Fig. 25.4).

(*ii*) **Tillage operation and keeping the land fallow.** There are several diverse opinions as to whether deep ploughing gives good result or shallow ploughing. A number of researches support the view that in dry areas, shallow ploughing gives comparatively good crop yields. Shallow ploughing removes the weeds and enables the soil to absorb water. Deep ploughing often leads to soil erosion but in the areas where rainfall is sufficiently high, deep ploughing (upto 15–30 cm deep) is effective in removing weeds and increasing crops yields.

If the land is left uncultivated and sheep, goats and other cattle are allowed to graze and sit over it for some time, the soil becomes fertile. Though this practice is useful yet it is not possible in the countries like India where exists severe problem of cereals because of thick human population.

(*iii*) **Crop rotation, sowing of legumes and mixed cropping.** When the same crop is grown in the field every year, the soil becomes depleted in certain minerals. The soil loses its fertility even after the use of fertilizers and ultimately erosion sets in. Rotation of crops is an important method for checking erosion and maintaining productivity of soil. After 2 years crop should be changed in the fields. A good rotation should include a cultivated row crop, densely planted small grasses and a spreading legume or a legume and grass mixture. Selection of crops for rotation should be made taking into consideration the climate, economic condition, soil types, soil texture, slopes, nature of erosion, etc. Deep-rooted crops should be rotated by shallow-rooted crops. Deep-rooted crops absorb nutrients from the deeper strata of the soil. Thus, the minerals of top soil remain stored for future use by shallow-rooted plants. When the deep rooted crops die, they add humus in the soil which is future storehouse of plant nutrients. Growing of *Phaseolus* *mungo* (Mung in Hindi) in Bihar, *P. radiatus* (Urad in Hindi) and *P. mungo* in U.P. in kharif season before wheat cropping improves soil fertility as well as yields of cereal crops.

In Vidarbha (Maharashtra), the rotation of cotton, jowar, and groundnut has been proved useful in maintaining the productivity of the soil. Leguminous plants play very active part in increasing the nitrogenous contents in the soil because of the fact that bacteria inhabiting their root nodules fix nitrogen of atmosphere into nitrogenous compounds such as nitrates, nitrites, ammonium salts, amino acids and proteins. The nitrogenous compounds return to the soil by way of death and decay of underground nodulated roots of these legumes. Thus, growing of leguminous plants is one of the important methods for maintaining soil fertility.

The rotation of crop serves the following purposes :

(*a*) Enriches the soil,

(*b*) Improves the soil texture,

(*c*) Improves water holding capacity of the soil,

(*d*) Improves crop production,

(*e*) Controls the recurrence of weeds and diseases.

Mixed cropping is another important method for increasing productivity of the soil. In this practice, one main crop and one or two subsidiary crops are grown together on the same land, as for example, growing of Arhar, Urad, Til along with millet. This practice checks the soil erosion and avoids the risk of crop failure. If one crop fails due to diseases or any other factor, the others remain ensured.

(*iv*) **Mulching.** It means covering the soil surface by straw, leaves or grasses. Mulches of different kinds check soil erosion, increase soil fertility and also minimise moisture evaporation from the top soils. Various types of surface tillers and crop residues are helpful in obstructing the movement of soil particles.

(*v*) **Strip cropping.** It is an important method which employs all the advanced cultivation practices such as contour farming, proper tillage, crop rotation, mulching, cover cropping, etc. Strip cropping is very effective and practical means for controlling soil erosion. It is of the following types:

(*a*) Contour strip cropping,

(*b*) Field strip cropping,

(*c*) Wind strip cropping, and

(*d*) Permanent or temporary buffer strip cropping.

(*a*) ***Contour strip cropping.*** It is special type of contour farming in which soil exposing crops are grown on the strips across the slope on the level of contour and in the following season soil protecting crops are sown on the strips on which soil exposing crops were grown in the previous season and again in place of these soil protecting crops some soil exposing crops are sown. This practice is useful because it checks the flow of run off water, increases the infiltration of water in the soil and prevents soil erosion. (Fig. 25.5)

Fig. 25.5 Strip cropping on hills

(*b*) ***Field strip cropping.*** It is a farming in more or less parallel strips across fairly uniform slopes but not on the exact contour.

(*c*) *Wind strip cropping.* In this method tall growing plants (*e.g.* bajra, jowar, etc.) alternating with the short growing crops, such as 'arhar', 'urad' are sown in long straight strips right across the direction of wind regardless of contour.

(*d*) *Permanent or temporary buffer strip cropping.* This is a special type of contour strip cropping in which care is taken to check the soil erosion. In this, crop rotation practice is not applied and on the strip perennial legumes and grasses are planted on permanent or temporary basis.

2. **Agrostological methods.** The following are the important agrostological practices that check soil erosion :

(*i*) Cultivation of grasses (ley farming).

(*ii*) Retiring the land.

(*iii*) Afforestation and Reforestation.

(*iv*) Checking of overgrazing.

(*i*) **Cultivation of grasses (Ley farming)**. This method consists in growing grasses in rotation with agricultural crops. This practice improves the fertility of soil and helps in binding of the soil, thus preventing the soil erosion. This practice is recommended for Nilgiris and similar places which are subjected to very severe soil erosion.

(*ii*) **Retiring the land.** Areas subjected to heavy soil erosion should necessarily be put under thick cover of grasses. Under favourable climatic conditions grazing should also be allowed for short periods. Researches conducted at Solapur in Maharashtra have shown that grasses have good soil binding capacity. In Nilgiri hills, Tamil Nadu doobgrass (*Cynodon dactylon*), *Dectylis glomerata, Eragrostis amabitis* and *E. cerbula* are proved to be most effective in soil binding and in stabilizing the reserves of the bench terrace and sodding water channels.

(*iii*) **Afforestation and reforestation.** *Afforestation* means growing forests at places where there were no forests before due to lack of trees or due to adverse factors such as unstable soil, aridity, or swampiness. *Reforestation* means replanting of forests at places where they have been destroyed by uncontrolled forest fires, excessive felling and lopping. Plantation of trees in short blocks is known as a **wind break** and extensive plantation of trees is called **shelter belts**. Wind is a problem in dry areas. Soil erosion by wind can be checked or reduced if the soil is covered by vegetation. Wind breaks are planting of trees or other plants at right angles to prevailing wind that protect bare soil from full force of wind. Wind breaks reduce the velocity of wind, thereby decrease the soil erosion. In some cases, rows of trees are angles to prevailing wind. Wind break and shelter belt type of plantations are being done in some regions of Uttar Pradesh where desert is encroaching. The plantation is usually done in two or three belts. Small sized plants are planted on windward side and tall trees on leeward side. Wind breaks and shelter belts reduce the wind velocity considerably and also check the transport of lifted sand and soil particles. Afforestation is the best means to check the soil erosion. Lutz and Chandler (1946) cited the following points in support of the vegetational check of erosion :

(*a*) Infiltration of water is favoured due to high porosity of soil under vegetation. Percolation of water helps in preserving the soil moisture which accelerates growth of vegetation.

(*b*) Surface accumulation of organic matter increases the water holding capacity of soil.

(*c*) Root systems of vegetation hold the soil mechanically and provide stability to the soil.

(*d*) It gives protection to soil against wind. The forest vegetation shields the soil from direct effects of drought, snow and rain.

Destruction of vegetational cover usually results in accelerated erosion, flooding and silting. The past victims of soil erosion tell the story of drastic deforestation (destruction of forests). In our country deforestation has been recognised as the most potent and powerful cause of soil erosion and efforts are being made to check it.

State forest departments in India have already undertaken this practice, although in very restricted areas and have met with good success. Forest departments in Baramulla, Gulmarg, and Banihal in Kashmir afford good examples. In Baramulla region, mixed planting of *Pinus, Cedrus* and other species has been done. The plants are thriving well and have checked erosional losses of soil to a good degree. In the area where plantations have not been made the soil is actively eroding. Another example is afforded by Sankaracharya Hills in the vicinity of Srinagar which was heavily eroded in the past. It has now been improved by afforestation. Forest department took up the planting of conifer species like *Pinus wallichiana P. roxburghii, P. sylvestris, P. insignis, P. gerardiana, Cupressus arizonica, C. sempervirens, Juniperous* species, *Cedrus deodara,* and broad leaved species, like *Aesculus indica, Fraxinus, Juglans regia.* These plants check erosion to some extent.

The efforts are being made to check the spread of Rajasthan desert towards Uttar Pradesh and the plantation is being done in that region in two or three belts. Small-sized plants are planted on the windward side and tall trees on the leeward side. On the windward side *Leptadenia spartium, Cenchrus ciliaris, Balanites roxburghii, Calligonum polygonoides, Sachharum munja, Kochia indica,* etc. are commonly planted and on leeward side *Acacia senegal, A. leucophloea, Ricinus communis, Prosopis spicigera, P. juliflora, Parkinsonia,* etc. are being grown.

The problem of afforestation is the selection of suitable species for a given area. This can be accomplished by dividing the whole area into different catchment zones according to climate, soil and biota. Suitable species for each zone should be selected from those already growing there. Knowledge of successional trends of the vegetation will be of great help in afforestation practices. Lutz and Chandler (1946) have stated that climax vegetation in any region is most effective agent to prevent accelerated erosion. Afforestation also checks the erosion of unstable rocks probably by checking the leakage of water (Puri, 1951).

Along with afforestation and reforestation, forest management practices should also be strictly enforced which require

(*a*) Control on forest fires,

(*b*) Extension of forest areas,

(*c*) Checking extensive felling and lopping,

(*d*) Restricted grazing in forest areas, and

(*e*) Formation of reserve forests.

(*iv*) **Checking of overgrazing.** Since grazing in all the areas subjected to soil erosion cannot be completely stopped, a system of restricted and rotational grazing may be helpful in checking soil erosion to some extent. The area open to grazing for sometimes should be closed for the following years to facilitate regeneration of forest and to maintain thick ground vegetation.

3. **Dry farming practices.** Dry farming is useful in those areas where rainfall is very low, uncertain and uneven in distribution. In such areas maximum conservation of water in the soil and reduction of soil erosion are the two main problems.

The improved dry farming practice developed as result of researches carried out in India is an integrated rational system of cultivation of purely rainfed and drought enduring crops in dry areas. This practice consists in bunding, both contour and compartmental, and all other advanced agricultural practices to conserve maximum amount of rain water in the soil so as to make it available for crops during their growth period. In this, usually early maturing crop varieties are selected for cultivation. Important dry farming regions of India are:

(1) Punjab,

(2) Rajasthan,

(3) South-west part of U.P., north-west part of M.P., and

(4) Some parts of peninsular India including dry tracts of Maharashtra, Mysore, Andhra Pradesh and Tamil Nadu.

Dry farming methods and selection of crops vary in different regions as per requirement of local conditions.

B. Mechanical Methods

It is only in recent years that soil erosion problems have received attention of engineers. The mechanical practices of soil conservation include various engineering techniques and structures which are adopted to supplement the biological methods when the latter alone are not sufficiently effective. These practices aim at the following objectives :

1. To reduce the velocity of run-off water and to retain it for long period so as to allow maximum water to be absorbed and held in the soil.

2. To divide a long slope into several small parts so as to reduce the velocity of run-off water to the minimum, and

3. Protection against erosion by wind and water.

Mechanical methods for soil conservation are :

(*i*) Basin leaching,
(*ii*) Pan breaking,
(*iii*) Subsoiling,
(*iv*) Contour terracing,
(*v*) Contour trenching,
(*vi*) Terrace outlets,
(*vii*) Gully control,
(*viii*) Digging of ponds and reservoirs, and
(*ix*) Stream bank protection.

(*i*) **Basin leaching.** In this method, a number of small basins (water reservoirs) are made along the contour by means of an implement called basin blister. Basins collect and retain rain water for long period and also catch and stabilize downwardly moving soils of the slopes.

(*ii*) **Pan breaking.** In some areas, soils become impervious to water and are less productive because of formation of hard sheet of clay a few feet below the surface. Such areas can be made productive and water permeable by breaking hard clay pans by means of pan breaker on contour at a distance of about 5 feet. By pan breaking, drainage and percolation of rain water is improved and soil is saved from residual run-off and erosion.

(*iii*) **Sub-soiling.** In this method hard subsoil is broken deeply by means of an implement called subsoiler. This process promotes absorption of rain water in the soil and makes the soil more loose and fit to allow luxuriant growth of vegetation.

(*iv*) **Contour terracing.** Sometimes drainage channels or properly spaced ridges or soil mounds are formed along the contour (at right angles to the slope) to retain water in the soil and check the soil erosion. These are called terraces. Terraces are levelled areas constructed at right angles to the slope to reduce soil erosion. Terracing may be of the following four types :

(*a*) *Channel terracing.* This is concerned with making of wide but shallow channels on contours at suitable distance. In this process, the excavated soil is deposited along the lower edge of channel in the form of low ridge.

(*b*) *Narrow based ridge terracing.* This process is commonly called bunding. In this a number of narrow based ridges or bunds are constructed at distance of 1 m to 2 m across the slope along the contour.

(*c*) *Broad-based ridge terracing.* In this, wide but low bunds are made on contour by excavating soils from both the sides of ridge.

(*b*) *Bench terracing.* This method involves making of wide steplike platforms, the so called bench terraces, having suitable drops along contours (Fig. 25.6). Along the outer edges of bench terraces bunds of about one foot height are raised to check the downward flow of rain water and also soil erosion. The vertical drops may vary from 1 m to 2 m. Bench terracing is very costly process and so it should be applied in the area of land scarcity for growing money crops.

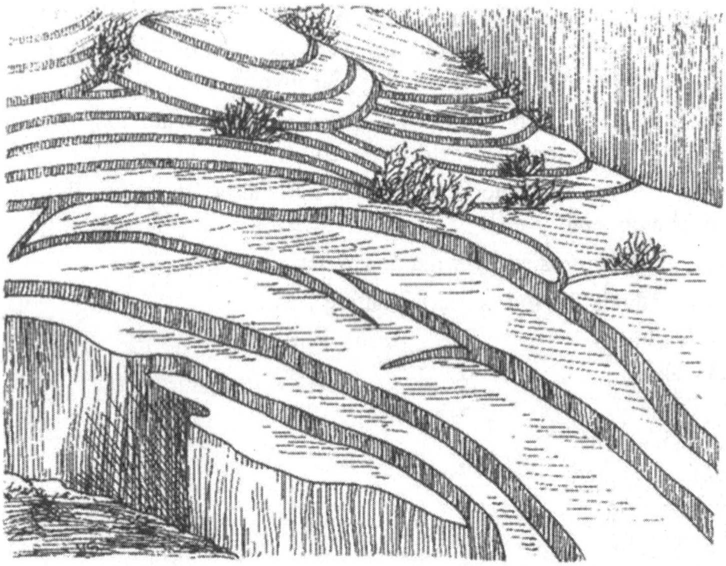

Fig. 25.6 Bench terraces on a steep slope.

Fig. 25.7. Contour trenching on a slopy land.

Terracing is the only practical method of soil conservation on steep land. It is an expensive method or reducing soil erosion since it requires moving soil to construct the levelled areas, protecting the steep areas between terraces and constant repair and maintenance. Many factors such as length, the steepness of slopes, type of soil, and amount of precipitation determine the feasibility of terracing.

(*v*) **Contour Trenching.** This method involves making a series of deep pits (2′ × 1′) or trenches across the slope at convenient distance (Fig. 25.7). The soil excavated from the trenches is deposited along the lower edge in the form of bund. On the ridges tree seeds are sown (Fig. 25.7).

(*vi*) **Terrace outlet.** In order to reduce soil erosion and to remove excess of rain water safely from the contour terraces pipe outlets are used or channels are made which are thickly covered by grasses.

(*vii*) **Gully and ravine control.** Gully formation can be checked by the following methods :

(*a*) By making perimeter bunds around gullies to check flow of water through it.

(*b*) By growing suitable soil-binding vegetation on the gullies to check soil erosion.

(*c*) Diversion trenches should be made around gullies.

(*viii*) **Ponds and reservoirs.** Small ponds and water reservoirs or dams should also be made at suitable places for irrigation and some other purposes. Various types of dams have been devised to arrest and plug gullies and thus to check soil erosion (Fig. 25.8). These dams may be (*a*) brush dams (*b*) earth dams, (*c*) concrete dams or (*d*) woven wire dams.

Fig. 25.8. A dam.

(*ix*) **Stream bank protection.** Banks of ravines and rivers with high vertical drops are subjected to heavy soil erosion. The bank erosions can be checked by making the drop slopy and by growing vegetation on the slopes or by constructing stone or concrete pitch.

The problem of soil conservation is very complicated and is receiving attention of scientists and engineers. The erosion of Siwaliks has been studied by Glover and Hamilton (1935) and Gorrie (1951) and various methods have been suggested for soil conservation. Recently Puri (1949, 1954) analysed the problems from different angles and suggested that soil should be kept under tree vegetation in order to check erosional losses and landslides. The Central Arid Zone Research Institute in collaboration with UNESCO is attempting to check erosional losses in Rajasthan. The Indian Council of Agricultural Research (ICAR), especially the Waste Land Reclamation Committee, is giving attention to the problem in collaboration with some State forest and agriculture departments. Central Soil Conservation Board has established 9 research centres, one each at Dehradun, Kota, Ootacamund, Bellary, Vasad, Agra, Chandigarh, Jodhpur and Chhatra to organise, co-ordinate and initiate researches on soil conservation, to meet the paucity of trained personnels for manning soil conservation schemes and to assist States and River valley projects technically.

QUESTIONS

1. Comment upon the following statements :
   (*i*) Presence of micro-organisms in soils is harmful to soil fertility.
   (*ii*) Acid soils are better for plant growth than alkali soils.
   (*iii*) Base exchange causes soil erosion.
2. Define Soil erosion. Give an account of various factors causing erosion of soil.
3. What are the aims of soil conservation? Discuss in brief the various biological and mechanical methods of soil conservation.
4. Write short notes on :
   (*i*) Crop rotation.
   (*ii*) Afforestation and Reforestation.
   (*iii*) Contour farming.

26

Soil Fertility—Its Meaning, Causes and Maintenance

The soil is a natural medium for plant growth and it supplies nutrients to plants. Some soils are productive and they support luxuriant growth of plants with very little human effort whereas others may be unproductive which support almost no useful plant life regardless of every human effort. In order for soil to be productive, it must (*i*) be easily tillable and fertile, (*ii*) contain all essential elements in the forms readily available to plants in sufficient amount, and (*iii*) physically good to support plants and contain just the right amount of water and air for proper root growth. The soil must supply these essentials every day in the life of the plant.

Soil fertility and soil productivity appear to be synonymous but in soil science these two terms bear different meanings. **Soil fertility** may be defined as the ability of soil to provide all essential plant nutrients in available forms and in a suitable balance whereas **soil productivity** is the resultant of several factors such as soil fertility, good soil management practices, availability of water supply and suitable climate. A soil can be highly fertile, *i.e.,* it has ready supply of nutrients in available form, yet it may not be highly productive. Water-logged soils may be highly fertile but may not produce good crop because of the unfavourable physical conditions. A fertile soil may be highly saline or alkaline which may not be good for agriculture. Sandy soil may be poor in fertility but with the use of fertilizers and water it may be made productive. **Soil fertility** thus denotes the status of plant nutrients in the soil whereas the soil productivity is the resultant of various factors influencing crop production.

In fact, there is no standard for either fertility or productivity because both depend upon the crops to be grown. Soil that is productive for potatoes may not necessarily be productive for certain other crops.

PLANT NUTRIENTS

Of the 90 or so chemical elements forming the earth's crust, 16 are known to be essential for plant growth and reproduction. Seven elements needed in good quantity (*macro nutrients*) are : hydrogen, oxygen, nitrogen and carbon from air and water and phosphorus, potassium and calcium from mineral particles in the soil. The other 9 elements needed only in small amount (micronutrients) are : magnesium, sulphur, boron, copper, iron, manganese, zinc, molybdenum and chlorine.

Jacob and Uexkull have listed cobalt and sodium as essential elements for plant growth. With the exception of hydrogen, carbon, and oxygen all other inorganic plant requirements are obtained directly or indirectly from the soil minerals, hence these elements are called mineral nutrients. Strictly speaking, nitrogen is not a mineral element but it has been included in the list because it can also be obtained by plants from soil.

The mineral elements are taken by plants from soils mostly in the form of ions. Plants obtain nutrients from the following four devices :

(*i*) from the soil solutions through roots,

(*ii*) from exchangeable ions on the surface of clay and humus particles through roots,

(*iii*) from readily decomposable minerals, and

(*iv*) through the leaves.

The essentiality of an element is proved by the following criteria :

(*a*) The element may be considered essential if its exclusion from the nutrient medium inhibits or drastically reduces the growth and reproduction of plant.

(*b*) Acute deficiency of the element produces certain well defined disease symptoms which are not produced by the deficiency of any other element.

(*c*) Deficiency disease symptoms will disappear if the particular element is supplied before the living system has been damaged beyond repairs.

The capacity of soil to supply the essential elements is a fundamental edaphological problem. In natural habitat the plant matter is returned to ground through decay and unless there is extensive leaching and percolation the inorganic nutrients become available again for new growth. Many farm crops are removed from ground in *toto* and the supply of essential elements thereby becomes depleted. To replace these elements taken away from ground by crops, the manures or fertilizers are added. Unless it is done the crop will suffer from deficiencies in the essential elements. Deficiencies become reflected in the growth of plants in several ways; some may cause a reduction in yield as a result of poor plant growth and some may delay maturation of crop, a function that may be very vital to crop yield in the places where the growing season is short. The symptoms of mineral deficiencies may be dwarf, spotted, distorted, curled or wilted leaves or rotting of the centre of fruits. Different mineral nutrients have certain specific deficiency symptoms. The physiological roles of various mineral elements in the life of plants and deficiency symptoms caused by them can be found in any plant physiology book.

SOIL FERTILITY FACTORS

Several factors are know to govern the fertility of soil. Some of the important factors are discussed below :

1. As a result of cropping, a large amount of organic matter and soil minerals are removed and if the normal cycling of mineral elements is retarded, loss in soil fertility may result.

2. Besides cropping, soil erosion and loss of water also causes tremendous loss of plant nutrients from the top soil.

Generally, water is lost through leaching, drainage, evapo-transpiration and runoff. The following adverse affects are observed due to water loss.

(*i*) Soil becomes very hard.

(*ii*) The seed germination percentage decreases.

(*iii*) The nutrients in the soil leach or evaporate.

(*iv*) The root growth retards, so that plants become stunted and, as a result, the yield is reduced.

(*v*) Stomata become closed, as a result of which accumulation of gases or metabolic wastes increases in plant tissues leading to death of the plant.

(*vi*) The activity of soil-microogranisms decreases.

3. Conversion of organic forms of nitrogen locked in humus into ammonia gas and nitrogen gas and leaching out of soluble nitrates and nitrites from surface soil greatly affect the fertility status of soil.

4. Like deficiency, the abundance of certain nutrient elements in soluble form may also be toxic and even the elements, say alkalies, essential for plant growth may be toxic if present in excess. Flowering plants do not grow in the soil containing more than 6 per cent NaCl and other salts. The elements are not equally toxic and the various species of plants differ in their susceptibility to different elements.

5. **Toxic chemicals and pesticides in soil.** Several agricultural chemicals being used for controlling various diseases and insect pests are highly toxic and their application adversely affects the soil microflora and fauna. Prolonged persistence of these pesticides in soil is bound to lower the soil fertility both directly and indirectly.

6. **Soil reaction.** The soils may be alkaline or neutral or acidic in their reaction. Some plants find acid soil unsuitable for growth and other plants find alkaline ground unfavourable. pH value of soil solution determines the availability of certain plant nutrients and thus it has bearing on soil fertility problem. Increase in the acidity of the soil makes mineral salts more soluble in soil solution and thus salts may become available in concentrations that may be highly toxic or may damage plants growing in such soils. Janick *et al.* (1969) have demonstrated that high concentration of both iron and aluminium may damage plants growing in acid soils.

MAINTENANCE OF SOIL FERTILITY

Soil fertility is the most important asset of a nation. Maintenance of soil fertility is an important aspect of agriculture. The soil fertility problem has been studied in many countries and scientists have brought to light several facts concerning soil fertility and its maintenance. Soil fertility is of two types;

(*a*) **Permanent fertility.** It is derived from the soil itself. It can be improved, maintained or corrected by soil management practices.

(*b*) **Temporary fertility.** It is acquired by suitable soil management but the response of built up soil fertility is highly dependent on the degree of permanent fertility which is already there.

Several methods are known for controlling the loss of soil fertility. Here only the important methods are discussed.

1. Application of Organic Manures and Chemical Fertilizers

Plants absorb water and minerals from the soil, which is essential for growth, flowering, crop yield, and other vital activities. Soil is a store house for organic and inorganic plant nutrients. Some soils are rich in organic and humus content and are considered to be fertile and more productive while others that are deficient in humus and minerals are less productive. The soil is subjected to a continuous depletion of nutrients due to its continuous use by crops. This requires the addition of mineral resources. The various soil components are being removed by living organisms and are returned to the soil by death and decay of organisms. If the rate of removal or loss of minerals is greater than the rate of addition, the soil will naturally become less fertile. The minerals of the soils are lost due to crops, leaching or soil erosion. The minerals are often removed from the top layer by rainwater. Cultivation of crops regularly, year after year, makes the soil less productive. In intensive cultivation there are little chances for the restoration of lost nutrients in the soil until they are supplied from outside. The leguminous plants, however, compensate the loss of nitrogenous compounds. Besides this, manure and fertilizers are to be supplemented to restore the fertility of the soil.

The deficiency of mineral nutrients in the soil either can be compensated through organic manures such as green manuring, compost etc. or it can be supplemented by the application of chemical fertilizers from outside sources.

A. Organic Manures

The organic content of the soil which is a good source of plant nutrients contributes most to the fertility of the soil. Organic manures improve soil fertility in the following ways :

(*i*) They modify the physical properties as increase in granulation of the soil and increase in permeability and moisture holding capacity of soil.

(*ii*) They provide food for soil microbes and thus enhance microbial activities.

(*iii*) Decomposition products of organic manures help to bring mineral constituents of soil into solution.

(*iv*) They improve physico-chemical properties of soil, such as cation exchange and buffering action.

Organic manures are of several kinds some of which are discussed below :

(*i*) **Farmyard manures.** Solid and liquid excreta as dung and urine of all farm animals are termed farmyard manures. They are ready made manures and contain nitrogen, phosphorus and potassium. The farmyard manures of different animals vary greatly in their composition but they are good for all types of soils and all the crops. Farmyard manure when collected in field in exposed condition for several months shows considerable loss of fertilizing value as upon decomposition a considerable amount of ammonia is lost by volatilization. Therefore, it is important to keep manure protected from weather and manure preparation should be carried out in trenches of about a metre depth. When the trenches are filled with dung etc, the surface is covered with a cowdung-earth slurry. In about 3 months the manure becomes ready for use.

(*ii*) **Compost.** Compost manure can be prepared from a variety of refuse materials, such as straw, sugar cane refuse, rice hulls, forest, litter, weeds, leaves, kitchen wastes. It is prepared in pits usually 6–8 m long, 1½ to 2 m wide and one metre deep. In the pits, 30 cm thick layer of plant residues moistened with dung, urine and water is formed and then a second layer of about 30 cm thickness of mixed refuse is spread over it and moistened with slurry. The operation is repeated until the heap rises to a height of about 50 cm above the ground level. The top is then covered with a thin layer of moist earth. After three months of decomposition the material is well mixed and again covered. After a couple of months the manure is ready for use. There are two types of composts :

(*a*) *Farmyard compost* which is obtained from animal excreta and plant residues.

(*b*) *Town compost* which is obtained by decomposition of kitchen wastes and garbage of towns and cities.

Compost manures are rich in all plant nutrients.

(*iii*) **Green manures.** Green manuring is the practice of growing, ploughing and mixing of green crops with soil to improve soil fertility and productivity. Its effects on soils are similar to those of farmyard manures. It is cheap and the best method to increase soil fertility as it can supplement farmyard and other organic manures without involving much cost. Green manures add nitrogen and organic matter to the soil for the improvement of crop yield. Through green manuring mobilization of minerals, reduction of organic nutrient losses due to erosion, leaching and percolation, and improvement in physical, chemical and biological activities of the soil can be achieved. Green manuring also improves soil aeration and drainage conditions. For green manuring both leguminous and non-leguminous crops are used. In India, leguminous crops such as sannhemp **(sanai),** dhaincha, berseem, clover, *Phaseolus mungo,* cowpea, are generally used for green manuring.

(*iv*) **Sawdust.** Sawdust can be used as bedding material to conserve animal urine or for making compost. It is a low fertilizing material but it is definitely richer than wheat straw in calcium.

(*v*) **Sewage.** In modern system of sanitation, water is used for removal of human excreta and other wastes. Sewage consists of two components :

(*a*) the solid part, called *sludge* and (*b*) the liquid part, called *effluent or sewage water.* Sewage is quite rich in several plant nutrients and can be used for fertilizing the crop by irrigating the soil directly with sewage water but there is a danger for the spread of several human diseases.

B. Chemical Fertilizers

Of the elements known to be essential for plant growth, nitrogen (N), phosphorus (P) and potassium (K) are required by plants in pretty large amounts, and are therefore, designated as *major* or *primary nutrients* while calcium, magnesium and sulphur are secondary nutrients. For acid soils, use of Ca and Mg is necessary. Seven elements iron, manganese, boron, molybdenum, copper, zinc and chlorine are required in trace amount and hence called *micro-nutrients.* Under continuous cultivation our soils are losing organic matter and mineral nutrients faster than they can be replaced. Regular loss of nutrients from the soil results in compact soil, shallow roots, increased drought, claddy and poorly productive soil. So, for the maintenance of soil fertility quick replacement of the organic matter and mineral nutrients removed from the soils is necessary (Table 26.1).

Nitrogen (N), phosphorus (P) and potassium (K), i.e., the primary plant nutrients are commonly applied to soils in the form of commercial fertilizers and hence they are often referred to as *fertilizer elements.*

Table 26.1. Fertilizers and their composition

	Fertilizer	Composition	Source	Nitrogen %
1.	Ammonium sulphate	$(NH_4)_2SO_4$	By-produce from coke and also as synthetic	26.6
2.	Ammonium nitrate	NH_4NO_3	Synthetic	33
3.	Ammonium chloride	NH_4Cl	Synthetic	25
4.	Ammonia liquor	Dilute NH_4OH	Synthetic	20—25
5.	Ammonium phosphate	$NH_4H_2PO_4$	Synthetic	11% N and 48% P_2O_5
6.	Diammonium phosphate (DAP)	$(NH_4)_2HPO_4$	Synthetic	21% N and 53% P_2O_5
7.	Sodium nitrate	$NaNO_3$	Chile saltpetre and synthetic	16
8.	Calcium ammonium nitrate	NH_4NO_3 and Dolomite	Synthetic	20.5
9.	Calcium cyanamide	$CaCN_2$	Synthetic	21
10.	Calcium nitrate	$Ca(NO_3)_2$	Synthetic	15
11.	Urea	$CO(NH_2)_2$	Synthetic	45

Chemical fertilizers are classified into the following three groups on the basis of materials supplied : (*i*) Nitrogenous fertilizers, (*ii*) Phosphorus fertilizers, and (*iii*) Potassium fertilizers.

Classification is not so simple as this grouping would imply, several such fertilizer materials as contain two of these elements, as for example, potassium nitrate and ammonium phosphate.

(*i*) **Nitrogenous fertilizers.** Crops usually take nitrogen from the soil in the form of nitrate (NO_3^-) and ammonium ions (NH_4^+). Nitrogen fertilizers may be divided for convenience into two groups: (*a*) Organic nitrogen fertilizers, and (*b*) Inorganic nitrogen fertilizers.

(*a*) **Organic nitrogen fertilizers.** Organic materials such as cotton seed meal, guano and fish tankage are nitrogen carriers but because they supply less than 2% of total nitrogen added in commercial fertilizers, their use is costly and hence they are not used extensively. Nitrogen of organic fertilizers is released slowly by microbial action. They are used as special fertilizers for gardens, lawns and potted plants.

(*b*) **Inorganic nitrogen fertilizers.** Several inorganic chemicals are used to supply nitrogen to plants. The most important of these are presented in Table 26.1.

(*ii*) **Phosphorus fertilizers.** Phosphorus has rightly been called 'master key' to agriculture as low crop production is due more often to lack of phosphate than to the deficiency of any other element except nitrogen. In phosphorus fertilizers this element is present in the form of phosphate or superphosphate salts and it is available to the plants when it is combined with organic matter or with calcium and magnesium. Phosphorus is also found in combination with iron and aluminium and is present in certain rock minerals as apatite. Plants take up phosphorus chiefly as phosphate $(PO_4)^-$, HPO_4^- and $H_2PO_4^-$ ions and the availability of these ions depends chiefly upon the acidity of ground. They become nearly insoluble in strongly acid or strongly alkaline soils. Release of phosphorus from phosphate rocks is slow. The breakdown of phosphate fertilizers produces phosphoric acid (P_2O_5) in soluble form that is absorbed by plants. Use of phosphate fertilizers on alkaline soils is not suitable. Phosphorus fertilizers are classified into (*i*) water soluble, (*ii*) citrate soluble, and (*iii*) insoluble. When the term P_2O_5 is used, it means water soluble plus citrate soluble P_2O_5. The following are the important phosphate fertilizers being used in all parts of the world including tropical Asia (Table 26.2).

Table 26.2. Phosphate fertilizers

Fertilizer with chemical composition	P_2O_5 available percentage
(*a*) **Fertilizers containing water soluble phosphorus :**	
(*i*) Superphosphate Ca $(H_2PO_4)_2$ and (ordinary grade) $CaHPO_4$	16—20
(*ii*) Superphosphate $Ca(H_2PO_4)_2$ and $CaHPO_4$ (concentrated)	40—45
(*iii*) Ammonium phosphate $NH_4H_2PO_4$	48 (11% N)
(*iv*) Diammonium phosphate $(NH_4)_2HPO_4$	46—53 (4% N)
(*b*) **Citrate soluble phosphorus fertilizers :**	
(*i*) Dicalcium phosphate	35—40
(*ii*) Basic slag (Indian) $(CaO)_5P_2O_5 . SiO_2$	3—5
(*c*) **Insoluble phosphorus fertilizers :**	
(*i*) Rock phosphate Fluor and chloraphatite	20—30
(*ii*) Bone Meal $Ca_3(PO_4)_2$	18—20

Superphosphate. It is water soluble fertilizer. It does not affect the soil adversely. It contains monocalcium phosphate, dicalcium phosphate and tricalcium phosphate, gypsum, silica, iron, aluminium sulphate and calcium fluoride and water.

Ammonium phosphate. It is a fertilizer containing both nitrogen and phosphorus. It is rich in phosphoric acid content but comparatively low in nitrogen content.

Ammonium superphosphate ($NH_4H_2PO_4$). $Ca_3(PO_4)_2(NH_4)_2SO_4$. It is the cheapest fertilizer and is a mixture of nitrogen and phosphorus fertilizers. It contains nitrogen 3 to 4 per cent and phosphorus pentaoxide (P_2O_5) 16 to 18%.

Nitrophosphate. It is highly hygroscopic. It contains nitrogen 13—18 per cent and phosphorus 20 per cent. It is suitable for acid soils.

Bone meal. It is derived from bone. Bone ash and bone char are the bone products. It is suitable for acidic soils.

Basic slag. Basic slag, a by-product in the manufacture of steel, is one of the cheapest sources of phosphorus. It is a double compound of silicate and phosphate of lime. It is dark brown in colour and alkaline in reaction.

Rock phosphate. It occurs in natural deposits. It is light grey in colour. It is a very cheap fertilizer suitable for acid soils.

(*iii*) **Potassium fertilizers.** These fertilizers are soluble in water which means that potassium is readily available to plants. Total potassium of potassium fertilizers is usually expressed in terms of water soluble potassium (K) or potash (K_2O). Soils of arid and semiarid areas are generally well supplied with potassium. Acid soils usually need potassium fertilizers more than neutral or alkaline soils because acid soils develop in the areas of high rainfall that leaches out available potassium. All potassium fertilizers are physiologically neutral in reaction. The following are some common potassium fertilizers.

(*i*) **Potassium chloride.** It is also called muriate of potash. It contains 48—62% K_2O. It is also cheap and neutral in reaction.

(*ii*) **Potassium sulphate.** It contains about 50% potash (K_2O). It is expensive fertilizer.

(*iii*) **Kainite.** It is natural potassium mineral which contains 14—20% potassium. It is suitable for alkaline soils.

(*iv*) **Wood ashes.** It is used in the form of ash as a manure. Potassium occurs in the form of potassium carbonate and the percentage of potash is from 2 to 6.5. It is suitable for alkaline soils.

Application of Micronutrients

In order to correct the deficiency of micronutrients, especially if it is very necessary, micronutrients should be added only after ascertaining the amount required. Copper, manganese, iron, zinc are supplied generally as their sulphate and boron is applied as borax. Molybdenum is supplied as sodium molybdate. In recent years there has been an increase in use of chelates to supply iron, zinc, manganese and copper.

Soils vary in their ability to supply available nutrients. Some soils may be deficient in nitrogen, some may be deficient in nitrogen and phosphorus, and still some others may be deficient in nitrogen, phosphorus and potassium. To suit the variable requirements of different soils and crops, *fertilizer mixtures* are prepared. Fertilizer mixtures or mixed fertilizers contain two or more fertilizer materials. If the ingredients and their amounts are known, the formula is referred to as **open** and if they are not disclosed the formula is termed **closed** one.

The kinds and amounts of fertilizers to be applied to soil are determined considering the following points :

(*a*) Kinds of crop to be grown—particularly its economic value, nutrient removal and absorbing ability.

(*b*) Chemical condition of soil in respect of total nutrients and available nutrients.

(*c*) Physical state of the soil, especially its moisture content and aeration. For recommending the kind and amount of fertilizers or soil amendments, the analysis of soil is essential.

2. Application of Soil Conservation Practices

Loss of plant nutrients and water from the soils due to soil erosion can be checked effectively and the fertility of soil can be maintained by application of various biological and engineering methods of soil conservation. A detailed account of these methods has already been given in soil conservation topic.

3. Water management Practices (Irrigation and Drainages)

Water supply is critical factor in crop production in most areas of the world. Soil moisture greatly affects the availability of mineral nutrients in the soil. It has been proved beyond doubt that fertilizer response is much higher with adequate irrigation.

Drainage and moisture control influence micronutrient availability in soils. Improving the drainage of acid soils encourages the formation of less toxic oxidized forms of iron and manganese.

4. Prevention and Elimination of Inorganic Chemical Contamination of Soil

Loss of soil fertility due to application of toxic chemicals as pesticides can be eliminated if (*i*) application of toxic chemicals to soil is reduced and (*ii*) the soil and crop are so managed as to prevent cycling of toxic chemicals.

5. Stabilization of Soil pH

The stabilization of pH through application of soil amendments and buffering seems to be an effective guard against the problems of non-availability of certain plant nutrients and radical changes in microbial activities arising due to change in soil pH. Several methods are discussed for correction of acidity and alkalinity of soils in the chapter on "Properties of Soil".

IRRIGATION SYSTEMS

Water is a very important natural resource, which is the basis of all life forms. For sustainable agricultural production, water is one of the most precious important inputs. Plants need water in huge amount throughout their life. Water is also one of the main factors that influence most of the metabolic process such as photosynthesis, respiration, adsorption, opening and closing of stomata and translocation of food material. The growth and yield of crop plants is very much affected by the availability of water.

Water is used in two ways:

1. Withdrawal or off-channel use
2. Non-withdrawal or on-channel use.

The amount of water taken out of a streams or pumped out from underground water reservoir or surface water reservoirs to be supplied to the points of major use, such as public water supply systems, irrigation and industries, is referred to as **offchannel** use. **In non-withdrawal use,** water is used without being removed from its natural source such as for navigation, swimming bodies, wildlife habitats and other recreation purposes.

Need for Irrigation

Irrigation is the artificial watering of soil to sustain plant growth. Irrigation has become necessary because of the limitation of using natural rainfall as the reliable source of water for agriculture. It is also difficult to store the rainwater for immediate irrigation purposes.

In India, monsoon is usually erratic; sometimes delayed, sometimes scarce or sometimes in excess. Through the monsoon season usually lasts for four months, most of the rains occur only during two months. Therefore, it is very difficult to plan the cropping pattern depending solely on the rains. The sufficient recharging of ground water is essential for sustainable farming, but usually this objective is not fulfilled due to unpredictable and fluctuating rains.

The efficient utilisation of water resources is essential for better crop production. It includes the suitability of land and water for irrigation, planning of crops and suitable water management practices. Water management includes irrigation and drainage.

The suitable irrigation depends on:

Time of irrigation

Amount of irrigation, and

Efficient method of irrigation

Similarly, suitable drainage depends on:

How much to drain

How best to drain

How rapidly to drain

Stages of Crop when Irrigation is Required

The growth span of a crop plant passes through various phases and stages of growth. The rate of irrigation varies during the different stages of plant growth, i.e., from seedling to maturity stage. The growth period of irrigated crops can generally be divided into three phases, namely: vegetative, flowering and maturity stage. At vegetative stage, light and intermittent irrigation is required whereas at flowering stage moderate and frequent irrigation is needed and during crop maturation stage again light irrigation is required.

Systems of Irrigation:

There are different water sources from which irrigation is carried out, for example, canals, wells, open wells, tube wells, tanks etc. Depending upon the crops, soil types, water resources, climate conditions and costs involved, several systems of irrigation are used which are as follows:

1. Surface irrigation systems
2. Subsoil irrigation systems
3. Drip irrigation systems

Surface Irrigation System

In this system, water is directly used on the surface of the soil and water follows the slope of the land. the surface irrigation system may be of following types:

Flood irrigation : Flood irrigation is used for close-grown crops such as rice and where farm fields are levelled and water is abundant. A sheet of water is allowed to advance from different sources and remains on a field for a given period, depending on the crop, the porosity of the soil and its drainage (Fig. 26.1 a).

(*a*) (*b*) (*c*)

Fig. 26.1 Different surface irrigation systems :
(*a*) Flood irrigation (*b*) Furrow irrigation (*c*) Basin irrigation

Basin flooding : It is used in widely spaced trees, such as in orchards, with basisns built around trees and filled with water (Fig. 26.1c).

Check basin : This system of irrigation is widely used in India, as it is more suitable to all types of soils and to a variety of crops.

Furrow irrigation : In furrow irrigation, water moves in the fields in furrows, between two ridges (Fig. 26.1b). It is employed in the fields with row crops such as cotton and vegetables. Parallel furrows, called corrugations, are used to spread water over fields that are too irregular to flood.

Sprinkler Irrigation System

In this system, irrigation is done through pressure, to the surface of any crop or soil, in the form of a thin spray. Sprinkler irrigation system uses less water and provides better control. Each sprinkler, spaced along a pipe, sprays water in a continuous circle until the moisture reaches the root level of the crop. Centre-pivot irrigation uses long lines of sprinklers that revolve in a circular field. It is used especially for feed crops such as alfalfa, tall crops and orchards (Fig 26.2*a*).

1. Sprinkler irrigation system conveys water from the source to the field, through pipes under pressure, and disributes over the field in the form of spray of 'rain like' droplets (Fig. 26.2b).

2. 10–20% of area is not irrigated at the corners of square of rectangular plot.

3. This system requires high energy and involves huge cost of the equipment.

This system can be followed in conditions when

1. the soil is too shallow.

2. the land is too steep.

3. less and frequent irrigations are needed.

4. the soils are very sandy.

This system is disadvantageous in the following conditions when:

1. strong winds cause improper distribution of water.

2. evaporation losses are high from sprinkler irrigation, especially under high temperature and low relative humidity conditions.

3. the initial cost is high.

Fig. 26.2 Sprinkler irrigation: a. centre-pivot irrigation with long lines of sprinkler, b. sprinkler irrigation system in which water is conveyed to field through pipes under pressure.

Sub-Soil Irrigation System

In this system, irrigation is done into a series of ditches in the field deeply to the impervious layer. Then it moves laerally and vertically through capillaries saturating the root zone. In this system, continuous supply of water in the root zone is assured from the artificial water table created by the ponding of irrigation water on the impervious layer. This system is very efficient because the water losses through evaporation from surface can be reduced. This system is more common in Gujarat and Jammu and Kashmir, for cash crops growing on sandy loam soils.

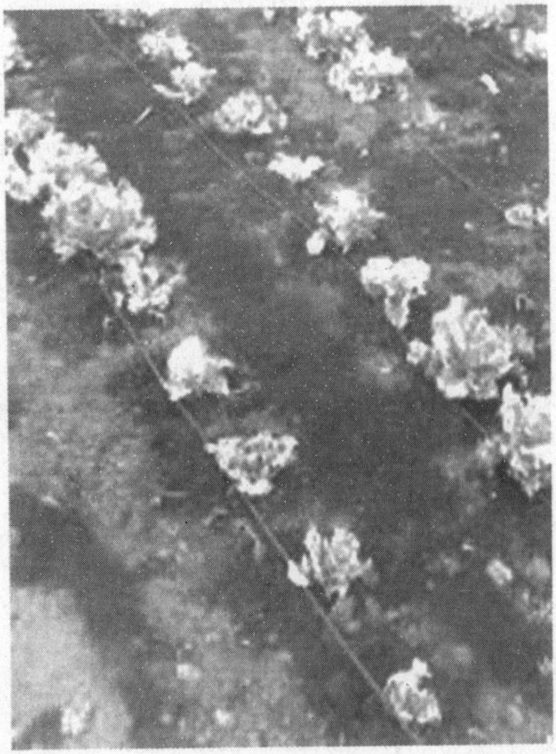

Fig. 26.3 Drip irrigation

Drip Irrigation System

It is also called **trickle irrigation.** This irrigation system involves the slow application of water, drop by drop, to the root zone of a crop. It is done through mechanical devices called emitters, located at selected points along water delivery lines. Drip irrigation is extensively used in areas of acute water scarcity and especially for crops such as coconut, grape, banana, citrus, sugarcane, cotton, maize, tomato and plantation crops (Fig. 26.3).

Advantages of Drip Irrigation

1. Water loss through transpiration is low.
2. It is possible to obtain better yield and quality of crops by controlling soil moisture-air-nutrient levels.
3. We can save the fertilisers by monitoring the supply of nutrients as per the need of the crop.
4. Improvements in biological fertility can be achieved by avoiding pollution.

Disadvantages of Conventional Irrigation Methods

- Major part of the water goes waste, only small quantity of water is utilised by the plants.
- Water is not uniformly distributed due to uneven and poor levelling of the field.
- Crops are usually subjected to cyclic changes of flooding and water stress situations by proividing heavy irrigation at one time and leaving the fields to dry up for about 10 to 15 days.
- The low lying fields always get excess water that causes prolonged water logging, due to lack of levelling of fields.
- In the fields, about 10-15% of land is utilised for making channels and distributaries etc. which decreases the area of cultivation.
- Excessive irrigation and poor water management leads to water logging and accumulation of salty in the upper layer of the soil.
- In conventional irrigation, unmanageable and undesirable weeds grow.

Problems of Excess Irrigation

- Excess irrigation causes several changes in the soil and plants, resulting in reduced growth and some times even death of plants.
- Germinating seeds are sensitive to water logging, since they are totally dependent on the surrounding soil space for oxygen supply.
- Yield of cereals is reduced if excess irrigation is given.
- Excess water causes injury to the plant due to low oxygen supply to the root system and accumulation of toxic substances in the soil.
- Leaching of nitrates and denitrification occur which result in nitrogen deficiency.
- Permeability of roots decrease due to shortage of oxygen. It results in decreased uptake of water and nutrient.

QUESTIONS

1. What do you mean by soil fertility? How does it differ from soil productivity? Discuss in brief various factors affecting loss of soil fertility.
2. Define soil fertility and describe the methods of checking loss of soil fertility.
3. Write brief notes on :
   (*a*) Soil productivity
   (*b*) Biological fertilizers
   (*c*) Green manuring.
4. How does the soil moisture help in maintaining the soil fertility? Discuss in brief different methods of irrigation used by Indian farmers.

27

Soil Degradation and Wasteland

Soil degradation has been defined as a process that leads to decline in the fertility or future productive capacity of soil as a result of human activity (United Nations Environment Programme, 1993). It occurs whenever the natural balances in the landscape are changed by human activity through misuse or overuse of soil. Degraded soils which result in poor or no production are also called *problem soils.*

Waste lands are those which for one or the other reason have poor life sustaining property. Out of 100 per cent potentially active lands only 44 per cent are available for cultivation and 56 per cent of land are non-available for cultivation.

The wasteland can be made useful by increasing productivity of land by using some useful methods as afforestation or by using biofertilizers.

Soil degradation is a complex phenomenon derived by interaction between natural and socio economic factors. The degradation or deterioration of soil may be caused by the following factors :

1. *Physical factors, e.g.* loss of fertile top soil due to water or wind erosion.
2. *Chemical factors e.g.* depletion of nutrients or the toxicity due to acidity or alkalinity (salinization) or water logging.
3. *Biological factors* which affect the microflora and reduce the microbial activity of the soil. These factors reduce the yield.

Some other factors as deforestation, extensive cultivation on marginal land, improper cultivation practices like monocropping, poor manuring, misuse of fertilizers or excess use of fertilizers, excessive irrigation, over-grazing, fragility of soil, adverse weather and mining may accelerate the process of soil degradation.

During last decade the nutrients deficiency has been considered as the main cause of poor productivity and crop failure. A study of the current trends in agronomic practices has suggested that the nutrients deficiency is further aggravated by continued use of high yielding crop varieties, intensive cropping pattern and relatively poor fertilizers.

Among the major causes of degradation, water erosion is considered to be the most severe one which covers almost 87% of the affected area. The main cause of water erosion is removal of vegetation, over exploitation of vegetation, over grazing and improper agricultural practices. The latest data revealed that erosion has rendered 200 million hectares or 36% of the total area of the country barren (Table 27.1).

Soil degradation is a global phenomenon. Of the world's total land area of 13.5 billion hectares, only 3.03 billion hectares (22 per cent) is actually cultivable and about 2 billion hectares is degraded. The annual loss of land is expected to go up to 10 million hectares by 2000 A.D.

(Yadava, 1996). In India alone, about 188 million hectares or almost 57% of total land area, is degraded (Sehgel and Abrol, 1994).

Table 27.1. Soil degradation (in million ha) in India

Causes of degradation	National commission on Agriculture (1976)	Sehgal and Abrol 1994	1997
Erosion	150.0	162.4	167
Saline & alkali soils	7.0	10.1	11
Water logging	6.0	11.6	13
Shifting cultivation	—	—	9
Decline in soil fertility	—	3.7	2
Total	163.0	187.8	202

The severity of soil degradation can be measured by the quantity of soil lost every year from each hectare of soil (Tables 27.2 & 27.3).

Table 27.2. Severity of soil degradation
(Slight, 5-10 tonnes; moderate, 11-20 tonnes; high, 21-40 tonnes; and severe, more than 40 tonnes)

	Severity of Degradation				Total area
Degradation	Slight	Moderate	High	Severe	(Million ha)
Water erosion	5.0	24.3	107.2	12.7	148.9
Wind erosion	0.0	0.0	10.8	2.7	13.5
Loss of nutrients	0.0	0.0	3.7	0.0	3.7
Salinization	2.8	2.0	5.3	0.0	10.1
Water-logging	6.4	5.2	0.0	0.0	11.6
Total	14.2	31.5	127.0	15.1	187.7

Table 27.3. Estimated annual costs of soil degradation (billion Rupees)

Degradation (billion Rs.)	Annual estimated cost (Per cent)	Share of Total
Erosion	61	69
Salinity	19	21
Water-logging	8	9
Depletion of nutrients	1	1
Total	89	100

Overall damage of some important crops due to soil degradation is given in Table 27.4.

Table 27.4. Losses in some important crops due to soil degradation
(According to Brandon and Hommann (1995)

Crops	Loss Percentage	Losses in billion Rs.
Paddy	9.7	29.2
Wheat	10.8	26.4

Barley	9.1	0.4
Groundnut	8.1	5.8
Rape seed	8.3	3.9
Cotton	10.2	2.2
Maize	8.0	2.4
Sugar cane	12.1	11.7
Total	100	88.7

Causes of Soil Degradation

The main reasons for unproductiveness or degradation of soils are as follows :

1. Nutrient disorder
2. Water-logging
3. Salinity
4. Erosion
5. Biological degradation
6. Other causes

Nutrient disorder. Most of the Indian soils are deficient in nutrients and organic matter. Organic matter is rapidly decomposed and leached or eroded by heavy rains. In addition to these causes, intensive cultivation using high-yielding short-duration and fertilizer-responsive cultivars of crops has further accelerated the loss of plant nutrients which is much greater than what is supplemented through fertilizers. According to an estimate of 1992, every year 20.2 million tonnes of NPK is removed by growing crops. The data published by National Bureau of Soil Survey and Land Use Planning (Sehgal and Abrol, 1994) show that about 3.7 million ha land suffers from nutrient loss or depletion of organic matter. The problem is more severe in the cultivated areas of the subtropical belt. Out of 20.2 million tonnes NPK removed by the plants, only 2.66 million tonnes comes from fertilizers and 3 million tonnes from organic sources. If the loss of nutrients due to soil erosion is included, the loss of nutrients from top soil is 43 million tonnes.

Water-logging. Soils become water-logged when the water balance of an area is disturbed because of excess recharge. Important sources of water are heavy rains, overland water flow towards basin, seepage from canals and distribution system and tidal flooding. Natural basins without outlet for water, low permeability of subsurface horizons, internal drainage, low intake rate of surface soils and obstructions to natural flow of rain water etc. are conditions cause water-logging. In highly productive areas, canal irrigation is responsible for a rapid rise in water table. Expansion of canal irrigation is also directly concerned with widespread water-logging and salinity problems in arid and semiarid areas. Disturbances in the hydrologic cycle due to inefficient use of surface irrigation water, poor land development, seepage and poor drainage have resulted in higher water tables. Most of the canal areas in arid and semiarid regions are rich in soluble salts. In irrigation these salts are dissolved in soil water and rise to the surface through capillary action. When the water dries up, the salts are left on the upper surface as a crust or layer. According to National Commission on Agriculture (1976), about 6 million ha area is under water-logged condition. Data of World Bank Survey (1995) reveal that India loses 1.2 to 6 million tonnes of food grains production every year due to water-logging of soil. The water-logging and salinity cause a loss of Rs. 12 billion to 27 billion annually.

Salinity (Saline and alkali soils). Salinity directly affects the productivity by making the soil unsuitable for crop growth. Indirectly it lowers productivity through its adverse effects on the availability of nutrients. The adverse effect of alkalinity on availability of nutrients is due to deflocculation effect of sodium ions. An area of about 21.7 million hectares of soil is rendered unproductive due to salinity and water-logging. The saline degradation is due to natural causes and poor irrigation practices which disturb the water cycle in areas. Most of the crops in India are affected due to salinity. Productivity loss of some crops is given in Table 27.5. The detail account of saline, alkaline soils and acidic soils are given in 'Soil Property' Chapter of this book.

Table 27.5. Loss (in million tonnes) of production of some crops due to salinity

Crops	FAO (1993) estimate	Indian (1986) estimate
Paddy	0.54	1.31
Wheat	0.68	0.95
Barley	0.03	0.03
Groundnut	0.08	0.08
Gram	0.12	0.09
Cotton	0.28	0.18
Maize	0.11	0.13
Sugarcane	3.76	6.49
Total	6.2	9.65

Erosion. Soil erosion is the major cause of soil degradation. In the soil erosion uppermost fertile layer of soil which contains essential nurients is lost. Thus soil becomes deficient in essential minerals and this results in productivity loss. Deforestation or destruction of forests accompanied by reduced frequency of rainfall leads to soil erosion and causes damage to agriculture property. Deforestation causes fast degradation when the soil is steep slopy or easily erodible. Destruction of natural vegetation cover is a major factor responsible for erosion of soils by water and wind. According to Global Assessment of Soil Degradation (GLASOD), deforestation is the main cause of soil erosion by wind in about 98% of the area. Overgrazing, cutting of timber trees, collection of fuel wood, shifting cultivation and encroachment of forest areas are some of the important factors responsible for the loss of vegetation cover on the soil which ultimately causes soil erosion. The latest data provided by Sehgal and Abrol (1994) show that the total degraded land in India is 187.8 million ha, of which 162.4 million ha is degraded due to soil erosion alone (Table 27.6). Table 27.6 presents the area under different types of soil degradation in different years:

Table 27.6. Area under different types of soil degradation in different years (in million ha)

	Different types of degradation			
	Salinity	*Water-logging*	*Erosion*	*Total*
1980	8.0	6.0	150.0	168.4
1985	9.4	8.5	141.5	166.3
1990	9.8	10.1	152.8	180.6
1994	10.1	11.6	162.4	191.8

Biological degradation. The factors which affect soil microflora and fauna also reduce the biological or microbial activity of soil adversely. These factors reduce the yield. It is well known that monocropping (growing the same crop on the same land year after year) often leads to increasing attack of pests and diseases. The fatal nematodes threaten potato cultivation in the Nilgiris and, if not controlled they may pose threat to potato cultivation in that area. Excess use of pesticide reduces microbial activity and biomass. Applications of some pesticide chemicals (*e.g.*, amitrole, atrazine, bromacil, picloram, etc.) inhibit nitrification. The nodulation and growth of some leguminous crops and nitrogen fixation are inhibited by different pesticides. Disposal of oil shales, heavy metal contamination of soil and spillage of crude oils adversely affect soil microflora which ultimately affect soil productivity and cause soil degradation.

Other Causes of Soil Degradation

Extension of cultivation to marginal land. Due to tremendous population increase the use of land is increasing day by day. Marginal lands though sustainable for farming are less fertile and more prone to degradation. Examples of marginal lands are : steep slopy lands, shallow or sandy soils and the lands in dry and semi-dry areas.

Improper crop rotation. Due to shortage of land, increase of population and economic pressure, the farmers have adopted intensive cropping patterns of commercial crops in place of more balanced cereal-legume rotations. During last two decades the area under food crops decreased and that under non-food crops increased. Intensive cultivation leads to removal of large quantities of nutrients from the soil which results to in loss of soil fertility.

Fertilizer misuse. Soil fertility is reduced due to prolonged intensive cultivation. The farmers maintain productivity of soil by applying chemical fertilizers but make less use of organic manures. Although the yield can be maintained by using fertilizers that provide deficient minerals yet their use often results in deficiencies of other nutrients.

Overgrazing. In India pasture land area is decreasing day by day due to expansion of agricultural land. Recent satellite data show that the area under pasture land is severely degraded. This poor condition of pasture lands is due to excessive grazing. The unchecked and indiscriminate grazing on forest land also leads to degradation of forest soils. Overgrazing directly leads to disappearance of vegetation which is one of the important causes of wind and water erosion in dry lands.

Mining. Mining disturbs the physical, chemical and biological features of the soil. The impact of mining on soil depends on the physical, chemical properties of the waste generated. The soil profile is changed, the top soil is turned deep inside the dumps. The erodible material is almost devoid of organic matter and lacks in mineral plant nutrients. According to an estimate, about 0.8 million ha soil is degraded due to mining activity.

Impact of Soil Degradation

The following are the impacts of soil degradation :

1. Degradation leads to reduction in crop yield in the affected lands and a possible decline in cropping intensity.
2. In extreme cases, soil becomes unfit for cultivation.
3. Silting of drainage, canals, rivers and reservoirs results in increased floods and droughts.
4. In some cases farmers use more fertilizer inputs to compensate reduced soil productivity while in other cases, they use excess fertilizers.
5. The rate of siltation in many water reservoirs are significantly high. According to Central Water Commission (1991), nearly 11 per cent of the total capacity of water reservoirs has been silted.
6. Soil degradation has several adverse impacts on the environment. It affects global climate through alterations in water cycle and energy balances and disruptions of carbon, nitrogen and sulphur cycles.

The estimated annual loss of different crops due to soil degradation rangs from Rs. 89 billion to 232 billion which represent a loss of 11 to 26 per cent yield (Table 27.6).

NATIONAL WASTELAND DEVELOPMENT BOARD

For wasteland and its management a separate board was established in 1985 which deals with the land degradation, reclamation of ravines, usar lands and arid tracts and deforestation which is called National Wasteland Development Board (NWDB).

The main objectives of National Development Board is to check land degradation, to convert wastelands for sustainable use, to increase biomass and to restore ecological balance. At present in 146 districts of India wastelands have been identified. The NWDB has converted 2.7 million hectares of wasteland for afforestation in 1989-90 costing Rs. 700 crores. Now this department has been merged under Ministry of Rural Development. There are several regional centres of NWDB. These centres assist the central and state governments in evaluation and formulation of projects to provide training and reorientation programmes. These centres are :

North India	—	University of Horticulture & Forestry Solon (H.P.).
South India	—	PME of Social Forestry Programme UAS, GKVK Campus, Bangalore (Karnataka).
West India	(*i*)	Indian Institute of Management, Vastrapur, Ahmedabad (Gujarat).
	(*ii*)	Agriculture Finance Consultants Ltd. Chhatrapati Shivaji Maharaj Marg, Mumbai (Maharashtra).
East India	—	Jadavpur University, Calcutta (West Bengal).
Central India	—	Indian Institute of Forest Management, Nehru Nagar, Bhopal (MP).
North-Eastern India	—	North-Eastern Hill University (NEHU), Shillong (Meghalaya).

QUESTIONS

1. Define wasteland or problem soil. What are the corrective measures for problem soils?
2. Discuss the various causes of soil degradation.
3. Write short notes on :
   (*i*) Impacts of soil degradation on agriculture.
   (*ii*) National Wasteland Development Board (NWDB).
   (*iii*) Biological degradation

INDEX

PLANT ECOLOGY & ETHNOBOTANY

A

B

C

D

E

F

G

H

I

J

K

L

M

Q

R

S

SOIL SCIENCE

A

B

C

D

E

T

W